behavioral genetics

FIFTH EDITION

behavioral genetics

Robert Plomin | Institute of Psychiatry, London

John C. DeFries | University of Colorado, Boulder

Gerald E. McClearn | Pennsylvania State University, University Park

Peter McGuffin | Institute of Psychiatry, London

Worth Publishers
New York

PUBLISHER: Catherine Woods

ACQUISITIONS EDITOR: Erik Gilg

SENIOR MARKETING MANAGER: Katherine Nurre

SENIOR PROJECT EDITOR: Mary Louise Byrd

ART DIRECTOR: Barbara Reingold

SENIOR DESIGNER: Kevin Kall

PRODUCTION MANAGER: Sarah Segal

COMPOSITION AND ILLUSTRATIONS: Northeastern Graphic, Inc.

MANUFACTURING: RR Donnelley

Library of Congress Control Number: 2007940103

ISBN-13: 978-1-4292-0577-1
ISBN-10: 1-4292-0577-6

Printed in the United States of America
First printing

COVER PHOTOS: (*Foreground*) Twin boys, Roger Ressmeyer/CORBIS.
(*Background*) DNA, Digital Vision/Veer.

Worth Publishers
41 Madison Avenue
New York, NY 10010
www.worthpublishers.com

C O N T E N T S

CHAPTER SIX

CHAPTER SEVEN

CHAPTER EIGHT

CHAPTER TWELVE

Developmental Psychopathology   225

CHAPTER THIRTEEN

Personality and Personality Disorders   238

CHAPTER FOURTEEN

CHAPTER FIFTEEN

CHAPTER SIXTEEN

The Interplay between Genes and Environment   305

CHAPTER SEVENTEEN

Evolution and Behavior   334

CHAPTER EIGHTEEN

APPENDIX

Robert Plomin is MRC Research Professor of Behavioral Genetics at the Institute of Psychiatry in London, where he is also director of the Social, Genetic and Developmental Psychiatry Centre. The goal of the centre is to bring together genetic and environmental research strategies to investigate behavioral development, a theme that characterizes Plomin's research. Plomin is conducting a study of all twins born in England during the period 1994 to 1996, focusing on developmental delays in early childhood. After receiving his doctorate in psychology from the University of Texas, Austin, in 1974, he worked with John DeFries and Gerald McClearn at
the Institute for Behavioral Genetics at the University of Colorado, Boulder. Together, they initiated several large longitudinal twin and adoption studies of behavioral development throughout the life span. From 1986 until 1994, he worked with McClearn at Pennsylvania State University. They launched a study of elderly twins reared apart and twins reared together to study aging, and they developed mouse models to identify genes in complex behavioral systems. Plomin's current interest is in harnessing the power of molecular genetics to identify genes for behavioral traits. He is a past president of the Behavior Genetics Association.

John C. DeFries is professor of psychology and faculty fellow of the Institute for Behavioral Genetics, University of Colorado, Boulder. After receiving his doctorate in agriculture (with specialty training in quantitative genetics) from the University of Illinois in 1961, he remained on the faculty of the University of Illinois for six years. In 1962, he began research on mouse behavioral genetics, and the following year he was a research fellow in genetics at the University of California, Berkeley, where he conducted research in the laboratory of Gerald McClearn. After returning to Illinois in 1964, DeFries initiated an extensive genetic analysis of open-
field behavior in laboratory mice. Three years later, he joined the Institute for Behavioral Genetics, and he served as its director from 1981 to 2001. DeFries and Steve G. Vandenberg founded the journal *Behavior Genetics* in 1970, and DeFries and Robert Plomin founded the Colorado Adoption Project in 1975. For over three decades, DeFries's major research interest has concerned the genetics of reading disabilities, and he is currently associate director of the Colorado Learning Disabilities Research Center. He served as president of the Behavior Genetics Association in 1982 and 1983, receiving the association's Th. Dobzhansky Award for Outstanding Research in 1992; and he became a Fellow of the American Association for the Advancement of Science (Section J, Psychology) in 1994.

Gerald E. McClearn is Evan Pugh Professor in the College of Health and Human Development at Pennsylvania State University, University Park. After receiving his doctorate from the University of Wisconsin in 1954, he taught at Yale University, Allegheny College, and the University of California, Berkeley, before moving to the University of Colorado in 1965. There he founded the Institute for Behavioral Genetics in 1967. In 1981, McClearn moved to Penn State, where he has served as associate dean for research and dean of the College of Health and Human Development. He was also founding head of the Program in Biobehavioral Health and founding director of the Center for Developmental and Health Genetics. His research with colleagues at Penn State on mice has two main emphases: drug-related processes and behavioral and physiological aging. With Robert Plomin and other colleagues at Penn State and in Sweden, he has been involved for the past 17 years in large-scale studies of genetic and environmental influences on pattern and rate of aging in Swedish twins.

Peter McGuffin is dean of the Institute of Psychiatry (IoP), Kings College, London. He was previously director of the Medical Research Council (UK) Social, Genetic and Developmental Psychiatry Research Centre at the IoP. He graduated from Leeds University Medical School in 1972 and underwent a period of postgraduate training in internal medicine before specializing in psychiatry at the Bethlem Royal and Maudsley Hospitals, London. In 1979, he was awarded a Medical Research Council Fellowship to train in genetics at the Institute of Psychiatry in London and at Washington University Medical School, St. Louis, Missouri. During this time, he completed the work for his doctoral dissertation, which constituted one of the first genetic linkage studies on schizophrenia. He went on to carry out family and twin studies of depression and other psychiatric disorders, attempting to integrate the investigation of genetic and environmental influences. His current work continues with this general theme, while at the same time incorporating molecular genetic techniques and their applications in the study of both normal and abnormal behaviors. McGuffin has been president of the International Society of Psychiatric Genetics since 1995 and is a founder fellow of Britain's Academy of Medical Sciences.

PREFACE

Genetics was one of the major scientific accomplishments of the twentieth century beginning with the rediscovery of Mendel's laws of heredity and ending with the first draft of the complete DNA sequence of the human genome. The pace of discoveries has continued to accelerate in the first part of the twenty-first century. One of the most dramatic developments in the behavioral sciences during the past few decades is the increasing recognition and appreciation of the important contribution of genetic factors. Genetics is not a neighbor chatting over the fence with some helpful hints—it is central to the behavioral sciences. In fact, genetics is central to all the life sciences and gives the behavioral sciences a place in the biological sciences. Genetics includes diverse research strategies such as twin and adoption studies (called quantitative genetics) that investigate the influence of genetic and environmental factors as well as strategies to identify specific genes (called molecular genetics). Behavioral genetics is a specialty that applies these genetic research strategies to the study of behavior, such as psychiatric genetics (the genetics of mental illness) and psychopharmacogenetics (the genetics of behavioral responses to drugs).

The goal of this book is to share with you our excitement about behavioral genetics, a field in which we believe some of the most important discoveries in the behavioral sciences have been made in recent years. This is the fifth edition of a textbook (Plomin, DeFries, & McClearn, 1980) that followed an earlier version (McClearn & DeFries, 1973). The earlier editions focused on the methods of behavioral genetics. The third edition (Plomin, DeFries, McClearn, & Rutter, 1997) was entirely rewritten, with a topical rather than a methodological focus. The fourth edition added as a coauthor Peter McGuffin, who is an expert in the quantitative and molecular genetic analysis of psychopathology. The fifth edition continues to emphasize what we know about genetics in psychology and psychiatry rather than how we know it. Its goal is not to train students to become behavioral geneticists but rather to introduce students in the behavioral, biological, and social sciences to the field of behavioral genetics. The fifth edition brings the book up to date in this fast-moving field with more than 800 new references. It also expands one chapter on psychopathology into three chapters, reflecting the enormous growth of genetic research on mental illness. It also adds a new chapter called Pathways between Genes and Behavior.

We begin with an introductory chapter that will, we hope, whet your appetite for learning about genetics in the behavioral sciences. The next few chapters present the basic rules of heredity, its DNA basis, and the methods used to find genetic influence and to identify specific genes. The rest of the book highlights what is known about genetics in psychology and psychiatry. The areas about which most is known are cognitive disabilities and abilities, psychopathology, and personality. We also consider areas of psychology more recently introduced to genetics, such as health psychology and aging. These topics are followed by three chapters, one on pathways between genes and behavior, one on environment, and one on evolution. The future of genetic research lies in what we call behavioral genomics—understanding the pathways between genes and behavior, from the expression of genes to the brain. At first, a chapter on the environment might seem odd in a textbook on genetics, but, in fact, the environment is crucial at every step in the pathways between genes, brain, and behavior. One of the oldest controversies in the behavioral sciences, the

enetics) versus nurture (environment) controversy, has given way to a
ure and nurture are important for complex behavioral traits. Moreover,
has made important discoveries about how the environment affects behav-
nt. The chapter on the environment is followed by a chapter on evolution
ion can be viewed as environmental change at the macro rather than the
el of the population. The last chapter looks to the future of behavioral genet-
out these chapters, quantitative genetics and molecular genetics are inter-
of the most exciting developments in behavioral genetics is the ability to begin
specific genes that influence behavior.

se behavioral genetics is an interdisciplinary field that combines genetics and
avioral sciences, it is complex. We have tried to write about it as simply as possi-
hout sacrificing honesty of presentation. Although our coverage is representative,
y no means exhaustive or encyclopedic. History and methodology are relegated to
s and an appendix to keep the focus on what we now know about genetics and behav-
The appendix by Shaun Purcell presents an overview of statistics, quantitative genetic
ory, and a type of quantitative genetic analysis called model fitting. In this edition we
ave retained an interactive Web site that brings the appendix to life with demonstrations:
http://statgen.iop.kcl.ac.uk/bgim. The Web site was designed and written by Shaun Purcell. A list of other useful Web sites, including those of relevant associations, databases,
and other resources, is included after the reference section.

The text is sprinkled with 24 brief, autobiographical "close-ups" of researchers in the
field in order to personalize the research. The close-ups are meant to be representative
of researchers in the field rather than an honor roll of the most illustrious researchers. To
reinforce this point, many of the close-ups in this edition are different from those in the
previous edition. In this edition, we have especially tried to add close-ups about younger
scientists. We are grateful to our colleagues for contributing autobiographical statements
and photographs.

This edition benefited greatly from the advice of many colleagues, far too many to
name. We are grateful to Helena Kiernan, who helped us organize the revision, in a race
to see what would come first: the book or her baby. As in the previous edition, Mary
Louise Byrd, our project editor, eased the publication process. We also appreciate the
effort and support of our editor, Erik E. Gilg, who encouraged us to undertake this revision and accelerated its publication.

CHAPTER ONE

Overview

S ome of the most important recent discoveries about behavior involve genetics. For example, autism (Chapter 12) is a rare but severe disorder beginning early in childhood in which children withdraw socially, not engaging in eye contact or physical contact, with marked communication deficits and stereotyped behavior. Until the 1980s, autism was thought to be environmentally caused by cold, rejecting parents or by brain damage. But genetic studies comparing the risk for identical twins, who are identical genetically (like clones), and fraternal twins, who are only 50 percent similar genetically, indicate substantial genetic influence. If one member of an identical twin pair is autistic, the risk that the other twin is also autistic is very high, about 60 percent. In contrast, for fraternal twins, the risk is very low. Molecular genetic studies are attempting to identify individual genes that contribute to the genetic susceptibility to autism.

Later in childhood, a very common concern especially in boys is a cluster of attention-deficit and disruptive behavior problems, called attention-deficit hyperactivity disorder (ADHD) (Chapter 12). Several twin studies have shown that ADHD is highly heritable (genetically influenced). ADHD is one of the first behavioral areas in which specific genes have been identified. Although many other areas of childhood psychopathology show genetic influence, none are as heritable as autism and ADHD. Some behavior problems, such as childhood anxiety and depression, are only moderately heritable, and others, such as antisocial behavior in adolescence, show little genetic influence.

More relevant to college students are personality traits such as risk-taking (often called sensation-seeking) (Chapter 13), drug use and abuse (Chapter 14), and learning abilities (Chapters 8 and 9). All these domains have consistently shown substantial genetic influence in twin studies and have recently begun to yield clues concerning individual genes that contribute to their heritability. These domains are also examples of an important general principle: Not only do genes contribute to disorders such as autism and ADHD, they also play an important role in normal variation. For example, you might be surprised to learn that differences in weight are almost as heritable as height (Chapter 14). Even though we can control how much we eat and are free to go on crash diets, differences among us in weight are much more a matter of nature (genetics) than nurture (environment). Moreover, normal variation in weight is as highly heritable as overweight or obesity. The same story can be told for behavior. Genetic differences do not just make some of us abnormal; they contribute to differences among all of us in normal variation for personality and cognitive abilities.

One of the greatest genetic success stories involves the most common behavioral disorder in later life, the terrible memory loss and confusion of Alzheimer's disease, which strikes as many as one in five individuals in their eighties (Chapter 7). Although Alzheimer's disease rarely occurs before the age of 60, some early-onset cases of dementia run in families in a simple manner that suggests the influence of only one gene. In 1992, a single gene on chromosome 14 was found to be responsible for many of these early-onset cases.

The gene on chromosome 14 is not responsible for the much more common form of Alzheimer's disease that occurs after 60 years of age. Like most behavioral disorders, late-onset Alzheimer's disease is not caused by a single gene. Still, twin studies indicate genetic influence. If you have a twin who has Alzheimer's disease, your risk of developing it is twice as great if you are an identical twin rather than a fraternal twin. These findings suggest genetic influence.

Even for complex disorders like late-onset Alzheimer's, it is now possible to identify genes that contribute to the risk for the disorder. In 1993, a gene was identified that predicts risk for late-onset Alzheimer's disease far better than any other known risk factor. If you inherit one copy of a particular form (allele) of the gene, your risk for Alzheimer's disease is about four times greater than if you have another allele. If you inherit two copies of the allele (one from each of your parents), your risk is much greater. Finding these genes for early-onset and late-onset Alzheimer's disease has greatly increased our understanding of the brain processes that lead to dementia.

Another example of recent genetic discoveries involves mental retardation (Chapter 7). It has been known for decades that the single most important cause of mental retardation is the inheritance of an entire extra chromosome 21. (Our DNA, the basic hereditary molecule, is packaged as 23 pairs of chromosomes, as explained in Chapter 4.) Instead of inheriting only one pair of chromosomes 21, one from the mother and one from the father, an entire extra chromosome is inherited, usually from the mother. Often called Down syndrome, trisomy-21 is one of the major reasons why women worry about pregnancy later in life. Down syndrome occurs much more frequently when mothers are over 40 years old. The extra chromosome can be detected in the first few months of pregnancy by a procedure called amniocentesis.

In 1991, researchers identified a single gene that is the second most common cause of mental retardation. This form of mental retardation is called fragile X retardation. The gene that causes the disorder is on the X chromosome. Fragile

X mental retardation occurs nearly twice as often in males as in females because males have only one X chromosome. If a boy has the fragile X allele on his X chromosome, he will develop the disorder. Females have two X chromosomes, and it is necessary to inherit the fragile X allele on both X chromosomes in order to develop the disorder. However, females with one fragile X allele can also be affected to some extent. The fragile X gene is especially interesting because it involves a newly discovered type of genetic defect in which a short sequence of DNA mistakenly repeats hundreds of times. This type of genetic defect is now also known to be responsible for several other previously puzzling diseases (Chapter 3).

Genetic research on behavior goes beyond just demonstrating the importance of genetics to the behavioral sciences and allows us to ask questions about how genes influence behavior. For example, does genetic influence change during development? Consider cognitive ability, for example; you might think that as time goes by we increasingly accumulate the effects of Shakespeare's "slings and arrows of outrageous fortune." That is, environmental differences might become increasingly important during one's life span, whereas genetic differences might become less important. However, genetic research shows just the opposite: Genetic influence on cognitive ability increases throughout the individual's life span, reaching levels later in life that are nearly as great as the genetic influence on height (Chapter 8). This finding is an example of developmental genetic analysis.

School achievement and the results of tests you took to apply to college are influenced almost as much by genetics as are the results of tests of cognitive abilities such as intelligence (IQ) tests (Chapter 9). Even more interesting, the substantial overlap between such achievement and the ability to perform well on tests is nearly all genetic in origin. This finding is an example of what is called multivariate genetic analysis.

Genetic research is also changing the way we think about the environment (Chapter 16). For example, we used to think that growing up in the same family makes brothers and sisters similar psychologically. However, for most behavioral dimensions and disorders, it is genetics that accounts for similarity among siblings. Although the environment is important, environmental influences can make siblings growing up in the same family different, not similar. This genetic research has sparked an explosion of environmental research looking for the environmental reasons why siblings in the same family are so different.

Recent genetic research has also shown a surprising result that emphasizes the need to take genetics into account when studying the environment: Many environmental measures used in the behavioral sciences show genetic influence! For example, research in developmental psychology often involves measures of parenting that are assumed to be measures of the family environment. However, genetic research during the past decade has convincingly shown genetic influence on parenting measures. How can this be? One way is that genetic differences influence parents' behavior toward their children. Genetic differences among children can also make a contribution. For example, parents who have more books in their home have children who do better in school, but this correlation does not necessarily mean that having more books in the home is an environmental cause for children performing well in school. Genetic factors could affect parental traits that relate both to the number of books parents have in their home and to their children's achievement at school. Genetic involvement has also been found for many other ostensible measures of the environment, including childhood accidents, life events, and social support. To some extent, people create their own experiences for genetic reasons.

These are examples of what you will learn about in this book. The simple message is that genetics plays a major role in behavior. Genetics integrates the behavioral sciences into the life sciences. Although research in behavioral genetics has been conducted for many years, the field-defining text was published only in 1960 (Fuller & Thompson, 1960). Since that date, discoveries in behavioral genetics have grown at a rate that few other fields in the behavioral sciences can match. For example, one of the major accomplishments of the life sciences has been the sequencing of the human genome, that is, identifying each of the more than 3 billion steps in the spiral staircase that is DNA. This knowledge and new DNA tools are leading to the identification of all the genetic differences between us that are responsible for the heritability of normal and abnormal behavior.

Recognition of the importance of genetics is one of the most dramatic changes in the behavioral sciences during the past two decades. Almost eighty years ago, Watson's (1930) behaviorism detached the behavioral sciences from their budding interest in heredity. A preoccupation with the environmental determinants of behavior continued until the 1970s, when a shift began toward the more balanced contemporary view that recognizes genetic as well as environmental influences. This shift toward genetics in the behavioral sciences can be seen in the increasing numbers of publications in the behavioral sciences that involve genetics, as illustrated in Figure 1.1 for twin studies of behavior. Other ways of documenting the rise of behavioral genetics, especially molecular genetic studies, would show similarly striking increases.

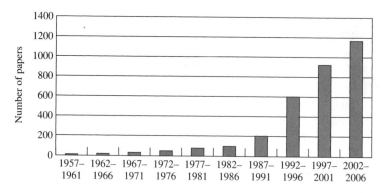

▌ **Figure 1.1** Fifty years of behavioral genetic twin studies.

CHAPTER TWO

Mendel's Laws of Heredity

Huntington disease (HD) begins with personality changes, forgetfulness, and involuntary movements. It typically strikes in middle adulthood; during the next 15 to 20 years, it leads to complete loss of motor control and intellectual function. No treatment has been found to halt or delay the inexorable decline. This is the disease that killed the famous depression-era folksinger Woody Guthrie. Although it affects only about 1 in 20,000 individuals, a quarter of a million people in the world today will eventually develop Huntington disease.

When the disease was traced through many generations, it showed a consistent pattern of heredity. Afflicted individuals had one parent who also had the disease, and approximately half the children of an affected parent developed the disease. (See Figure 2.1 for an explanation of symbols traditionally used to describe family trees, called *pedigrees*. Figure 2.2 shows an example of a Huntington disease pedigree.) What rules of heredity are at work? Why does this lethal condition persist in the population? We will answer these questions in the next section, but first, consider another inherited disorder.

In the 1930s, a Norwegian biochemist discovered an excess of phenylpyruvic acid in the urine of a pair of mentally retarded siblings and suspected that the condition was due to a disturbance in the metabolism of phenylalanine. Phenylalanine is one of the essential amino acids, which are the building blocks of proteins, and is present in many foods in the normal human diet. Other retarded individuals were soon found with this same excess. This type of mental retardation came to be known as phenylketonuria (PKU).

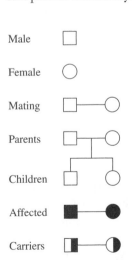

Male	□
Female	○
Mating	□—○
Parents	□—○
Children	□ ○
Affected	■—●
Carriers	◨—◐

Figure 2.1 Symbols used to describe family pedigrees.

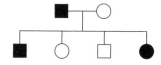

Figure 2.2 Huntington disease. HD individuals have one HD parent. About 50 percent of the offspring of HD parents will have HD.

Figure 2.3 Phenylketonuria. PKU individuals do not typically have parents with PKU. If one child has PKU, the risk for other siblings is 25 percent. As explained later, parents in such cases are carriers for one allele of the PKU gene, but a child must have two alleles in order to be afflicted with recessive disorders such as PKU.

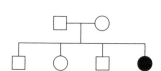

Although the frequency of PKU is only about 1 in 10,000, PKU once accounted for about 1 percent of the mentally retarded institutionalized population. PKU has a pattern of inheritance very different from that of Huntington disease. PKU individuals do not usually have affected parents. Although this might make it seem at first glance as if PKU is not inherited, PKU does in fact "run in families." If one child in a family has PKU, the risk for siblings to develop it is about 25 percent, even though the parents themselves may not be affected (Figure 2.3). One more piece of the puzzle is the observation that when parents are genetically related ("blood" relatives), typically in marriages between cousins, they are more likely to have children with PKU. How does heredity work in this case?

Mendel's First Law of Heredity

Although Huntington disease and phenylketonuria, two examples of hereditary transmission of mental disorders, may seem complicated, they can be explained by a simple set of rules about heredity. The essence of these rules was worked out more than a century ago by Gregor Mendel (1866).

Mendel studied inheritance in pea plants in the garden of his monastery in what is now the Czech Republic (Box 2.1). On the basis of his many experiments, Mendel concluded that there are two "elements" of heredity for each trait in each individual and that these two elements separate, or segregate, during reproduction. Offspring receive one of the two elements from each parent. In addition, Mendel concluded that one of these elements can "dominate" the other, so that an individual with just one dominant element will display the trait. A nondominant, or *recessive*, element is expressed only if both elements are recessive. These conclusions are the essence of Mendel's first law, the *law of segregation*.

No one paid any attention to Mendel's law of heredity for over 30 years. Finally, in the early 1900s, several scientists recognized that Mendel's law is a general law of inheritance, not one peculiar to the pea plant. Mendel's "elements" are now known as *genes*, the basic units of heredity. Some genes have only one form throughout a species, for example, in all pea plants or in all people. Heredity focuses on genes that have different forms: differences that cause some pea seeds to be wrinkled or smooth, or that cause some people to have Huntington disease or PKU. The alternative forms of a gene are called *alleles*. An individual's combination of alleles is its *genotype*, whereas the observed traits are its *phenotype*. The fundamental issue of heredity in the behavioral sciences is the extent to which differences in genotype account for differences in phenotype, observed differences among individuals.

This chapter began with two very different examples of inherited disorders. How can Mendel's law of segregation explain both examples?

Huntington Disease

Figure 2.4 shows how Mendel's law explains the inheritance of Huntington disease. HD is caused by a dominant allele. Affected individuals have one dominant allele (*H*) and one recessive, normal allele (*h*). (It is rare that an HD individual has two *H* alleles, an event that would require both parents to have HD.) Unaffected individuals have two normal alleles.

As shown in Figure 2.4, a parent with HD whose genotype is *Hh* produces gametes (egg or sperm) with either the *H* or the *h* allele. The unaffected (*hh*) parent's gametes all have an *h* allele. The four possible combinations of these gametes from the mother and father result in the offspring genotypes shown at the bottom of Figure 2.4. Offspring will always inherit the normal *h* allele from the unaffected parent, but they have a 50 percent chance of inheriting the *H* allele from the HD parent. This pattern of inheritance explains why HD individuals always have a parent with HD and why 50 percent of the offspring of an HD parent develop the disease.

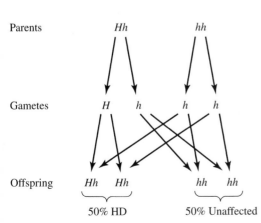

Parents   *Hh*   *hh*

Gametes   *H*   *h*   *h*   *h*

Offspring   *Hh*   *Hh*   *hh*   *hh*

50% HD   50% Unaffected

Figure 2.4 Huntington disease is due to a single gene, with the allele for HD dominant. *H* represents the dominant HD allele and *h* is the normal recessive allele. Gametes are sex cells (eggs and sperm); each carries just one allele. The risk of HD in the offspring is 50 percent.

BOX 2.1

Gregor Mendel's Luck

Before Mendel (1822–1884), much of the research on heredity involved crossing plants of different species. But the offspring of these matings were usually sterile, which meant that succeeding generations could not be studied. Another problem with research before Mendel was that features of the plants investigated were complexly determined. Mendel's success can be attributed in large part to the absence of these problems.

Mendel crossed different varieties of pea plants of the same species; thus the offspring were fertile. In addition, he picked simple either-or traits, qualitative traits, that happened to be due to single genes. He was also lucky that in the traits he chose, one allele completely dominated the expression of the other allele, which is not always the case. However, one feature

Gregor Johann Mendel. A photograph taken at the time of his research. (Courtesy of V. Orel, Mendel Museum, Brno, Czech Republic.)

of Mendel's research was not due to luck. Over seven years, while raising over 28,000 pea plants, he counted all offspring rather than being content, as researchers before him had been, with a verbal summary of the typical results.

Mendel studied seven qualitative traits of the pea plant, such as whether the seed was smooth or wrinkled. He obtained 22 varieties of the pea plant that differed in these seven characteristics. All the varieties were true-breeding plants: those that always yield the same result when crossed with the same kind of plant. Mendel presented the results of eight years of research on the pea plant in his 1866 paper. This paper, *Experiments with Plant Hybrids*, now forms the cornerstone of genetics and is one of the most influential publications in the history of science.

In one experiment, Mendel crossed true-breeding plants with smooth seeds and true-breeding plants with wrinkled seeds. Later in the summer, when he opened the pods containing their offspring (called the F_1, or first filial generation), he found that all of them had smooth seeds. This result indicated that the then-traditional view of blending inheritance was not correct. That is, the F_1 did not have seeds that were even moderately wrinkled. These F_1 plants were fertile, which allowed Mendel to take the next step of allowing plants of the F_1 generation to self-fertilize and then studying their offspring, F_2. The results were striking: Of the 7324 seeds from the F_2, 5474 were smooth and 1850 were wrinkled. That is, $\frac{3}{4}$ of the offspring had smooth seeds and $\frac{1}{4}$ had wrinkled seeds. This result indicates that the factor responsible for wrinkled seeds had not been lost in

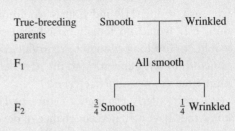

the F_1 generation but had merely been dominated by the factor causing smooth seeds. The figure above summarizes Mendel's results.

Given these observations, Mendel deduced a simple explanation involving two hypotheses. First, each individual has two hereditary "elements," now called alleles (alternate forms of a gene). For Mendel's pea plants, these alleles determined whether the seed was wrinkled or smooth. Thus, each parent has two alleles (either the same or different) but transmits only one of the alleles to each offspring. The second hypothesis is that, when an individual's alleles are different, one allele can dominate the other. These two hypotheses neatly explain the data (see the figure below).

The true-breeding parent plant with smooth seeds has two alleles for smooth seeds (SS). The true-breeding parent plant with wrinkled seeds has two alleles for wrinkled seeds (ss). First-generation (F_1) offspring receive one allele from each parent and are therefore Ss. Because S dominates s, F_1 plants will have smooth seeds. The real test is the F_2 population. Mendel's theory predicts that when F_1 individuals are self-fertilized or crossed with other F_1 individuals, $\frac{1}{4}$ of the F_2 should be SS, $\frac{1}{2}$ Ss, and $\frac{1}{4}$ ss. Assuming S dominates s, then Ss should have smooth seeds like the SS. Thus, $\frac{3}{4}$ of the F_2 should have smooth seeds and $\frac{1}{4}$ should have wrinkled, which is exactly what Mendel's data indicated. Mendel also discovered that the inheritance of one trait is not affected by the inheritance of another trait. Each trait is inherited in the expected 3:1 ratio.

Mendel was not so lucky in terms of acknowledgment of his work during his lifetime. When Mendel published the paper about his theory of inheritance in 1866, reprints were sent to scientists and libraries in Europe and the United States. However, for 35 years, Mendel's findings on the pea plant were ignored by most biologists, who were more interested in evolutionary processes that could account for change rather than continuity. Mendel died in 1884 without knowing the profound impact that his experiments would have during the twentieth century.

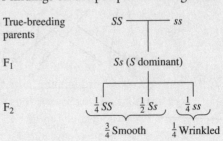

Why does this lethal condition persist in the population? If HD had its effect early in life, HD individuals would not live to reproduce. In one generation, HD would no longer exist because any individual with the HD allele would not live long enough to reproduce. The dominant allele for HD is maintained from one generation to the next because its lethal effect is not expressed until after the reproductive years.

A particularly traumatic feature of HD is that offspring of HD parents know they have a 50 percent chance of developing the disease and of passing on the HD gene. In 1983, DNA markers were used to show that the gene for HD is on chromosome 4, as will be discussed in Chapter 4. In 1993, the HD gene itself was identified. Now it is possible to determine for certain whether a person has the HD gene.

This genetic advance raises its own problems. If one of your parents had HD, you would be able to find out whether or not you have the HD allele. You would have a 50 percent chance of finding that you do not have the HD allele, but you would also have a 50 percent chance of finding that you do have the HD allele and will eventually die from it. In fact, most people at risk for HD decide *not* to take the test. Identifying the gene does, however, make it possible to determine whether a fetus has the HD allele and holds out the promise of future interventions that can correct the HD defect (Chapter 6).

Phenylketonuria

Mendel's law also explains the inheritance of PKU. Unlike HD, PKU is due to the presence of two recessive alleles. For offspring to be affected, they must have two copies of the allele. Those offspring with only one copy of the allele are not afflicted with the disorder. They are called *carriers* because they carry the allele and can pass it on to their offspring. Figure 2.5 illustrates the in-

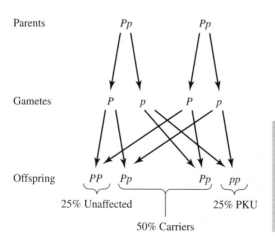

Figure 2.5 PKU is inherited as a single gene. The allele that causes PKU is recessive. *P* represents the normal dominant allele, and *p* is the recessive allele for PKU. Parents are carriers; the risk of PKU for their offspring is 25 percent.

heritance of PKU from two unaffected carrier parents. Each parent has one PKU allele and one normal allele. Offspring have a 50 percent chance of inheriting the PKU allele from one parent and a 50 percent chance of inheriting the PKU allele from the other parent. The chance of both these things happening is 25 percent. If you flip a coin, the chance of heads is 50 percent. The chance of getting two heads in a row is 25 percent (i.e., 50 percent times 50 percent).

This pattern of inheritance explains why unaffected parents have children with PKU and why the risk of PKU in offspring is 25 percent when both parents are carriers. For PKU and other recessive disorders, identification of the genes makes it possible to determine whether potential parents are carriers. Identification of the PKU gene also makes it possible to determine whether a particular pregnancy involves an affected fetus. In fact, all newborns in most countries are screened for elevated phenylalanine levels in their blood, because early diagnosis of PKU can help parents prevent retardation by serving low-phenylalanine diets to their affected children.

Figure 2.5 also shows that 50 percent of children born of two carrier parents are likely to be carriers, and 25 percent will inherit the normal allele from both parents. If you understand how a recessive trait such as PKU is inherited, you should be able to work out the risk for PKU in offspring if one parent has PKU and the other parent is a carrier. (The risk is 50 percent.)

We have yet to explain why recessive traits like PKU are seen more often in offspring whose parents are genetically related. Although PKU is rare (1 in 10,000), about 1 in 50 individuals are carriers of one PKU allele (Box 2.2). If you are a PKU carrier, your chance of marrying someone who is also a carrier is 2 percent. However, if you marry someone genetically related to you, the PKU allele must be in your family, so the chances are much greater than 2 percent that your spouse will also carry the PKU allele.

It is very likely that we all carry at least one harmful recessive gene of some sort. However, the risk that our spouses are also carriers for the same disorder is small unless we are genetically related to them. In contrast, about half the children born to incestuous relationships between father and daughter show severe genetic abnormalities, often including childhood death or mental retardation. This pattern of inheritance explains why most severe genetic disorders are recessive: Because carriers of recessive alleles do not show the disorder, they escape eradication by natural selection.

It should be noted that even single-gene disorders such as PKU are not so simple, because many different mutations of the gene occur and these have different effects (Scriver & Waters, 1999). New PKU mutations emerge in individuals with no family history. Some single-gene disorders are largely caused by new mutations. In addition, age of onset may vary for single-gene disorders, as it does in the case of HD.

BOX 2.2

How Do We Know That 1 in 50 People Are Carriers for PKU?

I f you randomly mate F_2 plants to obtain an F_3 generation, the frequencies of the S and s alleles will be the same as in the F_2 generation, as will the frequencies of the SS, Ss, and ss genotypes. Shortly after the rediscovery of Mendel's law in the early 1900s, this implication of Mendel's law was formalized and eventually called the *Hardy-Weinberg equilibrium:* The frequencies of alleles and genotypes do not change across generations unless forces such as natural selection or migration change them. This rule is the basis for a discipline called population genetics, whose practitioners study forces that change gene frequencies (see Chapter 17).

Hardy-Weinberg equilibrium also makes it possible to estimate frequencies of alleles and genotypes. The frequencies of the dominant and recessive alleles are usually referred to as p and q, respectively. Eggs and sperm have just one allele for each gene. The chance that any particular egg or sperm has the dominant allele is p. Because sperm and egg unite at random, the chance that a sperm with the dominant allele fertilizes an egg with the dominant allele is the product of the two frequencies, $p \times p = p^2$. Thus, p^2 is the frequency of offspring with two dominant alleles (called the *homozygous dominant* genotype). In the same way, the *homozygous recessive* genotype has a frequency of q^2. As shown in the diagram at the right, the frequency of offspring with one dominant allele and one recessive allele (called the *heterozygous* genotype) is $2pq$. In other words, if a population is in Hardy-Weinberg equilibrium, the frequency of the offspring genotypes is $p^2 + 2pq + q^2$. In populations with random mating, the expected genotypic frequencies are merely the product of $p + q$ for the mothers' alleles and $p + q$ for the fathers' alleles. That is, $(p + q)^2 = p^2 + 2pq + q^2$.

		Eggs	
Frequencies		p	q
Sperm	p	p^2	pq
	q	pq	q^2

For PKU, q^2, the frequency of PKU individuals (homozygous recessive) is 0.0001. If you know q^2, you can estimate the frequency of the PKU allele and PKU carriers, assuming Hardy-Weinberg equilibrium. The frequency of the PKU allele is q, which is the square root of q^2. The square root of 0.0001 is 0.01, so that 1 in 100 alleles in the population are the recessive PKU alleles. If there are only two alleles at the PKU locus, then the frequency of the dominant allele (p) is $1 - 0.01 = 0.99$. What is the frequency of carriers? Because carriers are heterozygous genotypes with one dominant allele and one recessive allele, the frequency of carriers of the PKU allele is 1 in 50 (that is, $2pq = 2 \times 0.99 \times 0.01 = 0.02$).

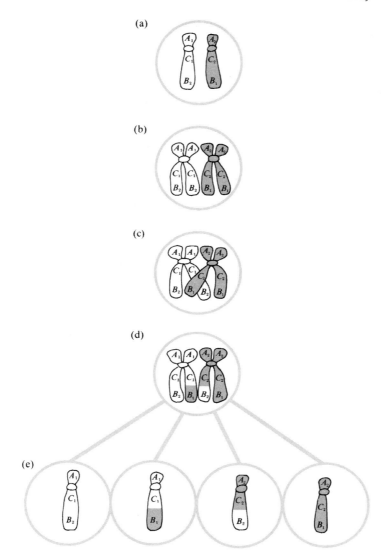

Figure 2.7 Illustration of recombination. The maternal chromosome, carrying the alleles A_1, C_1, and B_2, is represented in white; the paternal chromosome, with alleles A_2, C_2, and B_1, is gray. The right chromatid (the duplicated chromosome produced during meiosis) of the maternal chromosome crosses over (recombines) with the left chromatid of the paternal chromosome.

meiosis. During this stage, the chromatids can break and rejoin (Figure 2.7d). Each of the chromatids will be transmitted to a different gamete (Figure 2.7e). Consider only the A and B loci for the moment. As shown in Figure 2.7e, one gamete will carry the genes A_1 and B_2, as in the mother, and one will carry A_2 and B_1, as in the father. The other two will carry A_1 with B_1 and A_2 with B_2. For the latter two pairs, recombination has taken place—these combinations were not present on the parental chromosomes.

The probability of recombination between two loci on the same chromosome is a function of the distance between them. In Figure 2.7, for example, the A and C loci have not recombined. All gametes are either A_1C_1 or A_2C_2, as in the parents, because the crossover did not occur between these loci. Crossover could occur between the A and C loci, but it would happen less frequently than between A and B.

These facts have been used to "map" genes on chromosomes. The distance between two loci can be estimated by the number of recombinations per 100 gametes. This distance is called a map unit or *centimorgan*, named after T. H. Morgan, who first identified linkage groups in the fruit fly *Drosophila* (Morgan, Sturtevant, Muller, & Bridges, 1915). If two loci are far apart, like the A and B loci, recombination will separate the two loci as often as if the loci were on different chromosomes, and they will not appear to be linked.

To identify the location of a gene on a particular chromosome, *linkage analysis* can be used. Linkage analysis refers to techniques that use information about violations of independent assortment to identify the chromosomal location of a gene. DNA markers serve as signposts on the chromosomes, as discussed in Chapter 6. Since 1980, the power of linkage analysis has greatly increased with the discovery of millions of these markers. Linkage analysis looks for a violation of independent assortment between a trait and a DNA marker. In other words, linkage analysis assesses whether the DNA marker and the trait co-assort in a family more often than expected by chance.

SUMMING UP

Mendel also showed that the inheritance of one gene is not affected by the inheritance of another gene. This is Mendel's second law, the law of independent assortment. Violation of Mendel's second law indicates that genes are inherited together on the same chromosome. This inheritance pattern is the basis for linkage analysis, which makes it possible to locate genes on specific chromosomes.

In 1983, the gene for Huntington disease was shown to be linked to a DNA marker near the tip of one of the larger chromosomes (chromosome 4; see Chapter 6) (Gusella et al., 1983). This was the first time that the new DNA markers had been used to demonstrate a linkage for a disorder for which no chemical mechanism was known. DNA markers that are closer to the Huntington gene have since been developed and have made it possible to pinpoint the gene. As noted earlier, the gene itself was finally located precisely in 1993.

Once a gene has been found, two things are possible. First, the DNA variation responsible for the disorder can be identified. This identification provides a DNA test that is directly associated with the disorder in individuals and is more than just a risk estimate calculated on the basis of Mendel's laws. That is, the DNA test can be used to diagnose the disorder in individuals regardless of information about other family members. Second, the protein coded by the gene can be studied; this investigation is a major step toward understanding how the gene has its effect and thus can possibly lead to a therapy. In the case of Huntington disease, the gene codes for a previously unknown protein, called huntingtin. This protein interacts with many other proteins, which has hampered efforts to develop drug therapies (Ross, 2004).

Although the disease process of the Huntington gene is not yet fully understood, Huntington disease, like fragile X mental retardation mentioned in Chapter 1, also involves a type of genetic defect in which a short sequence of DNA is repeated many times (see Chapter 3). The defective gene product slowly has its effect over the life course by contributing to neural death in the cerebral cortex and basal ganglia. This leads to the motoric and cognitive problems characteristic of Huntington disease.

Finding the PKU gene was easier because its enzyme product was known, as described in Chapter 1. In 1984, the gene for PKU was found and shown to be on chromosome 12 (Lidsky et al., 1984). For decades, PKU infants have been identified by screening for the physiological effect of PKU—high blood phenylalanine levels—but this test is not highly accurate. Developing a DNA test for PKU has been hampered by the discovery that there are many different mutations at the PKU locus and that these mutations differ in the magnitude of their effects. This diversity contributes to the variation in blood phenylalanine levels among PKU individuals.

Of the several thousand single-gene disorders known (about half of which involve the nervous system), the precise chromosomal location has been identified for several hundred genes. The gene itself and the specific mutation have been found for more than a hundred disorders, and this number is increasing. One of the goals of the Human Genome Project is to identify all genes. Rapid progress toward this goal holds the promise of identifying genes even for complex behaviors influenced by multiple genes as well as environmental factors.

Summary

Huntington disease (HD) and phenylketonuria (PKU) are examples of dominant and recessive disorders, respectively. They follow the basic rules of heredity described by Mendel more than a century ago. A gene may exist in two or more different forms (alleles). One allele can dominate the expression of the other. The two alleles, one from each parent, separate (segregate) during gamete formation. This rule is Mendel's first law, the law of segregation. The law explains many features of inheritance: why 50 percent of the offspring of an HD parent are eventually afflicted, why this lethal gene persists in the population, why PKU children usually do not have PKU parents, and why PKU is more likely when parents are genetically related.

Mendel's second law is the law of independent assortment: The inheritance of one gene is not affected by the inheritance of another gene. However, genes that are closely linked on the same chromosome can co-assort, thus violating Mendel's law of independent assortment. Such violations make it possible to map genes to chromosomes by using linkage analysis. For Huntington disease and PKU, linkage has been established and the genes responsible for the disorders have been identified.

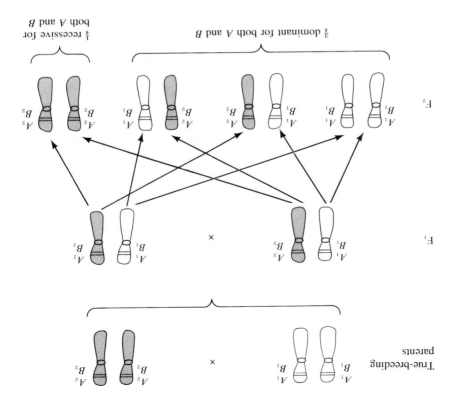

Figure 2.6 An exception to Mendel's second law occurs if two genes are closely linked on the same chromosome. The A_1 allele and the B_1 allele are dominant; the A_2 and B_2 alleles are recessive.

The reason why such violations of Mendel's second law are important is that they make it possible to map genes to chromosomes. If the inheritance of a particular pair of genes violates Mendel's second law, then it must mean that they tend to be inherited together and thus reside on the same chromosome. This phenomenon is called *linkage*. However, it is actually not sufficient for two linked genes to be on the same chromosome; they must also be very close together on the chromosome. Unless genes are near each other on the same chromosome, they will recombine by a process in which chromosomes exchange parts. Recombination occurs during meiosis in the ovaries and testes, when gametes are produced.

Figure 2.7 illustrates recombination for three loci (A, C, B) on a single chromosome. The maternal chromosome, carrying the alleles A_1, C_1, and B_1, is represented in white; the paternal chromosome with alleles A_2, C_2, and B_2, is gray. During meiosis, each chromosome duplicates to form sister chromatids (Figure 2.7b). These sister chromatids may cross over one another, as shown in Figure 2.7c. This overlap happens an average of one time for each chromosome during

---

S U M M I N G   U P

Mendel's theory of heredity can explain dominant (Huntington disease) and recessive (PKU) patterns of inheritance. A gene may exist in two or more different forms (alleles). The two alleles, one from each parent, separate (segregate) during gamete formation. This is Mendel's first law, the law of segregation.

Mendel's Second Law of Heredity

Not only do the alleles for Huntington disease segregate independently during gamete formation, they are also inherited independently from the alleles for PKU. This finding makes sense, because Huntington disease and PKU are caused by different genes, each of the two genes is inherited independently. Mendel experimented systematically with crosses between varieties of pea plants that differed in two or more traits. He found that alleles for the two genes assort independently. In other words, the inheritance of one gene is not affected by the inheritance of another gene. This is Mendel's *law of independent assortment*.

Most important about Mendel's second law are its exceptions. We now know that genes are not just floating around in eggs and sperm. They are carried on *chromosomes*. The term *chromosome* literally means "colored body," because in certain laboratory preparations the staining characteristics of these structures are different from those of the rest of the nucleus of the cell. Genes are located at places called *loci* (singular, *locus*, from the Latin, meaning "place") on chromosomes. Eggs contain just one chromosome from each pair of the mother's set of chromosomes, and sperm contain just one from each pair of the father's set. An egg fertilized by a sperm thus has the full chromosome complement, which, in humans, is 23 pairs of chromosomes. Chromosomes are discussed in more detail in Chapter 4.

When Mendel studied the inheritance of two traits at the same time (let's call them A and B), he crossed true-breeding parents that showed the dominant trait for both A and B with parents that showed the recessive forms for A and B. He found second-generation (F_2) offspring of all four possible types: dominant for A and B, dominant for A and recessive for B, recessive for A and dominant for B, and recessive for A and B. The frequencies of the four types of offspring were as expected if A and B were inherited independently. Mendel's law is violated, however, when genes for two traits are close together on the same chromosome. If Mendel had studied the joint inheritance of two such traits, the results would have surprised him. The two traits would not have been inherited independently. Figure 2.6 illustrates what would happen if the genes for traits A and B were very close together on the same chromosome. Instead of finding all four types of F_2 offspring, Mendel would have found only two types: dominant for both A and B and recessive for both A and B.

CHAPTER THREE

Beyond Mendel's Laws

C olor blindness shows a pattern of inheritance that does not appear to conform to Mendel's laws. The most common color blindness involves difficulty in distinguishing red and green, a condition caused by a lack of certain color-absorbing pigments in the retina of the eye. It occurs more frequently in males than in females. More interesting, when the mother is color blind and the father is not, all of the sons but none of the daughters are color blind (Figure 3.1a). When the father is color blind and the mother is not, offspring are seldom affected (Figure 3.1b). But something remarkable happens to these apparently normal daughters of a color-blind father: Half of their sons are likely to be color blind. This is the well-known skip-a-generation phenomenon—fathers have it, their daughters do not, but some of the grandsons do. What could be going on here in terms of Mendel's laws of heredity?

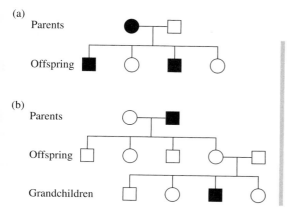

(a)
Parents
Offspring

(b)
Parents
Offspring
Grandchildren

Figure 3.1 Inheritance of color blindness. (a) A color-blind mother and unaffected father have color-blind sons but unaffected daughters. (b) An unaffected mother and color-blind father have unaffected offspring, but daughters have sons with 50 percent risk for color blindness. (See Figure 2.1 for symbols used to describe family pedigrees.)

Genes on the X Chromosome

There are two chromosomes called the sex chromosomes because they differ for males and females. Females have two X chromosomes, and males have one X chromosome and a smaller chromosome called Y.

Color blindness is caused by a recessive allele on the X chromosome. But males have only one X chromosome; so, if they have one allele for color blindness (*c*) on their single X chromosome, they are color blind. For females to be color blind, they must inherit the *c* allele on both of their X chromosomes. For this reason, the hallmark of a sex-linked (meaning *X-linked*) recessive gene is a greater incidence in males. For example, if the frequency of an X-linked recessive allele (*q* in Chapter 2) for a disorder were 10 percent, then the expected frequency of the disorder in males would be 10 percent, but the frequency in females (q^2) would be only 1 percent (i.e., $0.10^2 = 0.01$).

Figure 3.2 illustrates the inheritance of the sex chromosomes. Both sons and daughters inherit one X chromosome from their mother. Daughters inherit their father's single X chromosome and sons inherit their father's Y chromosome. Sons cannot inherit an allele on the X chromosome from their father. For this reason, another sign of an X-linked recessive trait is that father-son resemblance is negligible. Daughters inherit an X-linked allele from their father, but they do not express a recessive trait unless they receive another such allele on the X chromosome from their mother.

Inheritance of color blindness is further explained in Figure 3.3. In the case of a color-blind mother and unaffected father (Figure 3.3a), the mother has the *c* allele on both of her X chromosomes and the father has the normal allele (*C*)

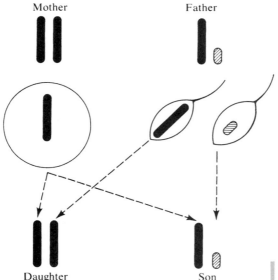

Figure 3.2 Inheritance of X and Y chromosomes.

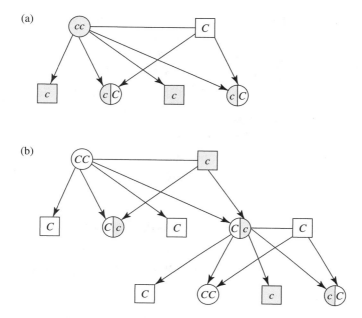

Figure 3.3 Color blindness is inherited as a recessive gene on the X chromosome. *c* refers to the recessive allele for color blindness, and *C* is the normal allele. (a) Color-blind mothers are homozygous recessive (*cc*). (b) Color-blind fathers have a *c* allele on their single X chromosome, which is transmitted to daughters but not to sons.

on his single X chromosome. Thus, sons always inherit an X chromosome with the *c* allele from their mother and are color blind. Daughters carry one *c* allele from their mother but are not color blind because they have inherited a normal, dominant *C* allele from their father. They carry the *c* allele without showing the disorder, so they are called *carriers*, a status indicated by the half-shaded circles in Figure 3.3.

In the second example (Figure 3.3b), the father is color blind but the mother is neither color blind nor a carrier of the *c* allele. None of the children are color blind, but the daughters are all carriers because they must inherit their father's X chromosome with the recessive *c* allele. You should now be able to predict the risk of color blindness for offspring of these carrier daughters. As shown in the bottom row of Figure 3.3b, when a carrier daughter (*Cc*) has children by an unaffected male (*C*), half of her sons but none of her daughters are likely to be color blind. Half of the daughters are carriers. This pattern of inheritance explains the skip-a-generation phenomenon. Color-blind fathers have no color-blind sons or daughters (assuming normal, noncarrier mothers), but their daughters are carriers of the *c* allele. The daughters' sons have a 50 percent chance of being color blind.

The sex chromosomes are inherited differently for males and females, so detecting X linkage is much easier than identifying a gene's location on other chromosomes. Color blindness was the first reported human X linkage. About 1000 genes have been identified on the X chromosome, as well as a disproportionately high number of single-gene diseases (Ross et al., 2005). The Y chromosome has genes for determining maleness plus only about 50 other genes and the smallest number of genes associated with disease of any chromosome.

SUMMING UP

Recessive genes on the X chromosome, such as the gene for color blindness, result in more affected males than females and appear to skip a generation.

Other Exceptions to Mendel's Laws

Several other genetic phenomena do not appear to conform to Mendel's laws in the sense that they are not inherited in a simple way through the generations.

New Mutations

The most common type of exceptions to Mendel's laws involves new DNA mutations that do not affect the parent because they occur during the formation of the parent's eggs or sperm. But this situation is not really a violation of Mendel's laws, because the new mutations are passed on according to Mendel's laws, even though affected individuals have unaffected parents. Many genetic diseases involve such spontaneous mutations, which are not inherited from the preceding generation. An example is Rett syndrome, an X-linked dominant disorder that has a prevalence of about 1 in 10,000 in girls. Although girls with Rett syndrome develop normally during the first year of life, they later regress and eventually become both mentally and physically disabled. Boys with this mutation on their single X chromosome die either before birth or in the first two years after birth. (See Chapter 7.)

In addition, DNA mutations frequently occur in cells other than those that produce eggs or sperm and are not passed on to the next generation. This mutation type is the cause of many cancers, for example. Although these mutations affect DNA, they are not heritable because they do not occur in the eggs or sperm.

Changes in Chromosomes

Changes in chromosomes are an important source of mental retardation, as discussed in Chapter 7. For example, Down syndrome occurs in about 1 in 1000 births and accounts for more than a quarter of individuals with mild to moderate retardation. It was first described by Langdon Down in 1866, the same year that Mendel published his classic paper. For many years, the origin of Down syndrome defied explanation because it does not "run in families." Another puzzling feature is that it occurs much more often in the offspring of women who gave birth after 40 years of age. This relationship to maternal age suggests environmental explanations.

Instead, in the late 1950s, Down syndrome was shown to be caused by the presence of an entire extra chromosome with its thousands of genes. As explained in Chapter 4, during the formation of eggs and sperm (called *gametes*), each of the 23 pairs of chromosomes separates, and egg and sperm carry just one member of each pair. When the sperm fertilizes the egg, the pairs are reconstituted, with one chromosome of each pair coming from the father and the other coming from the mother. But sometimes the initial division in gamete formation is not even. When this accident happens, one egg or sperm might have both members of a particular chromosome pair and another egg or sperm might have neither. This failure to apportion the chromosomes equally is called *nondisjunction* (Figure 3.4). Nondisjunction is a major reason why so many conceptions abort spontaneously in the first few weeks of prenatal life. However, in the case of certain chromosomes, some fetuses with chromosomal anomalies are able to survive, though with developmental abnormalities. A prominent example is that of Down syndrome, which is caused by the presence of three copies (called *trisomy*) of one of the smallest chromosomes (chromosome 21). No individuals have been found with only one of these chromosomes (*monosomy*), which might occur when nondisjunction leaves an egg or sperm

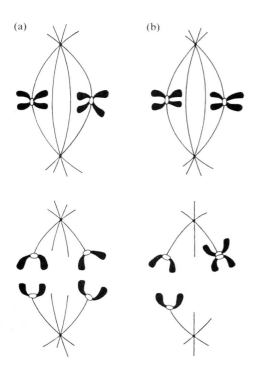

(a)   (b)

Figure 3.4 An exception to Mendel's laws of heredity: nondisjunction of chromosomes. (a) When eggs and sperm are formed, chromosomes for each pair line up and then split, and each new egg or sperm has just one member of each chromosome pair. (b) Sometimes this division does not occur properly, so one egg or sperm has both members of a chromosome pair and the other egg or sperm has neither.

with no copy of the chromosome and another egg or sperm with two copies. It is assumed that this monosomy is lethal. Apparently, too little genetic material is more damaging than extra material. Because most cases of Down syndrome are created anew by nondisjunction, Down syndrome generally is not familial.

Nondisjunction also explains why the incidence of Down syndrome is higher among the offspring of older mothers. All the immature eggs of a female mammal are present before birth. These eggs have both members of each pair of chromosomes. Each month, one of the immature eggs goes through the final stage of cell division. Nondisjunction is more likely to occur as the female grows older and activates immature eggs that have been dormant for decades. In contrast, fresh sperm are produced all the time. For this reason, the incidence of Down syndrome is not affected by the age of the father.

Many women worry about reproducing later in life because of chromosomal abnormalities such as Down syndrome. Much of the worry of pregnancies later in life can be relieved by *amniocentesis*, a procedure that examines the chromosomes of the fetus.

Expanded Triplet Repeats

We have known about mutations and chromosomal abnormalities for a long time. Two other exceptions to Mendel's rules were discovered more recently. One is in effect a special form of mutation that involves *repeat sequences* of DNA. Although we do not know why, some very short segments of DNA—two, three, or four nucleotide bases of DNA (Chapter 4)—repeat a few times or up to a few dozen times. Different repeat sequences can be found in as many as 50,000 places in the human genome. Each repeat sequence has several, often a dozen or more, alleles that consist of various numbers of the same repeat sequence; these alleles are usually inherited from generation to generation according to Mendel's laws. For this reason, and because there are so many of them, repeat sequences are widely used as DNA markers in linkage studies.

Sometimes the number of repeats at a particular locus increases and causes problems (Wells & Warren, 1998). About 20 diseases are now known to be associated with such expansions of repeat sequences; all involve the brain and thus lead to behavioral problems. For example, most cases of Huntington disease involve a repeat in the Huntington gene on chromosome 4. It is called a triplet repeat because the repeated unit is a certain sequence of three nucleotide bases of DNA. All combinations of the four nucleotide bases of DNA (see Chapter 4) are possible but certain combinations are more common, such as CGG and CAG. Normal Huntington alleles contain between 11 and 34 copies of the triplet repeat, but Huntington alleles have more than 40 copies. The expanded number of triplet repeats is unstable and can increase in subsequent generations. This phenomenon explains a previously mysterious non-Mendelian process called *genetic anticipation*, in which symptoms appear at an earlier age and with greater severity in successive generations. For Huntington disease, longer expansions lead to earlier onset of the disorder and greater severity. The expanded triplet repeat is CAG, which codes for the amino acid glutamine and results in a protein with an expanded number of glutamines in the middle of the protein. The additional glutamines change the conformation of the protein and confer new and toxic properties to the protein. This leads to neural death, especially in the cerebral cortex and basal ganglia. Despite this non-Mendelian twist of genetic anticipation, Huntington disease generally follows Mendel's laws of heredity as a single-gene dominant disorder.

Genetic anticipation was originally described early in the century as a phenomenon generally occurring in "insanity." Recent studies of schizophrenia and manic-depressive psychosis (bipolar disorder) do indeed find evidence of genetic anticipation, an observation suggesting the possibility that expanded triplet repeats might also affect these disorders. However, one of the problems is that sampling artifacts can mimic true anticipation. For example, having schizophrenia reduces the chances of finding a partner and having children. Therefore individuals who suffer from schizophrenia but still become parents are likely to have had a late onset of their illness after having already produced their children. If any of the children become affected, their age of onset is likely to be close to the average for schizophrenia and hence will be earlier than their affected parent. Despite such difficulties in pinning down anticipation, the possible clue it provides toward identifying genes seems worth following up. Several laboratories have now reported evidence of trinucleotide repeat expansions somewhere within the genomes of patients with schizophrenia and bipolar disorder, but as yet the genes containing these expansions have not been identified (Tsutsumi et al., 2004).

Fragile X mental retardation, the most common cause of mental retardation after Down syndrome, is also caused by an expanded triplet repeat that violates Mendel's laws. Although this type of mental retardation was known to occur almost twice as often in males as in females, its pattern of inheritance did not conform to sex linkage because it is caused by an unstable expanded repeat. As explained in Chapter 7, the expanded triplet repeat makes the X chromosome fragile in a certain laboratory preparation, which is how fragile X received its name. Parents who inherit X chromosomes with a normal number of repeats (5 to 40 repeats) at a particular locus sometimes produce eggs or sperm with an expanded number of repeats (up to 200 repeats), called a *premutation*. This premutation does not cause retardation in the offspring, but it is unstable and often leads to much greater expansions (200 or more repeats) in the next generation, which do cause retardation (Figure 3.5). Unlike the expanded repeat responsible for Huntington disease, the expanded repeat sequence (CGG) for fragile X mental retardation interferes with transcription of the DNA into messenger RNA (Garber, Smith, Reines, & Warren, 2006; see Chapter 7).

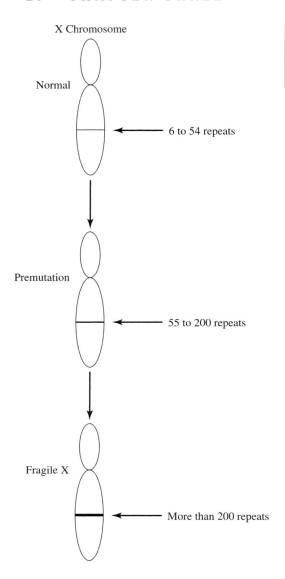

X Chromosome

Normal

← 6 to 54 repeats

Premutation

← 55 to 200 repeats

Fragile X

← More than 200 repeats

Figure 3.5 Fragile X mental retardation involves a triplet repeat sequence of DNA on the X chromosome that can expand over generations.

Genomic Imprinting

Another example of exceptions to Mendel's laws is called *genomic imprinting* (Reik & Walter, 2001). In genomic imprinting, the expression of a gene depends on whether it is inherited from the mother or from the father, even though, as usual, one allele is inherited from each parent. The precise mechanism by which one parent's allele is imprinted is not known, but it usually involves inactivation of a part of the gene by a process called *methylation* (Delaval & Feil, 2004; see Chapter 15); recently, attention has been directed toward RNA-mediated silencing involving non-coding RNA (Pauler & Barlow, 2006; see Chapter 4).

About two dozen such genes have been described in both mice and humans (Morison, Ramsay, & Spencer, 2005). The most striking example of genomic imprinting in humans involves deletions of a small part of chromosome 15 that lead to two very different disorders, depending on whether a deletion is inherited from the mother or the father. When it is inherited from the mother, it causes what is known as Angelman syndrome, which causes severe mental retardation and other manifestations, such as an awkward gait and frequent inappropriate laughter. When a deletion is inherited from the father, it causes other behavioral problems, such as overeating, temper outbursts, and depression, as well as physical problems such as obesity and short stature (Prader-Willi syndrome).

In addition to genes acting differently depending on whether they are inherited from the mother or the father, the expansion of repeat sequences discussed in the previous section sometimes depends on the sex of the individual making the gamete. For example, the fragile X repeat sequence expands in females (mothers), whereas the Huntington disease repeat sequence expands in males (fathers).

SUMMING UP

Other exceptions to Mendel's laws include new mutations, changes in chromosomes, expanded triplet sequences, and genomic imprinting. Many genetic diseases involve spontaneous mutations that are not inherited from generation to generation. Changes in chromosomes include nondisjunction, which is the single most important cause of mental retardation and produces the trisomy of Down syndrome. Expanded triplet repeats are responsible for the next most important cause of mental retardation, fragile X, and for Huntington disease. Genomic imprinting occurs when the expression of a gene depends on whether it is inherited from the mother or from the father, as in Angelman and Prader-Willi syndromes.

Complex Traits

Most psychological traits show patterns of inheritance that are much more complex than those of Huntington disease or PKU. Consider schizophrenia and general cognitive ability.

Schizophrenia

Schizophrenia (Chapter 10) is a severe mental disorder characterized by thought disorders. Nearly 1 in 100 people throughout the world are afflicted by this disorder at some point in life, 100 times more than Huntington disease or PKU. Schizophrenia shows no simple pattern of inheritance like Huntington disease, PKU, or

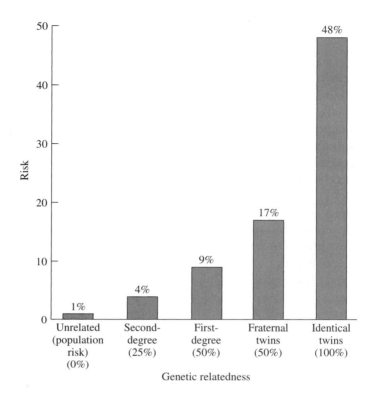

Figure 3.6 Risk for schizophrenia increases with genetic relatedness. (Data adapted from Gottesman, 1991.)

color blindness, but it is familial (Figure 3.6). A special incidence figure used in genetic studies is called a *morbidity risk estimate* (also called the *lifetime expectancy*), which is the chance of being affected during an entire lifetime. The estimate is "age-corrected" for the fact that some as yet unaffected family members have not yet lived through the period of risk. If you have a second-degree relative (grandparent or aunt or uncle) who is schizophrenic, your risk for schizophrenia is about 4 percent, four times greater than the risk in the general population. If a first-degree relative (parent or sibling) is schizophrenic, your risk is about 9 percent. If several family members are affected, the risk is greater. If your fraternal twin has schizophrenia, your risk is higher than for other siblings, about 17 percent, even though fraternal twins are no more similar genetically than siblings. Most striking, the risk is about 48 percent for an identical twin whose co-twin is schizophrenic. Identical twins develop from one embryo, which in the first few days of life splits into two embryos, each with the same genetic material (Chapter 5).

Clearly, the risk of developing schizophrenia increases systematically as a function of the degree of genetic similarity which an individual has to another who is affected. Heredity appears to be implicated, but the pattern of affected individuals does not conform to Mendelian proportions. Are Mendel's laws of heredity at all applicable to such a complex outcome?

General Cognitive Ability

Many psychological traits are quantitative dimensions, as are physical traits such as height and biomedical traits such as blood pressure. Quantitative dimensions are often continuously distributed in the familiar bell-shaped curve, with most people in the middle and fewer people toward the extremes.

For example, as discussed in Chapter 8, an intelligence test score from a general test of intelligence is a composite of diverse tests of cognitive ability and is used to provide an index of general cognitive ability. Intelligence test scores are normally distributed for the most part.

Because general cognitive ability is a quantitative dimension, it is not possible to count "affected" individuals. Nonetheless, it is clear that general cognitive ability runs in families. For example, parents with high intelligence test scores tend to have children with higher than average scores. Like schizophrenia, transmission of general cognitive ability does not seem to follow simple Mendelian rules of heredity.

The statistics of quantitative traits are needed to describe family resemblance. (An overview is provided in the Appendix.) Over a hundred years ago, Francis Galton, the father of behavioral genetics, tackled this problem of describing family resemblance for quantitative traits. He developed a statistic that he called co-relation and that has become the widely used correlation coefficient. More formally, it is called the Pearson product-moment correlation, named after Karl Pearson, Galton's colleague. The *correlation* is an index of resemblance that ranges from .00, indicating no resemblance, to 1.0, indicating perfect resemblance.

Correlations for intelligence test scores show that resemblance of family members depends on the closeness of the genetic relationship (Figure 3.7). The correlation of intelligence test scores for pairs of individuals taken at random from the population is .00. The correlation for cousins is about .15. For half siblings, who have just one parent in common, the correlation is about .30. For full siblings, who have both parents in common, the correlation is about .45; this correlation is similar to that between parents and offspring. Scores for fraternal twins correlate about .60, which is higher than the correlation of .45 for full siblings but lower than the correlation for identical twins, which is about .85. In addition, husbands and wives correlate about .40, a result that has implications for interpreting sibling and twin correlations, as discussed in Chapter 8.

How do Mendel's laws of heredity apply to continuous dimensions such as general cognitive ability?

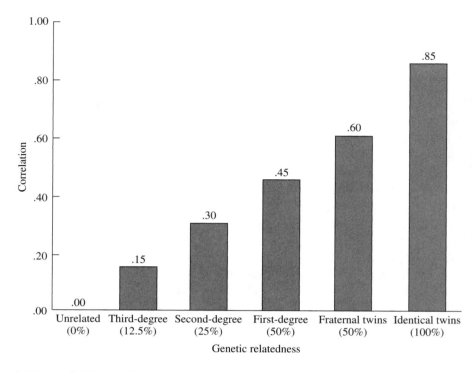

Figure 3.7 Resemblance for general cognitive ability increases with genetic relatedness. (Data adapted from Bouchard & McGue, 1981, as modified by Loehlin, 1989.)

Pea Size

Although pea plants might not seem relevant to schizophrenia or cognitive ability, they provide a good example of complex traits. A large part of Mendel's success in working out the laws of heredity came from choosing simple traits that are either-or qualitative traits. If Mendel had studied, for instance, size of the pea seed as indexed by its diameter, he would have found very different results. First, pea seed size, like most traits, is continuously distributed. If he had taken plants with big seeds and crossed them with plants with small seeds, the seed size of their offspring would be neither big nor small. In fact, the seeds would have varied in size from small to large, with most offspring seeds of average size.

Only ten years after Mendel's report, Francis Galton studied pea seed size and concluded that it is inherited. For example, parents with large seeds were likely to have offspring with larger than average seeds. In fact, Galton developed the fundamental statistics of regression and correlation mentioned above in order to describe the quantitative relationship between pea seed size in parents and offspring. He plotted parent and offspring seed sizes and drew the regression line that best fits the observed data (Figure 3.8). The slope of the

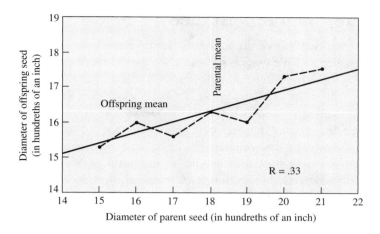

Figure 3.8 First regression line (solid line), drawn by Galton in 1877 to describe the quantitative relationship between pea seed size in parents and offspring. The dashed line connects actual data points. (Courtesy of the Galton Laboratory.)

regression line is .33. This means that, for the entire population, as parental size increases by one unit, the average offspring size increases one-third of one unit.

Galton also demonstrated that human height shows the same pattern of inheritance. Children's height correlates with the average height of their parents. Tall parents have taller than average children. Children with one tall and one short parent are likely to be of average height. Inheritance of this trait is quantitative rather than qualitative. Quantitative inheritance is the way in which nearly all complex behavioral as well as biological traits are inherited.

Does quantitative inheritance violate Mendel's laws? When Mendel's laws were rediscovered in the early 1900s, many scientists thought this must be the case. They thought that heredity must involve some sort of blending, because offspring resemble the average of their parents. Mendel's laws were dismissed as a peculiarity of pea plants or of abnormal conditions. However, recognizing that quantitative inheritance does *not* violate Mendel's laws is fundamental to an understanding of behavioral genetics.

SUMMING UP

Schizophrenia and general cognitive ability are examples of complex traits which determine how much relatives resemble each other by the number of genes they share. Identical twins are more similar than fraternal twins and first-degree relatives are more similar than second-degree relatives. Such complex, quantitative traits typical of behavioral disorders and dimensions do not violate Mendel's laws, as explained in the following section.

Multiple-Gene Inheritance

The traits that Mendel studied, as well as Huntington disease and PKU, are examples in which a single gene is necessary and sufficient to cause the disorder. That is, you will have Huntington disease only if you have the H allele (necessary); if you have the H allele, you will have Huntington disease (sufficient). Other genes and environmental factors have little effect on its inheritance. In such cases, a dichotomous (either-or) disorder is found: You either have the specific allele, or not, and thus you have the disorder, or not. More than 2000 such single-gene disorders are known definitely and again as many are considered probable (http://www.ncbi.nlm.nih.gov/Omim/mimstats.html).

In contrast, more than just one gene is likely to affect complex disorders such as schizophrenia and continuous dimensions such as general cognitive ability. When Mendel's laws were rediscovered in the early 1900s, a bitter battle was fought between so-called Mendelians and biometricians. Mendelians looked for single-gene effects, and biometricians argued that Mendel's laws could not apply to complex traits because they showed no simple pattern of inheritance. Mendel's laws seemed especially inapplicable to quantitative dimensions.

In fact, both sides were right and both were wrong. The Mendelians were correct in arguing that heredity works the way Mendel said it worked, but they were wrong in assuming that complex traits will show simple Mendelian patterns of inheritance. The biometricians were right in arguing that complex traits are distributed quantitatively, not qualitatively, but they were wrong in arguing that Mendel's laws of inheritance are particular to pea plants and do not apply to higher organisms.

The battle between the Mendelians and biometricians was resolved when biometricians realized that Mendel's laws of inheritance of single genes also apply to complex traits that are influenced by *several* genes. Such a complex trait is called a *polygenic* trait. Each of the influential genes is inherited according to Mendel's laws.

Figure 3.9 illustrates this important point. The top distribution shows the three genotypes of a single gene with two alleles that are equally frequent in the population. As discussed in Box 2.1, 25 percent of the genotypes are homozygous for the A_1 allele (A_1A_1), 50 percent are heterozygous (A_1A_2), and 25 percent are homozygous for the A_2 allele (A_2A_2). If the A_1 allele were dominant, individuals with the A_1A_2 genotype would look just like individuals with the A_1A_1 genotype. In this case, 75 percent of individuals would have the observed trait (phenotype) of the dominant allele. For example, as discussed in Box 2.1, in Mendel's crosses of pea plants with smooth or wrinkled seeds, he found that in the F_2 generation, 75 percent of the plants had smooth seeds and 25 percent had wrinkled seeds.

However, not all alleles operate in a completely dominant or recessive manner. Many alleles are additive in that they each contribute something to the phenotype. In Figure 3.9a, each A_2 allele contributes equally to the pheno-

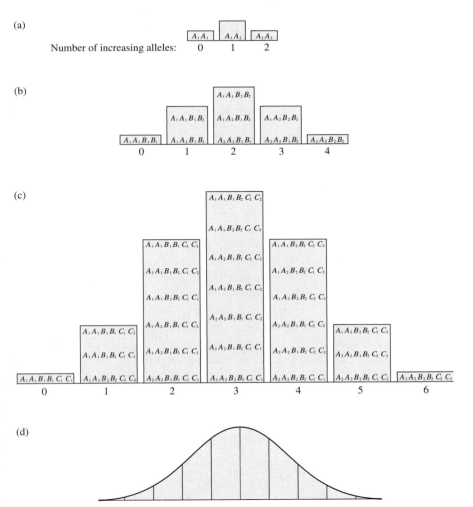

Figure 3.9 Single-gene and multiple-gene distributions for traits with additive gene effects. (a) A single gene with two alleles yields three genotypes and three phenotypes. (b) Two genes, each with two alleles, yield nine genotypes and five phenotypes. (c) Three genes, each with two alleles, yield twenty-seven genotypes and seven phenotypes. (d) Normal bell-shaped curve of continuous variation.

type, so if you have two A_2 alleles, you would have a higher score than if you had just one A_2 allele. Figure 3.9b adds a second gene (B) that affects the trait. Again, each B_2 allele makes a contribution. Now there are nine genotypes and five phenotypes. Figure 3.9c adds a third gene (C), and there are 27 genotypes. Even if we assume that the alleles of the different genes equally affect the trait and that there is no environmental variation, there are still seven different phenotypes.

So, even with just three genes and two alleles for each gene, the phenotypes begin to approach a normal distribution in the population. When we consider environmental sources of variability and the fact that the effects of alleles are not likely to be equal, it is easy to see that the effects of even a few genes will lead to a quantitative distribution. Moreover, the complex traits that interest behavioral geneticists may be influenced by dozens or even hundreds of genes. Thus, it is not surprising to find continuous variation at the phenotypic level, even though each gene is inherited in accord with Mendel's laws.

Quantitative Genetics

The notion that multiple-gene effects lead to quantitative traits is the cornerstone of a branch of genetics called *quantitative genetics.*

Quantitative genetics was introduced in papers by R. A. Fisher (1918) and by Sewall Wright (1921). Their extension of Mendel's single-gene model to the multiple-gene model of quantitative genetics (Falconer & Mackay, 1996) is described in the Appendix. This multiple-gene model adequately accounts for the resemblance of relatives. If genetic factors affect a quantitative trait, phenotypic resemblance of relatives should increase with increasing degrees of genetic relatedness. First-degree relatives (parents, offspring, full siblings) are 50 percent similar genetically. The simplest way to think about this is that offspring inherit half their genetic material from each parent (X linkage aside). If one sibling inherits a particular allele from a parent, the other sibling has a 50 percent chance of inheriting that same allele. Other relatives differ in their degree of genetic relatedness.

Figure 3.10 illustrates degrees of genetic relatedness for the most common types of relatives, using male relatives as examples. Relatives are listed in relation to an individual in the center, the index case. The illustration goes back three generations and forward three generations. First-degree relatives (e.g., fathers, sons), who are 50 percent similar genetically, are each one step removed from the index case. Second-degree relatives (e.g., uncles) are two steps removed and are only half as similar genetically (i.e., 25 percent) as first-degree relatives are. Third-degree relatives (e.g., cousins) are three steps removed and half as similar genetically (i.e., 12.5 percent) as second-degree relatives are. Identical twins are a special case, because they are the same person genetically.

For schizophrenia and general cognitive ability, phenotypic resemblance of relatives increases with genetic relatedness. For schizophrenia, as noted earlier (see Figure 3.6), the chance that an individual picked at random from the population will develop schizophrenia is about 1 percent. If a second-degree relative (grandparent, aunt, or uncle) is affected, the risk is 4 percent. If a first-degree relative (parent, sibling, or offspring) is affected, the risk doubles to about 9 percent. Finally, the risk shoots up to 48 percent for identical twins, which is much greater than the risk of 17 percent for fraternal twins. How can there be a

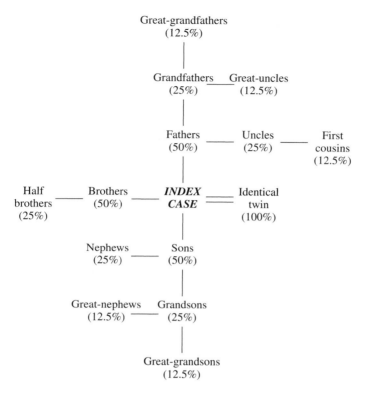

Figure 3.10 Genetic relatedness: Male relatives of male index case (proband), with degree of genetic relatedness in parentheses.

dichotomous disorder if many genes cause schizophrenia? One possible explanation is that genetic risk is normally distributed but that schizophrenia is not seen until a certain threshold is reached. Another explanation is that disorders are actually dimensions artificially established on the basis of a diagnosis. That is, there may be a continuum between what is normal and abnormal. These alternatives are described in Box 3.1.

For general cognitive ability, phenotypic resemblance also increases with genetic relatedness, as shown in Figure 3.7. Cousins (third-degree relatives) are only half as similar as half siblings (second-degree relatives), and half siblings are less similar than full siblings (first-degree relatives). Identical twins are more similar than fraternal twins.

These data for schizophrenia and general cognitive ability are consistent with the hypothesis of genetic influence, but they do not *prove* that genetic factors are important. It is possible that familial resemblance increases with genetic relatedness for environmental reasons. First-degree relatives might be more similar because they live together. Second-degree and third-degree relatives might be less similar because of less similarity of rearing.

BOX 3.1

Liability-Threshold Model of Disorders

If complex disorders such as schizophrenia are influenced by many genes, why are they diagnosed as qualitative disorders rather than assessed as quantitative dimensions? Theoretically, there should be a continuum of genetic risk, from people having none of the alleles that increase risk for schizophrenia to those having most of the alleles that increase risk. Most people should fall between these extremes, with only a moderate susceptibility to schizophrenia.

One model assumes that risk, or liability, is distributed normally but that the disorder occurs only when a certain threshold of liability is exceeded, as represented in the accompanying figure by the shaded area in (a). Relatives of an affected person have a greater liability, that is, their distribution of liability is shifted to the right, as in (b). For this reason, a greater proportion of the relatives of affected individuals exceed the threshold and manifest the disorder. If there is such a threshold, familial risk can be high only if genetic or shared environmental influence is substantial, because many of an affected individual's relatives will fall just below the threshold and not be affected.

Liability and threshold are hypothetical constructs. However, it is possible to use the liability-threshold model to estimate correlations from family risk data (Falconer, 1965; Smith, 1974). For example, the correlation estimated for first-degree relatives for schizophrenia is .45, an estimate based on a population base rate of 1 percent and risk to first-degree relatives of 9 percent.

Although correlations estimated from the liability-threshold model are widely reported for psychological disorders, it should be emphasized that this statistic refers to hypothetical constructs of a threshold and an underlying liability derived from diagnoses, not to the risk for the actual diagnosed disorder. In the previous example, the actual risk for schizophrenia for first-degree relatives is 9 percent, even though the liability-threshold correlation is .45.

Alternatively, a second model assumes that disorders are actually continuous phenotypically; the disorder might not just appear after a certain threshold is reached. Instead, symptoms might increase continuously from the normal to the abnormal. A continuum from normal to abnormal seems likely for common disorders such as depression and alcoholism. For example, people vary in the frequency and severity of their depression. Some people rarely get the blues; for others, depression completely disrupts their lives. Individuals diagnosed as depressed might be extreme cases that differ quantitatively, not qualitatively, from the rest of the population. In such cases, it may be possible to assess the continuum directly, rather than assuming a continuum from dichotomous diagnoses using the liability-threshold model. Even for less common disorders like schizophrenia, there is increasing interest in the possibility that there may be no sharp threshold dividing the normal from the abnormal, but rather a continuum from normal to abnormal thought processes. A method called *DF extremes analysis* can be used to investigate the links between the normal and abnormal (see Box 7.1).

The relationship between dimensions and disorders is key, as discussed in later chapters. The best evidence for genetic links between dimensions and disorders will come as specific genes are found for behavior. For example, will a gene associated with diagnosed depression also relate to differences in mood within the normal range?

(a)

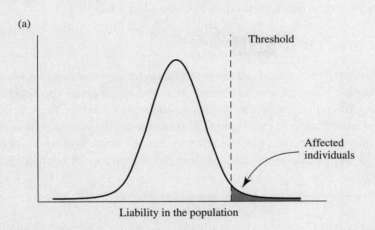

Liability in the population

(b)

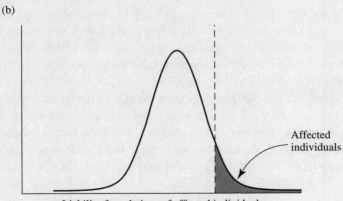

Liability for relatives of affected individuals

Two experiments of nature are the workhorses of human behavioral genetics that help to disentangle genetic and environmental sources of family resemblance. One is the *twin study*, which compares the resemblance within pairs of identical twins, who are genetically identical, to the resemblance within pairs of fraternal twins, who, like other siblings, are 50 percent similar genetically. The second is the *adoption study*, which separates genetic and environmental influences. For example, when biological parents relinquish their children for adoption at birth, any resemblance between these parents and their adopted-away offspring can be attributed to shared heredity rather than to shared environment, if there is no selective placement. In addition, when these children are adopted, any resemblance between the adoptive parents and their adopted children can be attributed to shared environment rather than to shared heredity. The twin and adoption methods are discussed in Chapter 5.

SUMMING UP

The battle between the Mendelians and biometricians was resolved when it was realized that Mendel's laws of the inheritance of single genes also apply to those complex traits that are influenced by several genes. Each of these genes is inherited according to Mendel's laws. This concept is the cornerstone of the field of quantitative genetics, which is a theory and set of methods (such as the twin and adoption methods) for investigating the inheritance of complex, quantitative traits.

As described in Chapter 5, results obtained from twin and adoption studies indicate that genetic factors play a major role in familial resemblance for schizophrenia and cognitive ability. The present chapter explains how the pattern of inheritance for complex disorders like schizophrenia and continuous dimensions like cognitive ability differs from that seen for single-gene traits, because multiple genes are involved. However, each gene is inherited according to Mendel's laws.

Chapter 4 briefly describes the DNA basis for Mendel's laws of heredity. Chapter 5 returns to research using the twin and adoption methods, which aim to assess the net effects of multiple genes and multiple experiences on complex behavioral traits of interest to psychologists. DNA markers are currently being used to find the genes that influence complex traits such as schizophrenia and cognitive abilities. The techniques used to do this are described in Chapter 6.

Summary

Mendel's laws of heredity do not explain all genetic phenomena. For example, genes on the X chromosome, such as the gene for color blindness, require an extension of Mendel's laws. Other exceptions to Mendel's laws include new mutations, changes in chromosomes such as the chromosomal nondisjunction that causes Down syndrome, expanded DNA triplet repeat sequences responsible for Huntington disease and fragile X mental retardation, and genomic imprinting.

Most psychological dimensions and disorders show more complex patterns of inheritance than do single-gene disorders such as Huntington disease, PKU, or color blindness. Complex disorders such as schizophrenia and continuous dimensions such as cognitive ability are likely to be influenced by multiple genes as well as by multiple environmental factors. Quantitative genetic theory extends Mendel's single-gene rules to multiple-gene systems. The essence of the theory is that complex traits can be influenced by many genes but each gene is inherited according to Mendel's laws. Quantitative genetic methods, especially adoption and twin studies, can detect genetic influence for complex traits.

CHAPTER FOUR

DNA: The Basis of Heredity

Mendel was able to deduce the laws of heredity even though he had no idea of how heredity works at the chemical or physiological level. Quantitative genetics, such as twin and adoption studies, depends on Mendel's laws of heredity but does not require knowledge of the biological basis of heredity. However, it is important to understand the biological mechanisms underlying heredity for two reasons. First, understanding the biological basis of heredity makes it clear that the processes by which genes affect behavior are not mystical. Second, this understanding is crucial for appreciating the exciting advances in attempts to identify genes associated with behavior. This chapter describes the biological basis of heredity, how the process is regulated, how genetic variations arise, and how this genetic variation is detected, using the techniques of molecular genetics. There are many excellent genetics texts that provide great detail about these issues (e.g., Lewin, 2004; Watson et al., 2004). The biological basis of heredity includes the fact that genes are contained on structures called chromosomes. The linkage of genes that lie close together on a chromosome has made possible the mapping of the human genome. Moreover, abnormalities in chromosomes contribute importantly to behavioral disorders, especially mental retardation.

DNA

Nearly a century after Mendel did his experiments, it became apparent that DNA (deoxyribonucleic acid) is the molecule responsible for heredity. In 1953, James Watson and Francis Crick proposed a molecular structure for DNA that could explain how genes are replicated and how DNA codes for proteins. As shown in Figure 4.1, the DNA molecule consists of two strands that are held apart by pairs of four bases: adenine, thymine, guanine, and cytosine. As a result

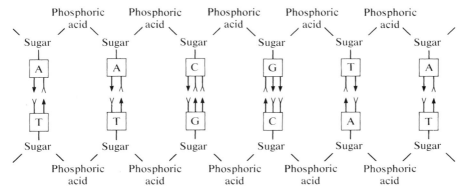

Figure 4.1 Flat representation of the four DNA bases in which adenine (A) always pairs with thymine (T) and guanine (G) always pairs with cytosine (C). (From *Heredity, Evolution, and Society* by I. M. Lerner. W. H. Freeman and Company. ©1968.)

of the structural properties of these bases, adenine always pairs with thymine and guanine always pairs with cytosine. The backbone of each strand consists of sugar and phosphate molecules. The strands coil around each other to form the famous double helix of DNA (Figure 4.2).

The specific pairing of bases in these two-stranded molecules allows DNA to carry out its two functions: to replicate itself and to direct the synthesis of proteins. Replication of DNA occurs during the process of cell division. The double helix of the DNA molecule unzips, separating the paired bases (Figure 4.3). The two strands unwind, and each strand attracts the appropriate bases to construct its complement. In this way, two complete double helices of DNA are created where there was previously only one. This process of replication is the essence of life that began billions of years ago when the first cells replicated themselves. It is also the essence of each of our lives that begins with a single cell and faithfully reproduces our DNA in trillions of cells.

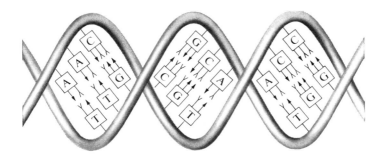

Figure 4.2 A three-dimensional view of a segment of DNA. (From *Heredity, Evolution, and Society* by I. M. Lerner. W. H. Freeman and Company. ©1968.)

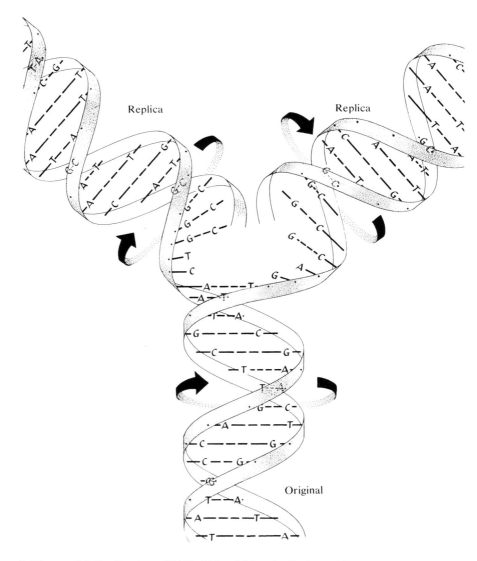

Figure 4.3 Replication of DNA. (After *Molecular Biology of Bacterial Viruses* by G. S. Stent. W. H. Freeman and Company. ©1963.)

The second major function of DNA is to direct the synthesis of proteins according to the genetic information that resides in the particular sequence of bases. DNA encodes the various sequences of the 20 amino acids making up the thousands of specific enzymes and proteins that are the stuff of living organisms. Box 4.1 describes this process, the so-called central dogma of molecular genetics.

What is the genetic code contained in the sequence of DNA bases, which is transcribed to messenger RNA (mRNA; see Box 4.1) and then translated into amino acid sequences? The code consists of various sequences of three bases, which are called *codons* (Table 4.1). For example, three adenines in a row (AAA) in the DNA molecule will be transcribed in mRNA as three uracils (UUU). This mRNA codon codes for the amino acid phenylalanine. Although there are 64 possible triplet codons ($4^3 = 64$), there are only 20 amino acids. Some amino acids are coded by as many as six codons. Any one of three particular codons signal the end of a transcribed sequence.

This same genetic code applies to all living organisms. Discovering this code was one of the great triumphs of molecular biology. The human set of DNA sequences (the genome) consists of about 3 billion base pairs, counting just one chromosome from each pair of chromosomes. The 3 billion base pairs contain about 25,000 protein-coding genes, which range in size from about 1000 bases to 2 million bases. The chromosomal locations of most genes are known. About

TABLE 4.1
The Genetic Code

Amino Acid*	DNA Code
Alanine	CGA, CGG, CGT, CGC
Arginine	GCA, GCG, GCT, GCC, TCT, TCC
Asparagine	TTA, TTG
Aspartic acid	CTA, CTG
Cysteine	ACA, ACG
Glutamic acid	CTT, CTC
Glutamine	GTT, GTC
Glycine	CCA, CCG, CCT, CCC
Histidine	GTA, GTG
Isoleucine	TAA, TAG, TAT
Leucine	AAT, AAC, GAA, GAG, GAT, GAC
Lysine	TTT, TTC
Methionine	TAC
Phenylalanine	AAA, AAG
Proline	GGA, GGG, GGT, GGC
Serine	AGA, AGG, AGT, AGC, TCA, TCG
Threonine	TGA, TGG, TGT, TGC
Tryptophan	ACC
Tyrosine	ATA, ATG
Valine	CAA, CAG, CAT, CAC
(Stop signals)	ATT, ATC, ACT

*The 20 amino acids are organic molecules that are linked together by peptide bonds to form polypeptides, which are the building blocks of enzymes and other proteins. The particular combination of amino acids determines the shape and function of the polypeptide.

BOX 4.1

The "Central Dogma" of Molecular Genetics

Genetic information flows from DNA to messenger RNA (mRNA) to protein. These protein-coding genes are DNA segments that are a few thousand to several million DNA base pairs in length. The DNA molecule contains a linear message consisting of four bases (adenine, thymine, guanine, and cytosine); in this two-stranded molecule, A always pairs with T and G with C. The message is decoded in two basic steps, shown in the figure: (a) transcription of DNA into a different sort of nucleic acid called ribonucleic acid, or RNA, and (b) translation of RNA into proteins.

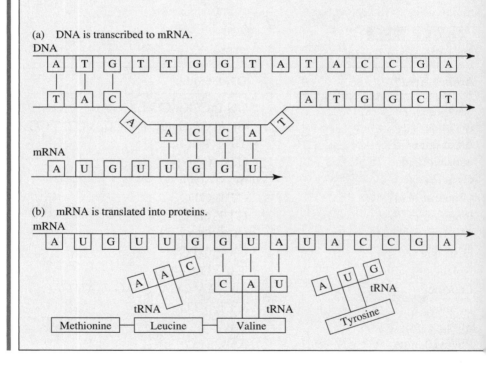

(a) DNA is transcribed to mRNA.

(b) mRNA is translated into proteins.

a third of our protein-coding genes are expressed only in the brain; these are likely to be most important for behavior. The human genome sequence is like an encyclopedia of genes with 3 billion letters, equivalent in length to about 3000 books of 500 pages each. Continuing with this simile, the encyclopedia of genes is written in an alphabet consisting of 4 letters (A, T, G, C), with 3-letter words (codons) organized into 23 volumes (chromosomes). This simile, however, does not comfortably extend to the fact that each encyclopedia is different; millions of letters (about 1 in 1000) differ for any two people. There is no single human genome; we each have a different genome, except for identical twins. Most of the

In the transcription process, the sequence of bases in one strand of the DNA double helix is copied to RNA, specifically a type of RNA called messenger RNA (mRNA) because it relays the DNA code. mRNA is single stranded and is formed by a process of base pairing similar to the replication of DNA, except that uracil substitutes for thymine (so that A pairs with U instead of T). In the figure, one DNA strand is being transcribed—the DNA bases ACCA have just been copied as UGGU in mRNA. mRNA leaves the nucleus of the cell and enters the cell body (cytoplasm), where it connects with ribosomes, which are the factories where proteins are built.

The second step involves translation of the mRNA into amino acid sequences that form proteins. Another form of RNA, called transfer RNA (tRNA), transfers amino acids to the ribosomes. Each tRNA is specific to 1 of the 20 amino acids. The tRNA molecules, with their attached specific amino acids, pair up with the mRNA in a sequence dictated by the base sequence of the mRNA, as the ribosome moves along the mRNA strand. Each of the 20 amino acids found in proteins is specified by a codon made up of three sequential mRNA bases. In the figure, the mRNA code has begun to dictate a protein that includes the amino acid sequence methionine-leucine-valine-tyrosine. Valine has just been added to the chain that already includes methionine and leucine. The mRNA triplet code GUA attracts tRNA with the complementary code CAU. This tRNA transfers its attached amino acid valine, which is then bonded to the growing chain of amino acids. The next mRNA codon, UAC, is attracting tRNA with the complementary codon, AUG, for tyrosine. Although this process seems very complicated, amino acids are incorporated into chains at the incredible rate of about 100 per second. Proteins consist of particular sequences of about 100 to 1000 amino acids. The sequence of amino acids determines the shape and function of proteins. Protein shape is subsequently altered in other ways called *posttranslational changes*. These changes affect its function and are not controlled by the genetic code.

Surprisingly, DNA that is transcribed and translated like this represents only about 2 percent of the genome. What is the other 98 percent doing? See the text for an answer.

---

life sciences focus on the generalities of the genome, but the genetic causes of diseases and disorders lie in these variations in the genome. These variations on the human theme are the focus of behavioral genetics.

The twentieth century has been called the century of the gene. The century began with the re-discovery of Mendel's laws of heredity. The word *genetics* was first coined in 1905. Almost fifty years later, Crick and Watson described the double helix of DNA, the premier icon of science. The pace of discoveries accelerated greatly during the next fifty years, culminating at the turn of the twenty-first century with the sequencing of the human genome. More than 90 percent of the human genome was sequenced by 2001 (International Human Genome Sequencing Consortium, 2001; Venter et al., 2001). Subsequent publications have presented the finished sequence for most chromosomes (e.g., Gregory et al., 2006).

Sequencing of the human genome and the technologies associated with it have led to an explosion of new findings in genetics. One of many examples is *alternative splicing*, in which mRNA is spliced to create different transcripts, which are then translated into different proteins (Brett, Pospisil, Valcarcel, Reich, & Bork, 2002). More than half of all genes are alternatively spliced, which more than doubles the number of gene products (Johnson et al., 2003).

The speed of discovery in genetics is now so great that it would be impossible to predict what will happen in the next five years, let alone the next fifty years. Most geneticists would agree with Francis Collins (2006), the director of the U.S. National Human Genome Research Institute and leader in the Human Genome Project, who expects that we will each possess an electronic chip containing our DNA sequence. Individual DNA chips would herald a revolution in personalized medicine in which treatment can be individually tailored rather than our present one-size-fits-all approach. The greatest value of DNA lies in its ability to predict genetic risk that could lead to preventative interventions. That is, rather than treating problems after they occur, DNA may allow us to predict problems and intervene to prevent them. This could involve genetic engineering that alters DNA, although so far gene therapy in the human species has proven difficult even for single-gene disorders (Rubanyi, 2001). To prevent behavioral problems affected by many genes as well as many environmental factors, behavioral and environmental engineering will be needed.

For behavioral genetics, the most important thing to understand about the DNA basis of heredity is that the process by which genes affect behavior is not mystical. Genes code for sequences of amino acids that form the thousands of proteins of which organisms are made. Proteins create the skeletal system, muscles, the endocrine system, the immune system, the digestive system, and, most important for behavior, the nervous system. Genes do not code for behavior directly, but DNA variations that create differences in these physiological systems can affect behavior.

SUMMING UP

DNA is a double helix that includes four different bases. Its structure allows DNA to replicate itself and to direct protein synthesis. DNA codes for the 20 amino acids by means of sequences of three bases, called codons. The codons make up the genetic code. The human genome consists of 3 billion nucleotide base pairs and has about 25,000 protein-coding genes, a third of which are expressed only in the brain. The DNA sequence of the human genome has recently been identified and has led to an explosion of new knowledge and new techniques.

Gene Expression

Genes do not blindly pump out their protein products. When the gene product is needed, many copies of its mRNA will be present, but otherwise very few copies of the mRNA are transcribed. You are changing the rates of transcription of genes for neurotransmitters by reading this sentence. Because mRNA exists for only a few minutes and then is no longer translated into protein, changes in the rate of transcription of mRNA are used to control the rate at which genes produce proteins. This is what is meant by *gene expression*.

RNA is no longer thought of as merely the messenger that translates the DNA code into proteins. In terms of evolution, RNA was the original genetic code, and it still is the genetic code for most viruses. Double-stranded DNA presumably had a selective advantage over RNA because the single strand of RNA left it vulnerable to predatory enzymes. DNA became the faithful genetic code that is the same in all cells, at all ages, and at all times. In contrast, RNA, which degrades quickly, is tissue-specific, age-specific, and state-specific. For these reasons, RNA can respond to environmental changes by regulating the transcription and translation of protein-coding DNA. In Chapter 15 we return to consider RNA and gene expression as a biological foundation for understanding how the environment works in the pathways between genes and behavior.

As mentioned in Box 4.1, only about 2 percent of the genome involves protein-coding DNA as described by the central dogma. What is the other 98 percent doing? It had been thought that it is "junk" that has just hitched a ride evolutionarily. A recent finding with far-reaching implications is that most human DNA is transcribed into RNA which is not the mRNA translated into amino acid sequences. It is called *non-coding RNA*. It seems unlikely that such a complicated mechanism would have evolved unless it served some function. In recent years, it has become clear that such non-coding RNA plays an important role in regulating the expression of protein-coding DNA, especially in humans.

One type of non-coding RNA has been known for 30 years. Embedded in protein-coding genes are DNA sequences, called *introns*, that are transcribed into RNA but are spliced out before the RNA leaves the nucleus. The remaining parts of genes are spliced back together and are called *exons*. Exons exit the nucleus and are translated into amino acid sequences. Exons usually consist of only a few hundred base pairs, but introns vary widely in length, from 50 to 20,000 base pairs. Only exons are translated into amino acid sequences that make up proteins. However, introns are not "junk." In many cases they regulate the transcription of the gene in which they reside and in some cases they also regulate other genes.

About a quarter of the human genome involves introns. A recent exciting finding is that another quarter of the human genome involves non-coding RNA other than introns (Mattick, 2004). Like introns, this non-coding RNA can regulate protein-coding genes, leading the way to a new world of regulatory networks across the genome and to many new targets for potential sources of genetic variation (Mattick, 2005; Mattick & Makunin, 2006).

One class of non-coding RNA is called *microRNA*, because they are usually only 21 base pairs long (although the DNA coding for them is about 80 base pairs). Even though they are tiny, microRNA play a big role in gene regulation, particularly in the development of the nervous system (Kosik, 2006) and especially that of primates (Berezikov et al., 2006). More than 500 classes of microRNA have been identified and shown to regulate protein-coding genes by binding to (and thus silencing) mRNA (Lim et al., 2005). Amazingly, the 500 microRNA identified to date appear to regulate the expression of more than a third of all coding mRNA. Moreover, microRNA are likely to be just the tip of the iceberg of non-coding RNA effects on gene regulation (Mendes Soares & Valcárcel, 2006). The list of novel mechanisms by which non-coding RNA can regulate gene expression is growing rapidly (Costa, 2005; Huttenhofer, Schattner, & Polacek, 2005).

Some gene regulation is short term and responsive to the environment. Other gene regulation leads to long-term changes in development. Figure 4.4 shows how regulation often works for classical protein-coding genes. Many of these genes include regulatory sequences that normally block the gene from being transcribed. If a particular molecule binds with the regulatory sequence, it will free the gene for transcription. Most gene regulation involves several mechanisms that act like a committee voting on increases or decreases in transcription. That is, several transcription factors act together to regulate the rate of specific mRNA transcription.

Similar mechanisms can also lead to long-term developmental changes. The key question about development is how differentiation occurs, how we start life as a single cell and end up with trillions of cells, all of which have the same DNA but many different functions. Some basic aspects of development are programmed in genes. For example, we have 38 *homeobox* genes similar to genes in most animals that act as master switches to control the timing of development of different parts of the body, by coding for a protein that switches on cascades of other genes. However, for the most part, development is not hard wired in the genes. For example, a thousand different molecules must be synthesized in a specific sequence during the half-hour life cycle of bacteria. It was formerly assumed that this sequential synthesis was programmed genetically,

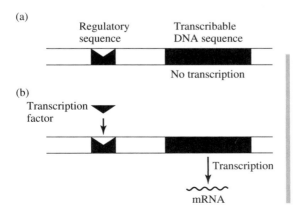

(a)
Regulatory sequence   Transcribable DNA sequence

No transcription

(b)
Transcription factor

Transcription

mRNA

Figure 4.4 Transcription factors can regulate protein-coding genes by controlling mRNA transcription. (a) A regulatory sequence normally shuts down transcription of its gene; (b) but when a particular transcription factor comes along and binds to the regulatory sequence, the gene is freed for transcription.

with genes orchestrated to turn on at the right moment. However, the sequence of steps is not programmed in DNA. Instead, transcription rates of DNA depend upon the products of earlier DNA transcription and upon experiences. Consider another example: When songbirds are first exposed to their species' songs, the experience causes changes in expression of a set of brain-cell genes encoding proteins that regulate the transcription of other brain-cell genes (Mello, Vicario, & Clayton, 1992). Tweaking a rat's whiskers causes changes in gene expression in the cells of the sensory cortex (Mack & Mack, 1992).

Rather than our looking at the expression of a few genes, DNA *microarrays* have made it possible to assess the degree of expression of all genes in the genome simultaneously, including non-coding RNA. Because this is the RNA parallel of the genome, it is called the *transcriptome*. Although microarrays have only been in use for about 10 years (Lander, 1999), there are thousands of studies that compare gene expression profiles in individuals before versus after administration of drugs and that compare gene expression profiles in two groups of individuals—"cases" (affected individuals such as schizophrenics) versus control (comparison) individuals. The importance of DNA microarrays for gene expression profiling for behavioral genetics lies in the transcriptome's position as the first step in the correlation between genes and behavior. This huge area of research is described in Chapter 15.

For now, the important point is that RNA regulates the transcription of genes in response to both the internal and the external environments. Many times more DNA is invested in regulatory genes, including introns and other non-coding RNA, than in protein-coding genes.

SUMMING UP

Many genes are involved in regulating the transcription of other genes rather than in synthesizing proteins. This includes DNA that is transcribed into RNA but not translated into proteins, such as introns, and DNA that is transcribed into non-coding RNA such as microRNAs. Gene regulation is also responsible for long-term developmental changes.

Mutations

Behavioral genetics asks why people are different behaviorally—for example, why people differ in cognitive abilities and disabilities, psychopathology, and personality. For this reason, it focuses on genetic and environmental differences that can account for these observed differences among people. New DNA differences occur when mistakes, called mutations, are made in copying DNA. These mutations result in different alleles (called polymorphisms) such as the alleles responsible for the variations that Mendel found in pea plants, for Huntington disease and PKU, and for complex behavioral traits such as schizophrenia and cognitive abilities. Mutations that occur in the creation of eggs and sperm will be transmitted faithfully unless natural selection intervenes (Chapter 17). The effects that count in terms of natural selection are effects on survival and reproduction. Because evolution has so finely tuned the genetic system, most new mutations in regions of DNA that are translated into amino acid sequences have deleterious effects. However, once in a great while a mutation will make the system function a bit better. In evolutionary terms, this outcome means that individuals with the mutation are more likely to survive and reproduce.

A single-base mutation can result in the insertion of a different amino acid into a protein. Such a mutation can alter the function of the protein. For example, in the figure in Box 4.1, if the first DNA codon TAC is miscopied as TCC, the amino acid arginine will be substituted for methionine. (Table 4.1 indicates that TAC codes for methionine and TCC codes for arginine.) This single amino acid substitution in the hundreds of amino acids that make up a protein might have no noticeable effect on the protein's functioning; then again, it might have a small effect; or it might have a major, even lethal, effect. A mutation that leads to the loss of a single base is likely to be more damaging than a mutation causing a substitution, because the loss of a base shifts the reading frame of the triplet code. For example, if the second base in the box figure were deleted, TAC-AAC-CAT becomes TCA-ACC-AT. Instead of the amino acid chain containing methionine (TAC) and leucine (AAC), the mutation would result in a chain containing serine (TCA) and tryptophan (ACC).

Mutations are often not so simple. For example, a particular gene can have mutations at several locations. As an extreme example, over 60 different mutations have been found in the gene responsible for PKU, and some of these different mutations have different effects (Scriver & Waters, 1999). Another example involves triplet repeats, mentioned in Chapter 3. Most cases of Huntington disease are caused by three repeating bases (CAG). Normal alleles have from 11 to 34 CAG repeats in a gene that codes for a protein found throughout the brain. For the many individuals with Huntington disease, the number of CAG repeats varies from 37 to more than 100. Because triplet repeats involve three bases, the presence of any number of repeats does not shift the reading frame of transcription. However, the CAG repeat responsible for Huntington disease is transcribed into mRNA and translated into protein, which means that multiple repeats of an amino acid are inserted into the protein. Which amino acid? CAG is the mRNA code, so the DNA code is GTC. Table 4.1 shows that GTC codes for the amino acid glutamine. Having a protein encumbered with many extra copies of glutamine reduces the protein's normal activity; therefore the lengthened protein would show loss of function. However, although Huntington disease is a dominant disorder, the other allele should be operating normally, producing enough of the normal protein to avoid trouble. This possibility suggests that the Huntington allele, which adds dozens of glutamines to the protein, might confer a new property (gain of function) that creates the problems of Huntington disease.

Although many millions of our 3 billion base pairs differ among some people, about 2 million base pairs differ among at least 1 percent of the population. As described in the following section, these DNA polymorphisms have made it possible to identify genes responsible for the heritability of traits, including complex behavioral traits.

Detecting Polymorphisms

Much of the success of molecular genetics comes from the availability of millions of DNA polymorphisms. Previously, genetic markers were limited to the products of single genes, such as the red blood cell proteins that define the blood groups. In 1980, new genetic markers that are the actual polymorphisms in the DNA were discovered. Because millions of DNA base sequences are polymorphic, these DNA polymorphisms can be used in genomewide linkage studies to determine the chromosomal location of single-gene disorders, as described in Chapter 6. As noted earlier, in 1983 such DNA markers were first used to localize the gene for Huntington disease at the tip of the short arm of chromosome 4. The latest development is the use of hundreds of thousands of DNA markers to conduct genomewide association studies to identify genes associated with complex disorders including behavioral disorders (Hirschhorn & Daly, 2005).

The time is coming soon when we will be able to detect every single DNA polymorphism by sequencing each individual's entire genome. The race is on to sequence all 3 billion bases of DNA of an individual for less than $1000 (Service, 2006). Until then, two types of DNA polymorphisms are used that account for most of the polymorphisms in the genome: *microsatellite markers*, which have many alleles, and *single nucleotide polymorphisms* (*SNPs*), which have just two alleles (Weir, Anderson, & Hepler, 2006). Box 4.2 describes how microsatellite markers and SNPs are detected, and explains the technique of polymerase chain reaction (PCR). This is fundamental for detection of all DNA markers, because PCR makes millions of copies of a small stretch of DNA. The triplet repeats mentioned in relation to Huntington disease are an example of a microsatellite repeat marker, which can involve two, three, or four base pairs that are repeated up to a hundred times and which have been found at as many as 50,000 loci throughout the genome. The number of repeats at each locus differs among individuals and is inherited in a Mendelian manner. For example, a microsatellite marker might have three alleles, in which the two-base sequence C-G repeats 14, 15, or 16 times.

SNPs (called "snips") are by far the most common type of DNA polymorphisms. As their name suggests, a SNP involves a mutation in a single nucleotide, for example, a mutation that changes the first codon in Box 4.1 from TAC to TCC, thus substituting arginine for methionine when the gene is transcribed and translated into a protein. SNPs that involve a change in an amino acid sequence are called *nonsynonymous* and are thus likely to be functional: The resulting protein will contain a different amino acid. Most SNPs in coding regions are *synonymous:* They do not involve a change in amino acid sequence because the

BOX 4.2

DNA Markers

Microsatellite repeats and SNPs are genetic polymorphisms in DNA. They are called DNA markers rather than genetic markers because they can be identified directly in the DNA itself rather than attributed to a gene product, such as the red blood cell proteins responsible for blood types. Investigations of both of these DNA markers are made possible by a technique called *polymerase chain reaction* (PCR). In a few hours, millions of copies of a particular small sequence of DNA a few hundred to two thousand base pairs in length can be created. To do this copying, the sequence of DNA surrounding the DNA marker must be known. From this DNA sequence, 20 bases on both sides of the polymorphism are synthesized. These 20-base DNA sequences, called *primers*, are unique in the genome and identify the precise location of the polymorphism.

Polymerase is an enzyme that begins the process of copying DNA. It begins to copy the DNA on each strand of DNA at the point of the primer. One strand is copied from the primer on the left in the right direction and the other strand is copied from the primer on the right in the left direction. In this way, PCR results in a copy of the DNA between the two primers. When this process is repeated many times, even the copies are copied and millions of copies of the double-stranded DNA between the two primers are produced. (For an animation, see http://www.maxanim.com/genetics/PCR/PCR.swf.)

The simplest way to identify a polymorphism from the PCR-amplified DNA fragment is to sequence the fragment. Sequencing would indicate how many repeats are present for microsatellite markers and which allele is present for SNPs. Because we have two alleles for each locus, we can have two different alleles (heterozygous) or have two copies of the same allele (homozygous). A more cost-effective approach is used for microsatellite markers that sorts DNA fragments by length; this indicates the number of repeats. For SNPs, the DNA fragments can be made single-stranded and allowed to find their match (hybridize) to a single-stranded probe for one or the other SNP allele. For example, in the figure below, the target probe is ATCATG with a SNP at the third nucleotide base. The PCR-amplified DNA fragment TAGTAC has hybridized successfully with the probe. In high-throughput approaches, a fluorescent molecule is attached to the DNA fragments so that the fragments light up if they successfully hybridize with the probe. (The TATGAC allele is unable to hybridize with the probe.)

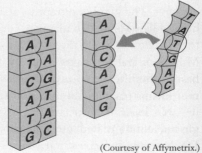

(Courtesy of Affymetrix.)

SNP involves one of the alternate DNA codes for the same amino acid (see Table 4.1). Although nonsynonymous SNPs are more likely to be functional because they change the amino acid sequence of the protein, it is possible that synonymous SNPs might have an effect by changing the rate at which mRNA is translated into proteins. The field is just coming to grips with the functional effects of other SNPs throughout the genome, such as SNPs in non-coding RNA regions of the genome. More than 10 million SNPs have been reported in populations around the world; more than 5 million SNPs have been validated (http://www.ncbi.nlm.nih.gov/SNP/), and about 2 million meet the criterion of occurring in at least 1 percent of a population. This work is being systematized by the International HapMap Project (http://www.hapmap.org/index.html.en), which has genotyped more than a million SNPs for 269 individuals from four ethnic groups (International HapMap Consortium, 2005). The project is called HapMap because its aim was to create a map of correlated SNPs throughout the genome. SNPs close together on a chromosome are unlikely to be separated by recombination, but recombination does not occur evenly throughout the genome. There are blocks of SNPs that are very highly correlated with one another and are separated by so-called *recombinatorial hotspots*. These blocks are called *haplotype blocks*. (In contrast to *genotype*, which refers to a pair of chromosomes, the DNA sequence on one chromosome is called a *haploid genotype*, which has been shortened to *haplotype*.) By identifying a few SNPs that tag a haplotype block, it may be necessary to genotype only half a million SNPs rather than many millions of SNPs in order to scan the entire genome for associations with phenotypes.

A new type of polymorphism has recently attracted considerable attention: *copy number variants* that involve duplication of long stretches of DNA, often encompassing protein-coding genes as well as non-coding genes (Redon et al., 2006). A recent survey of the human genome identified more than 3000 autosomal copy number variants, 800 of which appeared at a frequency of at least 3 percent (Wong et al., 2007). Remarkably, copy number variants result in the genomes of individuals differing by as much as 10 million base pairs of the 3 billion base pairs in the average genome. Given the speed of discoveries in molecular genetics following from the Human Genome Project, it is likely that many other surprises are in store.

SUMMING UP

Mutations are the source of genetic variability. About 3 million of our 3 billion base pairs differ from one individual to the next. Detecting these DNA polymorphisms has been the key to success in molecular genetics. Polymorphisms in DNA itself include multi-allelic microsatellite repeat markers, bi-allelic single nucleotide polymorphisms (SNPs), and copy number variants. Sequencing DNA is the ultimate way to detect all polymorphisms.

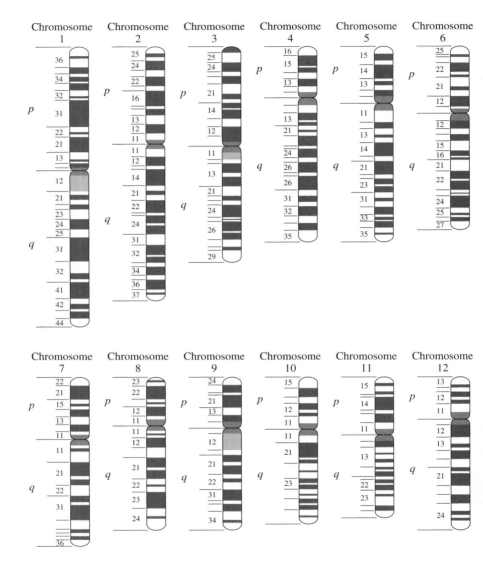

Chromosomes

As discussed in Chapter 2, Mendel did not know that genes are grouped together on chromosomes, so he assumed that all genes are inherited independently. However, Mendel's second law of independent assortment is violated
when two genes are close together on the same chromosome. In this case, the
two genes are not inherited independently; and, on the basis of this noninde-
pendent assortment, linkages between DNA markers have been identified and
used to produce a map of the genome. With the same technique, mapped DNA
markers are used to identify linkages with disorders and dimensions, including
behavior, as described in Chapter 6.

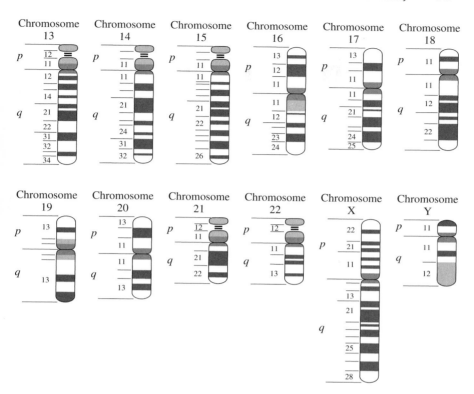

Figure 4.5 The 23 pairs of human chromosomes. The short arm above the centromere is called *p*, and the long arm below the centromere is called *q*. The bands, created by staining, are used to identify the chromosomes and to describe the location of genes. Chromosomal regions are referred to by chromosome number, arm of chromosome, and band. Thus, *1p36* refers to band 6 in region 3 of the *p* arm of chromosome 1. For more details about each chromosome and the locus of major genetic disorders, see http://www.ornl.gov/sci/techresources/Human_Genome/posters/chromosome/chooser.shtml.

Our species has 23 pairs of chromosomes, for a total of 46 chromosomes. The number of chromosome pairs varies widely from species to species. Fruit flies have 4, mice have 20, dogs have 39, and butterflies have 190. Our chromosomes are very similar to those of the great apes (chimpanzee, gorilla, and orangutan). Although the great apes have 24 pairs, two of their short chromosomes have been fused to form one of our large chromosomes.

As noted above, one pair of our chromosomes is the *sex chromosomes* X and Y. Females are XX and males are XY. All the other chromosomes are called *autosomes*. As shown in Figure 4.5, chromosomes have characteristic banding patterns when stained with a particular chemical. The *bands*, whose function is not known, are used to identify the chromosomes. At some point in each chromosome, there is a *centromere*, a region of the chromosome without genes, where the chromosome is attached to its new copy when cells reproduce. The short arm of the chromosome above the centromere is called p and the long arm below the centromere is called q. The location of genes is described in relation to the bands. For example, the gene for Huntington's disease is at $4p16$, which means the short arm of chromosome 4 at a particular band, number 6 in region 1.

In addition to providing the basis for gene mapping, chromosomes are important in behavioral genetics because mistakes in copying chromosomes during cell division affect behavior. There are two kinds of cell division. Normal cell division, called *mitosis*, occurs in all cells not involved in the production of gametes. These cells are called *somatic cells*. The sex cells produce eggs and sperm, the *gametes*. In mitosis, each chromosome in the somatic cell duplicates and divides to produce two identical cells. A special type of cell division called *meiosis* occurs in the sex cells of the ovaries and testes to produce eggs and sperm, both of which have only one member of each chromosome pair. Each egg and each sperm have 1 of over 8 million (2^{23}) possible combinations of the 23 pairs of chromosomes. Moreover, crossover (recombination) of members of each chromosome pair (see Figure 2.7) occurs about once per meiosis and creates even more genetic variability. When a sperm fertilizes an egg to produce a zygote, one chromosome of each pair comes from the mother's egg and the other from the father's sperm, thereby reconstituting the full complement of 23 pairs of chromosomes.

As indicated in Chapter 3, a common copying error for chromosomes is an uneven split of the pairs of chromosomes during meiosis, called nondisjunction (see Figure 3.4). The most common form of mental retardation, Down syndrome, is caused by nondisjunction of one of the smallest chromosomes, chromosome 21. Many other chromosomal problems occur, such as breaks in chromosomes that lead to inversion, deletion, duplication, and translocation (Figure 4.6). About half of all fertilized human eggs have a chromosomal abnormality. Most of these abnormalities result in early spontaneous abortions (miscarriages). At birth, about 1 in 250 babies has an obvious chromosomal abnormality. (Small abnormalities such as deletions have been difficult to detect but are being made much easier by DNA microarrays, which are described in Chapter 6.) Although chromosomal abnormalities occur for all chromosomes, only fetuses with the least severe abnormalities survive to birth. Some of these babies die soon after they are born. For example, most babies with three chromosomes (trisomy) of chromosome 13 die in the first month, and most of those with trisomy-18 die within the first year. Other chromosomal abnormalities are less lethal but result in behavioral and physical problems. Nearly all major chromosomal abnormalities influence cognitive ability, as expected if cognitive ability is affected by many genes. Because the behavioral effects of chromosomal abnormalities often involve mental retardation, they are discussed in Chapter 7.

Missing a whole chromosome is lethal, except for the X and Y chromosomes. Having an entire extra chromosome is also lethal except for the smallest chromosomes and the X chromosome, which is one of the largest. The reason

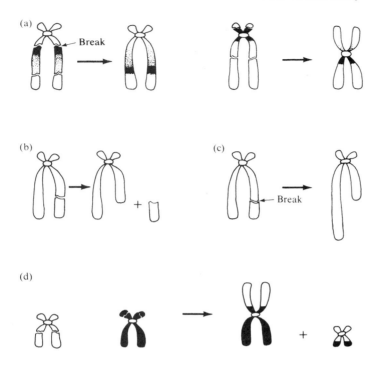

Figure 4.6 Common types of chromosomal abnormalities: (a) inversion;
(b) deletion; (c) duplication; (d) translocation between different chromosomes.

why the X chromosome is the exception is also the reason why half of all chromosomal abnormalities that exist in newborns involve the sex chromosomes. In females, one of the two X chromosomes is inactivated, in the sense that most of its genes are not transcribed. In males and females with extra X chromosomes, the extra X chromosomes also are inactivated. For this reason, even though X is a large chromosome with many genes, having an extra X in males or females or only one X in females is not lethal. The most common sex chromosome abnormalities are XXY (males with an extra X), XXX (females with an extra X), and XYY (males with an extra Y), each with an incidence of about 1 in 1000. The incidence of XO (females with just one X) is lower than expected, 1 in 2500 at birth, because 98 percent of such conceptuses abort spontaneously.

SUMMING UP

Our species has 23 pairs of chromosomes, including the pair of sex chromosomes X and Y. During meiosis, when eggs and sperm are produced, copying errors are occasionally made, leading to an uneven split of pairs of chromosomes, called nondisjunction, and other errors. Such chromosomal abnormalities contribute importantly to behavioral, especially cognitive, disorders.

The effects of most chromosomal abnormalities are manifold, often involving diverse behavioral and physical traits, which is not surprising because so many genes are involved.

Summary

One of the most exciting advances in biology has been understanding Mendel's "elements" of heredity. The double helix structure of DNA relates to its dual functions of self-replication and protein synthesis. The genetic code consists of a sequence of three DNA bases that codes for amino acids. DNA is transcribed to mRNA, which is translated into amino acid sequences. Many genes—especially DNA that is transcribed into RNA but not translated into DNA—are involved in regulating the transcription of other genes. Gene regulation is responsible for long-term developmental changes as well as short-term responses to environmental conditions.

Mutations are the source of genetic variability. Much of the success of molecular genetics comes from the availability of millions of polymorphisms in DNA itself. Microsatellite repeat polymorphisms and single nucleotide polymorphisms (SNPs) are the most widely used DNA markers. A recently discovered source of DNA polymorphisms is copy number variants.

Genes are inherited on chromosomes. Linkage between DNA markers and behavior can be detected by looking for exceptions to Mendel's law of independent assortment, because a DNA marker and a gene for behavior are not inherited independently if they are close together on the same chromosome. Our species has 23 pairs of chromosomes. Mistakes in duplicating chromosomes often affect behavior directly. About 1 in 250 newborns has a major chromosomal abnormality, and about half of these abnormalities involve the sex chromosomes.

CHAPTER FIVE

Nature, Nurture, and Behavior

Most behavioral traits are much more complex than single-gene disorders such as Huntington disease and PKU (see Chapter 2). Complex dimensions and disorders are influenced by heredity, but not by one gene alone. Multiple genes are usually involved, as well as multiple environmental influences. The purpose of this chapter is to describe ways in which we can study genetic effects on complex behavioral traits. The words *nature* and *nurture* have a rich and contentious history in the field, but are used here simply as broad categories representing genetic and environmental influences, respectively. They are not distinct categories–Chapter 16 discusses the interplay between them.

The first question that needs to be asked about behavioral traits is whether heredity is at all important. For single-gene disorders, this is not an issue, because it is usually obvious that heredity is important. For example, for dominant genes such as the gene for Huntington disease, you do not need to be a geneticist to notice that every affected individual has an affected parent. Recessive gene transmission is not as easy to observe, but the expected pattern of inheritance is clear. For complex behavioral traits in the human species, an experiment of nature—twinning—and an experiment of nurture—adoption—are widely used to assess the net effect of genes and environment. More direct genetic experiments are available to investigate animal behavior. The theory underlying these methods is called *quantitative genetics*. Quantitative genetics estimates the extent to which observed differences among individuals are due to genetic differences of any sort and to environmental differences of any sort without specifying what the specific genes or environmental factors are. When heredity is important—and it almost always is for complex traits like behavior—it is now possible to identify specific genes by using the methods of molecular genetics, the topic of Chapter 6. Behavioral genetics uses the methods of both quantitative genetics and molecular genetics to study behavior. Using genetically sensitive designs also facilitates the identification of specific environmental factors, which is the topic of Chapter 16.

Genetic Experiments to Investigate Animal Behavior

Dogs provide a dramatic yet familiar example of genetic variability within species (Figure 5.1). Despite their great variability in size and physical appearance—from a height of six inches for the Chihuahua to three feet for the Irish wolfhound— they are all members of the same species. Recent molecular genetic research suggests that dogs, which originated from wolves more than 100,000 years ago as they were domesticated by humans, may have enriched their supply of genetic variability by repeated intercrossing with wolves (Ostrander & Wayne, 2005). The genome of the domestic dog has been sequenced (Lindblad-Toh et al., 2005; Ostrander, Giger, & Lindblad-Toh, 2006), which creates a genetic basis for dog breeds and suggests that there are four basic genetic clusters of dogs: wolves and Asian dogs (the earliest domesticated dogs such as Akita and Lahsa Apso), mastiff-type dogs (e.g., mastiff and boxer), working dogs (e.g., collies and sheepdogs), and hunting dogs (e.g., hounds and terriers) (Parker et al., 2004).

Dogs also illustrate the extent of genetic effects on behavior. Although physical differences are most obvious, dogs have been bred for centuries as much for

Figure 5.1 Dog breeds illustrate genetic diversity within species for behavior as well as physical appearance.

Irish Setter   Boxer   Beagle

Dalmation   Afghan

their behavior as for their looks. In 1576, the earliest English-language book on dogs classified breeds primarily on the basis of behavior. For example, terriers (from *terra*, which is Latin for "earth") were bred to creep into burrows to drive out small animals. Another book, published in 1686, described the behavior for which spaniels were originally selected. They were bred to creep up on birds and then spring to frighten the birds into the hunter's net, which is the origin of the *springer spaniel*. With the advent of the shotgun, different spaniels were bred to point rather than to spring. The author of the 1686 work was especially interested in temperament: "Spaniels by Nature are very loveing, surpassing all other Creatures, for in Heat and Cold, Wet and Dry, Day and Night, they will not forsake their Master" (cited by Scott & Fuller, 1965, p. 47). These temperamental characteristics led to the creation of spaniel breeds selected specifically to be pets, such as the King Charles spaniel, which is known for its loving and gentle temperament.

Behavioral classification of dogs continues today. Sheepdogs herd, retrievers retrieve, trackers track, pointers point, and guard dogs guard with minimal training. Breeds also differ strikingly in intelligence and in temperamental traits such as emotionality, activity, and aggressiveness, although there is also substantial variation in these traits within each breed (Coren, 2005). The selection process can be quite fine tuned. For example, in France, where dogs are used chiefly for farm work, there are 17 breeds of shepherd and stock dogs specializing in aspects

Standard Poodle   Cocker Spaniel   Miniature Schnauzer

Fox Terrier   Pomeranian   Collie

Figure 5.2 J. P. Scott with the five breeds of dogs used in his experiments with J. L. Fuller. Left to right: wire-haired fox terrier, American cocker spaniel, African basenji, Shetland sheepdog, and beagle. (From *Genetics and the Social Behavior of the Dog* by J. P. Scott & J. L. Fuller. ©1965 by The University of Chicago Press. All rights reserved.)

of this work. In England, dogs have been bred primarily for hunting, and there are 26 recognized breeds of hunting dogs. Dogs are not unusual in their genetic diversity, although they are unusual in the extent to which different breeds have been intentionally bred to accentuate genetic differences in behavior.

An extensive behavioral genetic research program on breeds of dogs was conducted over two decades by J. Paul Scott and John Fuller (1965). They studied the development of pure breeds and hybrids of the five breeds pictured in Figure 5.2: wire-haired fox terriers, cocker spaniels, basenjis, Shetland sheepdogs, and beagles. These breeds are all about the same size, but they differ markedly in behavior. Although considerable genetic variability remains within each breed, average behavioral differences among the breeds reflect their breeding history. For example, as their history would suggest, terriers are aggressive scrappers; spaniels are nonaggressive and people oriented. Unlike the other breeds, Shetland sheepdogs have been bred, not for hunting, but for performing complex tasks under close supervision from their masters. They are very responsive to training. In short, Scott and Fuller found behavioral breed differences just about everywhere they looked—in the development of social relationships, emotionality, and trainability, as well as many other behaviors. They also found evidence for interactions between breeds and training. For example, scolding that would be brushed off by a terrier could traumatize a sheepdog.

Selection Studies

Laboratory experiments that select for behavior provide the clearest evidence for genetic influence on behavior. As dog breeders and other animal breeders have known for centuries, if a trait is heritable, you can breed selectively for it. Research in Russia aimed to understand how our human ancestors had domesticated dogs from wolves by selecting for tameness in foxes, which are notoriously wary of humans. Foxes that were the tamest when fed or handled were bred for more than 40 generations. The result of this selection study is a new breed of foxes that are like dogs in their friendliness and eagerness for human contact (Figure 5.3), so much so that these foxes have now become popular house pets in Russia (Trut, 1999).

Laboratory experiments typically select high and low lines in addition to maintaining an unselected control line. For example, in one of the largest and

Figure 5.3 Foxes are normally wary of humans and tend to bite. After selecting for tameness for 40 years, a program involving 45,000 foxes has developed animals that are not only tame but friendly. This one-month-old fox pup not only tolerates being held but is licking the woman's face. (Trut, 1999. Reprinted with permission.)

Figure 5.4 Mouse in an open field. The holes near the floor transmit light beams that electronically record the mouse's activity.

longest selection studies of behavior (DeFries, Gervais, & Thomas, 1978), mice were selected for activity in a brightly lit box called an open field, a measure of fearfulness that was invented more than 60 years ago (Figure 5.4). In the open field, some animals become immobile, defecate, and urinate, whereas others actively explore it. Lower activity scores are presumed to index fearfulness.

The most active mice were selected and mated with other high-active mice. The least active mice were also mated with each other. From the offspring of the high-active and low-active mice, the most and least active mice were again selected and mated in a similar manner. This selection process was repeated for 30 generations. (In mice, a generation takes only about three months.)

The results are shown in Figures 5.5 and 5.6 for replicated high, low, and control lines. Over the generations, selection was successful: The high lines became increasingly more active and the low lines less active (see Figure 5.5). Successful selection can occur only if heredity is important. After 30 generations of such selective breeding, a 30-fold average difference in activity has been achieved. There is no overlap between the activity of the low and high lines (see Figure 5.6). Mice from the high-active line now boldly run the equivalent total distance of the length of a football field during the six-minute test period, whereas the low-active mice quiver in the corners.

Another important finding is that the difference between the high and low lines steadily increases each generation. This outcome is a typical finding from

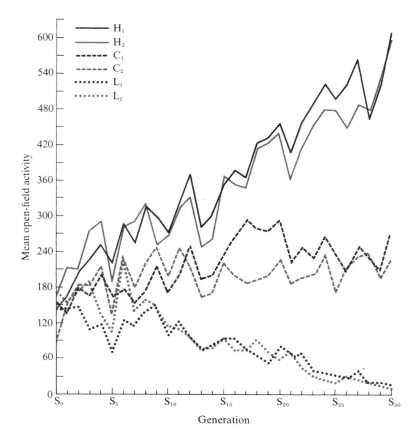

Figure 5.5 Results of a selection study of open-field activity. Two lines were selected for high open-field activity (H$_1$ and H$_2$), two lines were selected for low open-field activity (L$_1$ and L$_2$), and two lines were randomly mated within each line to serve as controls (C$_1$ and C$_2$). (From "Response to 30 generations of selection for open-field activity in laboratory mice" by J. C. DeFries, M. C. Gervais, & E. A. Thomas. *Behavior Genetics, 8,* 3–13. ©1978 by Plenum Publishing Corporation. All rights reserved.)

selection studies of behavioral traits, and strongly suggests that many genes contribute to variation in behavior. If just one or two genes were responsible for open-field activity, the two lines would separate after a few generations and would not diverge any further in later generations.

Despite the major investment required to conduct a selection study, the method continues to be used in behavioral genetics, in part because of the convincing evidence it provides for genetic influence on behavior, and also because it produces lines of animals that differ as much as possible genetically for a particular behavior (e.g., Gammie, Garland, & Stevenson, 2006; Stead et al., 2006).

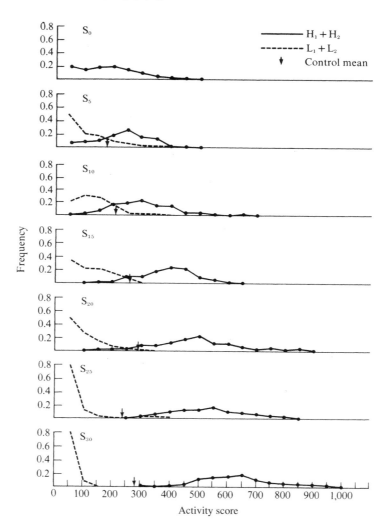

Figure 5.6 Distributions of activity scores of lines selected for high and low open-field activity for 30 generations (S_0 to S_{30}). Average activity of control lines in each generation is indicated by an arrow. (From "Response to 30 generations of selection for open-field activity in laboratory mice" by J. C. DeFries, M. C. Gervais, & E. A. Thomas. *Behavior Genetics, 8*, 3–13. ©1978 by Plenum Publishing Corporation. All rights reserved.)

Inbred Strain Studies

The other major quantitative genetic design for animal behavior compares *inbred strains*, in which brothers have been mated with sisters for at least 20 generations. This intensive inbreeding makes each animal within the inbred strain virtually a genetic clone of all other members of the strain. Because inbred strains differ genetically from one another, genetically influenced traits will show average differences between inbred strains reared in the same laboratory environment. Differences within strains are due to environmental influences. In animal behavioral genetic research, mice are most often studied; more than 100 inbred strains of mice are available. Some of the most frequently studied inbred strains are shown in Figure 5.7.

Studies of inbred strains suggest that most mouse behaviors show genetic influence. For example, Figure 5.8 shows the average open-field activity scores of two inbred strains called BALB/c and C57BL/6. The C57BL/6 mice are much more active than the BALB/c mice, an observation suggesting that genetics contributes to open-field activity. The mean activity scores of several crosses are also shown: F_1, F_2, and F_3 crosses (explained in Box 2.1) between the inbred strains,

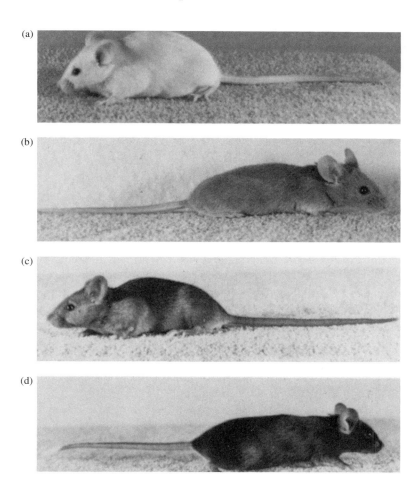

Figure 5.7 Four common inbred strains of mice: (a) BALB/c; (b) DBA/2; (c) C3H/2; (d) C57BL/6.

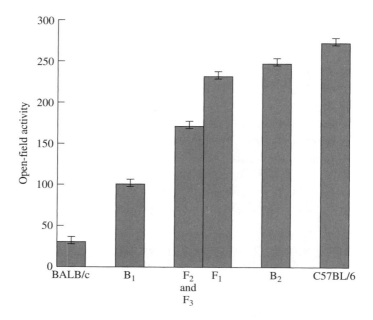

Figure 5.8 Mean open-field activity (± twice the standard error) of BALB/c and C57BL/6 mice and their derived F_1, backcross (B_1 and B_2), F_2, and F_3 generations. (From "Response to 30 generations of selection for open-field activity in laboratory mice" by J. C. DeFries, M. C. Gervais, & E. A. Thomas. *Behavior Genetics, 8,* 3–13. ©1978 by Plenum Publishing Corporation. All rights reserved.)

the backcross between the F_1 and the BALB/c strain (B_1 in Figure 5.8), and the backcross between the F_1 and the C57BL/6 strain (B_2 in Figure 5.8). There is a strong relationship between the average open-field scores and the percentage of genes obtained from the C57BL/6 parental strain, which again points to genetic influence.

Rather than just crossing two inbred strains, the *diallel design* compares several inbred strains and all possible F_1 crosses between them. Figure 5.9 shows the open-field results of a diallel cross between BALB/c, C57BL/6, and two other inbred strains (C3H/2 and DBA/2). C3H/2 is even less active than BALB/c, and DBA/2 is almost as active as C57BL/6. The F_1 crosses tend to correspond to the average scores of their parents. For example, the F_1 cross between C3H/2 and BALB/c is intermediate to the two parents in open-field activity.

Studies of inbred strains are also useful for detecting environmental effects. First, because members of an inbred strain are genetically identical, individual differences within a strain must be due to environmental factors. Large differences within inbred strains are found for open-field activity and most other behaviors studied, reminding us of the importance of prenatal and postnatal nurture as well as nature. Second, inbred strains can be used to assess the net effect of mothering by comparing F_1 crosses in which the mother is from either one strain or the other.

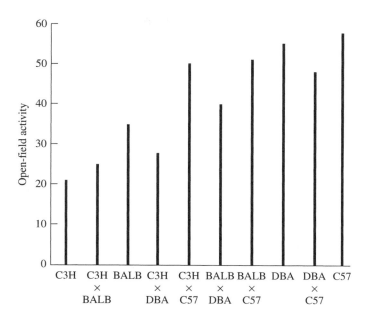

Figure 5.9 Diallel analysis of four inbred mouse strains for open-field activity. The F₁ strains are ordered according to the average open-field activity score of their parental inbred strains. (After Henderson, 1967.)

For example, the F_1 cross between BALB/c mothers and C57BL/6 fathers can be compared to the genetically equivalent F_1 cross between C57BL/6 mothers and BALB/c fathers. In a diallel study like that shown in Figure 5.9, these two hybrids had nearly identical scores, as was the case for comparisons between the other crosses as well. This result suggests that prenatal and postnatal maternal effects do not importantly affect open-field activity. If maternal effects are found, it is possible to separate prenatal and postnatal effects by cross-fostering pups of one strain with mothers of the other strain. Third, the environments of inbred strains can be manipulated in the laboratory to investigate interactions between genotype and environment, as discussed in Chapter 16. A type of genotype-environment interaction was reported in an influential paper in which genetic influence as assessed by inbred strains differed across laboratories, although the results for open-field activity were robust across laboratories (Crabbe, Wahlsten, & Dudek, 1999). Subsequent studies indicated that the rank order between inbred strains for behaviors showing large strain differences are stable across laboratories (Wahlsten et al., 2003). For example, comparisons over 50 years of research on inbred strains for locomotor activity and ethanol preference yield rank-order correlations of .85 to .98 across strains (Wahlsten, Bachmanov, Finn, & Crabbe, 2006). Another study of more than 2000 outbred mice also showed few interactions between open-field activity and experimental variables, such as who tests the mice and order of testing (Valdar, Solberg, Gauguier, Cookson et al., 2006). Nonetheless, there is value in multi-laboratory studies in terms of generalizability of inbred strain results (Kafkafi, Benjamini, Sakov, Elmer, & Golani, 2005).

More than 1000 behavioral investigations involving genetically defined mouse strains were published between 1922 and 1973 (Sprott & Staats, 1975), and the pace accelerated into the 1980s. Studies such as these played an important role in demonstrating that genetics contributes to most behaviors. Although inbred strain studies now tend to be overshadowed by more sophisticated genetic analyses, inbred strains still provide a simple and highly efficient test for the presence of genetic influence. For example, inbred strains have recently been used to screen for genetic mediation of associations between genomewide gene expression profiles and behavior (Letwin et al., 2006; Nadler et al., 2006), a topic to which we will return in Chapter 15.

SUMMING UP

Differences among breeds of dogs and selection studies of mice in the laboratory provide powerful evidence for the importance of genetic influence on behavior. Behavioral differences between inbred strains of mice, inbred by brother-sister matings for at least 20 generations, demonstrate the widespread contribution of genes to behavior. Differences within an inbred strain indicate the importance of environmental factors.

Investigating the Genetics of Human Behavior

Quantitative genetic methods to study human behavior are not as powerful or direct as selection studies or studies of inbred strains. Rather than using genetically defined populations such as inbred strains for mice or manipulating environments experimentally, human research is limited to studying naturally occurring genetic and environmental variation. Nonetheless, adoption and twinning provide experimental situations that can be used to test the relative influence of nature and nurture. As mentioned in Chapter 1, increasing recognition of the importance of genetics during the past two decades is one of the most dramatic shifts in the behavioral sciences. This shift is in large part due to the accumulation of adoption and twin research that consistently points to the important role played by genetics even for complex psychological traits.

Adoption Designs

Many behaviors "run in families," but family resemblance can be due to either nature or nurture. The most direct way to disentangle genetic and environmental sources of family resemblance involves adoption. Adoption creates pairs of genetically related individuals who do not share a common family environment. Their similarity estimates the contribution of genetics to family resemblance.

Lindon Eaves majored in genetics as an undergraduate at the University of Birmingham, England. He obtained his Ph.D. in human behavioral genetics in 1970. He taught at Oxford University for two years before moving to the United States in 1981, where he is now Distinguished Professor of Human Genetics and a professor of psychiatry at the Virginia Commonwealth University School of Medicine in Richmond. With Kenneth Kendler, he directs the Virginia Institute for Psychiatric and Behavioral Genetics. His research includes the study of genetic and environmental effects on personality and social attitudes, the genetic analysis of multiple variables, mate selection, genotype-environment interaction, segregation and linkage analysis, and the genetic analysis of developmental change. With Hans Eysenck and Nick Martin, he wrote *Genes, Culture and Personality: An Empirical Approach.* Eaves holds the James Shields award for twin research and the Dobzhansky award from the Behavior Genetics Association. He is a past president of the Behavior Genetics Association and past president of the International Society for Twin Studies. Currently, he directs the Virginia Twin Study of Adolescent Behavioral Development, which is analyzing the interaction of genetic and environmental effects in the development of adolescent behavioral problems.

Adoption also produces family members who share family environment but are not genetically related. Their resemblance estimates the contribution of family environment to family resemblance. In this way, the effects of nature and nurture can be inferred from experiments such as the adoption design. As mentioned earlier, quantitative genetic research does not in itself identify specific genes or environments. An important direction for future behavioral genetic research is to incorporate direct measures of genes (Chapter 6) and of environment (Chapter 16) into quantitative genetic designs.

For example, consider parents and offspring. Parents in a family study are "genetic-plus-environmental" parents in that they share both heredity and environment with their offspring. The process of adoption results in "genetic" parents and "environmental" parents (Figure 5.10). "Genetic" parents are birth parents who relinquish their child for adoption shortly after birth. Resemblance between birth parents and their adopted-away offspring directly assesses the genetic contribution to parent-offspring resemblance. "Environmental" parents are adoptive parents who adopt children genetically unrelated to them. When children are placed randomly into adoptive families, resemblance between adoptive parents

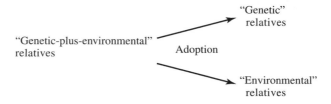

Figure 5.10 Adoption is an experiment of nurture that creates "genetic" relatives (biological parents and their adopted-away offspring; siblings adopted apart) and "environmental" relatives (adoptive parents and their adopted children; genetically unrelated children adopted into the same adoptive family). Resemblance for these "genetic" and "environmental" relatives can be used to test the extent to which resemblance between the usual "genetic-plus-environmental" relatives is due to either nature or nurture.

and their adopted children directly assesses the postnatal environmental contribution to parent-offspring resemblance. Data on birth parents are rare, but genetic influence can also be assessed by comparing "genetic-plus-environmental" families with adoptive families who share only family environment.

"Genetic" siblings and "environmental" siblings can also be studied. "Genetic" siblings are full siblings adopted apart early in life and reared in different homes. "Environmental" siblings are pairs of genetically unrelated children adopted early in life into the same adoptive home. As described in the Appendix, these adoption designs can be depicted more precisely as path models that are used in model-fitting analyses to test the fit of the model, to compare alternative models, and to estimate genetic and environmental influences (see the Appendix; Neale, 2004; Neale & Maes, 2003).

For most psychological traits that have been assessed in adoption studies, genetic factors appear to be important. For example, Figure 5.11 summarizes adoption results for general cognitive ability (see Chapter 8 for details). "Genetic" parents and offspring and "genetic" siblings significantly resemble each other even though they are adopted apart and do not share family environment. You can see that genetics accounts for about half of the resemblance for "genetic-plus-environmental" parents and siblings. The other half of familial resemblance appears to be explained by shared family environment, assessed directly by the resemblance between adoptive parents and adopted children, and between adoptive siblings. Chapter 8 describes a recent important finding that the influence of shared environment on cognitive ability decreases dramatically from childhood to adolescence.

One of the most surprising results from genetic research is that, for most psychological traits other than cognitive ability, resemblance between relatives is accounted for by shared heredity rather than by shared environment. For example, the risk of schizophrenia is just as great for offspring of schizophrenic parents whether they are reared by their biological parents or adopted away at

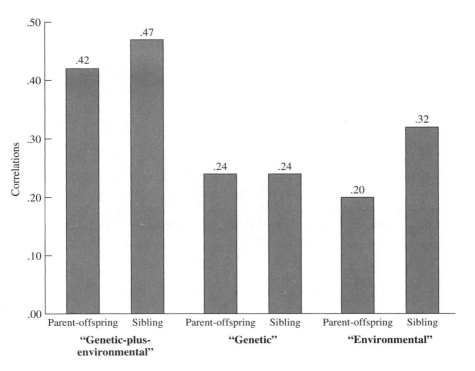

Figure 5.11 Adoption data indicate that family resemblance for cognitive ability is due both to genetic resemblance and to environmental resemblance. "Genetic" relatives refer to genetically related relatives adopted apart. "Environmental" relatives refer to genetically unrelated individuals adopted together. (Data adapted from Loehlin, 1989.)

birth and reared by adoptive parents. This finding implies that sharing a family environment does not contribute importantly to family resemblance. It does not mean that the environment or even the family environment is unimportant. As discussed in Chapter 16, quantitative genetic research such as adoption studies provides the best available evidence for the importance of environmental influence. The risk for first-degree relatives of schizophrenic probands who are 50 percent similar genetically is only about 10 percent, not 50 percent. Furthermore, although family environment does not contribute to the resemblance of family members, such factors could contribute to *differences* among family members, the *nonshared environment* (Chapter 16).

The first adoption study of schizophrenia, reported by Leonard Heston in 1966, is a classic study that was highly influential in turning the tide from assuming that schizophrenia was completely caused by early family experiences to recognizing the importance of genetics (Box 5.1). Box 5.2 considers some methodological issues in adoption studies.

BOX 5.1

The First Adoption Study of Schizophrenia

Environmentalism, which assumes that we are what we learn, dominated the behavioral sciences until the 1960s, when a more balanced view emerged that recognized the importance of nature as well as nurture. One reason for this major shift was an adoption study of schizophrenia reported by Leonard Heston in 1966. Although twin studies had for decades suggested genetic influence, schizophrenia was generally assumed to be environmental in origin, caused by early interactions with parents. Heston interviewed 47 adult adopted-away offspring of hospitalized schizophrenic women. He compared their incidence of schizophrenia with that of matched adoptees whose birth parents had no known mental illness. Of the 47 adoptees whose biological mothers were schizophrenic, 5 had been hospitalized for schizophrenia. Three were chronic schizophrenics hospitalized for several years. None of the adoptees in the control group were schizophrenic.

The incidence of schizophrenia in these adopted-away offspring of schizophrenic biological mothers was 10 percent. This risk is similar to the risk for schizophrenia found when children are reared by their schizophrenic parents. Not only do these findings indicate that heredity makes a major contribution to schizophrenia, they also suggest that shared rearing environment has little effect. When a biological parent is schizophrenic, the risk for schizophrenia is just as great for the offspring when they are adopted away at birth as it is when the offspring are reared by their schizophrenic parents.

Several other adoption studies have confirmed the results of Heston's study. His study is an example of what is called the *adoptees' study method* because the incidence of schizophrenia was investigated in the adopted-away offspring of schizophrenic biological mothers. A second major strategy is called the *adoptees' family method*. Rather than beginning with parents, this method begins with adoptees who are affected (probands) and adoptees who are unaffected. The incidence of the disorder in the biological and adoptive families of the adoptees is assessed. Genetic influence is suggested if the incidence of the disorder is greater for the biological relatives of the affected adoptees than for the biological relatives of the unaffected control adoptees. Environmental influence is indicated if the incidence is greater for the adoptive relatives of the affected adoptees than for the adoptive relatives of the control adoptees.

These adoption methods and their results for schizophrenia are described in Chapter 10.

Twin Design

The other major method used to disentangle genetic from environmental sources of resemblance between relatives involves twins (Segal, 1999). Identical twins, also called *monozygotic* (MZ) twins because they derive from one fertilized egg (zygote), are genetically identical (Figure 5.12). If genetic factors are

Figure 5.12 Twinning is an experiment of nature that produces identical twins, who are genetically identical, and fraternal twins, who are only 50 percent similar genetically. If genetic factors are important for a trait, identical twins must be more similar than fraternal twins. DNA markers can be used to test whether twins are identical or fraternal, although for most pairs it is easy to tell because identical twins (top photo) are usually much more similar physically than fraternal twins (bottom photo).

BOX 5.2

Issues in Adoption Studies

The adoption design is like an experiment that untangles nature and nurture as causes of family resemblance. The first adoption study, which investigated IQ, was reported in 1924 (Theis, 1924). The first adoption study of schizophrenia was reported in 1966 (see Box 5.1). Adoption studies have become more difficult to conduct as the number of adoptions has declined. In the 1960s, as many as 1 percent of all children were adopted. Adoption became much less frequent as contraception and abortion increased, and more unmarried mothers kept their infants.

One issue about adoption studies is representativeness. If biological parents, adoptive parents, or adopted children are not representative of the rest of the population, the generalizability of adoption results could be affected. However, means are more likely to be affected than variances, and genetic estimates rely primarily on variance. In the population-based Colorado Adoption Project (Petrill et al., 2003), for example, biological and adoptive parents appear to be quite representative of nonadoptive parents, and adopted children seem to be reasonably representative of nonadopted children. Other adoption studies, however, have sometimes shown less representativeness. Restriction of range can also limit generalizations from adoption studies (Stoolmiller, 1999).

Another issue concerns prenatal environment. Because birth mothers provide the prenatal environment for their adopted-away children, the resemblance between them might reflect prenatal environmental influences. A strength of

important for a trait, these genetically identical pairs of individuals must be more similar than first-degree relatives, who are only 50 percent similar genetically. Rather than comparing identical twins with nontwin siblings or other relatives, nature has provided a better comparison group: fraternal (*dizygotic*, or DZ) twins. Unlike identical twins, fraternal twins develop from separately fertilized eggs. They are first-degree relatives, 50 percent genetically related like other siblings. Half of fraternal twin pairs are same-sex pairs and half are opposite-sex pairs. Twin studies usually focus on same-sex fraternal twin pairs because they are a better comparison group for identical twin pairs, who are always same-sex pairs. If genetic factors are important for a trait, identical twins must be more similar than fraternal twins. (See Box 5.3 for more details about the twin method.)

How can you tell whether same-sex twins are identical or fraternal? DNA markers can tell. If a pair of twins differs for any DNA marker (excluding laboratory error), they must be fraternal because identical twins are identical genetically. If many markers are examined and no differences are found, the adoption studies is that prenatal effects can be tested independently from post-natal environment by comparing correlations for birth mothers and birth fathers. Although it is more difficult to study birth fathers, results for small samples of birth fathers show results similar to those for birth mothers for IQ and for schizophrenia. Another approach to this issue is to compare adoptees' biological half siblings related through the mother (maternal half siblings) with those related through the father (paternal half siblings). For schizophrenia, paternal half siblings of schizophrenic adoptees show the same risk for schizophrenia as maternal half siblings do, an observation suggesting that prenatal factors may not be of great importance (Kety, 1987).

Finally, selective placement could cloud the separation of nature and nurture by placing adopted-apart "genetic" relatives into correlated environments. For example, selective placement would occur if the adopted-away children of the brightest biological parents are placed with the brightest adoptive parents. If selective placement matches biological and adoptive parents, genetic influence could inflate the correlation between adoptive parents and their adopted children, and environmental influence could inflate the correlation between biological parents and their adopted-away children. If data are available on biological parents as well as adoptive parents, selective placement can be assessed directly. If selective placement is found in an adoption study, its effects need to be considered in interpreting genetic and environmental results. Although some adoption studies show selective placement for IQ, other psychological dimensions and disorders show little evidence for selective placement.

twin pair has a high probability of being identical. Physical traits such as eye color, hair color, and hair texture can be used in a similar way to diagnose whether twins are identical or fraternal. Such traits are highly heritable and are affected by many genes. If a twin pair differs for one of these traits, they are likely to be fraternal; if they are the same for many such traits, they are probably identical. In fact, a single question works pretty well because it sums up many such physical traits: When the twins were young, how difficult was it to tell them apart? To be mistaken for another person requires that many heritable physical characteristics be identical. Using physical similarity to determine whether twins are identical or fraternal is generally more than 95 percent accurate when compared with the results of DNA markers (e.g., Christiansen et al., 2003; Gao et al., 2006). DNA markers can be used to determine zygosity prenatally (Levy, Mirlesse, Jacquemard, & Daffos, 2002). In most cases, it is not difficult to tell whether twins are identical or fraternal (see Figure 5.12).

If a trait is influenced genetically, identical twins must be more similar than fraternal twins. However, when greater similarity of MZ twins is found, it is

BOX 5.3

The Twin Method

Francis Galton (1876) studied developmental changes in twins' similarity, but one of the first real twin studies was conducted in 1924 in which identical and fraternal twins were compared in an attempt to estimate genetic influence (Merriman, 1924). This twin study assessed IQ and found that identical twins were markedly more similar than fraternal twins, a result suggesting genetic influence. Dozens of subsequent twin studies of IQ confirmed this finding. Twin studies have also been reported for many other psychological dimensions and disorders; they provide the bulk of the evidence for the widespread influence of genetics in behavioral traits. Although most mammals have large litters, primates, including our species, tend to have single offspring. However, primates occasionally have multiple births. Human twins are more common than people usually realize—about 1 in 85 births are twins. Surprisingly, as many as 20 percent of fetuses are twins, but because of the hazards associated with twin pregnancies, often one member of the pair dies very early in pregnancy. Among live births, the numbers of identical and same-sex fraternal twins are approximately equal. That is, of all twin pairs, about one-third are identical twins, one third are same-sex fraternal twins, and one-third are opposite-sex fraternal twins.

Identical twins result from a single fertilized egg (called a zygote) that splits for unknown reasons, producing two (or sometimes more) genetically identical individuals. For about a third of identical twins, the zygote splits during the first five days after fertilization as it makes its way down to the womb. In this case, the identical twins have different sacs (called chorions) within the placenta. Two-thirds of the time, the zygote splits after it implants in the placenta and the twins share the same chorion. Identical twins who share the same chorion may be more similar for some psychological traits than identical twins who do not share the same chorion, although the evidence on this hypothesis is mixed (Fagard, Loos, Beunen, Derom, & Vlietinck, 2003; Gutknecht, Spitz, & Carlier, 1999; Jacobs et al., 2001; Phelps, Davis, & Schwartz, 1997; Riese, 1999; Sokol et al., 1995). When the zygote splits after about two weeks, the twins' bodies may be partially fused—so-called Siamese twins. Fraternal twins occur when two eggs are separately fertilized; they have different chorions. Like other siblings, they are 50 percent similar genetically.

The rate of fraternal twinning differs across countries, increases with maternal age, and may be inherited in some families. Increased use of fertility drugs results in greater numbers of fraternal twins because these drugs make it likely that more than one egg will ovulate. The numbers of fraternal twins have also increased since the early 1980s because of in vitro fertilization in which several fertilized eggs are implanted and two survive. The rate of identical twinning is not affected by any of these factors.

possible that the greater similarity is caused environmentally rather than genetically. The *equal environments assumption* of the twin method assumes that environmentally caused similarity is roughly the same for both types of twins reared in the same family. If the assumption were violated because identical twins experience more similar environments than fraternal twins, this violation would inflate estimates of genetic influence. The equal environments assumption has been tested in several ways and appears reasonable for most traits (Bouchard, & Propping, 1993; Derks, Dolan, & Boomsma, 2006).

Prenatally, identical twins may experience greater environmental *differences* than fraternal twins. For example, identical twins show greater birth weight differences than fraternal twins do. The difference may be due to greater prenatal competition, especially for the majority of identical twins who share the same *chorion*. To the extent that identical twins experience less similar environments, the twin method will underestimate heritability. Postnatally, the effect of labeling a twin pair as identical or fraternal has been studied by using twins who were misclassified by their parents or by themselves (e.g., Gunderson et al., 2006; Kendler, Neale, Kessler, Heath, & Eaves, 1993a; Scarr & Carter-Saltzman, 1979). When parents think that twins are fraternal but they really are identical, these mislabeled twins are as similar behaviorally as correctly labeled identical twins.

Another way in which the equal environments assumption has been tested takes advantage of the fact that differences within pairs of identical twins can only be due to environmental influences. The equal environments assumption is supported if identical twins who are treated more individually than others do not behave more differently. This is what has been found for most tests of the assumption in research on behavioral disorders and dimensions (e.g., Cronk et al., 2002; Kendler, Neale, Kessler, Heath, & Eaves, 1994; Loehlin & Nichols, 1976; Morris-Yates, Andrews, Howie, & Henderson, 1990).

A subtle, but important, issue is that identical twins might have more similar experiences than fraternal twins because identical twins are more similar genetically. That is, some experiences may be driven genetically. Such differences between identical and fraternal twins in experience are not a violation of the equal environments assumption because the differences are not caused environmentally (Eaves, Foley, & Silberg, 2003). This topic is discussed in Chapter 16.

As in any experiment, generalizability is an issue for the twin method. Are twins representative of the general population? Two ways in which twins are different are that twins are often born three to four weeks prematurely and intrauterine environments can be adverse when twins share a womb (Phillips, 1993). Newborn twins are also about 30 percent lighter at birth than the average singleton newborn, a difference that disappears by middle childhood (MacGillivray, Campbell, & Thompson, 1988). In childhood, language develops more slowly in twins and twins also perform less well on tests of verbal ability and IQ (Deary, Pattie, Wilson, & Whalley, 2005; Ronalds, De Stavola, &

Leon, 2005). These delays are similar for MZ and DZ twins and appear to be due to postnatal environment rather than prematurity (Rutter & Redshaw, 1991). Most of this cognitive deficit is recovered in the early school years (Christensen et al., 2006). Twins do not appear to be importantly different from singletons for personality (Johnson, Krueger, Bouchard, & McGue, 2002), psychopathology (Christensen, Vaupel, Holm, & Yashlin, 1995), or in motor development (Brouwer, van Beijsterveldt, Bartels, Hudziak, & Boomsma, 2006).

In summary, the twin method is a valuable tool for screening behavioral dimensions and disorders for genetic influence (Boomsma, Busjahn, & Peltonen, 2002; Martin, Boosma, & Machin, 1997). More than 5000 papers on twins were published during the five years from 2001 to 2006, and more than 500 of these involve behavior. The value of the twin method explains why most developed countries have twin registers (Bartels, 2007; Busjahn, 2002). The assumptions underlying the twin method are different from those of the adoption method, yet both methods converge on the conclusion that genetics is important in the behavioral sciences. Recall that for schizophrenia, the risk for a fraternal twin whose co-twin is schizophrenic is about 17 percent; the risk is 48 percent for identical twins (see Figure 3.6). For general cognitive ability, the correlation is about .60 for fraternal twins and .85 for identical twins (see Figure 3.7). The fact that identical twins are so much more similar than fraternal twins strongly suggests genetic influence. For both schizophrenia and general cognitive ability, fraternal twins are more similar than nontwin siblings, perhaps because twins shared the same uterus at the same time and are exactly the same age (Koeppen-Schomerus, Spinath, & Plomin, 2003).

Combination

During the past two decades, behavioral geneticists have begun to use designs that combine the family, adoption, and twin methods in order to bring more power to bear on these analyses. For example, it is useful to include nontwin siblings in twin studies to test whether twins differ statistically from singletons, and whether fraternal twins are more similar than nontwin siblings.

Two major combination designs bring the adoption design together with the family design and with the twin design. The adoption design comparing "genetic" and "environmental" relatives is made much more powerful by including the "genetic-plus-environmental" relatives of a family design. This is the design of one of the largest and longest ongoing genetic studies of behavioral development, the Colorado Adoption Project (Petrill, Plomin, DeFries, & Hewitt, 2003). This project has shown, for example, that genetic influence on general cognitive ability increases during infancy and childhood (Plomin, Fulker, Corley, & Defries, 1997).

The adoption-twin combination involves twins adopted apart and compares them with twins reared together. Two major studies of this type have been conducted, one in Minnesota (Bouchard, Lykken, McGue, Segal, & Tellegen, 1990;

CLOSE UP

Nancy Pedersen, professor of genetic epidemiology and psychology, is the chair of the Department of Medical Epidemiology and Biostatistics and, until recently, the director of the Swedish Twin Registry. A native of Minnesota, she has been at the Karolinska Institutet for over 25 years, where she came as a graduate student to work with the Swedish Twin Registry. As principal investigator of the Swedish Adoption Twin Study of Aging, which has been ongoing for over 20 years, and co-principal investigator on other studies of aging and the Study of Dementia in Swedish Twins, she has demonstrated how genetic influences decrease in importance late in life for cognitive abilities, whereas genetic influences for Alzheimer's disease remain substantial. Pedersen's current research efforts are focused on the etiology of chronic diseases of the elderly, including dementia, Parkinson's disease, and late-onset depression, as well as other diseases with neuropsychiatric components in midlife such as chronic fatigue. Key to her research is the study of comorbidity and the extent to which there are pleiotropic and epistatic effects explaining these comorbidities and associations. Through her efforts, the Swedish Twin Registry has been rejuvenated and expanded to include essentially all twins born in Sweden since 1886 and with prospective information that will continue to be a foundation for numerous research efforts in the field of genetic epidemiology.

Lykken, 2006) and one in Sweden (Kato & Pedersen, 2005; Pedersen, McClearn, Plomin, & Nesselroade, 1992a). These studies have found, for example, that identical twins reared apart from early in life are almost as similar in terms of general cognitive ability as are identical twins reared together, an outcome suggesting strong genetic influence and little environmental influence caused by growing up together in the same family (shared family environmental influence).

An interesting combination of the twin and family methods comes from the study of families of identical twins, which has come to be known as the families-of-twins method (D'Onofrio et al., 2003; Mendle et al., 2006). When identical twins become adults and have their own children, interesting family relationships emerge. For example, in families of male identical twins, nephews are as related genetically to their twin uncle as they are to their own father. That is, in terms of their genetic relatedness, it is as if the first cousins have the same father. Furthermore, the cousins are as closely related to one another as half siblings are. This design yields similar results in relation to cognitive ability.

Although not as powerful as standard adoption or twin designs, a design that has recently been used takes advantage of the increasing number of stepfamilies created as a result of divorce and remarriage (Reiss, Neiderhiser, Hetherington, & Plomin, 2000). Half siblings typically occur in stepfamilies because a woman brings a child from a former marriage to her new marriage and then has another child with her new husband. These children have only one parent (the mother) in common and are 25 percent similar genetically, unlike full siblings, who have both parents in common and are 50 percent similar genetically. Half siblings can be compared with full siblings in stepfamilies to assess genetic influence. Full siblings in stepfamilies occur when the mother brings full siblings from her former marriage or when she and her new husband have more than one child together. A useful test of whether stepfamilies differ from never-divorced families is the comparison between full siblings in the two types of families. This design has not yet been applied to general cognitive ability.

SUMMING UP

Adoption and twinning are like experiments that can be used to assess the relative contributions of nature and nurture to familial resemblance. For schizophrenia and cognitive ability, family members resemble one another even when they are adopted apart. Twin studies show that identical twins are more similar than fraternal twins. Results of family, adoption, and twin studies and of combinations of these designs converge on the conclusion that genetic factors contribute substantially to schizophrenia and cognitive ability.

Heritability

For the complex traits that interest behavioral scientists, it is possible to ask not only whether genetic influence is important but also *how much* genetics contributes to the trait. The question about whether genetic influence is important involves *statistical significance*, the reliability of the effect. For example, we can ask whether the resemblance between "genetic" parents and their adopted-away offspring is significant, or whether identical twins are significantly more similar than fraternal twins. Statistical significance depends on the size of the effect and the size of the sample. For example, a "genetic" parent-offspring correlation of .25 will be statistically significant if the adoption study includes at least 45 parent-offspring pairs. Such a result would indicate that it is highly likely (95 percent probability) that the true correlation is greater than zero.

The question about how much genetics contributes to a trait refers to *effect size*, the extent to which individual differences for the trait in the population can be accounted for by genetic differences among individuals. Effect size in this sense refers to individual differences for a trait in the entire population, not to certain individuals. For example, if PKU were left untreated, it would have a huge effect on the cognitive development of individuals homozygous for the recessive allele. However, because such individuals represent only 1 in 10,000 individuals in the population, this huge effect for these few individuals would have little effect overall on the variation in cognitive ability in the entire population. Thus, the size of the effect of PKU in the population is very small.

Many statistically significant environmental effects in the behavioral sciences involve very small effects in the population. For example, birth order is significantly related to intelligence test (IQ) scores (first-born children have higher IQs). This is a small effect in that the mean difference between first- and second-born siblings is less than two IQ points and their IQ distributions almost completely overlap. Birth order accounts for about 1 percent of the variance of IQ scores when other factors are controlled. In other words, if all you know about two siblings is their birth order, then you know practically nothing about their IQs.

In contrast, genetic effect sizes are often very large, among the largest effects found in the behavioral sciences, accounting for as much as half of the variance. The statistic that estimates the genetic effect size is called *heritability*. Heritability is the proportion of phenotypic variance that can be accounted for by genetic differences among individuals. As explained in the Appendix, heritability can be estimated from the correlations for relatives. For example, if the correlation for "genetic" (adopted-apart) relatives is zero, then heritability is zero. For first-degree "genetic" relatives, their correlation reflects half of the effect of genes because they are only 50 percent similar genetically. That is, if heritability is 100 percent, their correlation would be .50. In Figure 5.11, the correlation for "genetic" (adopted-apart) siblings is .24 for IQ scores. Doubling this correlation yields a heritability estimate of 48 percent, which suggests that about half of the variance in IQ scores can be explained by genetic differences among individuals.

Heritability estimates, like all statistics, include error of estimation, which is a function of the effect size and the sample size. In the case of the IQ correlation of .24 for adopted-apart siblings, the number of sibling pairs is 203. There is a 95 percent chance that the true correlation is between .10 and .38, which means that the true heritability is likely to be between 20 and 76 percent, a very wide range. For this reason, heritability estimates based on a single study need to be taken as very rough estimates surrounded by a large confidence interval, unless the study is very large. For example, if the correlation of .24 were based on a sample of 2000 instead of 200, there would be a 95 percent chance that the true heritability is between 40 and 56 percent. Replication across studies and across designs also allows more precise estimates.

If identical and fraternal twin correlations are the same, heritability is estimated as zero. If identical twins correlate 1.0 and fraternal twins correlate .50, a heritability of 100 percent is implied. In other words, genetic differences among individuals completely account for their phenotypic differences. A rough estimate of heritability in a twin study can be made by doubling the difference between the identical and fraternal twin correlations. As explained in the Appendix, because identical twins are identical genetically and fraternal twins are 50 percent similar genetically, the difference in their correlations reflects half of the genetic effect and is doubled to estimate heritability. For example, in Figure 3.7, IQ correlations for identical and fraternal twins are .85 and .60, respectively. Doubling the difference between these correlations results in a heritability estimate of 50 percent, which also suggests that about half of the variance of IQ scores can be accounted for by genetic factors. Because these studies include more than 10,000 pairs of twins, the error of estimation is small. There is a 95 percent chance that the true heritability is between .48 and .52.

Because disorders are diagnosed as either-or dichotomies, familial resemblance is assessed by concordances rather than by correlations. As explained in the Appendix, *concordance* is an index of risk. For example, if sibling concordance is 10 percent for a disorder, we say that siblings of probands have a 10 percent risk for the disorder.

If identical and fraternal twin concordances are the same, heritability must be zero. To the extent that identical twin concordances are greater than fraternal twin concordances, genetic influence is implied. For schizophrenia (see Figure 3.6), the identical twin concordance of .48 is much greater than the fraternal twin concordance of .17, a difference suggesting substantial heritability. The fact that in 52 percent of the cases identical twins are *dis*cordant for schizophrenia, even though they are genetically identical, implies that heritability is much less than 100 percent.

One way to estimate heritability for disorders is to use the liability-threshold model (see Box 3.1) to translate concordances into correlations on the assumption that a continuum of genetic risk underlies the dichotomous diagnosis. For schizophrenia, the identical and fraternal twin concordances of .48 and .17 translate into liability correlations of .86 and .57, respectively. Doubling the difference between these liability correlations suggests a heritability of about 60 percent. The five newest twin studies yield liability heritability estimates of about 80 percent (Cardno & Gottesman, 2000). As explained in Box 3.1, this statistic refers to a hypothetical construct of continuous liability as derived from a dichotomous diagnosis of schizophrenia rather than to the diagnosis of schizophrenia itself.

For combination designs that compare several groups, and even for simple adoption and twin designs, modern genetic studies are typically analyzed by using an approach called *model fitting*. Model fitting tests the significance of the fit between a model of genetic and environmental relatedness against the observed data. Different models can be compared, and the best-fitting model is used to estimate the effect size of genetic and environmental effects. Model fitting is described in the Appendix.

Interpreting Heritability

Heritability refers to the genetic contribution to individual differences (variance), *not* to the phenotype of a single individual. For a single individual, both genotype and environment are indispensable—a person would not exist without both genes and environment. As noted by Theodosius Dobzhansky (1964), the first president of the Behavior Genetics Association:

> The nature-nurture problem is nevertheless far from meaningless. Asking right questions is, in science, often a large step toward obtaining right answers. The question about the roles of genotype and the environment in human development must be posed thus: To what extent are the *differences* observed among people conditioned by the differences of their genotypes and by the differences between the environments in which people were born, grew and were brought up? (p. 55)

This issue is critical for the interpretation of heritability (Sesardic, 2005). You can still read in introductory textbooks that genetic and environmental effects on behavior cannot be disentangled because behavior is the product of genes and environment. An example sometimes given is the area of a rectangle. It is nonsensical to ask about the separate contributions of length and width to the area of a single rectangle because area is the product of length and width. Area does not exist without both length and width. However, if we ask, not about a single rectangle but about a population of rectangles (Figure 5.13), the variance in areas could be due entirely to length (b), entirely to width (c), or to

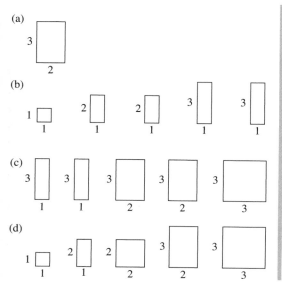

Figure 5.13 Individuals and individual differences. Genetic and environmental contributions to behavior do not refer to a single individual, just as the area of a single rectangle (a) cannot be attributed to the relative contributions of length and width, because area is the product of length and width. However, in a population of rectangles, the relative contribution of length and width to differences in area can be investigated. It is possible that length alone (b), width alone (c), or both (d) account for differences in area among rectangles.

both (d). Obviously, there can be no behavior without both an organism and an environment. The scientifically useful question is the origins of differences among individuals.

For example, the heritability of height is about 90 percent, but this does not mean that you grew to 90 percent of your height for reasons of heredity and that the other inches were added by the environment. What it means is that most of the height differences among individuals are due to the genetic differences among them. Heritability is a statistic that describes the contribution of genetic differences to observed differences among individuals in a particular population at a particular time. In different populations or at different times, environmental or genetic influences might differ, and heritability estimates in such populations could differ.

A counterintuitive example concerns the effects of equalizing environments. If environments were made the same for everyone in a particular population, heritability would be high in that population because individual differences that remained in the population would be due exclusively to genetic differences.

It should be emphasized that heritability refers to the contribution of genetic differences to observed differences among individuals for a particular trait in a particular population at a particular time. As noted in the previous chapter, 99.9 percent of our DNA does not vary from person to person. Because these genes are either the same or highly similar for everyone, they may not contribute to differences among individuals. However, if these nonvarying genes were disrupted by mutation, they could have a devastating, even lethal, effect on development, even though they may not normally contribute to variation in the population. Similarly, many environmental factors do not vary substantially; for example, the air we breathe and the essential nutrients we eat. Although at this level of analysis such nonvarying environmental factors do not contribute to differences among individuals, disruption of these essential environments could have devastating effects.

A related issue concerns average differences between groups, such as average differences between males and females, between social classes, or between ethnic groups. It should be emphasized that the causes of individual differences within groups have no implications for the causes of average differences between groups. Specifically, heritability refers to the genetic contribution to differences among individuals within a group. High heritability within a group does not necessarily imply that average differences between groups are due to genetic differences between groups. The average differences between groups could be due solely to environmental differences even when heritability within the groups is very high.

This point extends beyond the politically sensitive issues of gender, social class, and ethnic differences. As discussed in Chapter 11, a key issue in psychopathology concerns the links between the normal and the abnormal. Finding heritability for individual differences within the normal range of variation does not necessarily imply that the average difference between an extreme group and the rest of the population is also due to genetic factors. For example, if individual differences in depressive symptoms for an unselected sample are heritable, this finding does not necessarily imply that severe depression is also due to genetic factors. This point is worth repeating: The causes of average differences between groups are not necessarily related to the causes of individual differences within groups.

A related point is that heritability describes *what is* in a particular population at a particular time rather than *what could be*. That is, if either genetic influences change (e.g., changes due to migration) or environmental influences change (e.g., changes in educational opportunity), then the relative impact of genes and environment will change. Even for a highly heritable trait such as height, changes in the environment *could* make a big difference, for example, if an epidemic struck or if children's diets were altered. Indeed, the huge increase in children's heights during the past century is almost certainly a consequence of improved diet. Conversely, a trait that is largely influenced by environmental factors *could* show a big genetic effect. For example, genetic engineering can knock out a gene or insert a new gene that greatly alters the trait's development, something that can now be done in laboratory animals, as discussed in Chapter 15.

Although it is useful to think about what could be, it is important to begin with what is—the genetic and environmental sources of variance in existing populations. Knowledge about what is can sometimes help to guide research concerning what could be, as in the example of PKU. Most important, heritability has nothing to say about *what should be*. Evidence of genetic influence for a behavior is compatible with a wide range of social and political views, most of which depend on values, not facts. For example, no policies necessarily follow from finding genetic influence or even specific genes for cognitive abilities. It does not mean, for example, that we ought to put all our resources into educating the brightest children. Depending on our values, we might worry more about children falling off the low end of the bell curve in an increasingly technological society and decide to devote more public resources to those who are in danger of being left behind. Or we might decide that all citizens need to be computer literate so that they will not be left on the shore while everyone else is surfing the Internet.

A related point is that heritability does not imply genetic determinism. Just because a trait shows genetic influence does not mean that nothing can be done to change it. Environmental change is possible even for single-gene disorders. For example, when PKU was found to be a single-gene cause of mental retardation, it was not treated by means of eugenic (breeding) intervention or genetic engineering. An environmental intervention was successful in bypassing the genetic problem of high levels of phenolpyruvic acid: Administer a diet low in phenylalanine. This important environmental intervention was made possible by recognition of the genetic basis for this type of mental retardation.

For behavioral disorders and dimensions, the links between specific genes and behavior are weaker because behavioral traits are generally influenced by multiple genes and environmental factors. For this reason, genetic influence on behavior involves probabilistic propensities rather than predetermined programming. In other words, the complexity of most behavioral systems means that genes are not destiny. Although specific genes that contribute to complex disorders such as late-onset Alzheimer's disease are beginning to be identified, these genes only represent genetic risk factors in that they increase the probability of occurrence of the disorder but do not guarantee that the disorder will occur. An important corollary of the point that heritability does not imply genetic determinism is that heritability does not constrain environmental interventions such as psychotherapy.

We hasten to note that finding a gene that is associated with a disorder does not mean that the gene is "bad" and should be eliminated. For example, a gene associated with novelty seeking (Chapter 13) may be a risk factor for antisocial behavior, but it could also predispose individuals to scientific creativity. The gene that causes the flushing response to alcohol in Asian individuals protects them against becoming alcoholics (Chapter 14). The classic evolutionary example is a gene that causes sickle-cell anemia in the recessive condition but protects carriers against malaria in heterozygotes (Chapter 17). As we shall see, most complex traits are influenced by multiple genes, so we are all likely to be carrying many genes that contribute to risk for some disorders.

Finally, finding genetic influence on complex traits does not mean that the environment is unimportant. For simple single-gene disorders, environmental factors may have little effect. In contrast, for complex traits, environmental influences are usually as important as genetic influences. When one member of an identical twin pair is schizophrenic, for example, the other twin is not schizophrenic in about half the cases, even though members of identical twin pairs are identical genetically. Such differences within pairs of identical twins can only be caused by nongenetic factors. Despite its name, behavioral genetics is as useful in the study of environment as it is in the study of genetics. In providing a "bottom line" estimate of all genetic influence on behavior, genetic research also provides a "bottom line" estimate of environmental influence. Indeed, genetic research provides the best available evidence for the importance of the environment. Moreover, genetic research has made some of the most important discoveries in recent years about how the environment works in psychological development (Chapter 16).

In the field of quantitative genetics, the word *environment* includes all influences other than inherited factors. This use of the word *environment* is much broader than is usual in the behavioral sciences. In addition to environmental influences traditionally studied in the behavioral sciences, such as parenting, environment includes prenatal events and nongenetic biological events after birth, such as illnesses and nutrition. As mentioned in Chapter 3, environment even in-

Nick Martin had planned a career in politics and had begun a degree in the arts before becoming interested in the tension between political ideals of equality before the law and the biological reality of individual differences. As an undergraduate in Adelaide, South Australia, he began his first twin study—of school examination performance—with the encouragement of his father, who was also a geneticist. While writing up this study, he became aware of the radical new approach to behavioral genetic

analysis being forged by the biometrical geneticists in Birmingham, England. He went there for his Ph.D. studies and worked with Lindon Eaves and John Jinks for five years. Hans Eysenck and David Fulker, at the Institute of Psychiatry in London, were also great influences. The major achievements of his time with Eaves were the development of the genetic analysis of covariance structure on which multivariate genetic analysis is largely based, and the first power calculations for twin studies. These calculations showed that twin studies needed to be much larger, which prompted him to return to Australia and found the Australian Twin Registry. Much of Martin's subsequent work on the genetics of personality, alcoholism, and other psychiatric symptoms has been based on results from the Registry. His current interest is to use such large, well-phenotyped twin samples for genomewide association studies to discover major genes for behavioral traits.

cludes changes in DNA that are not inherited because they occur in cells other than testes and ovaries, where sperm and eggs are formed. For example, identical twins are not identical for such environmentally induced changes in DNA.

Beyond Heritability

As mentioned in Chapter 1, one of the most dramatic shifts in the behavioral sciences during the past few decades has been toward a balanced view that recognizes the importance of both nature and nurture in the development of individual differences in behavior. Behavioral genetic research has found genetic influence nearly everywhere it has looked. Indeed, it is difficult to find any behavioral dimension or disorder that reliably shows no genetic influence. On the other hand, behavioral genetic research also provides some of the strongest available evidence for the importance of environmental influences for the simple reason that heritabilities are seldom greater than 50 percent . This means that environmental factors are also important. This message of the importance of both nature and nurture is repeated throughout the following chapters. It is a message that seems to have gotten through to the public as well as academics. For example, a recent survey of parents and teachers of young children found that over 90 percent indicated that genetics is at least as important as environment for mental illness, learning difficulties, intelligence, and personality (Walker & Plomin, 2005).

As a result of the increasing acceptance of genetic influence on behavior, most behavioral genetic research reviewed in the rest of the book goes beyond merely estimating heritability. Estimating whether and how much genetics influences behavior is an important first step in understanding the origins of individual differences. But these are only first steps. As illustrated throughout this book, quantitative genetic research goes beyond heritability in three ways. First, instead of estimating genetic and environmental influence on the variance of one behavior at a time, multivariate genetic analysis investigates the origins of the covariance between behaviors. Some of the most important advances in behavioral genetics in recent years have come from multivariate genetic analyses. A second way in which behavioral genetic research goes beyond heritability is to investigate the origins of continuity and change in development. This is why so much recent behavioral genetic research is developmental, which is reflected in a new chapter (Chapter 12) on developmental psychopathology. Third, behavioral genetics has considered the interface between nature and nurture, which is discussed in Chapter 16. Moreover, the behavioral sciences are at the dawn of a new area in which molecular genetics will revolutionize genetic research by identifying specific genes responsible for the heritability of behavioral traits. Identifying genes will make it possible to address multivariate, developmental, and gene-environment interface issues with much greater precision. It will also facilitate the ultimate understanding of the pathways between genes and behavior (Chapter 15). Identifying genes is the topic of the next chapter.

SUMMING UP

The size of the genetic effect can be quantified by the statistic called heritability. Heritability estimates the proportion of observed (phenotypic) differences among individuals that can statistically be attributed to genetic differences. For schizophrenia and cognitive ability, genetic influence is not only significant but also substantial. Heritability describes *what is* in a particular population at a particular time, not *what could be* or *what should be*. Phenotypic differences not explained by genetic differences can be attributed to the environment. In this way, genetic studies provide the best available evidence for the importance of environment. Recognition of the importance of both nature and nurture has led behavioral genetic research to go beyond asking whether or how much genes influence behavior. We now ask questions about the multivariate relationship among behaviors, about developmental change and continuity, and about gene-environment interplay.

Equality

It was a self-evident truth to the signers of the American Declaration of Independence that all men are created equal. Does this mean that democracy rests on the absence of genetic differences among people? Absolutely not. The founding fathers of America were not so naive as to think that all people are created *identical*. The essence of a democracy is that all people should have legal equality *despite* their genetic differences.

The central message from behavioral genetics involves genetic individuality. With the exception of identical twins, each one of us is a unique genetic experiment, never to be repeated again. Here is the conceptualization on which to build a philosophy of the dignity of the individual! Human variability is not simply imprecision in a process that, if perfect, would generate unvarying representatives of an ideal person. Genetic diversity is the essence of life.

Summary

Quantitative genetic methods can detect genetic influence for complex traits. For animal behavior, selection studies and studies of inbred strains provide powerful tests of genetic influence. Adoption and twin studies are the workhorses for human quantitative genetics. They capitalize on the quasi-experimental situations caused by adoption and twinning to assess the relative contributions of nature and nurture. For schizophrenia and cognitive ability, for example, resemblance of relatives increases with genetic relatedness, an observation suggesting genetic influence. Adoption studies show family resemblance even when family members are adopted apart. Twin studies show that identical twins are more similar than fraternal twins. Results of such family, adoption, and twin studies converge on the conclusion that genetic factors contribute substantially to complex human behavioral traits, among other traits.

The size of the genetic effect is quantified by heritability, a statistic that describes the contribution of genetic differences to observed differences in a particular population at a particular time. For most behavioral dimensions and disorders, including cognitive ability and schizophrenia, genetic influence is not only detectable but also substantial, often accounting for as much as half of the variance in the population. Genetic influence in the behavioral sciences has been controversial in part because of misunderstandings about heritability.

Genetic influence on behavior is just that—an influence or contributing factor, not preprogrammed and deterministic. Environmental influences are usually as important as genetic influences. Behavioral genetics focuses on why people differ, that is, the genetic and environmental origins of individual differences that exist at a particular time in a particular population. Recognition of individual differences, regardless of their environmental or genetic origins, does not lessen the merits of the concept of equality.

CHAPTER SIX

Identifying Genes

Much more quantitative genetic research of the kind described in Chapter 5 is needed to identify the most heritable components and constellations of behavior, to investigate developmental change and continuity, and to explore the interplay between nature and nurture. However, one of the most exciting directions for research in behavioral genetics is the coming together of quantitative genetics and molecular genetics in attempts to identify specific genes responsible for genetic influence on behavior, even for complex behaviors for which many genes as well as many environmental factors are at work.

As illustrated in Figure 6.1, quantitative genetics and molecular genetics both began around the beginning of the twentieth century. The two groups, biometricians (Galtonians) and Mendelians, quickly came into contention, as described in Chapter 3. Their ideas and research grew apart as quantitative geneticists focused on naturally occurring genetic variation and complex quantitative traits, and molecular geneticists analyzed single-gene mutations, often those created artificially by chemicals or X-irradiation. During the past decade, however, quantitative genetics and molecular genetics have begun to come together to identify genes for complex, quantitative traits. Such a gene in multiple-gene systems is called a *quantitative trait locus (QTL)*. Unlike single-gene effects that are necessary and sufficient for the development of a disorder, QTLs contribute like probabilistic risk factors. QTLs are inherited in the same Mendelian manner as single-gene effects; but, if there are many genes that affect a trait, then each gene is likely to have a relatively small effect.

In addition to producing indisputable evidence of genetic influence, the identification of specific genes will revolutionize behavioral genetics by providing measured genotypes for investigating with greater precision the multivariate, developmental, and gene-environment interplay issues that have become the focus of quantitative genetic research. Moreover, once a gene is identified,

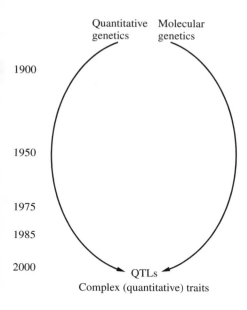

Figure 6.1 Quantitative genetics and molecular genetics are coming together in the study of complex quantitative traits and quantitative trait loci (QTLs).

it is possible to begin to explore the pathways between genes and behavior, which is the topic of Chapter 15. This new knowledge about specific genes associated with behavior will raise new ethical issues as well, as discussed at the end of this chapter and also in Chapter 18.

Animal Behavior

Chapter 5 indicated that inbred strain and selection studies with animals provide direct experiments to investigate genetic influence. In contrast, quantitative genetic research on human behavior is limited to adoption, the experiment of nurture, and twinning, the experiment of nature. Similarly, animal models provide more powerful means to identify genes than are available for our species because genes and genotypes can be manipulated experimentally.

Long before DNA markers became available in the 1980s, associations were found between single genes and behavior. The first example was discovered in 1915 by A. H. Sturtevant, inventor of the chromosome map. He found that a single-gene mutation that alters eye color in the fruit fly *Drosophila* also affects their mating behavior. Another example involves the single recessive gene that causes albinism and also affects open-field activity in mice. Albino mice are less active in the open field. It turns out that this effect is largely due to the fact that albinos are more sensitive to the bright light of the open field. With a red light that reduces visual stimulation, albino mice are almost as active as pigmented mice. These relationships are examples of what is called *allelic association*, the association between a particular allele and a phenotype. Rather than using genes like eye color and albinism that are known by their phenotypic effect, it is now possible to use hundreds of thousands of polymorphisms in DNA itself, either naturally occurring DNA polymorphisms, such as those determining eye color or albinism, or artificially created mutations.

Creating Mutations

In addition to studying naturally occurring genetic variation, geneticists have long used chemicals or X-irradiation to create mutations so that they could identify genes affecting complex traits including behavior. This section focuses on the use of mutational screening to identify genes that affect behavior.

During the past 40 years, hundreds of behavioral mutants have been created in organisms as diverse as bacteria, worms, fruit flies, zebrafish, and mice. (See Figure 6.2.) Information about these and other animal models for genetic research is available from www.nih.gov/science/models. This work illustrates the point that most normal behavior is influenced by many genes. Although any one of many single-gene mutations can seriously disrupt behavior, normal development is orchestrated by many genes working together. An analogy is an

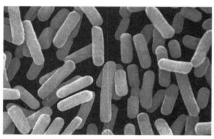

Bacteria (© Dr. Richard Kessel and Dr. Gene Shih/Visuals Unlimited.)

Zebrafish (© Inga Spence/Visuals Unlimited.)

Roundworm (© Wim van Egmond/ Visuals Unlimited.)

Mouse (© Redmond Durrell/Alamy.)

Figure 6.2 Behavioral mutants have been created in bacteria (shown magnified 25,000 times), roundworms (about 1 mm in length), fruit flies (about 2–4 mm), zebrafish (about 4 cm), and mice (about 9 cm without the tail).

Fruit fly (© BIOS Borrell Bartomeu/Peter Arnold, Inc.)

automobile, which requires thousands of parts for its normal functioning. If any one part breaks down, the automobile may not run properly. In the same way, if the function of any gene breaks down through mutation, it is likely to affect many behaviors. In other words, mutations in single genes can drastically affect behavior that is normally influenced by many genes. An important principle is *pleiotropy*, the effect of a single gene on many traits. The corollary is that any complex trait is likely to be *polygenic*, that is, influenced by many genes. Also, there is no necessary relationship between naturally occurring genetic variation and experimentally created genetic variation. That is, creating a mutation that affects a behavior does not imply that naturally occurring variation in that gene is associated with naturally occurring variation in the behavior.

Bacteria   Although the behavior of bacteria is by no means attention grabbing, they do behave. They move toward or away from many kinds of chemicals by rotating their propellerlike flagella. Since the first behavioral mutant in bacteria was isolated in 1966, the dozens of mutants that have been created emphasize the genetic complexity of an apparently simple behavior in a simple organism. For example, many genes are involved in rotating the flagella and controlling the duration of the rotation.

Roundworms   Among the 20,000 species of nematode (roundworm), *Caenorhabditis elegans* is about 1 mm in length and spends its three-week life span in the soil, especially in rotting vegetation, where it feeds on microbes such as bacteria. Conveniently, it also thrives in laboratory Petri dishes. Once viewed as an uninteresting, featureless tube of cells, *C. elegans* is now studied by thousands of researchers. It has 959 cells, of which 302 are nerve cells, including neurons in a primitive brain system called a nerve ring. A valuable aspect of *C. elegans* is that all its cells are visible with a microscope through its transparent body. The development of its cells can be observed, and it develops quickly because of its short life span.

Its behavior is more complex than that of single-celled organisms like bacteria, and many behavioral mutants have been identified (Hobert, 2003). For example, investigators have identified mutations that affect locomotion, foraging behavior, and learning (Rankin, 2002). *C. elegans* is especially important for functional genetic analysis because the developmental fate of each of its cells and the wiring diagram of its 302 nerve cells are known. In addition, most of its 20,000 genes are known, although we have no idea what half of them do. About half of the genes are known to match human genes. *C. elegans* has the distinction of being the first animal to have its genome of 100 million base pairs (3 percent of the size of the human genome) completely sequenced (Wilson, 1999). Despite these huge advantages for the experimental analysis of behavior, it has been difficult to connect the dots between genes, brain, and behavior (Schafer, 2005), which is a lesson to which we will return in Chapter 15.

Fruit flies   The fruit fly *Drosophila*, with about 2000 species, is the star organism in terms of behavioral mutants, with hundreds identified since the pioneering work of Seymour Benzer (Weiner, 1999). Its advantages include its small size (2–4 mm), the ease of growing it in a laboratory, its short generation time (about two weeks), and its high productivity (females can lay 500 eggs in 10 days). Its genome was sequenced in 1998.

The earliest behavioral research involved responses to light (phototaxis) and to gravity (geotaxis). Normal *Drosophila* move toward light (positive phototaxis) and away from gravity (negative geotaxis). Many mutants that were either negatively phototaxic or positively geotaxic were created. Attempts are continuing to identify the specific genes involved in these behaviors (Toma, White, Hirsch, & Greenspan, 2002).

The hundreds of other behavioral mutants included *sluggish* (generally slow), *hyperkinetic* (generally fast), *easily shocked* (jarring produces a seizure), and *paralyzed* (collapses when the temperature goes above 28°C). A *drop dead* mutant walks and flies normally for a couple of days and then suddenly falls on its back and dies. More complex behaviors have also been studied, especially courtship and learning. Behavioral mutants for various aspects of courtship and copulation have been found. One male mutant, called *fruitless*, courts males as well as females and does not copulate. Another male mutant cannot disengage from the female after copulation and is given the dubious title *stuck*. The first learning behavior mutant was called *dunce* and could not learn to avoid an odor associated with shock even though it had normal sensory and motor behavior.

Drosophila also offer the possibility of creating genetic *mosaics*, individuals in which the mutant allele exists in some cells of the body but not in others (Hotta & Benzer, 1970). As individuals develop, the proportion and distribution of cells with the mutant gene vary across individuals. By comparing individuals with the mutant gene in a particular part of the body—detected by a cell marker gene that is inherited along with the mutant gene—it is possible to localize the site where a mutant gene has its effect on behavior.

The earliest mosaic mutant studies involved sexual behavior and the X chromosome (Benzer, 1973). *Drosophila* were made mosaic for the X chromosome: Some body parts have two X chromosomes and are female, and other body parts have only one X chromosome and are male. As long as a small region toward the back of the brain is male, courtship behavior is male. Of course, sex is not all in the head. Different parts of the nervous system are involved in aspects of courtship behavior such as tapping, "singing," and licking. Successful copulation also requires a male thorax (containing the fly's version of a spinal cord between the head and abdomen) and, of course, male genitals (Greenspan, 1995).

Many other gene mutations in *Drosophila* have been shown to affect behaviors (Sokolowski, 2001). The future importance of *Drosophila* in behavioral research is assured by its unparalleled genomic resources (often called *bioinformatics*) (Matthews, Kaufman, & Gelbart, 2005).

Zebrafish   Although invertebrates like *C. elegans* and *Drosophila* are useful in behavioral genetics, many forms and functions are new to vertebrates. The zebrafish, named after its horizontal stripes, is common in many aquaria, grows to about 4 cm, and can live for 5 years. It has become a key vertebrate for studying early development because the developing embryo can be observed directly—it is not hidden inside the mother as are mammalian embryos. In addition, the embryos themselves are translucent. Nearly 1000 gene mutations that affect embryonic development have been identified. Behavior has recently become a focus of research on the zebrafish (Sison, Cawker, Buske, & Gerlai, 2006), including sensory and motor development (Guo, 2004), food and opiate preferences (Lau, Bretaud, Huang, Lin, & Guo, 2006), and novelty seeking and aggression (Blaser & Gerlai, 2006).

Mice and rats   The mouse is the main mammalian species for mutational screening (Kile & Hilton, 2005). Hundreds of lines of mice with mutations that affect behavior have been created (Godinho & Nolan, 2006). Many of these are preserved in frozen embryos that can be "reconstituted" on order. Resources describing the behavioral and biological effects of the mutations are available (e.g., www.informatics.jax.org/). Major initiatives are underway to use chemical mutagenesis to screen mice for mutations on a broad battery of measures of complex traits (Brown, Hancock, & Gates, 2006; Vitaterna, Pinto, & Takahashi, 2006). Behavioral screening is an important part of these initiatives because behavior can be an especially sensitive indicator of the effects of mutations (Crawley, 2003, 2007).

After the human, the mouse was the next mammalian target for sequencing the entire genome, which was accomplished in 2001 (Venter et al., 2001). The rat, whose larger size makes it the favorite rodent for physiological and pharmacological research, is also coming on strong in genomics research (Jacob & Kwitek, 2002; Smits & Cuppen, 2006). The rat genome was sequenced in 2004 (Gibbs et al., 2004). The bioinformatics resources for rodents are growing rapidly (DiPetrillo, Wang, Stylianou, & Paigen, 2005).

Targeted mutations in mice   In addition to mutational screening, the mouse is also the main mammalian species used to create *targeted mutations* that knock out the expression of specific genes. A targeted mutation is a process by which a gene is changed in a specific way to alter its function (Capecchi, 1994). Most often, genes are "knocked out" by deleting key DNA sequences that prevent the gene from being transcribed. Newer techniques produce more subtle changes that alter the gene's regulation; these changes lead to underexpression or overexpression of the gene rather than knocking it out altogether. In mice, the mutated gene is transferred to embryos (a technique called *transgenics* when the mutated gene is from another species). Once mice homozygous for the knock-out gene are bred, the effect of the knock-out gene on behavior can be investigated.

More than a thousand knock-out mouse lines have been created, many of which affect behavior. For example, since 1996 nearly 100 genes have been genetically engineered for their effect on alcohol responses (Crabbe, Phillips, Harris, Arends, & Koob, 2006). Another example is aggressive behavior in the male mouse for which 36 genetically engineered genes show effects (Maxson & Canastar, 2003). An ongoing project, called the Knockout Mouse Project, aims to create knock-outs in 8500 genes (www.nih.gov/science/models/mouse/knockout).

Gene targeting strategies are not without their limitations. One problem with knock-out mice is that the targeted gene is inactivated throughout the animal's life span. During development, the organism copes with the loss of the gene's function by compensating wherever possible. For example, deletion of a gene coding for a dopamine transporter protein (which is responsible for inactivating dopaminergic neurons by transferring the neurotransmitter back into the presynaptic terminal) results in a mouse that is hyperactive in novel environments (Giros, Jaber, Jones, Wightman, & Caron, 1996). These knock-out mutants exhibit complex compensations throughout the dopaminergic system that are not specifically due to the dopamine transporter itself (Jones et al., 1998). However, in most instances, compensations for the loss of gene function are invisible to the researcher, and caution must be taken to avoid attributing compensatory changes in the animals to the gene itself. These compensatory processes can be overcome by creating conditional knock-outs of regulatory elements; these conditional mutations make it possible to turn expression of the gene on or off at will at any time during the animal's life span, or the mutation can target specific areas of the brain.

To achieve maximum results, knock-out technology must overcome a major hurdle. Currently, there is no way to control the location of gene insertion in the mouse genome as the altered genes are taken up by the embryonic cells that will develop into the knock-out mutant. Neither can the number of inserted copies of the gene be controlled. Because so little is known about the effects of either precise location or copy number on gene function, the current technologies for producing mutants amount to making a large number of knock-outs, laboriously characterizing each of them to see where and to what degree the introduced construct is expressed, and choosing the best available model from the array of choices. Learning how to control the insertion site and copy number will greatly enhance the efficiency of knock-out technology.

Gene silencing   In contrast to knock-out studies, which alter DNA, another method uses double-stranded RNA to "knock down" expression of the gene that shares its sequence (Hannon, 2002). The gene silencing technique, which was discovered in 1997 and won the Nobel Prize in 2006 (Bernards, 2006), is called *RNA interference* (RNAi) or *small interfering RNA* (siRNA), because it degrades complementary RNA transcripts. siRNA kits are now available commercially that target nearly all the genes in the human and mouse genomes. More than a thousand papers on siRNA were published in 2006 alone, primarily using cell cultures where delivery of the siRNA to the cells is not a problem. However, in vivo animal model research necessary for behavioral analysis has begun, although delivery to the brain remains a problem (Thakker, Hoyer, & Cryan, 2006). For example, injecting siRNA in mouse brain has yielded knockdown results on behavior similar to results expected from knock-out studies (Salahpour, Medvedev, Beaulieu, Gainetdinov, & Caron, 2007). It is hoped that siRNA will soon have therapeutic applications (Kim & Rossi, 2007).

SUMMING UP

Many behavioral mutants have been identified in organisms as diverse as single-celled bacteria, roundworms, fruit flies, and rodents. This work shows that mutations in single genes can drastically affect behavior that is normally influenced by many genes. More than a thousand genetically engineered genes have been knocked out in mice, many of which have behavioral effects. Gene silencing using siRNA holds great promise for knocking down gene expression rather than knocking out gene function altogether.

Quantitative Trait Loci

Creating a mutation that has a major effect on behavior does not mean that this gene is specifically responsible for the behavior. Remember the automobile analogy in which any one of many parts can go wrong and prevent the automobile from running properly. Although the part that goes wrong has a big effect, that part is only one of many parts needed for normal functioning. Moreover, the genes changed by artificially created mutations are not necessarily responsible for the naturally occurring genetic variation detected in quantitative genetic research. Identifying genes responsible for naturally occurring genetic variation that affects behavior has only become possible in recent years. The difficulty is that, instead of looking for a single gene with a major effect, we are looking for many genes, each having a relatively small effect size—QTLs.

Animal models have been particularly useful in the quest for QTLs, because both genetics and environment can, and may, be manipulated and controlled in the laboratory, whereas for our species neither genetics nor environment may be manipulated. Animal model work on natural genetic variation and behavior has primarily studied the mouse and the fruit fly *Drosophila* (Kendler & Greenspan, 2006). Although this section emphasizes research on mice, similar methods have been used in *Drosophila* (Mackay & Anholt, 2006) and have been applied to many behaviors (Anholt & Mackay, 2004), including mating behavior (Moehring & Mackay, 2004), odor-guided behavior (Sambandan, Yamamoto, Fanara, Mackay, & Anholt, 2006), and locomotor behavior (Jordan, Morgan, & Mackay, 2006). In addition, as mentioned in the previous section, behavioral genetic research on the rat is also increasing rapidly for complex traits including behavior (Smits & Cuppen, 2006).

In animal models, linkage can be identified by using Mendelian crosses to trace the cotransmission of a marker whose chromosomal location is known and a single-gene trait, as illustrated in Figure 2.6. Linkage is suggested when the results violate Mendel's second law of independent assortment. However, as emphasized in previous chapters, behavioral dimensions and disorders are likely to be influenced by many genes; consequently, any one gene is likely to have only a small effect. If many genes contribute to behavior, behavioral traits will be distributed quantitatively. The goal is to find some of the many genes (QTLs) that affect these quantitative traits.

F_2 *crosses*   Although linkage techniques can be extended to investigate quantitative traits, most QTL analyses with animal models use association, which may be more powerful for detecting the small effect sizes expected for QTLs. Allelic association refers to the correlation or association between an allele and a trait. For example, the allelic frequency of DNA markers can be compared for groups of animals high or low on a quantitative trait. This approach has been applied to open-field activity in mice (Flint et al., 1995). F_2 mice were derived from a cross between high and low lines selected for open-field activity and subsequently inbred by using brother-sister matings for over 30 generations. Each F_2 mouse has a unique combination of alleles from the original parental strains because there is an average of one recombination in each chromosome inherited from the F_1 strain (see Figure 2.7). The most active and the least active F_2 mice were examined for 84 DNA markers spread throughout the mouse chromosomes in an effort to identify chromosomal regions that are associated with open-field activity (Flint et al., 1995b). The analysis simply compares the frequencies of marker alleles for the most active and least active groups.

Figure 6.3 shows that regions of chromosomes 1, 12, and 15 harbor QTLs for open-field activity. A QTL on chromosome 15 is related primarily to open-field activity and not to other measures of fearfulness, an observation suggesting the possibility of a gene specific to open-field activity. The QTL regions on chromosomes 1 and 12, on the other hand, are related to other measures of fearfulness, associations suggesting that these QTLs affect these diverse measures of fearfulness. QTLs were subsequently mapped in two large ($N = 815$ and 821) F_2 crosses from the replicate inbred lines of mice initially selected for open-field activity (Turri,

Figure 6.3 QTLs for open-field activity and other measures of fearfulness in an F_2 cross between high and low lines selected for open-field activity. The five measures are (1) open-field activity (OFA), (2) defecation in the open field, (3) activity in the Y maze, (4) entry in the open arms of the elevated plus maze, and (5) entry in the closed arms of the elevated plus maze, which is not a measure of fearfulness. LOD (logarithm to the base 10 of the odds) scores indicate the strength of the effect; and a LOD score of 3 or greater is generally accepted as significant. Distance in centimorgans (cM) indicates position on the chromosome, with each centimorgan roughly corresponding to 1 million base pairs. Below the distance scale are listed the specific short-sequence repeat markers for which the mice were examined and mapped. (Reprinted with permission from "A simple genetic basis for a complex psychological trait in laboratory mice" by J. Flint et al. *Science, 269*, 1432–1435. ©1995 American Association for the Advancement of Science. All rights reserved.)

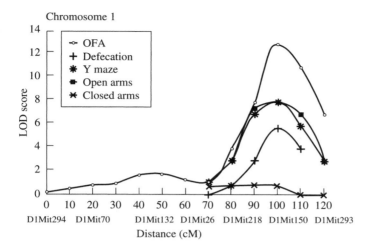

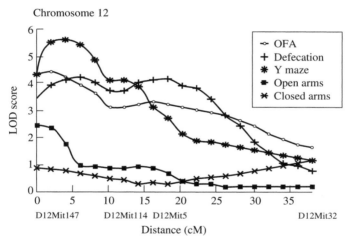

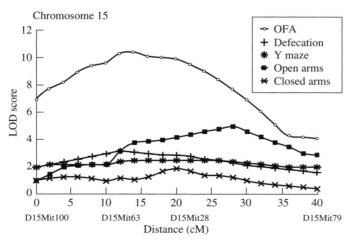

Jonathan Flint is in the Wellcome Trust Centre for Human Genetics at the University of Oxford. After studying history at Oxford University, he entered medical school at St. Mary's Hospital, London, where, largely due to the enthusiastic teaching in molecular genetics, he took a year out to work at a laboratory bench in Oxford. He witnessed at first hand one of the very first gene mapping experiments, for adult polycystic kidney disease. The excitement of the scientific environment, the apparent power of molecular techniques to make inroads into previously impregnable medical problems, led Flint to ask whether molecular genetics might be useful in behavioral disorders. After training in psychiatry at the Maudsley Hospital in London, he moved back to Oxford, where he demonstrated the importance of small chromosomal rearrangements as a new cause of mental retardation. At this time, the inconsistencies in the results of attempting to map psychiatric disease led Flint to ask whether behavior was just too complex to be dissected genetically with the currently available tools. He argued that only animal experiments could answer this question and has subsequently collaborated in work that has mapped the genetic basis of emotional behavior in mice.

Henderson, DeFries, & Flint, 2001). Results of this study both confirmed and extended the previous findings reported by Flint et al. (1995b). QTLs for open-field activity were replicated on chromosomes 1, 4, 12, and 15, and new evidence for additional QTLs on chromosome 7 and the X chromosome was also obtained. An exception is exploration in an enclosed arm of a maze (see Figure 6.3), which was included in the study as a control, because other research suggests that this measure is not genetically correlated with measures of fearfulness. Several studies have also reported associations between markers on the distal end of chromosome 1 and quantitative measures of emotional behavior, although it has been difficult to identify the specific gene responsible for the association (Fullerton, 2006).

HS crosses   Because the chromosomes of F_2 mice have only an average of one crossover between maternal and paternal chromosomes, the method has little resolving power to pinpoint a locus although it has good power to identify the chromosome on which a QTL resides. That is, QTL associations found by using F_2 mice only refer to general "neighborhoods," not specific addresses. The QTL neighborhood is usually very large, about 10 million to 20 million base pairs of DNA, and thousands of genes could reside there. One way to increase the resolving power is to use animals whose chromosomes are recombined to a greater

extent, either by breeding generations beyond the F_2 (Darvasi, 1998) or by crossing more than two strains together (Valdar, Solberg, Gauguier, Burnett et al.,
2006). The latter approach was used to increase 30-fold the resolving power of
the QTL study of fearfulness (Talbot et al., 1999). Mice in the top and bottom
20 percent of open-field activity scores were selected from 751 offspring of outbred crosses between eight different inbred strains (called heterogeneous stock,
HS). The results confirmed the association between emotionality and markers
on chromosome 1, although the association was closer to the 70-cM region than
the 100-cM region of chromosome 1 found in the earlier study (see Figure 6.3).
Some supporting evidence for a QTL on chromosome 12 was also found, but
none was found for chromosome 15. Work continues toward the identification
of the specific gene responsible for the chromosome 1 association using a new
commercially available HS cross that facilitates finer QTL mapping (Yalcin et
al., 2004) and other strategies (Willis-Owen & Flint, 2006).

Much QTL research in mice has been in the area of pharmacogenetics, a field
in which investigators study genetic effects on responses to drugs. At least
24 QTLs have been mapped for drug responses such as alcohol drinking, alcohol-
induced loss of righting reflex, acute alcohol and pentobarbital withdrawal, cocaine seizures, and morphine preference and analgesia (Crabbe, Phillips, Buck,
Cunningham, & Belknap, 1999). This achievement represented considerable
progress from the first effort to summarize QTLs for drug responses five years earlier (Crabbe, Belknap, & Buck, 1994), and progress has continued since (Bennett,
Downing, Parker, & Johnson, 2006). In some instances, the location of a mapped
QTL is close enough to a previously mapped gene of known function to make
studies of that gene informative. For example, several groups have mapped QTLs
for alcohol preference drinking to mouse chromosome 9 (Phillips, Brown et al.,
1998), in a region that includes the gene coding for the dopamine D_2 receptor subtype. Studies with D_2 receptor knock-out mice revealed that they showed reduced
alcohol preference drinking (Phillips, Belknap, Buck, & Cunningham, 1998).

Recombinant inbred strains   Another method used to identify QTLs for behavior involves special inbred strains called recombinant inbred (RI) strains. RI strains
are inbred strains derived from an F_2 cross between two inbred strains; this process
leads to recombination of parts of chromosomes from the parental strains. (See
Figure 6.4.) Thousands of DNA markers have been mapped in RI strains, thus en-
abling investigators to use these markers to identify QTLs associated with behavior without any additional genotyping (Plomin & McClearn, 1993b). The special
value of the RI QTL approach is that it enables all investigators to study essentially
the same animals, because the RI strains are extensively inbred. This feature of RI
QTL analysis means that each RI strain needs to be genotyped only once and that
genetic correlations can be assessed across measures, across studies, and across laboratories. The QTL analysis itself is much like the F_2 QTL analysis discussed earlier except that, instead of comparing individuals with recombined genotypes, the
RI QTL approach compares means of recombinant inbred strains. RI QTL work

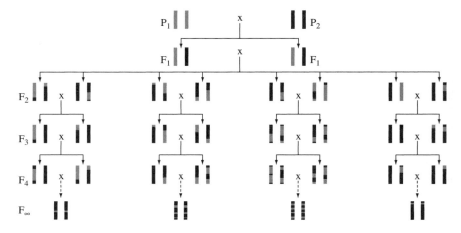

Figure 6.4 Construction of a set of recombinant inbred strains from the cross of two parental inbred strains. The F_1 is heterozygous at all loci that differ in the parental strains. Crossing F_1 mice produces an F_2 generation in which alleles from the parental strains segregate so that each individual is genetically unique. Comparing the quantitative trait scores of mice that differ genotypically for a particular DNA marker indicates QTL association, which is the F_2 cross discussed in the text. By inbreeding the F_2 with brother-sister matings for many generations, recombination continues until each RI strain is fixed homozygously at each gene for a single allele inherited from one or the other progenitor inbred strain. Unlike F_2 crosses, RI strains are stable genetically because each strain has been inbred. This means that a set of RI strains needs to be genotyped only once for DNA markers or phenotyped only once for behaviors and the data can be used in any other experiment using that set of RI strains. Similar to the F_2 cross, QTL association can be detected by comparing the quantitative trait scores of RI strains that differ genotypically for a particular DNA marker.

has also focused on pharmacogenetics. For example, RI QTL research has confirmed some of the associations for responses to alcohol found using F_2 crosses (Buck, Rademacher, Metten, & Crabbe, 2002). Although none of the specific genes have yet been identified, research is closing in on candidate QTLs for alcohol withdrawal (Rhodes & Crabbe, 2003). Research combining RI and F_2 QTL approaches are also making progress toward identifying genes for alcohol-related behaviors (Bennett, Carosone-Link, Zahniser, & Johnson, 2006).

One problem with the RI QTL method has been that only a few dozen RI strains were available, which means that only associations of large effect size could be detected. Also, it has been difficult to locate the specific genes responsible for associations. For example, during the past 15 years, more than 2000 QTL associations have been reported using crosses between inbred strains, but fewer than 1 percent have been localized (Flint, Valdar, Shifman, & Mott, 2005). A major new development is the creation of an RI series that includes 1000 RI strains from crosses between eight inbred strains (Churchill et al., 2004). By crossing eight inbred strains, the resulting RI strains will show greater recombination than seen in the two-strain RI example shown in Figure 6.4. Creating 1000 RI strains will also provide power to detect QTL associations of modest effect size. The "Collaborative Cross," as the project is known, will provide a valuable resource not only for identifying genes associated with complex traits but also for integrative analyses of complex systems that include gene expression as well as neural, pharmacological, and behavioral data (Chesler et al., 2005), as described in Chapter 15.

Synteny Homology

QTLs found in mice can be used as candidate QTLs for human research because nearly all mouse genes are similar to human genes. Moreover, chromosomal regions linked to behavior in mice can be used as candidate regions in human studies because parts of mouse chromosomes have the same genes in the same order as parts of human chromosomes, a relationship called *synteny homology*. It is as if about 200 chromosomal regions have been reshuffled onto different chromosomes from mouse to human. (See www.informatics.jax.org/ for details about synteny homology.) For example, the region of mouse chromosome 1 shown in Figure 6.3 to be linked with open-field activity has the same order of genes that happen to be part of the long arm of human chromosome 1, although syntenic regions are usually on different chromosomes in mouse and human. As a result of these findings, this region of human chromosome 1 has been considered as a candidate QTL region for human anxiety, and linkage with the syntenic region in human chromosome 1 has been reported in two large studies (Fullerton et al., 2003; Nash et al., 2004).

SUMMING UP

The mouse genome can be scanned for quantitative trait loci (QTLs) associated with behavior, even when many genes are involved for complex, quantitative traits. Mouse QTLs can point to candidate QTLs for human behavior because of the extensive synteny homology between mouse and human chromosomes.

Human Behavior

In our species, we cannot manipulate genes or genotypes as in knock-out studies or minimize environmental variation in a laboratory. Although this prohibition makes it more difficult to identify genes associated with behavior, this cloud has the silver lining of forcing us to deal with naturally occurring genetic and environmental variation. The silver lining is that results of human research will generalize to the world outside the laboratory and are more likely to translate to clinically relevant advances in diagnosis and treatment.

As described in Chapter 2, linkage has been extremely successful in locating the chromosomal neighborhood of single-gene disorders. For many decades, the actual residence of a single-gene disorder could be pinpointed when a physical

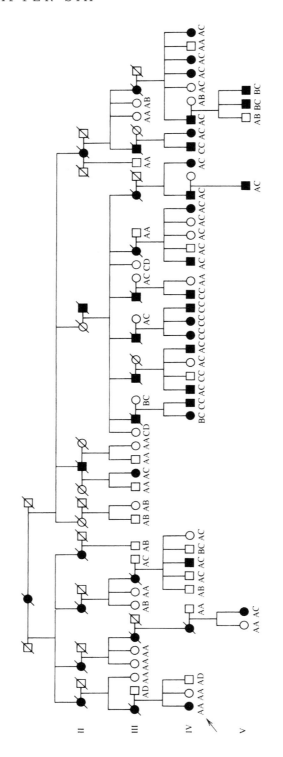

marker for the disorder was available, as was the case for PKU (high phenylala-nine levels), which led to identification of the culprit gene in 1984. With the dis-covery of DNA markers in the 1980s, screening the genome for linkage became possible for any single-gene disorder, which in 1993 led to the identification of the gene that causes Huntington disease (Bates, 2005).

During the past decade, attempts to identify genes responsible for the her-itability of complex traits have moved quickly from traditional linkage studies to QTL linkage to candidate gene association and, very recently, to genome-wide association studies, as it became apparent that genetic influence on com-plex traits is caused by many more genes of much smaller effect size than anticipated. This fast-moving journey is described in this section.

Linkage: Single-Gene Disorders

For single-gene disorders, linkage can be identified by using a few large family pedigrees, in which cotransmission of a DNA marker allele and a disorder can be traced. Because recombination occurs an average of once per chromosome in the formation of gametes passed from parent to offspring, a marker allele and an allele for a disorder on the same chromosome will be inherited together within a family. In 1983, the first DNA marker linkage was found for Huntington dis-ease in a single five-generation pedigree shown in Figure 6.5. In this family, the allele for Huntington disease is linked to the allele labeled C. All but one person with Huntington disease has inherited a chromosome that happens to have the C allele in this family. This marker is not the Huntington gene itself, because a recombination was found between the marker allele and Huntington disease for one individual; the leftmost woman with an arrow in generation IV had Hunt-ington disease but did not inherit the C allele for the marker. That is, this woman received that part of her affected mother's chromosome carrying the gene for Huntington disease, which is normally linked in this family with the C allele but in this woman is recombined with the A allele from the mother's other chromo-some. The farther the marker is from the disease gene, the more recombinations will be found within a family. Markers even closer to the Huntington gene were later found. Finally, in 1993, a genetic defect was identified as the CAG repeat sequence associated with most cases of Huntington disease, as described in Chapter 3. A similar approach was used to locate the genes responsible for other

Figure 6.5 Linkage between the Huntington disease gene and a DNA marker at the tip of the short arm of chromosome 4. In this pedigree, Huntington disease occurs in individuals who inherit a chromosome bearing the C allele for the DNA marker. A single individual shows a recombination (marked with an arrow) in which Huntington disease occurred in the absence of the C allele. (From "DNA markers for nervous system diseases" by J. F. Gusella et al. *Science, 225,* 1320–1326. ©1984. Used with permission of the American Association for the Advancement of Science.

single-gene disorders such as PKU on chromosome 12 and fragile X mental retardation on the X chromosome.

Although linkage analysis of large pedigrees has been very effective for locating genes for single-gene disorders, it is less powerful when several genes are involved. Early attempts to use this approach to investigate linkage for major psychiatric disorders led in the late 1980s to well-publicized reports of linkage, but these reports were later retracted. Even for complex disorders like schizophrenia, for which many genes may be involved in the population, the traditional large-pedigree design might be successful in identifying some families in which a particular gene has a large effect.

Linkage: Complex Disorders

Another linkage approach has greater power to detect genes of smaller effect size and can be extended to quantitative traits. Rather than studying a few families with many relatives as in traditional linkage, this method studies many families with a small number of relatives, usually siblings. The simplest method examines *allele sharing* for pairs of affected siblings in many different families, as explained in Box 6.1. As indicated in later chapters, the affected sib-pair linkage design is the most widely used linkage design for studying complex traits such as behavior.

Linkage based on allele sharing can also be investigated for quantitative traits by correlating allele sharing for DNA markers with sibling differences on a quantitative trait. That is, a marker linked to a quantitative trait will show greater than expected allele sharing for siblings who are more similar for the trait. The sib-pair QTL linkage design was first used to identify and replicate a linkage for reading disability on chromosome 6 (6p21; Cardon et al., 1994), a QTL linkage that has been replicated in several other studies (see Chapter 7). As seen in the following chapters, many genomewide linkage studies have been reported. However, replication of linkage results has generally not been as clear as in the case of reading disability, as seen, for example, in a review of 101 linkage studies of 31 human diseases (Altmuller, Palmer, Fischer, Scherb, & Wjst, 2001).

Association: Candidate Genes

A great strength of linkage approaches is that they systematically scan the genome with just a few hundred DNA markers looking for violations of Mendel's law of independent assortment between a disorder and a marker. However, a weakness of linkage approaches is that they cannot detect linkage for genes of small effect size expected for most complex disorders (Risch, 2000). Linkage is like using a telescope to scan the horizon systematically for distant mountains (large QTL effects). However, the telescope goes out of focus when trying to detect nearby hills (small QTL effects).

In contrast to linkage, which is systematic but not powerful, allelic association is powerful but, until recently, it was not systematic. Association is powerful because, rather than relying on recombination within families as in linkage, it simply compares allelic frequencies for groups such as individuals with the disorder (cases)

BOX 6.1

Affected Sib-Pair Linkage Design

The most widely used linkage design includes families in which two siblings are affected. "Affected" could mean that both siblings meet criteria for a diagnosis or that both siblings have extreme scores on a measure of a quantitative trait. The affected sib-pair linkage design is based on allele sharing—whether affected sibling pairs share 0, 1, or 2 alleles for a DNA marker (see the figure). For simplicity, assume that we can distinguish all four parental alleles for a particular marker. Linkage analyses require the use of markers with many alleles so that, ideally, all four parental alleles can be distinguished. The father is shown as having alleles A and B, and the mother has alleles C and D. There are four possibilities for sib-pair allele sharing: They can share no parental alleles; they can share one allele from father or one allele from mother; or they can share two parental alleles. When a marker is not linked to the gene for the disorder, each of these possibilities has a probability of 25 percent. In other words, the probability is 25 percent that sibling pairs share no alleles, 50 percent that they share one allele, and 25 percent that they share two alleles. Deviations from this expected pattern of allele sharing indicate linkage. That is, if a marker is linked to a gene that influences the disorder, more than 25 percent of the affected sibling pairs will share two alleles for the marker. Many examples of affected sib-pair linkage analyses are mentioned in later chapters. A recent example yielded evidence for linkage on chromosome 4 for alcohol dependence, especially for symptoms of tolerance to alcohol and out-of-control drinking (Prescott, Sullivan et al., 2006).

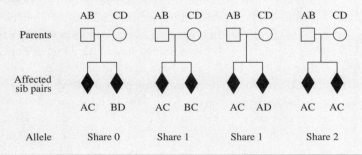

versus controls or low-scoring versus high-scoring individuals on a quantitative trait (Sham, Cherny, Purcell, & Hewitt, 2000). For example, as mentioned in Chapter 1, a particular allele of a gene (for apolipoprotein E on chromosome 19) involved in cholesterol transport is associated with late-onset Alzheimer's disease (Corder et al., 1993). In dozens of association studies, the frequency of allele 4 was found to be about 40 percent in individuals with Alzheimer's disease and about 15 percent in controls. In recent years, allelic associations have been reported for all domains of behavior, as discussed in later chapters, although none are nearly as large an effect as the association between apolipoprotein E and Alzheimer's disease.

The weakness of allelic association is that an association can only be detected if a DNA marker is itself the functional gene (called *direct association*) or very close to it (called *indirect association* or *linkage disequilibrium*). If linkage is a telescope, association is a microscope. As a result, hundreds of thousands of DNA markers need to be genotyped to scan the genome thoroughly. For this reason, until very recently, allelic association has been used primarily to investigate associations with genes thought to be candidates for association. For example, because the drug most commonly used to treat hyperactivity, methylphenidate, acts on the dopamine system, genes related to dopamine, such as the dopamine transporter and dopamine receptors, have been the target of candidate gene association studies. Evidence for QTL associations with hyperactivity involving the D_4 dopamine receptor (*DRD4*) and other dopamine genes is growing (Bobb, Castellanos, Addington, & Rappoport, 2006). For example, the frequency of the *DRD4* allele associated with hyperactivity is about 25 percent for children with hyperactivity and about 15 percent in controls. The problem with the candidate gene approach is that we often do not have strong hypotheses as to which genes are candidate genes. Indeed, as discussed earlier, pleiotropy makes it possible that any of the thousands of genes expressed in the brain could be considered as candidate genes.

Another less serious problem is that case-control comparisons can lead to spurious associations if the cases and controls are not matched. Within-family association designs have been developed for this reason because family members are perfectly matched. For example, siblings can be used to assess associations as well as linkage. Associations with quantitative traits can be assessed *between* families using sibling sums as well as *within* families using sibling differences (Fulker, Cherny, Sham, & Hewitt, 1999). For instance, sibling pairs with one or two copies of the *DRD4* allele associated with hyperactivity should have higher hyperactivity scores than sibling pairs who do not have the "hyperactivity" *DRD4* allele (between-family association). Even within sibling pairs, a sibling with the hyperactivity allele should have a higher activity score than a sibling partner who does not have the allele (within-family association) (Abecasis, Cardon, & Cookson, 2000).

The biggest problem is that reports of candidate gene associations have been difficult to replicate (Tabor, Risch, & Myers, 2002). This is a general problem for complex traits, not just for behavior (Ioannidis, Ntzani, Trikalinos, & Contopoulos-Ioannidis, 2001). For example, in a review of 600 reported associations with common medical diseases, only six have been consistently replicated (Hirschhorn, Lohmueller, Byrne, & Hirschhorn, 2002), although a follow-up meta-analysis indicated greater replication for larger studies (Lohmueller, Pearce, Pike, Lander, & Hirschhorn, 2003).

Much larger studies are needed because many genes of very small effect size appear to underlie the heritability of complex traits (Cardon & Bell, 2001). For case-control association studies, effect size is usually described by the *odds ratio* in which the odds of an allele in cases is divided by the odds of the allele in controls. An odds ratio of 1.0 means that there is no difference in allele frequency between cases and controls. An odds of 3.0 is considered a large effect, which has so far only been approached for Alzheimer's disease; as just mentioned, the frequency of allele 4 of the apolipoprotein E gene is about 40 percent in cases and 15 percent in controls, which is an odds ratio of 2.7. It was assumed that at least some odds ratios of 2 would be found for complex traits, which would require samples of about 500 cases and controls. However, few such associations with an odds ratio of 2 have been found in case-control studies of common disorders. In a review of 752 case-control studies, the odds ratio for replicated results in studies with large samples was only 1.2 (Ioannidis, Trikalinos, & Khoury, 2006). In recent years, in order to be able to detect odds ratios as small as 1.2, researchers have increased sample sizes tenfold (Zondervan & Cardon, 2004). If candidate gene associations are not found in such large samples, it is quite possible that researchers are looking at the wrong candidate genes. For example, only traditional protein-coding genes have been considered as candidate genes but, as discussed in Chapter 4, non-coding genes might be important.

Association: Genomewide

In summary, linkage is systematic but not powerful, and allelic association is powerful but not systematic. Allelic association can be made more systematic by using a dense map of markers. The problem with using a dense map of markers for a genome scan has been the amount of genotyping required and its expense. As many as 500,000 SNPs would be needed to scan the genome thoroughly for association, although selecting SNPs judiciously on the basis of haplotype blocks could reduce this number to about 350,000 SNPs (see Chapter 4; Cardon & Abecasis, 2003). Furthermore, these hundreds of thousands of DNA markers need to be genotyped on very large samples in order to detect associations of small effect size. Even genotyping 1000 individuals (for example, 500 cases and 500 controls) on the 350,000 SNPs comes to 350 million genotypings, which, until recently, would have cost tens of millions of dollars. This is why most association studies have been limited to considering a few candidate genes.

Very recently, a revolutionary advance has made genomewide association investigations possible (Hirschhorn & Daly, 2005). Microarrays, mentioned in Chapter 4 in relation to gene expression, can also be used to genotype hundreds of thousands of SNPs on a "chip" the size of a postage stamp. (See Box 6.2.) With microarrays, the cost of the experiment described above is less than half a million dollars instead of tens of millions. As a result of microarrays, genomewide association analysis has come to dominate attempts to identify genes for complex traits in just a couple of years. Although this is an exciting advance, it remains to be seen whether current genomewide scans of 500,000 SNPs on very large samples will be able to identify replicable associations (Collins, 2006b). If not, even larger samples may be needed to identify the variation in DNA that is responsible for

BOX 6.2

SNP Microarrays

Microarrays have made it possible to study the entire genome (DNA) and the entire transcriptome (RNA) (Plomin & Schalkwyk, 2007). A microarray is a tiny glass slide dotted with short DNA sequences called probes. Microarrays were first used about 10 years ago to assess gene expression, as mentioned in Chapter 4 and discussed again in Chapter 15. Recently, microarrays were developed to genotype SNPs. Microarrays detect SNPs using the same hybridization method described in Box 4.2 and repeated in the figure on the facing page. The difference is that microarrays probe for hundreds of thousands of SNPs on a platform the size of a postage stamp. This miniaturization requires little DNA and makes the method fast and inexpensive.

Several types of microarrays are available commercially (Syvanen, 2005); the figure shows the microarray manufactured by Affymetrix called GeneChip. As shown in the figure, many copies of a 25 nucleotide base sequence surrounding a SNP are used to probe reliably for each allele of the SNP. An individual's DNA is cut with restriction enzymes into tiny fragments which are all amplified by *polymerase chain reaction* (*PCR*; see Box 4.2). Using a single PCR to chop up and amplify the entire genome, called *whole genome amplification*, was the crucial trick that made microarrays possible. The PCR-amplified DNA fragments are made single stranded and washed over the probes on the microarrays so that the individual's DNA fragments will hybridize to the probes if they find exact matches. A fluorescent tag is added to the DNA fragments so that they will fluoresce if they hybridize with a probe, as shown in the figure. The microarray includes probes for both SNP alleles to indicate whether an individual is homozygous or heterozygous. An animation of DNA microarray methodology is available at www.bio.davidson.edu/courses/genomics/chip/chip.html.

Microarrays make it possible to conduct genomewide association studies with hundreds of thousands of SNPs. However, because microarrays cost several hundred dollars each and can be used only once, it is still expensive to conduct studies with the many thousands of subjects needed to detect associations of

the ubiquitous heritability found for complex traits. Recognition of the need for very large samples has led to ambitious projects like DeCode Genetics (www.decode.com), which is studying most of the population of Iceland, and the U.K. BioBank (www.ukbiobank.ac.uk), which will follow 500,000 individuals.

What good will come from identifying genes if they have such small effect sizes? One answer is that we can study pathways between each gene and behavior. Even for genes with very small effect on behavior, the road signs are clearly marked in a bottom-up analysis that begins with gene expression, although the pathways quickly divide and become more difficult to follow to higher levels of

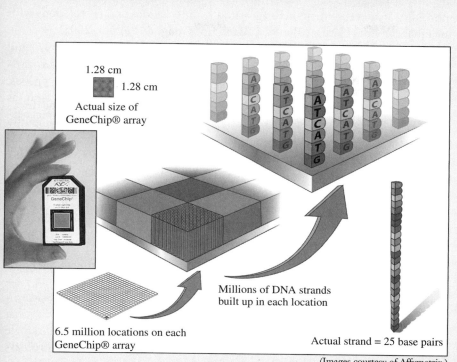

1.28 cm

1.28 cm

Actual size of GeneChip® array

Millions of DNA strands built up in each location

6.5 million locations on each GeneChip® array

Actual strand = 25 base pairs

(Images courtesy of Affymetrix.)

small effect size. Rather than individually genotyping each subject with a microarray and then averaging the results for cases and controls statistically, it is possible to average the groups biologically by pooling a few nanograms of DNA from each individual and genotyping the pooled DNA for each group on a microarray (Pearson et al., 2007). DNA pooling thus makes it possible to use microarrays to screen very large groups of cases and controls for a few hundred dollars rather than a few hundred thousand dollars.

analysis such as the brain and behavior. However, even if there are hundreds of genes that have small effects on a particular behavior, this set of genes will be useful in top-down analyses that begin with behavior and investigate multivariate, developmental, and genotype-environment interface issues, and translate these findings into gene-based diagnosis and treatment as well as prediction and prevention of disorders. These issues about pathways between genes and behavior are the topic of Chapter 15. With microarrays, it would not matter for top-down analyses if there were hundreds of genes that predict a particular trait. Indeed, for each trait, we can imagine microarrays with thousands of genes that include all the genes relevant to that trait's multivariate heterogeneity and co-morbidity, its developmental changes, and its interactions and correlations with the environment (Harlaar, Butcher, Meaburn, Craig, & Plomin, 2005).

SUMMING UP

Linkage studies of large family pedigrees can localize genes for single-gene disorders. Other linkage designs such as the affected sib-pair linkage design can identify linkages for more complex disorders. Techniques are also available to detect QTL linkage for quantitative traits and have been used to show linkage with reading disability. Allelic association, which can detect QTLs of smaller effect size, has until recently been limited to candidate genes such as dopamine-related genes for hyperactivity. With SNP microarrays, it is now possible to conduct systematic genomewide scans for allelic association with hundreds of thousands of SNPs.

BOX 6.3

Genetic Counseling

Genetic counseling is an important interface between the behavioral sciences and genetics. A huge area of research has published nearly 2000 papers since 2000. Genetic counseling goes well beyond simply conveying information about genetic risks and burdens. It helps individuals come to terms with the information by dispelling mistaken beliefs and allaying anxiety in a nondirective manner that aims to inform rather than to advise. In the United States, over 3000 health professionals have been certified as genetic counselors, and about half of these were trained in two-year master's programs (Mahowald, Verp, & Anderson, 1998). For more information about genetic counseling as a profession, including practice guidelines and perspectives, see the National Society of Genetic Counselors (www.nsgc.org/), which sponsors the *Journal of Genetic Counseling* and has a useful link called *How to Become a Genetic Counselor.* For more general information about professional education in genetic counseling, see the National Coalition for Health Professional Education in Genetics (www.nchpeg.org/).

Until recently, most genetic counseling was requested by parents who had an affected child and were concerned about risk for other children. Now genetic risk is often assessed directly by means of DNA testing. As more genes are identified for disorders, genetic counseling is increasingly involved in issues related to prenatal diagnoses, prediction, and intervention. This new information will create new ethical dilemmas. Huntington disease provides a good example. If you had a

Ethics and the Future

It is clear that we are at the dawn of a new era in which behavioral genetic research is moving beyond the demonstration of the importance of heredity to the identification of specific genes. In clinics and research laboratories, behavioral scientists of the future will routinely collect cells from the inside of the cheek and send them to a laboratory for DNA extraction (if we do not yet each have a memory key with our complete DNA sequence). Trait-specific sets of hundreds of genes will be available on microarrays that can genotype even large samples at a modest cost; these "gene set" data will be incorporated in behavioral research as genetic risk indicators. This is already happening for single-gene disorders like fragile X mental retardation as well as for the QTL association between apolipoprotein E and late-onset dementia.

As is the case with most important advances, identifying genes for behavior will raise new ethical issues. These issues are already beginning to affect genetic counseling (Box 6.3). Genetic counseling is expanding from the diagnosis

---

parent with the disease, you would have a 50 percent chance of developing the disease. However, with the discovery of the gene responsible for Huntington disease, it is now possible to diagnose in almost all cases whether a fetus or an adult will have the disease. Would you want to take the test? It turns out that the majority of people at risk choose not to take the test, largely because there is as yet no cure (Maat-Kievit et al., 2000). If you did take the test, the results are likely to affect knowledge of risk for your relatives. Do your relatives have the right to know, or is their right not to know more important? One generally accepted rule is that informed consent is required for testing; moreover, children should not be tested before they become adults, unless a treatment becomes available.

Another increasingly important problem concerns the availability of genetic information to employers and insurance companies. These issues are most pressing for single-gene disorders like Huntington disease, in which a single gene is necessary and sufficient to develop the disorder. For most behavioral disorders, however, genetic risks will involve QTLs that are probabilistic risk factors rather than certain causes of the disorder. A major new dilemma concerns the burgeoning industry of direct-to-consumer marketing of genetic tests (Biesecker & Marteau, 1999; Wade & Wilfond, 2006). Although genetic counseling has traditionally focused on single-gene and chromosomal disorders, increasingly the field is encompassing complex disorders including behavioral disorders (Finn & Smoller, 2006). Despite the ethical dilemmas that arise with the new genetic information, it should also be emphasized that these findings have the potential for profound improvements in the prediction, prevention, and treatment of diseases.

BOX 6.4

GATTACA?

Will DNA chips make the 1997 science fiction film *GATTACA* come true? In this story, individuals are selected as "valids" and "invalids" for education and employment on the basis of their DNA. DNA microarrays might also be used for "designer babies" by selecting embryos for in vitro fertilization. What about parents who want to use DNA chips to select egg or sperm donors?

What about using DNA chips for postnatal screening for purposes of interventions that avoid risks or enhance strengths? For decades, we have screened newborns for PKU because there is a relatively simple dietary intervention that prevents the mutation from damaging the developing brain. If similarly low-tech and inexpensive interventions such as dietary changes could make a difference for some behavioral genotypes, parents might want to take advantage of them even if the QTL only has a small effect. Expensive high-tech genetic engineering is unlikely to happen for a long time because it is proving very difficult for simple single-gene disorders (Goncalves, 2005), and it will be many orders of magnitude more difficult and less effective for complex traits influenced by many genes.

The most general fear is that finding genes associated with behavior will undermine support for social programs because it will legitimate social inequality as "natural." As indicated at the end of Chapter 5, the unwelcome truth is that equal opportunity will not produce equality of outcome because people differ in part for genetic reasons. Democracy is needed to ensure that all people are treated equally *despite* their differences. On the other hand, finding heritability or even specific genes associated with behavior does not imply

and prediction of rare untreatable single-gene conditions to the prediction of common, often treatable or preventable conditions (Karanjawala & Collins, 1998). Although there are many unknowns in this uncharted terrain, the benefits of identifying genes for understanding the etiology of behavioral disorders and dimensions seem likely to outweigh the potential abuses (Box 6.4). Forewarned of problems and solutions that have arisen with single-gene disorders, we should be forearmed as well to prevent abuses as genes that influence complex behavioral traits are discovered (Rothstein, 2005). The Nuffield Council on Bioethics has produced two relevant reports on behavioral genetics that are available online. One report is on genetics and severe mental disorders (www.nuffieldbioethics.org/go/ourwork/mentaldisorders/ introduction) and the other is more generally on behavioral genetics (www.nuffield-

(Columbia Pictures/Photofest.)

that behavior is immutable. Indeed, genetic research provides the best available evidence that nongenetic factors are important. PKU provides an example that even a single gene that causes mental retardation can be ameliorated environmentally.

"There is no gene for the human spirit" is the subtitle of the film *GATTACA*. It connotes the fear lurking in the shadows that finding genes associated with behavior will limit our freedom and our free will. In large part, such fears involve misunderstandings about how genes affect complex traits (Rutter & Plomin, 1997). Finding genes associated with behavior will not open a door to Huxley's brave new world where babies are engineered genetically to be alphas, betas, and gammas. The balance of risks and benefits to society of DNA chips is not clear—each of the problems identified above could also be viewed as potential benefits, depending on one's values. We need to be cautious and to think about societal implications and ethical issues. But there is also much to celebrate here in terms of the increased potential for understanding the behavior of our species.

bioethics.org/go/ourwork/behaviouralgenetics/introduction). See also a statement on behavioral genetics from the American Society for Human Genetics (Sherman et al., 1997).

Summary

Although much more quantitative genetic research is needed, one of the most exciting directions for genetic research in the behavioral sciences involves harnessing the power of molecular genetics to identify specific genes responsible for the widespread influence of genetics on behavior. Animal studies provide powerful designs to identify genes. Many behavioral mutants have been identified from studies of chemically induced mutations in organisms as diverse as

independently and in the past were usually institutionalized, today they often live in the community in special residences or with their families. People with IQs from 20 to 35 can learn some self-care skills and understand language, but they have trouble speaking and require considerable supervision. Individuals with IQs below 20 may understand a simple communication but usually cannot speak; they remain institutionalized.

General Cognitive Disability: Quantitative Genetics

In the behavioral sciences, it is now widely accepted that genetics substantially influences general cognitive ability; this belief is based on evidence presented in Chapter 8. Although one might expect that low IQ scores are also due to genetic factors, this conclusion does not necessarily follow. For example, cognitive disability can be caused by environmental trauma, such as birth problems, nutritional deficiencies, or head injuries. Given the importance of cognitive disability, it is surprising that no twin or adoption studies of moderate or severe cognitive disability have been reported. Nonetheless, one sibling study suggests that moderate and severe cognitive disability may be due largely to nonheritable factors. In a study of over 17,000 white children, 0.5 percent were moderately to severely disabled (Nichols, 1984). As shown in Figure 7.1, the siblings of these children

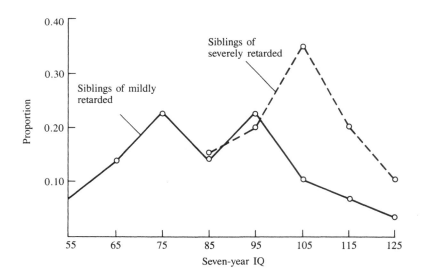

Figure 7.1 Siblings of children with mild cognitive disability tend to have lower than average IQs. In contrast, siblings of severely disabled children tend to have normal IQs. These trends suggest that mild disability is familial but severe disability is not. (From Nichols, 1984.)

showed no cognitive disability. The siblings' average IQ was 103, with a range of 85 to 125. In other words, moderate to severe cognitive disability showed no familial resemblance, a finding implying that it is not heritable. Although most moderate and severe cognitive disability may not be inherited from generation to generation, it is often caused by noninherited DNA events, such as new gene mutations and new chromosomal abnormalities, as discussed in the following sections.

In contrast, siblings of mildly disabled children tend to have lower than average IQ scores (Figure 7.1). The average IQ for these siblings of mildly disabled children (1.2 percent of the sample were mildly disabled) was only 85. This important finding, that mild cognitive disability is familial and moderate and severe disabilities are not familial, also emerged from the largest family study of mild cognitive disability, which considered 80,000 relatives of 289 mentally disabled individuals (Reed & Reed, 1965). This family study showed that mild mental disability is very strongly familial. If one parent is mildly disabled, the risk for cognitive disability in the children is about 20 percent. If both parents are mildly disabled, the risk is nearly 50 percent.

Although cognitive disability runs in families, it could do so for reasons of nurture rather than nature. Twin and adoption studies of mild cognitive disability are needed to disentangle the relative roles of nature and nurture. Three small twin studies suggest some genetic influence on mild cognitive disability (Nichols, 1984; Rosanoff, Handy, & Plesset, 1937; Wilson & Matheny, 1976). The largest study consisted of a U.S. population of 15 pairs of identical twins and 23 pairs of fraternal twins in which at least one member of the twin pair was mildly disabled (Nichols, 1984). The probandwise concordances were 75 percent for identical twins and 46 percent for fraternal twins, suggesting moderate genetic influence.

Two twin studies of unselected samples of twins have been used to investigate the origins of low IQ in infancy (Petrill et al., 1997) and in middle-aged adults (Saudino, Plomin, Pedersen, & McClearn, 1994). Both studies found that low IQ is at least as heritable as IQ in the normal range, suggesting that heritable factors might contribute to the familial resemblance found for mild cognitive disability. However, few subjects in these studies had IQs below 70. One study of over 3000 twin pairs assessed at 2, 3, and 4 years of age focused on those twins in the lowest 5 percent of the distribution of IQ and found that low IQ is at least as heritable as IQ in the normal range (Spinath, Harlaar, Ronald, & Plomin, 2004). Similar to the study mentioned above, probandwise twin concordances were 74 percent for identical twins and 45 percent for fraternal twins. Although these are not definitive studies of mild cognitive disability, they are consistent in pointing to a moderate role for genetic factors. In contrast to moderate to severe disability, mild disability seems likely to be at the lower end of the distribution of genetic and environmental factors that are responsible for variation in general cognitive ability (Plomin, 1999a), the topic of Chapter 8.

In studies considering disorders, the co-occurrence of several disorders in the same person is a common finding. For example, cognitive disability co-occurs with medical problems and with behavioral problems. About a third of children with mild cognitive disability have medical problems. Although it has been assumed that the medical problems caused the cognitive disability, another possibility is that genetic factors account for both the medical problems and the cognitive disability. That this might be so is suggested by studies that show that children with both mild cognitive disability and medical problems are more likely to have parents who have cognitive disabilities (Dykens & Hodapp, 2001). Similarly, about half of children with mild cognitive disability also have behavioral problems. Nothing is known as yet about whether the behavioral problems follow from the disability or whether there are genetic factors at work that affect both disability and behavioral problems.

General Cognitive Disability: Single-Gene Disorders

More than 250 genetic disorders, most extremely rare, include low IQ among their symptoms (Inlow & Restifo, 2004), and many more are likely to be discovered (Raymond & Tarpey, 2006). The classic disorder is PKU, discussed in Chapter 2; a newer discovery is fragile X, mentioned in Chapter 3. We will first discuss these two single-gene disorders, which are known for their effect on cognitive disability, as well as Rett syndrome, a common cause of cognitive disability in females. Then we will mention three single-gene disorders that also contribute to cognitive disability, even though their primary defect is something other than cognitive ability.

Until recently, much of what was known about these disorders, as well as the chromosomal disorders described in the next section, came from studies of patients in institutions. These earlier studies painted a gloomy picture. But more recent systematic surveys of entire populations show a wide range of individual differences, including individuals whose cognitive functioning is in the normal range. These genetic disorders shift the IQ distribution downward, but a wide range of individual differences remains.

Phenylketonuria

The most well known inherited form of moderate cognitive disability is phenylketonuria (PKU), which occurs in about 1 in 10,000 births, although its frequency varies widely from a high of 1 in 5000 in Ireland to a low of 1 in 100,000 in Finland. In the untreated condition, IQ scores are often below 50, although the range includes some near-normal IQs. As mentioned in Chapter 2, PKU is a single-gene recessive disorder that previously accounted for about 1 percent of mildly disabled individuals in institutions. PKU is the best example of the usefulness of finding genes for behavior. Knowledge that PKU is caused by a single gene led to an understanding of how the genetic defect causes cognitive disability. Mutations in the gene (*PAH*) that produces the enzyme phenylalanine hydroxylase lead to an enzyme that does not work properly, that is, one that is less efficient in breaking down phenylalanine. Phenylalanine comes from food, especially red meats; if it cannot be broken down properly, it builds up and damages the developing brain.

Although PKU is inherited as a simple single-gene recessive disorder, the molecular genetics of PKU is not so simple (Scriver & Waters, 1999). The *PAH* gene, which is on chromosome 12, shows more than 400 different mutations, some of which cause milder forms of cognitive disability (Zschocke, 2003). Similar findings have emerged for many classic single-gene disorders. Different mutations can do different things to the gene's product, and this variability makes understanding the disease process more difficult (Waters, 2003). It also makes DNA diagnosis more difficult, because DNA markers that identify all the mutations have to be used. A quantitative trait model that has recently been proposed considers the effects of *PAH* mutations in the context of normal variation in phenylalanine levels (Kaufmann, 1996). A mouse model of a mutation in the *PAH* gene shows similar phenotypic effects (Zagreda, Goodman, Druin, McDonald, & Diamond, 1999) and has been widely used in gene therapy research (Chen & Woo, 2005).

To allay fears about how genetic information will be used, it is important to note that knowledge about the single-gene cause of PKU did not lead to sterilization programs or genetic engineering. Instead, an environmental intervention—a diet low in phenylalanine—successfully prevented the development of cognitive disability. Widespread screening at birth for this genetic effect began in 1961, a program demonstrating that genetic screening can be accepted when a relatively simple intervention is available (Guthrie, 1996). However, despite screening and intervention, PKU individuals still tend to have a slightly lower IQ, especially when the low phenylalanine diet has not been strictly followed (Smith, Beasley, Wolff, & Ades, 1991). It is generally recommended that the diet be maintained as long as possible, at least through adolescence. PKU women must return to a strict low phenylalanine diet before becoming pregnant to prevent their high levels of phenylalanine from damaging the fetus (Lee, Ridout, Walter, & Cockburn, 2005).

Fragile X Syndrome

As mentioned in Chapters 1 and 3, fragile X is the second most common cause of cognitive disability after Down syndrome, and the most common inherited form. It is twice as common in males as in females. The frequency of fragile X is usually given as 1 in 5000 males and 1 in 10,000 females (Crawford, Acuna, & Sherman, 2001). At least 2 percent of the male residents of schools for cognitively disabled persons have the fragile X syndrome. Most cases of fragile X males are moderately disabled, but many are only mildly disabled and some have normal intelligence. Only about one-half of girls with fragile X are affected, because one of the two X chromosomes for girls inactivates, as mentioned in Chapter 4. Although fragile X is a major source of the greater incidence of cognitive disability in boys, other genes on the X chromosome have been implicated in cognitive disability (Ropers & Hamel, 2005).

For fragile X males, IQ declines after childhood. In addition to generally lower IQ, about three-quarters of fragile X males show large, often protruding ears, a long face with a prominent jaw, and, after adolescence, enlarged testicles. They also often show unusual behaviors such as odd speech, poor eye contact (gaze aversion), and flapping movements of the hand. Language difficulties range from an absence of speech to mild communication difficulties. Often observed is a speech pattern called "cluttering" in which talk is fast, with occasional garbled, repetitive, and disorganized speech. Spatial ability tends to be affected more than verbal ability. Comprehension of language is often better than expression and better than expected on the basis of IQ scores (Dykens, Hodapp, & Leckman, 1994; Hagerman, 1995). Parents frequently report overactivity, impulsivity, and inattention.

Until the gene for fragile X was found in 1991, its inheritance was puzzling (Verkerk et al., 1991). It did not conform to a simple X-linkage pattern because its risk increased across generations. The fragile X syndrome is caused by an expanded triplet repeat (CGG) on the X chromosome (Xq27.3). The disorder is called fragile X because the many repeats cause the chromosome to be fragile at that point and to break during laboratory preparation of chromosomes. The disorder is now diagnosed on the basis of DNA sequence. As mentioned in Chapter 3, parents who inherit X chromosomes with a normal number of repeats (6 to 40 repeats) can produce eggs or sperm with an expanded number of repeats (up to 200 repeats), called a *premutation*. This premutation does not cause cognitive disability in their offspring, but it is unstable and often leads to much greater expansions (more than 200 repeats) in later generations, especially when the premutated X chromosome is inherited through the mother. The risk that a premutation will expand to a full mutation increases over four generations from 5 to 50 percent, although it is not yet possible to predict when a premutation will expand to a full mutation. The mechanism by which expansion occurs is not known. The full mutation causes fragile X in almost all males but in only half of the females. Females are mosaics for fragile X in the sense that one X chromosome is inactivated, so some cells will have the full mutation and others will be normal (Kaufmann & Reiss, 1999). As a result, females with the full mutation have much more variable symptoms.

The triplet repeat is in an untranslated region at the beginning of a gene (*Fragile X Mental Retardation-1, FMR1*) that, when expanded to a full mutation, prevents that gene from being transcribed. The mechanism by which the full mutation prevents transcription is *DNA methylation*, the most well understood developmental mechanism of genetic regulation, as discussed in Chapter 15. DNA methylation prevents transcription by binding a methyl group to DNA, usually at CG repeat sites. The full mutation for fragile X, with its hundreds of CGG repeats, causes hypermethylation and thus shuts down transcription of the *FMR1* gene. The gene's protein product (FMRP) binds RNA, which means that the gene product regulates expression of other genes. FMRP facilitates translation of hundreds of neuronal RNAs; thus, the absence of FMRP causes diverse problems (Khandjian, Bechara, Davidovic, & Bardoni, 2005). The *FMR1* gene is found in species as diverse as yeast and rodents, and it is expressed in most tissues, with the highest levels in the brain and testes. In the mouse brain, the gene is expressed most in hippocampus, cerebellum, and cerebral cortex. Mouse knock-out studies have shown learning deficits and behavioral problems expected from the human disorder (Koukoui & Chaudhuri, 2007). Recent neurogenetic discoveries about the *FMR1* mutation in *Drosophila* have increased the pace of research in this area (Zhang & Broadie, 2005), including a possible drug treatment (Ropers, 2006). Research on fragile X is moving rapidly from molecular genetics to neurobiology (Garber et al., 2006). Researchers hope that, once the functions of FMRP are understood, it can be artificially supplied. In addition, methods for identifying carriers of premutations have improved; these screening tests will help people carrying premutations to avoid producing children who have a larger expansion and therefore suffer from fragile X syndrome.

Rett Syndrome

Rett syndrome is the most common single-gene cause of general cognitive disability in females (1 in 10,000). The disorder shows few effects in infancy, although the head, hands, and feet are slow to grow. Cognitive development is normal during infancy but, by school age, girls with Rett syndrome are generally unable to talk and about half are unable to walk (Weaving, Ellaway, Gecz, & Christodoulou, 2005). Although prone to seizures and gastrointestional disorders, women with Rett syndrome can live to 60 years. The single-gene disorder was mapped to the long arm of the X chromosome (Xq28), and then to a specific gene (*MECP2*, which encodes methyl-CpG-binding protein-2) (Amir et al., 1999). *MECP2* is a gene involved in the methylation process that silences other genes during development and thus has diffuse effects throughout the brain (Bienvenu & Chelly, 2006). The effects are variable in females because of random X-chromosome inactivation in females (see Chapter 4). Males with *MECP2* mutations usually die before or shortly after birth.

More than 250 other single-gene disorders, whose primary defect is something other than cognitive disability, also show effects on IQ (Inlow & Restifo, 2004). Three of the most common disorders are Duchenne muscular dystrophy, Lesch-Nyhan syndrome, and neurofibromatosis.

Duchenne Muscular Dystrophy

This syndrome is an X-linked ($Xp21$) recessive disorder that occurs in 1 in 3500 males. It is a neuromuscular disorder that causes progressive wasting of muscle tissue beginning in childhood and usually leads to death by age 20 years as a result of respiratory or cardiac failure. The average IQ of males with Duchenne muscular dystrophy (DMD) is 85. Verbal abilities are more severely impaired than nonverbal abilities, although effects on cognitive ability are highly variable.

The gene's product (dystrophin) encodes a cell membrane protein in muscle cells. Although it is not known how it affects cognitive function, it is found in the brain as well as in muscle tissue. The *DMD* gene is so huge—2.3 million base pairs with 79 exons—that it takes many hours to be transcribed (Tennyson, Klamut, & Worton, 1995). Dozens of different mutations in the *DMD* gene have been found, and at least a third of cases are due to new mutations. A mutation in the X-linked muscular dystrophy gene in mice arose spontaneously in the C57BL strain (Bulfield, Siller, Wight, & Moore, 1984). Although the mutation greatly reduces dystrophin in muscle and brain, few clinical signs can be found in these mice even though the mutation is devastating in the human species. This species difference suggests that mice have some compensatory mechanism that might be used to ameliorate the human condition (Deconinck & Dan, 2007). No cure is as yet available although there is hope for gene therapy (Foster, Foster, & Dickson, 2006).

Lesch-Nyhan Syndrome

Lesch-Nyhan syndrome is another X-linked ($Xq26$-27) recessive disorder, with an incidence of about 1 in 20,000 male births. The most striking feature of this disorder is compulsive self-injurious behavior, reported in over 85 percent of cases (Anderson & Ernst, 1994). Most typical is lip and finger biting, which is often so severe that it leads to extensive loss of tissue (Jinnah et al., 2006). The self-injurious behavior begins as early as in infancy or as late as in adolescence. The behavior is painful to the individual, yet uncontrollable. In terms of cognitive disability, most individuals have moderate or severe learning difficulties, and speech is usually impaired. Memory for both recent and past events appears to be unaffected. Many medical problems are also involved that lead to death before 30 years.

The gene (*HPRT1*) codes for an enzyme (hypoxanthine phosphoribosyl-transferase, HPRT) involved in production of nucleic acids. As in PKU and DMD, hundreds of different mutations are involved (Jinnah, Harris, Nyhan, & O'Neill, 2004). Nonetheless, the disorder can be diagnosed genetically by analyzing transcribed DNA (i.e., mRNA converted to cDNA) for the absence of one of the eight exons in *HPRT1*. Mutations in *HPRT1* have pervasive structural effects on dopamine systems, including abnormally few dopaminergic nerve terminals and cell bodies (Nyhan & Wong, 1996). An interesting study of six pairs of identical twins found that the twins were highly similar for mutations in

HPRT1 (Curry et al., 1997). This finding suggests that genetic factors might govern variation in the frequency of mutations, although shared environmental factors cannot be excluded until comparisons are made with fraternal twins.

One of the first transgenic knock-out mouse strains created involved the *HPRT1* gene responsible for Lesch-Nyhan (Kuehn, Bradley, Robertson, & Evans, 1987). Investigators found a response in mice similar to that of the *DMD* mutation, that is, knocking out the *HPRT1* gene seemed to have no effect on the brain or behavior of the mice (Engle et al., 1996). Cross-species comparisons suggest an evolutionary reason for phenotypic disparities between HPRT-deficient humans and mice (Keebaugh, Sullivan, & Thomas, 2007).

Neurofibromatosis Type 1

First described more than a century ago, neurofibromatosis type 1 involves skin tumors and tumors in nerve tissue (Ward & Gutmann, 2005). Its symptoms are highly variable, beginning with chocolate-colored spots that appear in early childhood. The tumorous lumps are not cancerous and primarily cause cosmetic disfigurement, although the tumors can cause more serious problems if they compress a nerve. The majority of affected individuals have learning difficulties (Coude, Mignot, Lyonnet, & Munnich, 2006; Hyman, Shores, & North, 2005), and the learning difficulties are usually general (Hyman, Arthur, & North, 2006). Most individuals with neurofibromatosis survive until middle age.

Neurofibromatosis type 1 is caused by a dominant allele in the neurofibromin gene (*NF1*) on chromosome 17 (17*q*11.2) and is surprisingly common (about 1 in 3000 births) for a dominant allele (Ferner, 2007). Although not nearly as big as the *DMD* gene, the *NF1* gene is large, with 59 exons spread over 350 kilobases. The *NF1* gene product, neurofibromin, is expressed in a wide variety of tissues including neurons, and accordingly has a wide variety of effects (Trovo-Marqui & Tajara, 2006). The allele that causes neurofibromatosis, which is involved in tumor suppression, is inherited from the father in more than 90 percent of the affected individuals. However, like other genes involved in cognitive disability, the *NF1* gene has a very high mutation rate; approximately half of all cases are new mutations. The mechanism by which mutations lead to tumors is not known, but most mutations lead to a truncation of the protein neurofibromin.

One early mouse model knocks out *NF1* and results in deficits in learning and memory in heterozygous mice (homozygous mutants do not survive) (Silva, Smith, & Giese, 1997). Although these heterozygous mice do not show tumors, chimeric mice with some cells with the homozygous knock-out mutation have tumors (Cichowski et al., 1999). Another model also knocks out the *NF1* gene but also knocks out a gene (*p53*) known to be responsible for tumor suppression and involved in many cancers (Vogel et al., 1999). Heterozygous *NF1* knock-out mice with the *p53* knock-out have tumors and deficits in learning and memory. Other knock-outs in mice and *Drosophila* have been created (Dasgupta & Gutmann, 2003), including a conditional knock-out which makes it possible to turn expression of the gene on and off during development (Zhu, Ghosh, Charnay, Burns, & Parada, 2002; see Chapter 6). *NF1* knock-out mice are being used as a model for tumor formation in the nervous system (Rubin & Gutmann, 2005). Gene expression profiling (see Chapter 15) has begun to trace the pathways between the *NF1* gene, brain, and behavior in mice (Donarum, Halperin, Stephan, & Narayanan, 2006).

SUMMING UP

Although little is known about the quantitative genetics of general cognitive disability, more than 250 single-gene disorders, most extremely rare, include cognitive disability among their symptoms and many more are likely to be discovered. The classic disorder is PKU, caused by a gene on chromosome 12. A newer discovery is a gene responsible for most of the excess of male cognitive disability: Fragile X syndrome is caused by a triplet repeat on the X chromosome that expands over several generations and prevents a nearby gene from being transcribed. An important cause of general cognitive disability in females is Rett syndrome. Other single-gene disorders that contribute to cognitive disability include Duchenne muscular dystrophy, Lesch-Nyhan syndrome, and neurofibromatosis type 1. These are typically large genes with dozens of different mutations. Mouse models have been helpful in understanding the function of these genes. Figure 7.2 shows the average IQ of individuals with the most common single-gene causes of cognitive disability. It should be remembered, however, that the

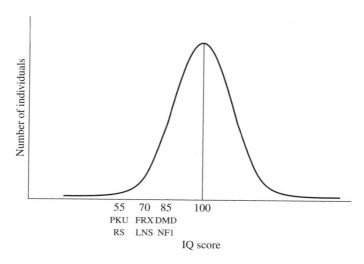

Figure 7.2 Single-gene causes of general cognitive disability: phenylketonuria (PKU), Rett syndrome (RS), fragile X syndrome (FRX), Lesch-Nyhan syndrome (LNS), Duchenne muscular dystrophy (DMD), and neurofibromatosis type 1 (NF1). Despite the lower average IQs, a wide range of cognitive functioning is found.

range of cognitive functioning is very wide for these disorders. The defective allele shifts the IQ distribution downward, but a wide range of individual IQs remains. Also, although there are hundreds of such rare single-gene disorders, together they account for only a tiny portion of cognitive disability. Most cognitive disability is mild, and may be the low end of the normal distribution of general cognitive ability and caused by many QTLs of small effect as well as multiple environmental factors. Molecular genetic research on general cognitive ability is discussed in the next chapter.

General Cognitive Disability: Chromosomal Abnormalities

Much more common than the single-gene causes of general cognitive disability are the chromosomal abnormalities that lead to cognitive disability. Most common are abnormalities that involve an entire extra chromosome, but deletions in parts of chromosomes also can contribute to cognitive disability. As the resolution of chromosomal analysis becomes finer, many more minor deletions are likely to be found. A study of children with unexplained moderate to severe cognitive disability found that 7 percent of them had subtle chromosomal abnormalities; these same abnormalities were found in only 0.5 percent of children with mild cognitive disability (Knight et al., 1999). The recent application of microarrays (see Chapter 6) has led to an explosion of research on minor deletions and rearrangements (de Vries et al., 2005; Knight & Regan, 2006; Mao & Pevsner, 2005; Ming et al., 2006; Rooms, Reyniers, & Kooy, 2005). About 5 percent of individuals with cognitive disability have been found to have subtle chromosomal abnormalities (Moog et al., 2005). As in the Knight et al. (1999) study, these subtle abnormalities are found in moderate to severe disability, not in mild disability, which is the low end of the normal distribution of general cognitive ability, as discussed in the next chapter.

Angelman syndrome (AS), mentioned in Chapter 3 as an example of genomic imprinting, involves a small deletion in chromosome 15*q*11 that usually occurs spontaneously in the formation of gametes rather than being an abnormality inherited in generation after generation (Cassidy & Schwartz, 1998). When the deletion comes from the mother (1 in 25,000 births), it causes moderate cognitive disability, abnormal gait, speech impairment, seizures, and an inappropriate happy demeanor that includes frequent laughing and excitability (Williams et al., 2006). When inherited from the father (1 in 15,000 births), the same chromosomal deletion causes *Prader-Willi syndrome* (PWS), which most noticeably involves overeating and temper outbursts but also leads to multiple learning difficulties and an IQ in the low normal range. In a quarter of cases, however, PWS occurs because two maternal copies of chromosome 15 are inherited but no paternal copy (Whittington, Butler, & Holland, 2007).

Most of the time, this chromosomal abnormality occurs spontaneously. The risk to siblings of probands with the disorder is low, less than 1 percent. AS and PWS may be caused by different genes in the 15q11 region (Cassidy & Schwartz, 1998). AS usually involves the ubiquitin-protein ligase gene, *UBE3A*. The ubiquitin ligase enzyme is a key part of a cellular protein degradation system and is crucial for brain development (Kishino, Lalande, & Wagstaff, 1997). Like the genes discussed above, *UBE3A* shows many different mutations, most spontaneous (Fang et al., 1999). PWS involves a cluster of seven genes that includes a gene, *SNRPN*, which affects mRNA processing (Schweizer, Zynger, & Francke, 1999).

Williams syndrome, with an incidence of about 1 in 10,000 births, is caused by a small deletion from chromosome 7 (7q11.2), a region that includes more than 20 genes. Most cases are spontaneous. Williams syndrome involves disorders of connective tissue that lead to growth retardation and multiple medical problems. General cognitive disability is common, and most affected individuals have learning difficulties that require special schooling. As adults, most affected individuals are unable to live independently. One of the most common behavioral symptoms is hypersensitivity to noise. When Williams syndrome was first studied, investigators thought that the expressive language skills of affected individuals were superior to their other cognitive abilities. However, later research indicated that language is affected as much as other cognitive abilities, although the effects are highly variable (Greer, Brown, Pai, Choudry, & Klein, 1997; Karmiloff-Smith et al., 1997). As is typical of chromosomal abnormalities that include several genes, no consistent brain pathology is found (Cherniske et al., 2004). Two of the 20 deleted genes are the gene for elastin (*ELN*) and the gene for an enzyme called LIM kinase (*LIMK*). In normal cells, elastin is a key component of connective tissue, conferring its elastic properties. Mutation or deletion of *ELN* leads to the vascular disease observed in Williams syndrome. *LIMK* is strongly expressed in the brain, and the absence of LIM kinase is thought to account for the cognitive disability in Williams syndrome.

The next sections include descriptions of the classic chromosomal abnormalities that involve cognitive disability. Chromosomes and chromosomal abnormalities, such as nondisjunction and the special case of abnormalities involving the X chromosome, were introduced in Chapters 3 and 4.

Down Syndrome

As described in Chapter 3, Down syndrome is caused by a trisomy of chromosome 21 (Roizen & Patterson, 2003). It was one of the first identified genetic disorders and its 150-year history parallels the history of genetic research (Patterson & Costa, 2005). It is the single most important cause of general cognitive disability and occurs in about 1 in 1000 births. It is so common that its general features are probably familiar to everyone (Figure 7.3). Although more than 300

Figure 7.3 Child with Down syndrome. (Laura Dwight/ Peter Arnold.)

abnormal features have been reported for Down syndrome children, a handful of specific physical disorders are diagnostic because they occur so frequently. These features include increased neck tissue, muscle weakness, speckled iris of eye, open mouth, and protruding tongue. Some symptoms, such as increased neck tissue, become less prominent as the child grows. Other symptoms, such as cognitive disability and short stature, are noted only as the child grows. About two-thirds of affected individuals have hearing deficits, and one-third have heart defects, leading to an average life span of 50 years. As first noted by Langdon Down, who identified the disorder in 1866, children with Down syndrome appear to be obstinate but otherwise generally amiable. Although it might be assumed that these diverse effects come from overexpression of specific genes on chromosome 21 (because there are three copies of the chromosome), it is possible that having so much extra genetic material creates a general instability in development. Animal models have played an important role in understanding gene dosage imbalances in Down syndrome (Antonarakis, Lyle, Dermitzakis, Reymond, & Deutsch, 2004) in general and in cognitive disorders in particular (Seregaza, Roubertoux, Jamon, & Soumireu-Mourat, 2006). Advances in genetics have stimulated a resurgence of research on Down syndrome with the hope of ameliorating at least some of its symptoms (Antonarakis & Epstein, 2006).

The most striking feature of Down syndrome is general cognitive disability. As is the case for all single-gene and chromosomal effects on general cognitive ability, affected individuals show a wide range of IQs. The average IQ among children with Down syndrome is 55, with only the top 10 percent falling within the lower end of the normal range of IQs. By adolescence, language skills are generally at about the level of a three-year-old child. Most individuals with Down syndrome who reach the age of 45 years suffer from the cognitive decline of dementia, which was an early clue suggesting that a gene related to dementia might be on chromosome 21 (see later).

In Chapter 3, Down syndrome was used as an example of an exception to Mendel's laws because it does not run in families. Because individuals with Down syndrome do not reproduce, most cases are created anew each generation by nondisjunction of chromosome 21. Another important feature of Down syndrome is that it occurs much more often in women giving birth later in life, for reasons explained in Chapter 3.

Sex Chromosome Abnormalities

Extra X chromosomes also cause cognitive disabilities, although the effect is highly variable, which is the reason why many cases remain undiagnosed (Lanfranco, Kamischke, Zitzmann, & Nieschlag, 2004). In males, an extra X chromosome causes *XXY male syndrome*. It is very common, occurring in about 1 in 500 male births. The major problems involve low testosterone levels after adolescence, leading to infertility, small testes, and breast development. Early detection and hormonal therapy are important to alleviate the condition, although infertility remains (Simm & Zacharin, 2006). Males with XXY male syndrome also have a somewhat lower than average IQ, and most have speech and language problems, and poor school performance (Mandoki, Sumner, Hoffman, & Riconda, 1991). In females, extra X chromosomes (called *triple X syndrome*) occur in about 1 in 1000 female births. Females with triple X show an average IQ of about 85, lower than for XXY males (Bender, Linden, & Robinson, 1993). Unlike XXY males, XXX females have normal sexual development and are able to conceive children; they have so few problems that they are rarely detected clinically. Verbal scores are lower than nonverbal scores and many require speech therapy. For both XXY and XXX individuals, head circumference at birth is smaller than average, a feature suggesting that the cognitive deficits may be prenatal in origin (Ratcliffe, 1994). As is generally the case for chromosomal abnormalities, structural brain imaging research indicates diffuse effects (Giedd et al., 2007).

In addition to having an extra X chromosome, it is possible for males to have an extra Y chromosome (XYY) and for females to have just one X chromosome (XO, called *Turner syndrome*). There is no equivalent syndrome of males with a Y chromosome but no X because this is fatal. XYY males, about 1 in 1000 male births, are taller than average after adolescence and have normal sexual development. More than 95 percent of XYY males do not even know they have an extra Y chromosome. Although XYY males have fewer cognitive problems than XXY males, about half have speech difficulties, and language and reading problems (Geerts, Steyaert, & Fryns, 2003). Their average IQ is about

10 points lower than that of their siblings with normal sex chromosomes. Juvenile delinquency is also associated with XYY. The XYY syndrome was the center of a furor in the 1970s, when it was suggested that such males are more violent, a suggestion possibly triggered by the notion of a "supermale" with exaggerated masculine characteristics caused by their extra Y chromosome, although not supported by research.

Turner syndrome females (XO) occur in about 1 in 2500 female births, although 98 percent of XO fetuses miscarry, accounting for 10 percent of the total number of spontaneous abortions. The main problems are short stature and abnormal sexual development; infertility is common. Puberty rarely occurs without hormone therapy; even with therapy, the individual is infertile because she does not ovulate. Hormonal treatment is now standard and many women with XO have conceived with in vitro fertilization (Stratakis & Rennert, 2005). Although verbal IQ is about normal, performance IQ is lower, about 90, after adolescence (Smith, Kimberling, & Pennington, 1991).

SUMMING UP

Small deletions of chromosomes can result in cognitive disability, as in Angelman syndrome, Prader-Willi syndrome, and Williams syndrome. The most common cause of cognitive disability is Down syndrome, which is due to the presence of three copies of chromosome 21. Mild cognitive disability generally occurs in individuals with extra X chromosomes: XXY males and XXX females. Some cognitive disability appears in males with an extra Y chromosome and in Turner females with a missing X chromosome. Although the average cognitive ability of these groups is generally lower than the average for the entire population, individuals in these groups show a wide range of cognitive functioning. An exciting area of research uses microarrays to detect subtle chromosomal abnormalities, which include about 5 percent of individuals with moderate to severe cognitive disability.

Figure 7.4 illustrates the most common chromosomal causes of general cognitive disability. Again, it should be emphasized that there is a wide range of cognitive functioning around the average IQ scores shown in the figure.

Learning Disorders

Learning disabilities involve cognitive processes related to achievement at school, especially reading but also communication and mathematics. Behavioral genetic research brings genetics to the field of educational psychology, which has been slow to recognize the importance of genetic influence (Plomin & Walker, 2003; Wooldridge, 1994), even though teachers in the classroom do (Walker & Plomin, 2005). This section focuses on low performance in cognitive processes

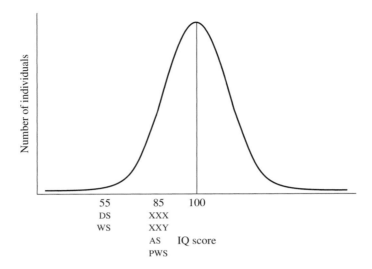

Figure 7.4 The most common chromosomal causes of general cognitive disability are Down syndrome (DS) and the sex chromosomal abnormalities XXX and XXY. The average IQs of individuals with XYY and XO are only slightly lower than normal and thus are not listed. Deletions of very small parts of chromosomes contribute importantly to general cognitive disability, but most are rare, such as Angelman syndrome (AS), Prader-Willi syndrome (PWS), and Williams syndrome (WS). For all these chromosomal abnormalities, a wide range of cognitive functioning is found.

related to academic achievement, whereas Chapter 9 focuses on normal variation in these processes. We begin with reading disability because reading is the primary problem in about 80 percent of children with a diagnosed learning disorder. We then consider communication disorders and mathematics and finally, their interrelationships.

Reading Disability

As many as 10 percent of children have difficulty learning to read. Children with reading disability (also known as *dyslexia*) read slowly and often with poor comprehension. When reading aloud, they perform poorly. For some, specific causes can be identified, such as cognitive disability, brain damage, sensory problems, and deprivation. However, many children without such problems find it difficult to read.

Family studies have shown that reading disability runs in families. The largest family study included 1044 individuals in 125 families with a reading-disabled child and 125 matched control families (DeFries, Vogler, & LaBuda, 1986). Siblings and parents of the reading-disabled children performed significantly worse on reading tests than did siblings and parents of control children. Earlier small twin studies suggested that familial resemblance for reading disability involves genetic factors (Bakwin, 1973; Decker & Vandenberg, 1985), al-

Stephen A. Petrill is a developmental psychologist interested in the gene-environment processes influencing the development of reading and math skills. He received his undergraduate training in psychology from the University of Notre Dame and graduate training from Case Western Reserve University; he did postgraduate research in London and at Penn State University. Petrill is currently professor of human development and family sciences at Ohio State University. Since graduate school, he has been drawn to behavioral genetics because of its interdisciplinary nature. He is convinced that integrating genetic, neuropsychological, cognitive, and psychometric perspectives will build better theories of development. His special interest is in the identification of the environmental and genetic influences that affect the development of reading and math skills. Are there some environmental influences that are consistent across reading and math development? How does the environment vary across development and how does this variation relate to individual differences in improvement or decline in academic performance over time? How do gene-environment correlations and interactions impact the development of reading and math skills? Despite their implications for theory and practical applications, surprisingly few answers are available to these important questions.

though one twin study showed little evidence of genetic influence (Stevenson, Graham, Fredman, & McLoughlin, 1987). The first major twin study confirmed genetic influence on reading disability (DeFries, Knopik, & Wadsworth, 1999). For more than 250 twin pairs in which at least one member of the pair was reading disabled, twin concordances were 66 percent for identical twins and 36 percent for fraternal twins, a result suggesting substantial genetic influence. A large twin study in the U.K. found similar results in the early school years for both reading disability and reading ability (Harlaar, Spinath, Dale, & Plomin, 2005c; Kovas, Haworth, Dale, & Plomin, 2007). In all of these studies, shared environmental influence is modest, typically accounting for less than 20 percent of the variance (Olson, 2007).

As part of the DeFries et al. twin study, a new method was developed to estimate the genetic contribution to the mean difference between the reading-disabled probands and the mean reading ability of the population. This type of analysis is called *DF extremes analysis* after its creators (DeFries & Fulker, 1985), and is described in Box 7.1. In a meta-analysis of all studies of reading disability,

BOX 7.1

DF Extremes Analysis

T he genetic and environmental causes of individual differences through-out the range of variability in a population can differ from the causes of the average difference between an extreme group and the rest of the population. For example, finding genetic influence on individual differences in reading ability in an unselected sample (Chapter 9) does not mean that the average difference in reading ability between reading-disabled individuals and the rest of the population is also influenced by genetic factors. Alternatively, it is possible that reading disability represents the extreme end of a continuum of reading ability, rather than a distinct disorder. That is, reading disability might be quantitatively rather than qualitatively different from the normal range of reading ability. DF extremes analysis, named after its creators (DeFries & Fulker, 1985, 1988), addresses these important issues concerning the links between the normal and abnormal.

DF extremes analysis takes advantage of quantitative scores of the relatives of probands rather than just assigning a dichotomous diagnosis to the relatives and assessing concordance for the disorder. The figure on the facing page shows hypothetical distributions of reading performance of an unselected sample of twins and of the identical (MZ) and fraternal (DZ) co-twins of probands (P) with reading disability (DeFries, Fulker, & LaBuda, 1987). The mean score of the probands is $\bar{P}$. The differential regression of the MZ and the DZ co-twin means $\bar{C}_{MZ}$ and $\bar{C}_{DZ}$ toward the mean of the unselected population μ provides a test of genetic influence. That is, to the extent that reading deficits of probands are heritable, the quantitative reading scores of identical co-twins will be more similar to those of the probands than will the scores of fraternal twins. In other words, the mean reading scores of identical co-twins will regress less far back toward the population mean than will those of fraternal co-twins.

The results for reading disability are similar to those illustrated in the figure. The scores of the identical co-twins regress less far back toward the population mean than do those of the fraternal co-twins. This finding suggests that genetics contributes to the mean difference between the reading-disabled probands and the population. Twin *group correlations* provide an index of how far the co-twins regress toward the population mean. For reading disability, the twin group correlations are .90 for identical twins and .65 for fraternal twins. Doubling the difference between these group correlations suggests a *group heritability* of 50 percent, similar to the results of more sophisticated DF extremes

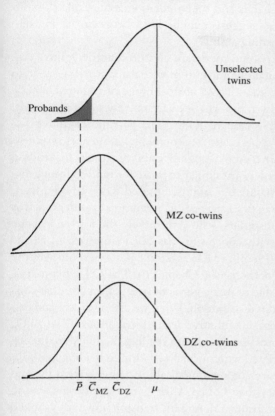

analysis (DeFries & Gillis, 1993). In other words, half of the mean difference between the probands and the population is heritable. This is called "group heritability" to distinguish it from the usual heritability estimate, which refers to differences between individuals rather than to mean differences between groups.

DF extremes analysis is conceptually similar to the liability-threshold model described in Box 3.1. The major difference is that the threshold model assumes a continuous dimension even though it assesses a dichotomous disorder. The liability-threshold analysis converts dichotomous diagnostic data to a hypothetical construct of a threshold with an underlying continuous liability. In contrast, DF extremes analysis assesses rather than assumes a continuum. If all the assumptions of the liability-threshold model are correct for a particular disorder, it will yield results similar to the DF extremes analysis to the extent that the quantitative dimension assessed underlies the qualitative disorder (Plomin, 1991). In the case of reading disability, a liability-threshold analysis of these twin data yields an estimate of group heritability similar to that of the DF extremes analysis.

In addition, DF extremes analysis can be used to examine the genetic and environmental origins of the co-occurrence between disorders. For example, attention-deficit hyperactivity disorder (see Chapter 12) is often found among reading-disabled children. Multivariate DF extremes analysis suggests that genetic factors are largely responsible for this overlap in the two disorders, especially for the inattention component (Willcutt, Pennington, & DeFries, 2000). In other words, the two disorders appear to share some genetic influences.

DF extremes analysis for reading disability estimates that about 60 percent of the mean difference between the probands and the population is heritable (Plomin & Kovas, 2005). The analysis also suggests genetic links between reading disability and normal variation in reading ability.

As described earlier in this chapter, moderate and profound cognitive disability is caused by single-gene mutations and chromosomal abnormalities that do not contribute importantly to variation in the normal range of cognitive ability. In contrast, mild cognitive disability appears to be quantitatively, not qualitatively, different from normal variation in cognitive ability. That is, mild cognitive disability is the low end of the same genetic and environmental influences responsible for variation in the normal distribution of cognitive ability. Results for reading disability and other common disorders are similar to those for mild cognitive disability rather than more severe cognitive disability. Phrased more provocatively, these findings from DF extremes analysis suggest that common disorders such as reading disability are not really disorders—they are merely the low end of the normal distribution. This important conclusion has been summarized in the phrase "the abnormal is normal" (Plomin & Kovas, 2005). This view fits with the quantitative trait locus (QTL) hypothesis (see Chapter 6). The QTL hypothesis assumes that genetic influence is due to many genes of small effect size that contribute to a normal quantitative trait distribution. What we call disorders and disabilities are the low end of these quantitative trait distributions. The QTL hypothesis predicts that when genes associated with reading disability are identified, the same genes will be associated with normal variation in reading ability.

Early molecular genetic research on reading disability assumed that the target was a single major gene rather than QTLs. Various modes of transmission have been proposed, especially autosomal dominant transmission and X-linked recessive transmission. The autosomal dominant hypothesis takes into account the high rate of familial resemblance but fails to account for the fact that about a fifth of reading-disabled individuals do not have affected relatives. An X-linked recessive hypothesis is suggested when a disorder occurs more often in males than in females, as is the case for reading disability. However, the X-linked recessive hypothesis does not work well as an explanation of reading disability. As described in Chapter 3, one of the hallmarks of X-linked recessive transmission is the absence of father-to-son transmission, because sons inherit their X chromosome only from their mother. Contrary to the X-linked recessive hypothesis, reading disability is transmitted from father to son as often as from mother to son. It is now generally accepted that, like most complex disorders, reading disability is caused by multiple genes as well as by multiple environmental factors (Fisher & DeFries, 2002).

One of the most exciting findings in behavioral genetics in the past decade is that the first quantitative trait locus for a human behavioral disorder was reported for reading disability, using sib-pair *QTL linkage analysis* (Cardon et al.,

1994). As explained in Chapter 6, siblings can share 0, 1, or 2 alleles for a particular DNA marker. If siblings who share more alleles are also more similar for a quantitative trait such as reading ability, then QTL linkage is likely. QTL linkage analysis is much more powerful when one sibling is selected because of an extreme score on the quantitative trait. When one sibling was selected for reading disability, the reading ability score of the co-sibling was also lower when the two siblings shared alleles for markers on the short arm of chromosome 6 (6p21). These QTL linkage results for four DNA markers in this region are depicted by the dotted line in Figure 7.5, showing significant linkage for the D6S105 marker. Significant linkage was also found for markers in this region in an independent sample of fraternal twins (see solid line in Figure 7.5) and in several replication studies in the broader region of 6p23-21.3 (Fisher et al., 1999, 2002; Gayán et al., 1999; Grigorenko et al., 1997; Grigorenko, Wood, Meyer, & Pauls, 2000; Kaplan et al., 2002; Turic et al., 2003). Most recently, the 6p linkage was reported in Swahili-speaking families in Tanzania (Grigorenko, Naples, & Chang, 2007). Despite these consistent linkage results, it has been difficult to identify the specific genes responsible for the QTL linkage among

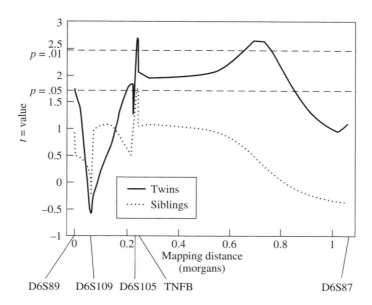

Figure 7.5 QTL linkage for reading disability in two independent samples in which at least one member of the pair is reading disabled: siblings (dotted line) and fraternal twins (solid line). D6S89, D6S109, D6S105, and TNFB are DNA markers in the 6p21 region of chromosome 6. The t-values are an index of statistical significance. The marker D6S105 is significant at the $p = .05$ level for siblings and at the $p = .01$ level for fraternal twins. (After Cardon et al., 1994; modified from DeFries & Alarcón, 1996; courtesy of Javier Gayán.)

the hundreds of genes in this gene-rich region of chromosome 6, but the search has narrowed to two genes very close together at *6p22*: *KIAA0319* and *DCDC2* (Fisher & Francks, 2006; McGrath, Smith, & Pennington, 2006; Williams & O'Donovan, 2006). Both genes provide plausible pathways between genes, brain, and behavior (Galaburda, LoTurco, Ramus, Fitch, & Rosen, 2006).

In 1983, traditional analyses of pedigrees were used to show linkage to chromosome 15 (15*q*21) (Smith, Kimberling, Pennington, & Lubs, 1983). Although two early studies failed to replicate the finding, many subsequent studies have reported linkage in the 15*q*15-21 region, which includes about 10 million nucleotide bases (Chapman et al., 2004; Fulker et al., 1991; Grigorenko et al., 1997; Marino et al., 2004; Morris et al., 2004; Schulte-Körne et al., 1998; Smith, Kimberling et al., 1991). A high-profile report identifying a gene responsible for the linkage (Taipale et al., 2003) was followed by many failures to replicate, and

CLOSE UP

Shelley Smith is a professor of pediatrics and director of the Hattie B. Munroe Molecular Genetics Center at the University of Nebraska Medical Center, Omaha.

Dr. Smith was trained as a medical geneticist at Indiana University. Her interest in specific reading disability (dyslexia) began with a large pedigree ascertained by Dr. Herbert Lubs in which the disorder appeared to be inherited in an autosomal dominant fashion. Further studies of similar multigeneration families as well as segregation analyses in other populations indicated that dyslexia is a complex quantitative trait influenced by at least several major genes. This led to a long-standing collaboration with the Institute for Behavioral Genetics under Dr. John DeFries, who had assembled an excellent group of researchers studying the phenotypic characteristics and genetic parameters of dyslexia in a large twin study. This well-characterized population was ideal for linkage studies, leading to the initial localizations on chromosomes 6*p* and 15*q* which have been replicated by numerous independent groups. While careful evaluation of the dyslexia phenotype was critical to the success of the linkage and association studies by reducing heterogeneity, further studies have indicated that at least some of these loci may have broader effects. Research with Dr. Bruce Pennington and Dr. Erik Willcutt has indicated that they may also be involved in a related condition, speech sound disorder, and also with ADHD.

the search continues (Fisher & Francks, 2006). Other linkages with reading disability on 1*p*34-*p*36, 2*p*16-*p*15, and 18*p*11 show greater consistency than do linkages for other complex disorders (Schumacher, Hoffmann, Schmael, Schulte-Korne, & Nothen, 2007).

Communication Disorders

DSM-IV includes four types of communication disorders: expressive language (putting thoughts into words) disorder, mixed receptive (understanding the language of others) and expressive language disorder, phonological (articulation) disorder, and stuttering (speech interrupted by prolonged or repeated words, syllables, or sounds). Hearing loss, cognitive disability, and neurological disorders are excluded. Genetic research suggests that all of these cognitive problems are genetically associated with clinically diagnosed language problems (Bishop, 2006).

Several family studies, examining communication disorders broadly, indicate that communication disorders are familial (Stromswold, 2001). For children with communication disorders, about a quarter of their first-degree relatives report similar disorders; these communication disorders appear in about 5 percent of the relatives of controls (Felsenfeld, 1994). Twin studies suggest that this familial resemblance is genetic in origin. A review of twin studies of language disability yields twin concordances of 75 percent for MZ twins and 43 percent for DZ twins (Stromswold, 2001). Using *DF extremes analysis*, the average weighted group heritability was 43 percent for language disabilities (Plomin & Kovas, 2005). A large twin study of language delay in infancy found high heritability, even at two years of age (Dale et al., 1998). The only adoption study of communication disorders confirms the twin results, suggesting substantial genetic influence (Felsenfeld & Plomin, 1997).

The high heritability of communication disorders has attracted attention from molecular genetics. A high-profile paper reported a mutation in a gene (*FOXP2*) that accounted for an unusual type of speech-language impairment that includes deficits in oro-facial motor control in one family (Lai, Fisher, Hurst, Vargha-Khadem, & Monaco, 2001). In the media, this finding was unfortunately trumpeted as "the" gene for language, whereas in fact the mutation is not found outside the original family (Meaburn, Dale, Craig, & Plomin, 2002; Newbury et al., 2002). QTL linkage studies have also pointed to linkages on 16*q* and 19*q* (SLI Consortium, 2004).

Family studies of stuttering over the past 50 years have shown that about a third of stutterers have other stutterers in their families. Most of our knowledge comes from the Yale Family Study of Stuttering, which includes nearly 600 stutterers and more than 2000 of their first-degree relatives (Kidd, 1983). Twin studies indicate that stuttering is highly heritable (Andrews, Morris-Yates, Howie, & Martin, 1991; Felsenfeld et al., 2000; Howie, 1981; Ooki, 2005). Although much remains to be learned about the genetics of stuttering, current evidence suggests substantial genetic influence (Yairi, Ambrose, & Cox, 1996). Molecular genetic research has begun using QTL linkage strategies (Suresh et al., 2006).

Mathematics Disability

For poor performance on tests of mathematics, the first twin study suggested moderate genetic influence (Alarcón, DeFries, Light, & Pennington, 1997). A study of 7-year-olds using U.K. National Curriculum scores for mathematics reported concordances of about 70 percent for MZ twins and 50 percent for DZ twins (Oliver et al., 2004). Using DF extremes analysis, the average weighted group heritability was .61 for mathematics disability (Plomin & Kovas, 2005). A recent study used Internet-administered tests of mathematics to select low-performing 10-year-old twins and reported a group heritability of .47 for low performance in mathematics (Kovas, Haworth, Petrill, & Plomin, in press). No molecular genetic research on mathematics disability has been reported.

Comorbidity between Learning Disabilities

Learning disabilities are distinguished from cognitive disability because they focus on what is thought to be specific disabilities as distinct from general cognitive disability. Nonetheless, two multivariate genetic analyses (see Appendix) suggest that there is substantial genetic overlap between reading and mathematics disabilities (Knopik, Alarcón, & DeFries, 1997; Kovas et al., 2007). Extending DF extremes analysis to bivariate analysis, genetic correlations of .53 and .67 between reading and mathematics disability were reported. In other words, many of the genes that affect reading disability also affect mathematics disability. Multivariate genetic research has been central to analyses of cognitive abilities; this research also suggests substantial genetic overlap among diverse cognitive abilities, as discussed in Chapter 9.

SUMMING UP

Twin studies suggest genetic influence for learning disorders, especially reading disability. Genetic influence is also found for communication disorders and mathematics disability. Genetic overlap has been found in multivariate genetic analyses of reading and mathematic disabilities. The first replicated QTL linkages for human behavioral disorders have been reported for reading disability.

Dementia

Although aging is a highly variable process, as many as 15 percent of individuals over 80 years of age suffer severe cognitive decline known as dementia; twice as many women as men are affected (Skoog, Nilsson, Palmertz, Andreasson, &

Svanborg, 1993). Prior to the age of 65 years, the incidence is less than 1 percent. Among the elderly, dementia accounts for more days of hospitalization than any other psychiatric disorder (Cumings & Benson, 1992). It is the fourth leading cause of death in adults.

At least half of all cases of dementia involve Alzheimer's disease (AD), which has been studied for more than a century (Goedert & Spillantini, 2006). AD occurs very gradually over many years, beginning with loss of memory for recent events. This mild memory loss affects many older individuals but is much more severe in individuals with AD. Irritability and difficulty in concentrating are also often noted. Memory gradually worsens to include simple behaviors, such as forgetting to turn off the stove or bath water, and wandering off and getting lost. Eventually—sometimes after 3 years, sometimes after 15 years—individuals with AD become bedridden. Biologically, AD involves extensive changes in brain nerve cells, including plaques and tangles (described later) that build up and result in death of the nerve cells. Although these plaques and tangles occur to some extent in most older people, they are usually restricted to the hippocampus. In individuals with AD, they are much more numerous and widespread.

Another type of dementia is the result of the cumulative effect of multiple small strokes in which blood flow to the brain becomes blocked, thus damaging the brain. This type of dementia is called multiple-infarct dementia (MID). (An infarct is an area damaged as a result of a stroke.) Unlike AD, MID is usually more abrupt and involves focal symptoms such as loss of language rather than general cognitive decline. Co-occurrence of AD and MID is seen in about a third of all cases. DSM-IV recognizes nine other kinds of dementias, such as dementias due to AIDS, to head trauma, and to Huntington's disease.

Like the situation for cognitive disability, surprisingly little is known about the quantitative genetics of either AD or MID. Family studies of AD probands estimate risk to first-degree relatives of nearly 50 percent by the age of 85, when the data are adjusted for age of the relatives (McGuffin, Owen, O'Donovan, Thapar, & Gottesman, 1994). Until recently, the only twin study of dementia was one reported 40 years ago. That twin study, which did not distinguish AD and MID, found concordances of 43 percent for identical twins and 8 percent for fraternal twins, results suggesting moderate genetic influence (Kallmann & Kaplan, 1955). More recent twin studies of AD also found evidence for genetic influence, with concordances two times greater for identical than for fraternal twins in Finland (Raiha, Kaprio, Koskenvuo, Rajala, & Sourander, 1996), Norway (Bergeman, 1997), Sweden (Gatz et al., 1997), and the United States (Breitner et al., 1995). In the largest twin study to date, liability to AD yielded a heritability estimate of .58 (Gatz et al., 2006).

Some of the most important molecular genetic findings for behavioral disorders have come from research on dementia (Pollen, 1993). Research has focused on a rare (1 in 10,000) type of Alzheimer's disease that appears before

65 years of age and shows evidence for autosomal dominant inheritance. Most of these early-onset cases are due to a gene (*PS1*, for presenilin-1) on chromosome 14 (St. George-Hyslop et al., 1992). In 1995, the *PS1* gene was identified (Sherrington et al., 1995), as well as a similar gene (*PS2*, for presenilin-2), which is on chromosome 1 (Hardy & Hutton, 1995). It is thought that these alleles lead to brain lesions surrounded by protein fragments called β-amyloid (Hardy & Selkoe, 2002). When β-amyloid builds up, it kills nerve cells. These genes are also involved in the other main feature of the brains of individuals with AD called neurofibrillary tangles, which are dense bundles of abnormal fibers that appear in the cytoplasm of certain nerve cells. As is often the case, dozens of different mutations in *PS1* (but not *PS2*) have been found, which will make screening difficult (Cruts, van Duijn, Backhovens, van den Broeck, & Wehnert, 1998). A small percentage of early-onset cases are linked to the amyloid precursor protein (*APP*) gene on chromosome 21.

The great majority of Alzheimer's cases occur after 65 years of age, typically in persons in their seventies and eighties. A major advance toward understanding late-onset Alzheimer's disease is the discovery of a strong allelic association with a gene (for apolipoprotein E) on chromosome 19 (Corder et al., 1993). This gene has three alleles (confusingly called alleles 2, 3, and 4). The frequency of allele 4 is about 40 percent in individuals with Alzheimer's disease and 15 percent in control samples. This result translates to about a sixfold increased risk for late-onset Alzheimer's disease for individuals who have one or two of these alleles. There is some evidence that allele 2, the least common allele, may play a protective role (Corder et al., 1994). Finding QTLs that protect rather than increase risk for a disorder is an important direction for genetic research. There is some evidence that apolipoprotein E is also weakly associated with cognitive aging in individuals without apparent dementia (Deary et al., 2004; Small, Rosnick, Fratiglioni, & Backman, 2004), although it has been argued that the association is the result of incipient AD (Savitz, Solms, & Ramesar, 2006).

Apolipoprotein E is a QTL in the sense that allele 4, although a risk factor, is neither necessary nor sufficient for developing dementia. For instance, nearly half of patients with late-onset Alzheimer's disease do not have that allele. Assuming a liability-threshold model, allele 4 accounts for about 15 percent of the variance in liability (Owen, Liddell, & McGuffin, 1994). Because apolipoprotein E was known for its role in transporting lipids throughout the body, its association with late-onset AD was puzzling at first. However, the product of allele 4 binds more readily with β-amyloid, leading to amyloid deposits, which in turn lead to plaques and, eventually, to death of nerve cells (Tanzi & Bertram, 2005). The product of allele 2 may block this buildup of β-amyloid. The product of allele 3 appears to buffer nerve cells against the other characteristic of AD, neurofibrillary tangles. Other roles for the gene product are also known, such as its increased production following injury to the nervous system, as in head injury, and, most important, its role in plaques (Hardy, 1997).

Because the gene for apolipoprotein E does not account for all the genetic influence on AD, the search is on for other QTLs. As is the case with other complex disorders, dozens of associations with candidate genes are reported each year but few of these replicate (Bertram & Tanzi, 2004). A recent meta-analysis of the hundreds of reports of associations with Alzheimer's disease finds evidence for significant associations for more than a dozen susceptibility QTLs (Bertram, McQueen, Mullin, Blacker, & Tanzi, 2007). Moreover, a gene (*SORL1*) involved in recycling amyloid precursor protein has shown significant associations in several case-control samples (Rogaeva et al., 2007).

More than a dozen knock-out mouse models of AD-related genes have been generated, and several of the mutants show β-amyloid deposits and plaques (Price, Sisodia, & Borchelt, 1998). However, no model has as yet been shown to have all the expected AD effects, including the critical effects on memory (McGowan, Eriksen, & Hutton, 2006).

Summary

More is known about specific genetic causes in cognitive disabilities than in any other area of behavioral genetics, although less is known about basic quantitative genetic issues. Surprisingly, no twin or adoption study has been reported for moderate or severe cognitive disability.

PKU has been known for decades as a single-gene recessive disorder that causes severe cognitive disability if untreated, although PKU is rare (1 in 10,000). The more recent discovery of fragile X syndrome is especially important. It is the most common cause of inherited cognitive disability (1 in several thousand males, half as common in females). It is caused by a triplet repeat (CGG) on the X chromosome that expands over several generations until it reaches more than 200 repeats, when it often causes moderate cognitive disability in males. Rett syndrome also involves a gene on the X chromosome that primarily causes cognitive disability in females, because males with the mutation die before or shortly after birth. Other single genes known primarily for other effects also contribute to cognitive disability, such as genes for Duchenne muscular dystrophy, Lesch-Nyhan syndrome, and neurofibromatosis.

Chromosomal abnormalities play an important role in cognitive disability. Small deletions of chromosomes can result in cognitive disability, as in Angelman syndrome, Prader-Willi syndrome, and Williams syndrome. The most common cause of cognitive disability is Down syndrome, caused by the presence of three copies of chromosome 21. Down syndrome occurs in about 1 in 1000 births and is responsible for about 10 percent of cognitively disabled individuals in institutions. Risk for cognitive disability is also increased by having an extra X chromosome (XXY males, XXX females). An extra Y chromosome (XYY males) and a missing X chromosome (Turner females) cause less disability. XYY males have speech and language problems; Turner females generally perform less well on nonverbal tasks such as spatial tasks. There is a wide range of cognitive functioning around the lowered average IQ scores found for all these genetic causes of cognitive disability.

Specific genes have also been localized for reading disability and for dementia. For reading disability, a replicated linkage on chromosome 6 has been reported, the first QTL linkage for human behavioral disorders. Quantitative genetic research indicates substantial genetic influence for reading disability. DF extremes analysis assesses genetic relationships between the normal and the abnormal. Much less is known about the genetics of other learning disorders. Substantial genetic influence has also been found for communication disorders and mathematics disability.

For dementia, several genes, especially that for presenilin-1 on chromosome 14, have been found that account for most cases of early-onset Alzheimer's disease, a rare (1 in 10,000) form of the disease that occurs before 65 years of age and often shows pedigrees consistent with autosomal dominant inheritance. Late-onset Alzheimer's disease is very common, striking as many as 15 percent of individuals over 85 years of age. The gene for apolipoprotein E is associated with late-onset Alzheimer's disease. Allele 4 of this gene increases risk about sixfold. The apolipoprotein E gene is a QTL in the sense that it is a probabilistic risk factor, not a single gene necessary and sufficient to develop the disorder.

CHAPTER EIGHT

General Cognitive Ability

G eneral cognitive ability is one of the most well studied domains in behavioral genetics. Nearly all this genetic research is based on a model in which cognitive abilities are organized hierarchically (Carroll, 1993, 1997), from specific tests to broad factors to general cognitive ability (often called *g*; Figure 8.1). There are hundreds of tests of diverse cognitive abilities. These tests measure several broad factors (specific cognitive abilities) such as verbal ability, spatial ability, memory, and speed of processing. Such tests are widely used in schools, industry, the military, and clinical practice.

These broad factors intercorrelate modestly. In general, people who do well on tests of verbal ability tend to do well on tests of spatial ability. *g*, that which is in common among these broad factors, was discovered by Charles Spearman over a century ago, about the same time that Mendel's laws of inheritance were rediscovered (Spearman, 1904). The phrase *general cognitive ability* to describe *g* is preferable to the word *intelligence* because the latter has so many

General cognitive
ability (*g*)

Specific cognitive
abilities

Tests

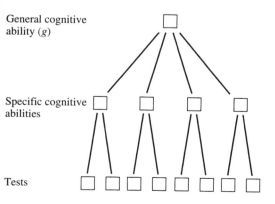

Figure 8.1 Hierarchical model of cognitive abilities.

different meanings in psychology and in the general language (Jensen, 1998). General texts on *g* are available (Brody, 1992; Deary, 2000; Mackintosh, 1998).

Most people are familiar with intelligence tests, often called IQ tests (intelligence quotient tests). These tests typically assess several cognitive abilities and yield total scores that are reasonable indices of *g*. For example, the Wechsler tests of intelligence, widely used clinically, include ten subtests such as vocabulary, picture completion (indicating what is missing in a picture), analogies, and block design (using colored blocks to produce a design that matches a picture). In research contexts, *g* is usually derived by using a technique called *factor analysis* that weights tests differently, according to how much they contribute to *g*. This weight can be thought of as the average of a test's correlations with every other test. This is not merely a statistical abstraction—one can simply look at a matrix of correlations among such measures and see that all the tests intercorrelate positively and that some measures (such as spatial and verbal ability) intercorrelate more highly than do other measures (such as nonverbal memory tests). A test's contribution to *g* is related to the complexity of the cognitive operations it assesses. More complex cognitive processes such as abstract reasoning are better indices of *g* than less complex cognitive processes such as simple sensory discriminations.

Although *g* explains about 40 percent of the variance among such tests, most of the variance of specific tests is independent of *g*. Clearly there is more to cognition than *g*. Specific cognitive abilities assessed in the psychometric tradition are the focus of the next chapter. As new tests of cognitive abilities are developed and their reliability and validity established, their relationship to *g* and their genetic and environmental origins can be investigated. For example, new measures of cognitive function are emerging from research on information processing and experimental cognitive psychology (Deary, 2001; Ramus, 2006), as well as measures of brain structure and function (de Geus, Wright, Martin, & Boomsma, 2001; Gray & Thompson, 2004; Toga & Thompson, 2005). In addition, just as there is more to cognition than *g*, there is clearly much more to achievement than cognition. Personality, motivation, and creativity all play a part in how well someone does in life. However, it makes little sense to stretch a word like *intelligence* to include all aspects of achievement such as emotional sensitivity (Goleman, 2005) and musical and dance ability (Gardner, 2006) that do not correlate with tests of cognitive ability (Visser, Ashton, & Vernon, 2006).

Despite the massive data pointing to the reality of *g*, considerable controversy continues to surround *g* and IQ tests, especially in the media. There is a wide gap between what lay people (including scientists in other fields) believe and what experts believe. Most notably, lay people often hear in the popular press that the assessment of intelligence is circular—intelligence is what intelligence tests assess. To the contrary, *g* is one of the most reliable and valid measures in the behavioral domain. Its long-term stability after childhood is greater than the stability of any other behavioral trait (Deary, Whiteman, Starr, Whalley, & Fox, 2004). It predicts important social outcomes such as educational and occupational levels far better than any other trait (Gottfredson, 1997; Neisser et al., 1996; Schmidt & Hunter, 2004). Although a few critics remain (Gould, 1996), *g* is widely accepted as a valuable concept by experts (Carroll, 1997). It is less clear what *g* is, whether *g* is due to a single general process such as executive function or speed of information processing, or whether it represents a concatenation of more specific cognitive processes (Deary, 2000; Mackintosh, 1998). The idea of a genetic contribution to *g* has produced controversy in the media, especially following the 1994 publication of *The Bell Curve* by Herrnstein and Murray (1994). In fact, these authors scarcely touched on genetics and did not view genetic evidence as crucial to their arguments. In the first half of the book, they showed, like many other studies, that *g* is related to educational and social outcomes. In the second half, however, they attempted to argue that certain conservative policies follow from these findings. But, as discussed in Chapter 18, public policy never necessarily follows from scientific findings; and on the basis of the same studies, it would be possible to present arguments that are the opposite of those of Herrnstein and Murray. Despite this controversy, there is considerable consensus among scientists—even those who are not geneticists—that *g* is substantially heritable (Brody, 1992; Mackintosh, 1998; Neisser, 1997; Snyderman & Rothman, 1988; Sternberg & Grigorenko, 1997). The evidence for a genetic contribution to *g* is presented in this chapter. (See also, Deary, Spinath, & Bates, 2006.)

Historical Highlights

The relative influences of nature and nurture on *g* have been studied since the beginning of the behavioral sciences. Indeed, a year before the publication of Gregor Mendel's seminal paper on the laws of heredity, Francis Galton (1865) published a two-article series on high intelligence and other abilities, which he later expanded into the first book on heredity and cognitive ability, *Hereditary Genius: An Enquiry into Its Laws and Consequences* (1869; see Box 8.1). The first twin and adoption studies in the 1920s also focused on *g* (Burks, 1928; Freeman, Holzinger, & Mitchell, 1928; Merriman, 1924; Theis, 1924).

Animal Research

Cognitive ability, at least problem-solving behavior and learning, can also be studied in other species. For example, in a well-known experiment in learning psychology, begun in 1924 by the psychologist Edward Tolman and continued by Robert Tryon, rats were selectively bred for their performance in learning a maze in order to find food. The results of subsequent selective breeding by Robert Tryon for "maze-bright" rats (few errors) and "maze-dull" rats (many

BOX 8.1

Francis Galton's life (1822–1911) as an inventor and explorer changed as he read the now-famous book on evolution written by Charles Darwin, his half cousin. Galton understood that evolution depends on heredity, and he began to ask whether heredity affects human behavior. He suggested the major methods of human behavioral genetics—family, twin, and adoption designs—and conducted the first systematic family studies showing that behavioral traits "run in families." Galton invented correlation, one of the fundamental statistics in all of science, in order to quantify degrees of resemblance among family members.

One of Galton's studies on mental ability was reported in an 1869 book, *Hereditary Genius: An Enquiry into Its Laws and Consequences*. Because there was no satisfactory way at the time to measure mental ability, Galton had to rely on reputation as an index. By "reputation," he did not mean notoriety for a single act, nor mere social or official position, but "the reputation of a leader of opinion, or an originator, of a man to whom the world deliberately acknowledges itself largely indebted" (1869, p. 37). Galton identified approximately 1000 "eminent" men and found that they belonged to only 300 families, a finding indicating that the tendency toward eminence is familial.

Taking the most eminent man in each family as a reference point, the other individuals who attained eminence were tabulated with respect to closeness of family relationship. As indicated in the diagram on the facing page, eminent status was more likely to appear in close relatives, with the likelihood of eminence decreasing as the degree of relationship became more remote.

errors) are shown in Figure 8.2. Substantial response to selection was achieved after only a few generations of selective breeding. There was practically no overlap between the maze-bright and maze-dull lines; all rats in the maze-bright line were able to learn to run through a maze with fewer errors than any of the rats in the maze-dull line. The difference between the bright and dull lines did not increase after the first half-dozen generations, possibly because

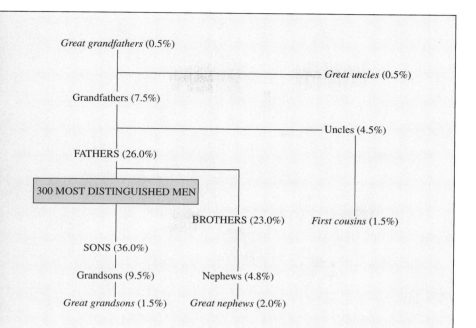

Galton was aware of the possible objection that relatives of eminent men share social, educational, and financial advantages. One of his counterarguments was that many men had risen to high rank from humble backgrounds. Nonetheless, such counterarguments do not today justify Galton's assertion that genius is solely a matter of nature (heredity) rather than nurture (environment). Family studies by themselves cannot disentangle genetic and environmental influences.

Galton set up a needless battle by pitting nature against nurture, arguing that "there is no escape from the conclusion that nature prevails enormously over nurture" (Galton, 1883, p. 241). Nonetheless, his work was pivotal in documenting the range of variation in human behavior and in suggesting that heredity underlies behavioral variation. For this reason, Galton can be considered the father of behavioral genetics.

brothers and sisters were often mated. Such inbreeding greatly reduces the amount of genetic variability within selected lines, a loss that inhibits progress in a selection study.

These maze-bright and maze-dull selected rats were used in one of the best-known psychological studies of genotype-environment interaction (Cooper & Zubek, 1958). Rats from the two selected lines were reared under one of three

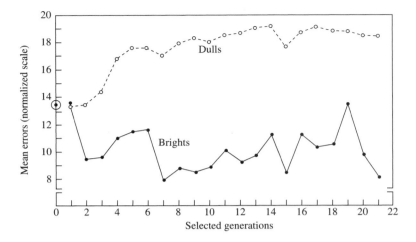

Figure 8.2 The results of Tryon's selective breeding for maze brightness and maze dullness in rats. (From "The inheritance of behavior" by G. E. McClearn. In L. J. Postman (Ed.), *Psychology in the Making.* © 1963. Used with permission of Alfred A. Knopf, Inc.)

conditions. One condition was "enriched," in that the cages were large and contained many movable toys. For the comparison condition, called "restricted," small gray cages without movable objects were used. In the third condition, rats were reared in a standard laboratory environment.

The results of testing the maze-bright and maze-dull rats reared in these conditions are shown in Figure 8.3. Not surprisingly, in the normal environment in which the rats had been selected, there was a large difference between the two selected lines. A clear genotype-environment interaction emerged for the enriched and restricted environments. The enriched condition had no

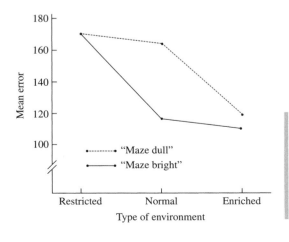

Figure 8.3 Genotype-environment interaction. The effects of rearing in a restricted, normal, or enriched environment on maze-learning errors differ for maze-bright and maze-dull selected rats. (From Cooper & Zubek, 1958.)

effect on the maze-bright rats, but it greatly improved the performance of the maze-dull rats. On the other hand, the restricted environment was very detrimental to the maze-bright rats but had little effect on the maze-dull ones. In other words, there is no simple answer concerning the effect of restricted and enriched environments in this study. It depends on the genotype of the animals. This example illustrates genotype-environment interaction, the differential response of genotypes to environments. Despite this persuasive example, other systematic research on learning generally failed to find widespread evidence of genotype-environment interaction (Henderson, 1972), although there is evidence for interactions with short-term factors as discussed in Chapter 5 (Crabbe et al., 1999).

In the 1950s and 1960s, studies of inbred strains of mice showed the important contribution of genetics to most aspects of learning. Genetic differences have been shown for maze learning as well as for other types of learning, such as active avoidance learning, passive avoidance learning, escape learning, lever pressing for reward, reversal learning, discrimination learning, and heart rate conditioning (Bovet, 1977). For example, differences in maze-learning errors among widely used inbred strains (Figure 8.4) confirm the evidence for genetic influence found in the maze-bright and maze-dull selection experiment. The DBA/2J strain learned quickly; the CBA animals were slow; and the BALB/c strain was intermediate. Similar results were obtained for active avoidance learning, in which mice learn to avoid a shock by moving from one compartment to

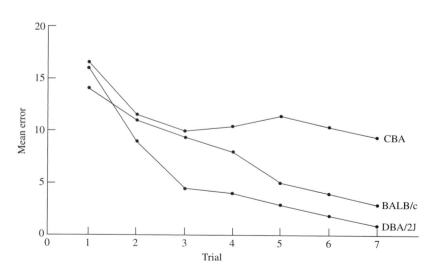

Figure 8.4 Maze-learning errors (Lashley III maze) for three inbred strains of mice. (From "Genetic aspects of learning and memory in mice" by D. Bovet, F. Bovet-Nitti, & A. Oliverio. *Science, 163*, 139–149. © 1969 by the American Association for the Advancement of Science.)

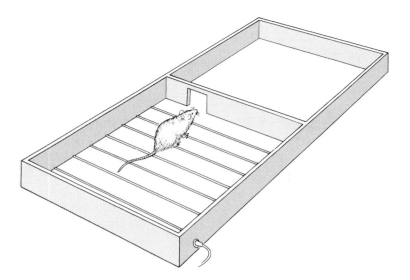

Figure 8.5 Avoidance learning in mice has been investigated by using a "shuttle box," which has two compartments and an electrified floor. The mouse is placed in one compartment, a light is flashed on, followed by a shock (delivered by an electrified grid on the floor) that continues until the mouse moves to the other compartment. Animals learn to avoid the shock by moving to the other compartment as soon as the light comes on. (From *The Experimental Analysis of Behavior* by Edmund Fantino & Cheryl A. Logan. © 1979 by W. H. Freeman and Company.)

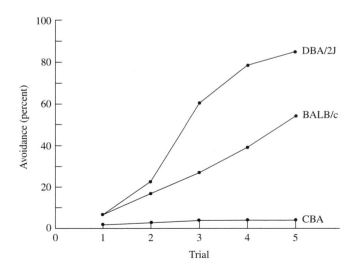

Figure 8.6 Avoidance learning for three inbred strains of mice. (From "Genetic aspects of learning and memory in mice" by D. Bovet, F. Bovet-Nitti, & A. Oliverio. *Science, 163,* 139–149. © 1969 by the American Association for the Advancement of Science.)

another whenever a light is flashed on (Figure 8.5). In this study, however, the CBA strain did not learn at all (Figure 8.6).

Although it has been assumed that *g* is not relevant to mouse learning (Macphail, 1993), a strong *g* factor runs through many learning tasks (Plomin, 2001). In half a dozen studies, intercorrelations among diverse learning tasks indicate that *g* accounts for at least 30 percent of the variance and appears to be moderately heritable (Galsworthy et al., 2005). *g* emerges even when other possible sources of intercorrelations among learning tasks such as emotional reactivity or sensory and motoric ability are controlled (Matzel et al., 2006). Very simple learning tasks such as avoidance learning are less likely to show the influence of *g* than are more complex cognitive tasks such as learning to run a maze or to classify objects (Thomas, 1996). Animal models of *g* will be useful for functional genomic investigations of the brain pathways between genes and *g* (see Chapter 15).

Human Research

Highlights in the history of human research on genetics and *g* include two early adoption studies which found that IQ correlations were greater in nonadoptive than in adoptive families, suggesting genetic influence (Burks, 1928; Leahy, 1935). The first adoption study that included IQ data for biological parents of adopted-away offspring also showed significant parent-offspring correlation, again suggesting genetic influence (Skodak & Skeels, 1949). Begun in the early 1960s, the Louisville Twin Study was the first major longitudinal twin study of IQ that charted the developmental course of genetic and environmental influences (Wilson, 1983).

In 1963, a review of genetic research on *g* was influential in showing the convergence of evidence pointing to genetic influence (Erlenmeyer-Kimling & Jarvik, 1963). In 1966, Cyril Burt summarized his decades of research on MZ twins reared apart, which added the dramatic evidence that MZ twins reared apart are nearly as similar as MZ twins reared together. After his death in 1973, Burt's work was attacked, with allegations that some of his data were fraudulent (Hearnshaw, 1979). Two subsequent books have reopened the case (Fletcher, 1990; Joynson, 1989). Although the jury is still out on some of the charges (Mackintosh, 1995; Rushton, 2002), it appears that at least some of Burt's data are dubious.

During the 1960s, environmentalism, which had been rampant until then in American psychology, was beginning to wane, and the stage was set for increased acceptance of genetic influence on *g*. Then, in 1969, a monograph on the genetics of intelligence by Arthur Jensen almost brought the field to a halt, because the monograph suggested that ethnic differences in IQ might involve genetic differences. Twenty-five years later, this issue was resurrected in *The Bell Curve* (Herrnstein & Murray, 1994) and caused a similar uproar. As we emphasized earlier, the causes of average differences between groups need not be related to the causes of individual differences within groups (see Chapter 5). The former question is much more difficult to investigate than the latter, which is the focus of the vast majority of genetic research on IQ. Although the question of the origins of ethnic differences in performance on IQ tests continues to be debated (e.g., Rushton & Jensen, 2005), the issue may not be resolved until the QTLs for IQ have been identified and the frequencies of their "increasing" and "decreasing" alleles (see Appendix) have been assessed in different groups. However, as discussed later in this chapter, no public policies would necessarily follow from such comparisons.

The storm raised by Jensen's monograph led to intense criticism of all behavioral genetic research, especially in the area of cognitive abilities (e.g., Kamin, 1974). These criticisms of older studies had the positive effect of generating a dozen bigger and better behavioral genetic studies that used family, adoption, and twin designs. These new projects produced much more data on the genetics of g than had been obtained in the previous 50 years. The new data contributed in part to a dramatic shift that occurred in the 1980s in psychology toward acceptance of the conclusion that genetic differences among individuals are significantly associated with differences in g (Snyderman & Rothman, 1988).

SUMMING UP

Selection and inbred strain studies indicate genetic influence on animal learning, such as the maze-bright and maze-dull selection study of learning in rats. Human twin and adoption studies of general cognitive ability have been conducted for more than 75 years. This research has led to widespread acceptance of the conclusion that genetic factors contribute to individual differences in general cognitive ability.

Overview of Genetic Research

In 1981 (Bouchard & McGue), a review of genetic research on g was published that summarized results from dozens of studies. Figure 8.7 is an expanded version of the summary of the review presented earlier in Chapter 3 (see Figure 3.7).

Genetic Influence

First-degree relatives living together are moderately correlated for g (about .45). As in Galton's original family study on hereditary genius (see Box 8.1), this resemblance could be due to genetic or to environmental influences, because such relatives share both. Adoption designs disentangle these genetic and environmental sources of resemblance. Because adopted-apart parents and offspring and siblings share heredity but not family environment, their similarity indicates that resemblance among family members is due in part to genetic factors.

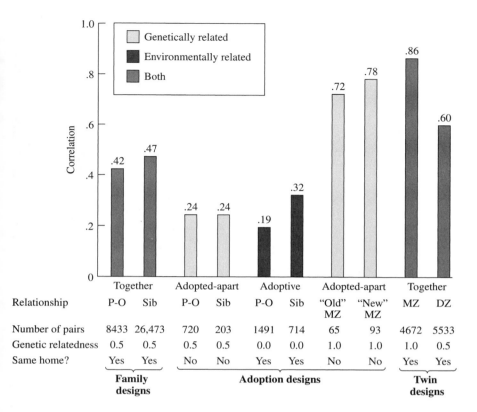

Relationship	Together		Adopted-apart		Adoptive		Adopted-apart		Together	
	P-O	Sib	P-O	Sib	P-O	Sib	"Old" MZ	"New" MZ	MZ	DZ
Number of pairs	8433	26,473	720	203	1491	714	65	93	4672	5533
Genetic relatedness	0.5	0.5	0.5	0.5	0.0	0.0	1.0	1.0	1.0	0.5
Same home?	Yes	Yes	No	No	Yes	Yes	No	No	Yes	Yes
	Family designs		**Adoption designs**						**Twin designs**	

Figure 8.7 Average IQ correlations for family, adoption, and twin designs. Based on reviews by Bouchard and McGue (1981), as amended by Loehlin (1989). "New" data for adopted-apart MZ twins include T. Bouchard et al. (1990) and Pedersen et al. (1992a).

For g, the correlation between adopted children and their genetic parents is .24. The correlation between genetically related siblings reared apart is also .24. Because first-degree relatives are only 50 percent similar genetically, doubling these correlations gives a rough estimate of heritability of 48 percent. As discussed in Chapter 5, this outcome means that about half of the variance in IQ scores in the populations sampled in these studies can be accounted for by genetic differences among individuals.

The twin method supports this conclusion. Identical twins are nearly as similar as the same person tested twice. (Test-retest correlations for g are generally between .80 and .90.) The average twin correlations are .86 for identical twins and .60 for fraternal twins. Doubling the difference between MZ and DZ correlations estimates heritability as 52 percent.

The most dramatic adoption design involves MZ twins who were reared apart. Their correlation provides a direct estimate of heritability. For obvious reasons, the number of such twin pairs is small. For several small studies

---

CLOSE UP

Thomas J. Bouchard, Jr., is a professor of psychology and director of the Minnesota Center for Twin and Adoption Research at the University of Minnesota. He received his doctorate in psychology from the University of California, Berkeley, in 1966. From 1966 to 1969, he held an appointment as an assistant professor in the Department of Psychology at the University of California, Santa Barbara. In the fall of 1969, he joined the faculty of the Department of Psychology at the University of Minnesota. In 1979, with colleagues in the Departments of Psychology and Psychiatry, and the Medical School, he began the Minnesota Study of Twins Reared Apart (MISTRA), a comprehensive medical and psychological study of monozygotic and dizygotic twins separated early in life and reared apart during their formative years. In 1983, Bouchard joined with other colleagues in the Department of Psychology to found the Minnesota Twin Family Registry, a resource for genetic research throughout the University of Minnesota. His research interests include genetic and environmental influences on personality, mental abilities, psychological interests, social values, and psychopathology, as well as the evolution of human behavior. He teaches differential psychology and evolutionary psychology.

---

published before 1981, the average correlation for MZ twins reared apart is .72 (excluding the suspect data of Cyril Burt). This outcome suggests higher heritability (72 percent) than the other designs. This high heritability estimate has been confirmed in two other studies of twins reared apart. In a report on 45 pairs of MZ twins reared apart, the correlation was .78 (T. Bouchard et al., 1990). A study of Swedish twins also included 48 pairs of MZ twins reared apart and reported the same correlation of .78 (Pedersen et al., 1992a). Possible explanations for this higher heritability estimate for adopted-apart MZ twins are discussed later.

Model-fitting analyses that simultaneously analyze all the family, adoption, and twin data summarized in Figure 8.7 yield heritability estimates of about 50 percent (Chipuer, Rovine, & Plomin, 1990; Loehlin, 1989). It is noteworthy that genetics can account for half of the variance of a trait as complex as general cognitive ability. In addition, the total variance includes error of measurement. Corrected for unreliability of measurement, heritability estimates would be higher. Regardless of the precise estimate of heritability, the point is that genetic influence on *g* is not only statistically significant, it is also substantial.

Although heritability could differ in different cultures, it appears that the level of heritability of g also applies to populations in countries other than American and Western European countries, where most studies have been conducted. Similar heritabilities have been found in twin studies in Russia (Lipovechaja, Kantonistowa, & Chamaganova, 1978; Malykh, Iskoldsky, & Gindina, 2005) and in the former East Germany (Weiss, 1982), as well as in rural India, urban India, and Japan (Jensen, 1998). Another interesting finding is that heritabilities for cognitive test scores are higher the more a test relates to g (Jensen, 1998). This result has been found in studies of older twins (Pedersen et al., 1992a), in individuals with cognitive disability (Spitz, 1988), and in a twin study using information-processing tasks (Vernon, 1989). These results suggest that g is the most highly heritable composite of cognitive tests.

Environmental Influence

If half of the variance of g can be accounted for by heredity, the other half is attributed to environment (plus errors of measurement). Some of this environmental influence appears to be shared by family members, making them similar to one another. Direct estimates of the importance of shared environmental influence come from correlations for adoptive parents and children and for adoptive siblings. Particularly impressive is the correlation of .32 for adoptive siblings. Because they are unrelated genetically, what makes adoptive siblings similar is shared rearing—having the same parents, the same diet, attending the same schools, etc. The adoptive sibling correlation of .32 suggests that about a third of the total variance can be explained by shared environmental influences. The correlation for adoptive parents and their adopted children is lower ($r = .19$) than that for adoptive siblings, a result suggesting that shared environment accounts for less resemblance between parents and offspring than between siblings.

Shared environmental effects are also suggested because correlations for relatives living together are greater than correlations for adopted-apart relatives. Twin studies also suggest shared environmental influence. In addition, shared environmental effects appear to contribute more to the resemblance of twins than to that of nontwin siblings, because the correlation of .60 for DZ twins exceeds the correlation of .47 for nontwin siblings. Twins may be more similar than other siblings because they shared the same womb and are exactly the same age. Because they are the same age, twins also tend to be in the same school, if not the same class, and share many of the same peers (Koeppen-Schomerus et al., 2003).

Model-fitting estimates of the role of shared environment for g based on the data in Figure 8.7 are about 20 percent for parents and offspring, about 25 percent for siblings, and about 40 percent for twins (Chipuer et al., 1990). One model-fitting analysis assumed that excess DZ similarity was due to prenatal effects and thus yielded an estimate of prenatal shared environment of about 40 percent (Devlin, Daniels, & Roeder, 1997). The rest of the environmental variance is attributed to nonshared environment and errors of measurement, factors that account for about 10 percent of the variance.

Assortative Mating

Several other factors need to be considered for a more refined estimate of genetic influence. One is *assortative mating*, which refers to nonrandom mating. Old adages are sometimes contradictory. Do "birds of a feather flock together" or do "opposites attract"? Research shows that, for some traits, "birds of a feather" do "flock together," in the sense that individuals who mate tend to be similar—although not as similar as you might think. For example, although there is some positive assortative mating for physical characters, the correlations between spouses are relatively low—about .25 for height and about .20 for weight (Spuhler, 1968). Spouse correlations for personality are even lower, in the .10 to .20 range (Vandenberg, 1972). Assortative mating for g is substantial, with average spouse correlations of about .40 (Jensen, 1978). In part, spouses select each other for g on the basis of education. Spouses correlate about .60 for education, which correlates about .60 with g.

Assortative mating is important for genetic research for two reasons. First, assortative mating increases genetic variance in a population. For example, if spouses mated randomly in relation to height, tall women would be just as likely to mate with short men as with tall men. Offspring of the matings of tall women and short men would generally be of moderate height. However, because there is positive assortative mating for height, children with tall mothers are also likely to have tall fathers, and the offspring themselves are likely to be taller than average. The same thing happens for short parents. In this way, positive assortative mating increases variance in that the offspring differ more from the average than they would if mating were random. Even though spouse correlations are modest, assortative mating can greatly increase genetic variability in a population, because its effects accumulate generation after generation.

Assortative mating is also important because it affects estimates of heritability. For example, it increases correlations for first-degree relatives. If assortative mating were not taken into account, it could inflate heritability estimates obtained from studies of parent-offspring (e.g., birth parents and their adopted-apart offspring) or sibling resemblance. For the twin method, however, assortative mating could result in underestimates of heritability. Assortative mating does not affect MZ correlations because MZ twins are identical genetically, but it raises DZ correlations because they are first-degree relatives. In this way, assortative mating lessens the difference between MZ and DZ correlations; it is this difference that provides estimates of heritability in the twin method. The model-fitting analyses described above took assortative mating into account in estimating the heritability of g to be about 50 percent. If assortative mating were not taken into account, its effects would be attributed to shared environment.

Nonadditive Genetic Variance

Nonadditive genetic variance also affects heritability estimates. For example, when we double the difference between MZ and DZ correlations to estimate heritability, we assume that genetic effects are largely additive. *Additive genetic effects* occur when alleles at a locus and across loci "add up" to affect behavior. However, sometimes the effects of alleles can be different in the presence of other alleles. These interactive effects are called *nonadditive*.

Dominance is a nonadditive genetic effect in which alleles at a locus interact rather than add up to affect behavior. For example, having one PKU allele is not half as bad as having two PKU alleles. Even though many genes operate with a dominant-recessive mode of inheritance, much of the effect of such genes can nonetheless be attributed to the average effect of the alleles. The reason is that, even though heterozygotes are phenotypically similar to the homozygote dominant, there is a substantial linear relationship between genotype and phenotype.

When several genes affect a behavior, the alleles at different loci can add up to affect behavior, or they can interact. This type of interaction between alleles at different loci is called *epistasis*. (See Appendix for details.)

Additive genetic variance is what makes us resemble our parents, and it is the raw material for natural selection. Our parents' genetic decks of cards are shuffled when our hand is dealt at conception. We and each of our siblings receive a random sampling of half of each parent's genes. We resemble our parents to the extent that each allele that we share with our parents has an average additive effect. Because we do not have exactly the same combination of alleles as our parents (we inherit only one of each of their pairs of alleles), we will differ from our parents for nonadditive interactions as a result of dominance or epistasis. The only relatives who will resemble each other for all dominance and epistatic effects are identical twins, because they are identical for all combinations of genes. For this reason, the hallmark of nonadditive genetic variation is that first-degree relatives are less than half as similar as MZ twins.

For *g*, the correlations in Figure 8.7 suggest that genetic influence is largely additive. For example, first-degree relatives are just about half as similar as MZ twins. However, there is evidence that assortative mating for *g* masks some nonadditive genetic variance. As indicated in the previous section, assortative mating, which is greater for *g* than for any other known trait, inflates correlations for first-degree relatives but does not affect MZ correlations. When assortative mating is taken into account in model-fitting analyses, some evidence appears for nonadditive genetic variance, although most genetic influence on *g* is additive (Chipuer et al.,1990; Fulker, 1979).

The presence of dominance can be seen from studies of inbreeding. (Inbreeding is mating between genetically related individuals.) If inbreeding occurs, offspring are more likely to inherit the same alleles at any locus. Thus, inbreeding makes it more likely that two copies of rare recessive alleles will be inherited, including those for harmful recessive disorders. In this sense,

for general cognitive ability from infancy through adolescence. As illustrated in Figure 8.9, correlations between parents and children for control (nonadoptive) families increase from less than .20 in infancy to about .20 in middle childhood and to about .30 in adolescence. The correlations between biological mothers and their adopted-away children follow a similar pattern, thus indicating that parent-offspring resemblance for g is due to genetic factors. Parent-offspring correlations for adoptive parents and their adopted children hover around zero, which suggests that family environment shared by parents and offspring does not contribute importantly to parent-offspring resemblance for g. These parent-offspring correlations for adoptive parents and their adopted children are slightly lower than those reported in other adoption studies (see Figure 8.7), possibly because selective placement was negligible in the Colorado Adoption Project (Plomin & DeFries, 1985).

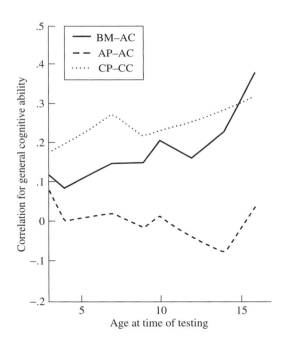

Figure 8.9 Parent-offspring correlations between parents' g scores and children's g scores for adoptive, biological, and control parents and their children at 3, 4, 7, 9, 10, 12, 14, and 16 years. Parent-offspring correlations are weighted averages for mothers and fathers to simplify the presentation. (The correlations for mothers and for fathers were similar.) The sample sizes range from 33 to 44 for biological fathers, 159 to 195 for biological mothers, 153 to 194 for adoptive parents, and 136 to 216 for control parents. (From "Nature, nurture and cognitive development from 1 to 16 years: A parent-offspring adoption study" by R. Plomin, D. W. Fulker, R. Corley, & J. C. DeFries. *Psychological Science, 8,* 442–447. © 1997.)

Figure 8.10 summarizes MZ and DZ twin correlations for *g* by age (McGue et al., 1993). The difference between MZ and DZ twin correlations increases slightly from early to middle childhood and then increases dramatically in adulthood (Loehlin, Horn, & Willerman, 1997). Because relatively few twin studies of *g* have included adults, summaries of IQ data (see Figure 8.7) rest primarily on data from childhood. Heritability in adulthood is higher; this conclusion is supported by five studies of MZ twins reared apart. These studies, unlike studies of twins reared together, almost exclusively include adults. The average heritability estimate from these studies of MZ twins reared apart is 75 percent (McGue et al., 1993). For example, one study included twins reared apart and matched twins reared together tested at the average age of 60 years as part of the Swedish Adoption-Twin Study of Aging (SATSA). These twins are much older than twins in other studies, and the study yielded a heritability estimate of 80 percent for *g* (Pedersen et al., 1992a), a result that was replicated when the SATSA twins were retested three years later (Plomin, Pedersen, Lichtenstein, & McClearn, 1994a). This is one of the highest heritabilities reported for any behavioral dimension or disorder. High heritability was even found in two studies of twins over 75 years of age (McClearn et al., 1997; McGue & Christensen, 2001), although some research suggests that heritability may decline in very old twins (Brandt et al., 1993; Finkel, Wille, & Matheny, 1998).

Why does heritability increase during the life span? Perhaps completely new genes come to affect *g* in adulthood. A more likely possibility is that relatively small genetic effects early in life snowball during development, creating

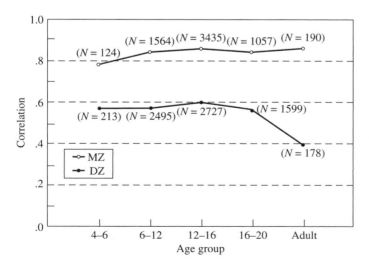

Figure 8.10 The difference between MZ and DZ twin correlations for *g* increases during adolescence and adulthood, a result suggesting increasing genetic influence. (From McGue et al., 1993, p. 63.)

larger and larger phenotypic effects. For the young child, parents and teachers contribute importantly to intellectual experience; but for the adult, intellectual experience is more self-directed. For example, it seems likely that adults with a genetic propensity toward high *g* keep active mentally by reading, arguing, and simply thinking more than other people do. Such experiences not only reflect but also reinforce genetic differences (Bouchard, Lykken, McGue, Segal, & Tellegen, 1997; Scarr, 1992; Scarr & McCartney, 1983).

Another important developmental finding is that the effects of shared environment appear to decrease. Twin study estimates of shared environment are weak because shared environment is estimated indirectly by the twin method; that is, shared environment is estimated as twin resemblance that cannot be explained by genetics. Nonetheless, the world's twin literature indicates that shared environment effects for *g* decline from adolescence to adulthood.

Adoption studies provide two types of evidence for the importance of shared environment in childhood. The classic adoption study of Skodak and Skeels (1949) found that adopted-away offspring had higher IQ scores than expected from the IQ scores of their biological parents, results confirmed in other studies (e.g., Capron & Duyme, 1989). This finding suggests that IQ scores are raised when children whose biological parents have lower than average IQ scores are adopted by adoptive parents whose IQ scores are higher than average. A study of 65 abused and neglected children with an average IQ of 78 when adopted at 4 or 5 years of age showed an average IQ of 91 at age 13 (Duyme, Dumaret, & Tomkiewicz, 1999). However, it is possible that adoption allowed an emergence of the children's normal IQ, which had been suppressed earlier by abuse and neglect.

The most direct evidence for the important effect of shared environment on individual differences in *g* comes from the resemblance of adoptive siblings, pairs of genetically unrelated children adopted into the same adoptive families. Figure 8.7 indicates an average IQ correlation of .32 for adoptive siblings. However, these studies assessed adoptive siblings when they were children. In 1978, the first study of older adoptive siblings yielded a strikingly different result: The IQ correlation was −.03 for 84 pairs of adoptive siblings who were 16–22 years of age (Scarr & Weinberg, 1978b). Other studies of older adoptive siblings have also found similarly low IQ correlations. The most impressive evidence comes from a ten-year longitudinal follow-up study of more than 200 pairs of adoptive siblings. At the average age of eight years, the IQ correlation was .26. Ten years later, their IQ correlation was near zero (Loehlin et al., 1989). Figure 8.11 shows the results of studies of adoptive siblings in childhood and in adulthood (McGue et al., 1993). In childhood, the average adoptive sibling correlation is .25; but in adulthood, the correlation for adoptive siblings is near zero.

These results represent a dramatic example of the importance of genetic research for understanding the environment. Shared environment is an important factor for *g* during childhood, when children are living at home. However, its importance fades in adulthood as influences outside the family become more

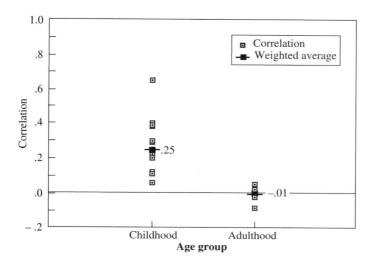

Figure 8.11 The correlation for adoptive siblings provides a direct estimate of the importance of shared environment. For *g*, the correlation is .25 in childhood and −.01 in adulthood, a difference suggesting that shared environment becomes less important after childhood. (From McGue et al., 1993, p. 67.)

salient. What are these mysterious nonshared environmental factors that make siblings growing up in the same family no more similar than individuals growing up in different families? This topic is discussed in Chapter 16.

In summary, from childhood to adulthood, heritability of *g* increases and the importance of shared environment decreases (Figure 8.12).

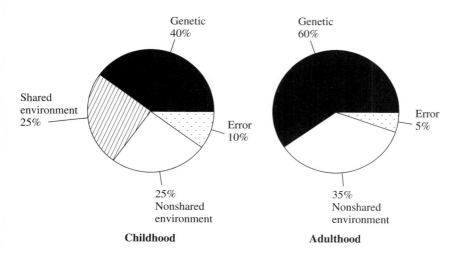

Figure 8.12 From childhood to adulthood, heritability of *g* increases and shared environment declines in importance.

Do Genetic Factors Contribute to Developmental Change?

The second type of genetic change in development refers to age-to-age change seen in longitudinal data in which individuals are assessed several times. It is important to recognize that genetic factors can contribute to change as well as to continuity in development. Change in genetic effects does not necessarily mean that genes are turned on and off during development, although this does happen. Genetic change simply means that genetic effects at one age differ from genetic effects at another age. For example, genes that affect cognitive processes involved in language cannot show their effect until language appears in the second year of life.

The issue of genetic contributions to change and continuity can be addressed by using longitudinal genetic data, in which twins or adoptees are tested repeatedly. The simplest way to think about genetic contributions to change is to ask whether change in scores from age to age show genetic influence. That is, although *g* is quite stable from year to year, some children's scores increase and some decrease. Genetic factors account for part of such changes, especially in childhood (Fulker, DeFries, & Plomin, 1988), and perhaps even in adulthood (Loehlin et al., 1989). Still, not surprisingly, most genetic effects on *g* contribute to continuity from one age to the next (Petrill et al., 2004; Rietveld, Dolan, van Baal, & Boomsma, 2003). Model-fitting analysis (see Appendix) is especially useful for longitudinal data because of the complexity of having multiple measurements for each subject. Several types of longitudinal genetic models have been proposed (Loehlin et al., 1989). A longitudinal model applied to twin and adoptive sibling data from infancy to middle childhood found evidence for genetic change at two important developmental transitions (Fulker, Cherny, & Cardon, 1993). The first is the transition from infancy to early childhood, an age when cognitive ability rapidly changes as language develops. The second is the transition from early to middle childhood, at seven years of age. It is no coincidence that children begin formal schooling at this age—all theories of cognitive development recognize this as a major transition.

Figure 8.13 summarizes these findings. Much genetic influence on *g* involves continuity. That is, genetic factors that affect infancy also affect early

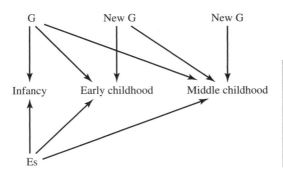

Figure 8.13 Genetic factors (G) contribute to change as well as continuity in *g* during childhood. Shared environment (Es) contributes only to continuity. (Adapted from Fulker, Cherny, & Cardon, 1993.)

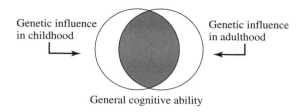

Figure 8.14 Although genetic influences on *g* in childhood are largely the same as those that affect *g* in adulthood, there is some evidence for genetic change.

childhood and middle childhood. However, some new genetic influence comes into play at the transition from infancy to early childhood. These new genetic factors continue to affect *g* throughout early childhood and into middle childhood. Similarly, new genetic influence also emerges at the transition from early to middle childhood. Still, a surprising amount of genetic influence on general cognitive ability in childhood overlaps with genetic influence even into adulthood, as illustrated in Figure 8.14.

As discussed earlier, shared environmental influences also affect *g* in childhood. Unlike genetic effects, which contribute to change as well as to continuity, longitudinal analysis suggests that shared environmental effects contribute only to continuity. That is, the same environmental factors shared by relatives affect *g* in infancy and in both early and middle childhood (see Figure 8.13). Socioeconomic factors, which remain relatively constant, might account for this shared environmental continuity.

SUMMING UP

Heritability of general cognitive ability increases during the life span. The effects of shared environment decrease during childhood to negligible levels after adolescence. Longitudinal genetic analyses of continuity and change indicate that much genetic influence on general cognitive ability contributes to continuity. However, some genetic influence affects change from age to age, especially during the transition from infancy to early childhood.

Identifying Genes

General cognitive ability is a reasonable candidate for molecular genetic research because it is one of the most heritable dimensions of behavior. As for most behaviors, many genes are likely to influence general cognitive ability. Conversely, no single gene is likely to account for a substantial proportion of the total genetic variance. The important implication is that molecular genetic strategies that can detect genes of small effect size are needed.

As mentioned in Chapter 6, knock-out genes have been shown to affect learning in mice. In our species, more than 250 single-gene disorders include cognitive disability among their symptoms (Inlow & Restifo, 2004). The major single-gene effects were described in Chapters 6 and 7. The classic example of a single-gene cause of severe cognitive disability is PKU. More recently, researchers have identified a gene causing the fragile X type of cognitive disability. A gene on chromosome 19 that encodes apolipoprotein E contributes substantially to risk for the dementia of late-onset Alzheimer's disease.

What about the normal range of general cognitive ability? Some evidence suggests that carriers for PKU show slightly lowered IQ scores (Bessman, Williamson, & Koch, 1978; Propping, 1987). However, differences in the number of fragile X repeats in the normal range do not relate to differences in IQ (Daniels et al., 1994). It is only when the number of repeats expands to more than 200 that cognitive disability occurs, as described in Chapter 7. As shown in a meta-analysis of 38 studies with more than 20,000 subjects, apolipoprotein E is associated with g as well as dementia in older people (Small et al., 2004), but it has been argued that this association is due to incipient dementia (Savitz et al., 2006). Apolipoprotein E is not associated with differences in g earlier in life (Deary et al., 2002).

In addition to investigating genes known to be involved in cognitive disability, one study employed a systematic allelic association strategy using DNA markers in or near candidate genes likely to be relevant to neurological functioning, such as genes involved in synaptic transmission and brain development. In the first report of this type, allelic association results were presented for 100 DNA markers for such candidate genes (Plomin et al., 1995). Although several significant associations were found in an original sample, only one association was replicated cleanly in an independent sample. This finding might well be a chance result, because 100 markers were investigated and follow-up studies failed to replicate the result (Petrill, Ball, Eley, Hill, & Plomin, 1998). Dozens of studies have subsequently explored other candidate gene associations with g but none have shown consistent results (see reviews by Payton, 2006; Plomin, Kennedy, & Craig, 2006). If, as suggested in Chapter 6, QTL effect sizes are very small for complex traits such as g, these studies were generally underpowered to detect small effects.

Another candidate gene strategy for identifying QTL associations for g is to focus on intermediate phenotypes—often called *endophenotypes*—that are presumed to be simpler genetically and thus more likely to yield QTLs of large effect size that can be detected with small samples (Goldberg & Weinberger, 2004; Winterer & Goldman, 2003). As discussed in Chapter 15, although all levels of analysis from genes to g are important to study in their own right and in terms of understanding pathways between genes and behavior, it seems unlikely that brain endophenotypes will prove to be simpler genetically or be more useful in identifying QTLs for g (Kovas & Plomin, 2006).

As mentioned in Chapter 6, attempts to find QTL associations with complex traits like *g* have begun to go beyond candidate genes to conduct systematic genome scans. Three QTL linkage reports on *g* have suggested several different linkage regions, including linkage near the region of 6*p*, which is the region that shows consistent linkage with reading disability, as discussed in Chapter 7 (Dick et al., 2006b; Luciano et al., 2006; Posthuma et al., 2005).

As explained in Chapter 6, association can detect QTLs of smaller effect size than can linkage. An early attempt to conduct a systematic association study of *g* using 2000 DNA markers came up empty-handed (Plomin et al., 2001). Microarrays have now made it possible to conduct genomewide association studies with hundreds of thousands of SNPs. Because it is very expensive to conduct microarray studies with the many thousands of subjects needed to detect associations of small effect size, DNA for low and high *g* groups can be averaged biologically by pooling a few nanograms of DNA from each individual (Sham, Bader, Craig, O'Donovan, & Owen, 2002) and genotyping the pooled DNA for each group on a microarray (Butcher et al., 2004; Kirov et al., 2006). DNA pooling thus makes it possible to use microarrays to screen very large groups of cases and controls for a few hundred dollars rather than a few hundred thousand dollars. The first study using this approach with pooled DNA and a microarray with 10,000 SNPs, focused on *g* and reported four SNPs associated with *g* at seven years of age, although the effect sizes were extremely small, less than 0.3 percent of the variance of *g* (Butcher et al., 2005b). Because many more SNPs are needed for a genomewide association scan, ongoing research is using a microarray with 500,000 SNPs (Butcher, Meaburn, Craig, Schalkwyk, & Plomin, 2006).

These *g* SNPs were aggregated in a composite "SNP set" and used in multivariate, developmental, and gene-environment analyses, as examples of behavioral genomic analyses that can be conducted when genes are identified for complex traits like *g* (Harlaar et al., 2005a). As an example of multivariate analysis, the *g* SNP set was also found to be significantly associated with reading. In terms of development, the SNP set for *g* at seven years yielded significant associations with *g* as early as two years of age, suggesting genetic continuity in development. In relation to gene-environment interplay, several examples of gene-environment interaction and correlation were found.

Finding QTLs for *g* has important implications for society as well as for science (Plomin, 1999b). The grandest implication for science is that QTLs for *g* will serve as an integrating force across diverse disciplines, with DNA as the common denominator, and will open up new scientific horizons for understanding learning and memory. In terms of implications for society, it should be emphasized that no public policies necessarily follow from finding genes associated with *g* because policy involves values. For example, finding genes for *g* does not mean that we ought to put all of our resources into educating the brightest children once we identify them genetically. Depending on our values, we might worry more about the children falling off the low end of the bell curve in an increasingly technological society and decide to devote more public resources to those who are in danger of being left behind. Potential problems related to finding genes associated with *g*, such as prenatal and postnatal screening, discrimination in education and employment, and group differences are already being considered (Newson & Williamson, 1999; Nuffield Council on Bioethics, 2002). As discussed in Box 6.4, we need to be cautious and to consider carefully societal implications and ethical issues, but there is also much to celebrate here in terms of increased potential for understanding our species' ability to think and learn.

Summary

The evidence for a strong genetic contribution to general cognitive ability (*g*) is clearer than for any other area of psychology. Although *g* has been central in the nature-nurture debate, few scientists now seriously dispute the conclusion that general cognitive ability shows significant genetic influence. The magnitude of genetic influence is still not universally appreciated, however. Taken together, this extensive body of research suggests that about half of the total variance of measures of *g* can be accounted for by genetic factors. Estimates of heritability are affected by assortative mating (which is substantial for *g*) and by nonadditive genetic variance (dominance and epistasis).

The heritability of *g* increases during an individual's life span, reaching levels in adulthood comparable to the heritability of height. The influence of shared environment diminishes sharply after adolescence. Longitudinal genetic analyses of *g* suggest that genetic factors primarily contribute to continuity, although some evidence for genetic change has been found, for example, in the transition from early to middle childhood.

Attempts to identify some of the genes responsible for the heritability of *g* have begun, including candidate gene studies, QTL linkage, and genomewide association studies.

CHAPTER NINE

Specific Cognitive Abilities

There is much more to cognitive functioning than general cognitive ability. As discussed in Chapter 8, cognitive abilities are usually considered in a hierarchical model (see Figure 8.1). General cognitive ability is at the top of the hierarchy, representing what all tests of cognitive ability have in common. Below general cognitive ability in the hierarchy are broad factors of specific cognitive abilities, such as verbal ability, spatial ability, memory, and speed of processing. These broad factors are indexed by several tests like the tests of verbal ability and spatial ability in Figure 9.1. The tests are at the bottom of the hierarchical model. Specific cognitive abilities correlate moderately with general cognitive ability, but they are also substantially different. In addition to specific tests, the bottom of the hierarchy can also be considered in terms of the elementary processes that are thought to be involved in processing information from input to storage and then from retrieval to output.

More is known about the genetics of the broad factors of specific cognitive abilities than about the elementary processes (Plomin & DeFries, 1998). This chapter presents genetic research on specific cognitive abilities, elementary processes, and their relationship to general cognitive ability. It also considers the genetics of a real-world aspect of cognitive abilities, school achievement.

Broad Factors of Specific Cognitive Abilities

The largest family study of specific cognitive abilities, called the Hawaii Family Study of Cognition, included more than a thousand families (DeFries et al., 1979). Like other work in this area, this study used a technique called factor analysis to identify the tightest clusters of intercorrelated tests. Four group factors were derived from 15 tests: verbal (including vocabulary and fluency), spatial (visualizing and rotating objects in two- and three-dimensional space), perceptual speed

(a) Tests of verbal ability

1. Vocabulary: In each row, circle the word that means the same or nearly the same as the underlined word. There is only one correct choice in each line.

      a. <u>arid</u>   coarse   clever   modest   dry
      b. <u>piquant</u> fruity   pungent   harmful upright

2. Word beginnings and endings: For the next three minutes, write as many words as you can that start with F and end with M.

3. Things: For the next three minutes, list all the things you can think of that are flat.

(b) Tests of spatial ability

1. Paper foam board: Draw a line or lines showing where the figure on the left should be cut to form the pieces on the right. There may be more than one way to draw the lines correctly.

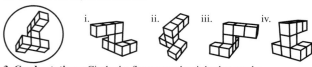

2. Mental rotations: Circle the two objects on the right that are the same as the object on the left.

3. Card rotations: Circle the figures on the right that can be rotated (without being lifted off the page) to exactly match the one on the left.

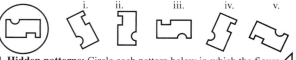

4. Hidden patterns: Circle each pattern below in which the figure appears. The figure must always be in this position, not upside down or on its side.

Figure 9.1 Tests of specific cognitive abilities such as those used in the Hawaii Family Study of Cognition include tasks resembling the ones shown here. (a) The answers for verbal test 1 are (i) dry and (ii) pungent. (b) For spatial test 1, the solution is that, in addition to the rectangle, only one line is needed: The two corners of a short side of the rectangle touch the circle and a single line extends the other short side to bisect the circle. The answers for the other spatial tests are 2. ii, iii; 3. i, iii, iv; 4. i, ii, vi.

(simple arithmetic and number comparisons), and visual memory (short-term and longer-term recognition of line drawings). Examples resembling some of the verbal and spatial tests used in the Hawaii Family Study of Cognition are shown in Figure 9.1.

Figure 9.2 summarizes parent-offspring resemblance for the four factors and the 15 cognitive tests for two ethnic groups. The most obvious fact is that familial resemblance differs for the four factors and for tests within each factor. The data were corrected for unreliability of the tests, so the differences in familial resemblance were not caused by reliability differences among the tests. For both groups, the verbal and spatial factors show more familial resemblance than

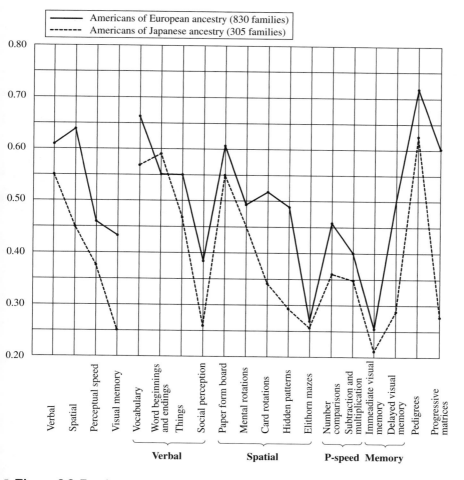

Figure 9.2 Family study of specific cognitive abilities. Regression of midchild on midparent for 4 group factors and 15 cognitive tests in 2 ethnic groups. (Data from DeFries et al., 1979.)

do the perceptual speed and memory factors. Other family studies also generally indicate that the greatest familial similarity occurs for verbal ability (DeFries, Vandenberg, & McClearn, 1976). It is not known why one group consistently shows greater parent-offspring resemblance than the other. This study is a good reminder of the principle that the results of genetic research can differ in different populations.

Figure 9.2 also makes another important point: Tests within each factor show dramatic differences in familial resemblance. For example, one spatial test, *Paper Form Board*, shows high familiality. The test involves showing how to cut a figure to yield a certain pattern—for example, how to cut a circle to yield a triangle and three crescents. Another spatial test, *Elithorn Mazes*, shows the lowest familial resemblance. This test involves drawing one line that connects as many dots as possible in a maze of dots. Although these tests correlate with each other and contribute to a broad factor of spatial ability, much remains to be learned about the genetics of the processes involved in each test.

The results of dozens of twin studies of specific cognitive abilities are summarized in Table 9.1 (Nichols, 1978). When we double the difference between the correlations for identical and fraternal twins to estimate heritability (see Chapter 5), these results suggest that specific cognitive abilities show slightly less genetic influence than general cognitive ability. Memory and verbal fluency show lower heritability, about 30 percent; the other abilities yield heritabilities of 40 to 50 percent. Although the largest twin studies do not consistently find greater heritability for particular cognitive abilities (Bruun, Markkananen, & Partanen, 1966; Schoenfeldt, 1968), it has been suggested that verbal and spatial abilities in general show greater heritability than do perceptual speed and especially memory

TABLE 9.1

Average Twin Correlations for Tests of Specific Cognitive Abilities

		Twin Correlations	
Ability	Number of Studies	Identical Twins	Fraternal Twins
Verbal comprehension	27	.78	.59
Verbal fluency	12	.67	.52
Reasoning	16	.74	.50
Spatial visualization	31	.64	.41
Perceptual speed	15	.70	.47
Memory	16	.52	.36

SOURCE: *Nichols (1978).*

abilities (Plomin, 1988). Earlier twin studies of specific cognitive abilities have been reviewed in detail elsewhere (DeFries et al., 1976). A study of 160 pairs of twins aged 15 to 19 found similar results for tests of verbal and spatial abilities. This study is notable because the sample population was Croatian (Bratko, 1997), thus broadening the population base of observations on this topic.

Two studies of identical and fraternal twins reared apart provide additional support for genetic influence on specific cognitive abilities. One is a U.S. study of 72 reared-apart twin pairs of a wide age range in adulthood (McGue & Bouchard, 1989), and the other is a Swedish study of older twins (average age of 65 years), including 133 reared-apart twins and 142 control twin pairs reared together (Pedersen, Plomin, Nesselroade, & McClearn, 1992b). Both studies show significant heritability estimates for all four specific cognitive abilities. As shown in Table 9.2, the heritability estimates are generally higher than those implied by the twin results summarized in Table 9.1. This discrepancy may be due to the trend, discussed in Chapter 8, for heritability for cognitive abilities to increase during the life span. In both studies, the lowest heritability is found for memory.

As described in Chapter 8, twin studies of general cognitive ability appear to indicate influence of shared environment in the sense that twin resemblance cannot be explained entirely by heredity. However, it was noted that both identical and fraternal twins experience more similar environments than do nontwin siblings. For this reason, twin studies inflate estimates of shared environment in studies of general cognitive ability. Adoption designs generally suggest less shared environmental influence, especially after childhood. The twin correlations in Table 9.1 also imply substantial influence of shared environment for specific cognitive abilities. In contrast, the two studies of twins reared apart, which also included control samples of twins reared together, found that shared environment has little influence. Studies of adoptive relatives can provide a direct test of shared environment, but only two adoption studies of specific cognitive abilities have been reported.

TABLE 9.2

Heritability Estimates for Specific Cognitive Abilities in Two Studies of Twins Reared Apart

	Heritability Estimate (%)	
Ability	McGue & Bouchard (1989)	Pedersen et al. (1992b)
Verbal	57	58
Spatial	71	46
Speed	53	58
Memory	43	38

SOURCE: Kovas et al. (2007).

One adoption study found little resemblance for adoptive parents and their adopted children or for adoptive siblings on subtests of an intelligence test, except for vocabulary (Scarr & Weinberg, 1978a). Thus, this study supports the results of the two twins-reared-apart adoption studies in suggesting that shared environment has little influence on specific cognitive abilities. Like the twin and twins-reared-apart studies, this adoption study found evidence for genetic influence, in that nonadoptive relatives showed greater resemblance than did adoptive relatives.

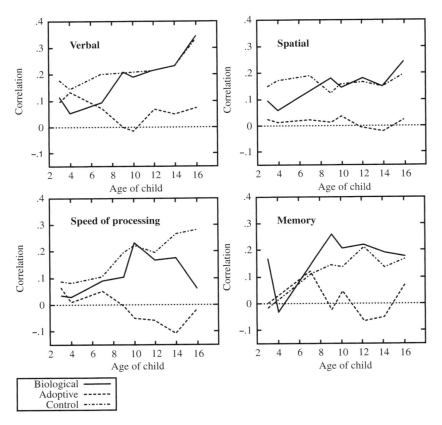

Figure 9.3 Parent-offspring correlations for factor scores for specific cognitive abilities for adoptive, biological, and control parents and their children at 3, 4, 7, 9, 10, 12, 14, and 16 years of age. Parent-offspring correlations are weighted averages for mothers and fathers. The N's range from 33 to 44 for biological fathers, 159 to 180 for biological mothers, 153 to 197 for adoptive parents, and 136 to 217 for control parents. (From "Nature, nurture and cognitive development from 1 to 16 years: A parent-offspring adoption study" by R. Plomin, D. W. Fulker, R. Corley, & J. C. DeFries. *Psychological Science, 8,* 442–447. © 1997. Used with permission of Psychological Science.)

Specific cognitive abilities are central to a 30-year longitudinal adoption study called the Colorado Adoption Project (Petrill et al., 2003). Figure 9.3 summarizes parent-offspring results for verbal, spatial, perceptual speed, and recognition memory abilities from early childhood through adolescence (Plomin et al., 1997b). Mother-child and father-child correlations were averaged for both adoptive and control (nonadoptive) families. For each ability, biological parent–adopted child and control parent–control child correlations tend to increase as a function of age. In contrast, adoptive parent–adopted child correlations do not differ substantially from zero at any age. These results indicate increasing heritability and no shared environment.

Developmental genetic analyses of Colorado Adoption Project data from adoptive and nonadoptive siblings indicate that genetically distinct specific cognitive abilities can be found as early as three years of age; they show increasing genetic differentiation from three to seven years of age (Cardon, 1994b). Like the findings for general cognitive ability (Chapter 8), new genetic effects are found at seven years, an observation hinting at a genetic transformation of cognitive abilities in the early school years (Cardon & Fulker, 1993). Shared environment shows little effect.

SUMMING UP

Family studies of specific cognitive abilities, most notably the Hawaii Family Study of Cognition, show greater familial resemblance for verbal and spatial abilities than for perceptual speed and memory. Tests within each ability vary in their degree of familial resemblance. Twin studies indicate that most of this familial resemblance is genetic in origin, as do studies of identical twins reared apart. Developmental analyses of adoption data indicate that heritability increases during childhood and that genetically distinct specific cognitive abilities can be found as early as three years of age. The results for family, twin, and adoption studies of verbal and spatial ability are summarized in Figure 9.4. These results converge on the conclusion that both verbal and spatial ability show substantial genetic influence but only modest influence of shared environment.

What about creativity? A review of ten twin studies of creativity yielded average twin correlations of .61 for identical twins and .50 for fraternal twins, results indicating only modest genetic influence and substantial influence of shared environment (Nichols, 1978). Some research implies that this modest genetic influence is entirely due to the overlap between tests of creativity and general cognitive ability. That is, when general cognitive ability is controlled, identical and fraternal twin correlations for tests of creativity are similar (Canter, 1973).

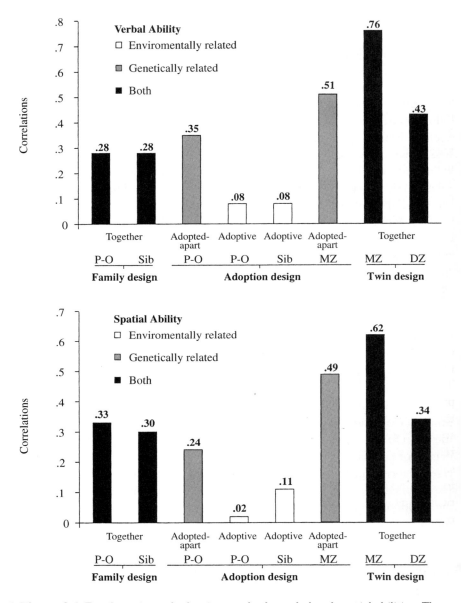

Figure 9.4 Family, twin, and adoption results for verbal and spatial abilities. The family study results are from the nearly 1000 Caucasian families in the Hawaii Family Study of Cognition, with parent-offspring correlations averaged for mothers and fathers rather than the regression of midchild on midparent shown in Figure 9.2 (De Fries et al., 1979). The adoption data are from the Colorado Adoption Project, with parent-offspring correlations shown when the adopted children were 16 years old and adoptive sibling correlations averaged across 9 to 12 years (Plomin et al., 1997b). The adopted-apart MZ twin data are averaged from the 95 pairs reported by T. Bouchard et al. (1990) and Pedersen et al. (1992b). The twin study correlations are based on more than 1500 pairs of wide age ranges in seven studies from four countries (Plomin, 1988). (From "Human behavioral genetics of cognitive abilities and disabilities" by R. Plomin & I. W. Craig (1997), *BioEssays, 19,* 1117–1124. Used with permission of BioEssays, ICSU Press.)

Information-Processing Measures

Future research on the genetics of specific cognitive abilities will capitalize on the laboratory tasks developed by experimental cognitive psychologists to assess how information is processed (Deary, 2000), an area called *mental chronometry* (Jensen, 2006). Twin studies using information-processing measures find some evidence for genetic influence. One early twin study focused on speed-of-processing measures, such as rapid naming of objects and letters (Ho, Baker, & Decker, 1988). These measures are similar to those used to assess the specific cognitive ability factor of perceptual speed. The results of this twin study yield evidence for moderate genetic influence. More traditional reaction-time measures of information processing also show genetic influence in twin studies (Finkel & McGue, 2007) and in a study of twins reared apart (McGue & Bouchard, 1989).

A study of 287 twin pairs aged 6 to 13 years (Petrill, Thompson, & Detterman, 1995) used a computerized battery of elementary cognitive tasks (Luo, Thompson, & Detterman, 2006) designed to test a theory that general cognitive ability is a complex system of independent elementary processes (Detterman, 1986). For example, a speed-of-processing factor was assessed by tasks such as decision time in stimulus discrimination. As shown in Figure 9.5, a probe stimulus is presented above an array of six stimuli, one of which matches the probe. The task is simply to touch as quickly as possible the stimulus that matches the probe. Information-processing tasks can subtract movement time from reaction time to obtain a purer measure of the time required to make the decision. In this study, a measure of decision time based on stimulus discrimination was highly reliable. Despite the simplicity of the task, it correlates −.42 with IQ. That is, shorter decision times are associated with higher IQ scores. Twin correlations for this measure of decision time were .61 for identical twins and .39 for fraternal twins, yielding a heritability of about 45 percent and about 15 percent influence of shared environment. The battery included other measures such as reaction

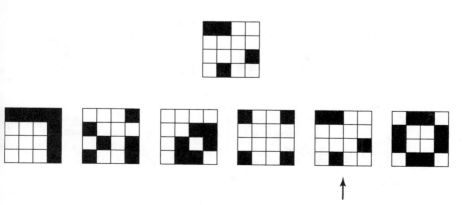

Figure 9.5 Stimulus discrimination display. Arrow indicates correct choice.

time, learning, and memory, most of which showed more modest heritability, ranging down to zero heritability for reaction time. Estimates of shared environment also varied widely for the various measures. Another computerized battery yielded similar results in a study of 278 adult twin pairs (Singer, MacGregor, Cherkas, & Spector, 2006).

In a study of 300 adult twin pairs, two classic elementary cognitive tasks were assessed: Sternberg's memory scanning and Posner's letter matching (Neubauer, Spinath, Riemann, Borkenau, & Angleitner, 2000). In the Sternberg measure, a random sequence of one, three, or five digits is presented. A target digit is shown, and the task is to indicate as quickly as possible whether the target digit was part of the previously shown set. Reaction time increases linearly from one to three to five digits and is assumed to index the added load for short-term memory. In the Posner task, pairs of letters are shown with the same physical and name identities (A-A), different physical but same name identity (A-a), or different physical and name identities (A-b). The task is to indicate whether the pairs of letters are exactly the same or different in some way. The difference in reaction times for name identity and physical identity is assumed to indicate the time needed for retrieval from long-term memory. These reaction time measures correlate about −.40 with IQ. MZ and DZ twin correlations for these five tasks are shown in Figure 9.6. An interesting result is that the more complex tasks such as the five-digit set of the Sternberg measure and the name identity task of the Posner measure showed heritabilities of about 50 percent. In contrast, the simpler tasks showed much lower heritabilities: 6 percent for the one-digit set of the Sternberg meas-

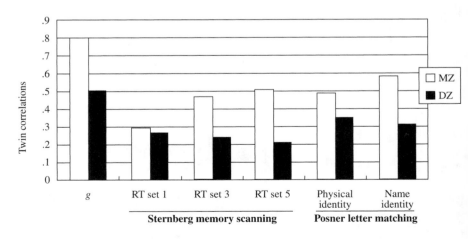

Figure 9.6 MZ and DZ correlations for two elementary cognitive tasks. See text for a description of the measures. (RT, reaction time.) The *g* measure was an unrotated principal component score derived from standard psychometric tests. (Adapted from Neubauer, Sange, & Pfurtscheller, 1999.)

ure and 28 percent for the physical identify task of the Posner measure. A meta-analysis of nine twin studies of reaction time measures supports the finding that heritability increases as complexity of the task increases (Beaujean, 2005).

Attempts to investigate even more basic processes have led to studies of speed of nerve conduction and brain wave measures of event-related potentials. Twin studies of speed of peripheral nerve conduction velocity show high heritability but little correlation with cognitive measures (Rijsdijk & Boomsma, 1997; Rijsdijk, Boomsma, & Vernon, 1995). Twin studies of event-related potentials yield widely varying heritability estimates across cortical sites, measurement conditions, and age, although much of this inconsistency could be due to the use of small samples (Hansell et al., 2005; van Baal, de Geus, & Boomsma, 1998). An EEG measure called *central coherence*, which assesses the connectivity between cortical regions, shows substantial heritability in childhood (van Baal et al., 1998) and adolescence (Van Beijsterveldt, Molenaar, de Geus, & Boomsma, 1998).

Multivariate genetic analysis of information-processing measures and their relationship to general and specific cognitive abilities is a special focus of research in this area, as discussed in the following section.

Multivariate Genetic Analysis: Levels of Processing

Genetic studies can go beyond the analysis of the variance of a single variable to consider genetic and environmental sources of covariance between traits. This technique is multivariate genetic analysis, mentioned in Chapter 5 and described in the Appendix. Multivariate genetic analysis yields a key statistic called the *genetic correlation* that indexes the extent to which genetic influences on one trait also affect another trait. A high genetic correlation implies that if a gene were associated with one trait, there is a good chance that this gene would also be associated with the other trait.

Multivariate genetic analyses of specific cognitive abilities (Figure 9.7) and their relationship to general cognitive ability have provided some important insights into the organization of cognitive abilities (Petrill, 1997). Although it is generally agreed that levels of abilities are related hierarchically, this conclusion is only a phenotypic description of the relationship among levels. In terms of genetics, three different models have been proposed, shown in simplified form in Figure 9.7. A bottom-up model (part c) assumes that different genes affect each basic element of information processing. These genetic influences on elementary processes feed into specific cognitive abilities, which in turn converge on general cognitive ability. The implication is that the influence of any gene found to be associated with higher levels of processing comes from that gene's association with a particular basic element of processing. In other words,

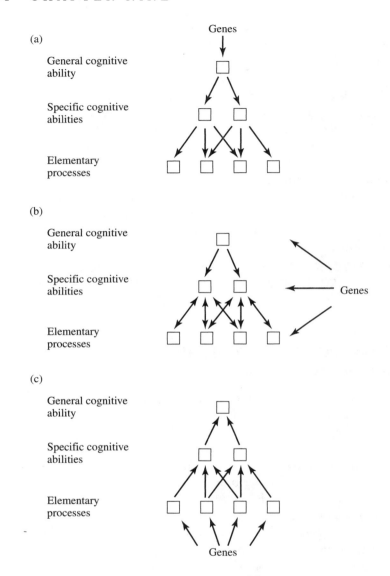

Figure 9.7 Genetic models of cognitive abilities: (a) top-down model; (b) levels-of-processing model; (c) bottom-up model.

when we control for genetic effects on basic elements, we find no additional genetic effects for higher levels of processing.

As reasonable as such a reductionistic model seems, it is possible that higher levels of processing involve new genetic effects not found at lower levels. The extreme version of this model is a top-down model (part a of Figure 9.7), which assumes that genes affecting cognitive abilities primarily affect general cognitive ability, perhaps as a result of some general mechanism such as neural speed. These genetic effects on general cognitive ability filter down to specific cognitive abilities and elementary processes. In the extreme, the top-down model implies that the influence of any gene associated with lower levels of processing comes from the association of that gene with general cognitive ability. That is, when we control for genetic effects on general cognitive ability, we find no additional genetic effects for lower levels of processing.

A compromise genetic model could be called a levels-of-processing model (part b). At each level of processing, there are unique genetic effects; but there are also genetic effects in common across levels of processing. In other words, as posited by the bottom-up model, there are genes specifically associated with each elementary process. In addition, as implied by the top-down model, there are genes associated with general cognitive ability that are not associated with lower levels of processing when general cognitive ability is controlled. In this sense, the levels-of-processing model suggests that both the bottom-up and the top-down models are correct. In addition, each level of processing has unique as well as common genetic effects. For example, some genetic effects will be found at the middle level of specific cognitive abilities that are not found at either the level of elementary processes or at the level of general cognitive ability. Chapter 15 focuses on related issues in the pathways between genes and behavior.

Multivariate genetic analyses have addressed the relationship between general cognitive ability and specific cognitive abilities and provide support for the levels-of-processing model (Alarcón, Plomin, Fulker, Corley, & DeFries, 1998; Cardon & Fulker, 1993; Casto, DeFries, & Fulker, 1995; Luo, Petrill, & Thompson, 1994; Pedersen, Plomin, & McClearn, 1994; Rijsdijk, Vernon, & Boomsma, 2002; Tambs, Sundet, & Magnus, 1986). Genes that affect one cognitive ability also affect other cognitive abilities, as assumed by the top-down model. However, there are some genetic effects unique to each cognitive ability, as hypothesized by the levels-of-processing model (Petrill, 2002; Plomin & Spinath, 2002).

The top-down model helps to explain an intriguing finding: Heritabilities of tests of specific cognitive abilities are strongly associated with the tests' correlations with general cognitive ability (Jensen, 1998). The higher the heritability of a test, the more that test correlates with general cognitive ability. For example, in the Swedish study of twins reared apart and twins reared together, tests of cognitive abilities differ in their heritability, and they differ in the extent to which they correlate with general cognitive ability. The correlation between the tests' heritabilities and their correlations with general cognitive ability was .77 after controlling for differential reliabilities of the tests (Pedersen et al., 1992b). The top-down model would predict this result in terms of the pervasive genetic effect of general cognitive ability.

Similar support is emerging for a levels-of-processing model for elementary cognitive processes and their genetic relationship with higher levels of processing (Luciano et al., 2004; Luciano et al., 2005; Wainwright, Wright, Luciano,

Geffen, & Martin, 2005). Although there are genetic effects in common across levels, unique genetic effects are found at each level. For example, in the Neubauer et al. twin study mentioned above, a general cognitive ability factor derived from the five elementary cognitive tasks listed in Figure 9.6 showed substantial genetic overlap with a general cognitive ability factor derived from psychometric tests, although some genetic variance was unique to the elementary tasks. The genetic correlation between reaction-time measures and *g* is very high, .90 in one review of a dozen studies (Jensen, 2006).

Although the levels-of-processing model best accounts for multivariate genetic results, what is most surprising is the extent to which genetic effects on cognitive abilities are general. These multivariate genetic results provide a challenging perspective on brain function in relation to learning and memory. According to the prevailing theory, the brain works in a modular fashion—cognitive processes are specific and independent. Implicit in this perspective is a bottom-up reductionistic view of genetics in which modules are the targets of

CLOSE UP

Juko Ando is a professor in the Department of Education, Faculty of Letters, at Keio University, Tokyo, Japan. He received his Ph.D. in education from Keio University in 1997. His Ph.D. dissertation was about behavioral genetic studies in educational settings using the co-twin control method. He established the Keio Twin Project (KTP) in 1998, with about 800 pairs of young adult twins in the Tokyo area, using a population-based twin registry of this area. KTP is a comprehensive behavioral genetic research project covering cognition, personality, mental health, neuropsychology, and molecular genetics. Ando collaborates with researchers in Australia, the Netherlands, Canada, Germany, Korea, and the United States. In 2003, he was a visiting scholar at the Institute for Behavioral Genetics at the University of Colorado, Boulder. In 2004, he started the Tokyo Twin Cohort Project (ToTCoP), a new large-scale longitudinal study of twins in infancy and childhood with over 1700 newborn twins as a part of the national project "Brain Science and Education." His current research interests are the genetic and environmental structures of cognition, personality, and sociability and their developmental processes. He has published many introductory books, chapters, and papers on behavioral genetics in Japan as well as scientific papers in international journals. Ando has served on the Membership Committee and is on the Board of Directors of the Behavior Genetics Association.

gene action. In contrast, the findings from multivariate genetic analyses are more compatible with a top-down view in which genetic effects operate primarily on general cognitive ability, rather than a bottom-up view in which genetic effects are specific to modules (Kovas & Plomin, 2006).

Holistic functioning of the brain seems reasonable given that the brain has evolved to learn from a variety of experiences and to solve a variety of problems. However, finding very high genetic correlations for psychometric tests typical of those used on intelligence tests does not prove that genetic effects are limited to a single general cognitive process. Another alternative is that specific cognitive abilities, as they are currently assessed with psychometric tests, might tap many of the same modular processes that are each affected by a different set of genes (Kovas & Plomin, 2006; Mackintosh, 1998). In other words, it is possible that different cognitive processes are independent but each is correlated with general cognitive ability. Such a finding would imply that general cognitive ability is not the result of some general process shared in common among these cognitive processes but that these independent processes all contribute to performance on tests of general cognitive ability. That is, general cognitive ability could be due to correlations among diverse cognitive processes or to independent component processes.

SUMMING UP

Information-processing measures also show genetic influence in the few available twin studies, especially for more complex tasks. Although information-processing research assumes a bottom-up model in which different genes affect basic elements of information processing, multivariate genetic analyses provide stronger support for a top-down model in which genes primarily affect general cognitive ability. Most research favors a compromise levels-of-processing model with unique genetic effects at each level of processing (bottom-up) but also genetic effects in common across levels of processing (top-down). The levels-of-processing model thus predicts that when genes associated with cognitive abilities are found, most of these genes will be associated with abilities throughout the hierarchy. Some genes, however, will be specific to certain abilities but not others. The considerable extent to which genetic effects are general across diverse cognitive abilities goes against the prevailing bottom-up model of cognitive neuroscience.

School Achievement

At first glance, tests of school achievement seem quite different from tests of specific cognitive abilities. School achievement tests focus on performance in specific domains such as grammar, history, and geometry. Moreover, the word *achievement* itself implies that such tests are due to dint of effort, assumed to be an environmental influence, in contrast to *ability*, for which genetic influence seems more reasonable. For the past half-century, environmental factors have been the focus of educational research on academic research, such as characteristics of schools, neighborhoods, and parents. Hardly any attention has been given to the possibility of genetic influences on the characteristics of children that affect learning in school (Plomin & Walker, 2003; Wooldridge, 1994). However, given the strong evidence for genetic influence on general cognitive ability, described in the previous chapter, and on specific cognitive abilities, described earlier in this chapter, it seems reasonable to expect that genetics plays a part in individual differences in learning in schools. Moreover, behavioral genetics can go beyond the rudimentary nature-nurture question to ask questions about "how" rather than "how much." For example, we can explore the genetic and environmental etiology of links between the normal (learning abilities) and abnormal (learning disabilities), links between ages, and links with general cognitive ability. Such questions about school achievement have been the target of much behavioral genetic research in the past decade. Reading and mathematics disabilities were discussed in Chapter 7, but the present discussion considers the normal range of individual differences in these and other aspects of school achievement.

The most well studied area by far is reading ability (Olson, 2007). As shown in Figure 9.8, a meta-analysis of a dozen twin studies indicates that reading-related processes such as word recognition, reading comprehension, and spelling show substantial genetic influence, with all heritability estimates within the narrow range of .54 to .63 (Harlaar, 2006). General reading based on composites of such tests yields a heritability estimate of .64. (See Figure 9.8.) Although it would

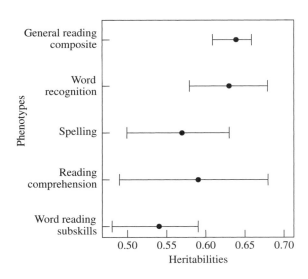

Figure 9.8 Meta-analysis of heritabilities of reading-related processes. The circles indicate the average heritability, and the lines around the circles indicate the 95 percent confidence intervals. (Adapted from Harlaar, 2006.)

TABLE 9.3

Twin Correlations for Report Card Grades for 13-Year-Olds

	Twin Correlation	
Subject Graded	Identical Twins	Fraternal Twins
History	.80	.51
Reading	.72	.57
Writing	.76	.50
Arithmetic	.81	.48

SOURCE: *Husén (1959).*

be reasonable to expect that learning to read (e.g., word recognition) might be less heritable than reading to learn (e.g., reading comprehension), reading in the early school years is also highly heritable (Harlaar et al., 2005b; Petrill et al., 2007). Even prereading skills such as phonological awareness, rapid naming, and verbal memory show substantial genetic influence (Samuelsson et al., 2007).

What about other academic subjects? One of the earliest studies used report card grades in a study of more than a thousand 13-year-old twins in Sweden (Husén, 1959). Twin correlations for history, reading, writing, and arithmetic (Table 9.3) suggest heritabilities of .58, .30, .52, and .66, respectively, and shared environment estimates of .22, .42, .24, and .15. Another early twin study of high school–age twins in the United States obtained data from the National Merit Scholarship Qualifying Test for 1300 identical and 864 fraternal twin pairs (Loehlin & Nichols, 1976). The twin correlations shown in Table 9.4 yield heritabilities of about .40 and shared environment

TABLE 9.4

Twin Correlations for School Achievement Test in High School

	Twin Correlation	
Test Subject	Identical Twins	Fraternal Twins
Social studies	.69	.52
Natural sciences	.64	.45
English usage	.72	.52
Mathematics	.71	.51

SOURCE: *Loehlin & Nichols (1976).*

TABLE 9.5

Twin Correlations for UK National Curriculum Ratings at 7, 9, and 10 Years

		Twin Correlation	
Subject		Identical Twins	Fraternal Twins
English	7 years	.82	.50
	9 years	.78	.46
	10 years	.80	.49
Math	7 years	.78	.47
	9 years	.76	.41
	10 years	.76	.48
Science	9 years	.76	.44
	10 years	.76	.57

SOURCE: *Kovas et al. (2007).*

estimates of about .30. Similar results in adolescence have been obtained in the Netherlands (Bartels, Rietveld, van Baal, & Boomsma, 2002b) and in Australia (Wainwright et al., 2005).

As with reading, early school achievement is also substantially heritable. In a longitudinal study of more than 2000 twin pairs in the UK, teachers assessed second-graders using criteria based on the UK National Curriculum for English, mathematics, and science at 7, 9, and 10 years (Kovas, Haworth, Dale, & Plomin, 2007). As shown in Table 9.5, twin correlations are remarkably consistent across subjects and across ages, suggesting heritabilities of about .60 and shared environment of only about .20, despite the fact that the twins grew up in the same family, attended the same school, and were often taught by the same teacher in the same classroom.

As mentioned earlier, behavioral genetics can go beyond the nature-nurture question of "how much." The first example concerns the genetic links between the normal (learning abilities) and abnormal (learning disabilities). This topic was addressed in relation to cognitive disability in Chapter 7, where DF extremes analysis was introduced (Box 7.1) and research using this method led to the conclusion that what we call abnormal may be part of the normal distribution. That is, mild cognitive disability is the low end of the same genetic and environmental influences responsible for variation in the normal distribution of general cognitive ability. In other words, mild cognitive disability is not really a disorder—it is the low end of the normal distribution. Similar results have been found for abilities and disabilities in reading, language, and mathematics

(Plomin & Kovas, 2005). DF extremes analyses of the data presented in Table 9.5 support the conclusion that the abnormal is normal across domains and ages (Kovas et al., 2007).

A second example involves developmental change and continuity in genetic and environmental influences. As discussed in Chapter 8 in relation to general cognitive ability, two types of developmental questions can be asked: Does heritability change during development? Do genetic factors contribute to developmental change? In the case of general cognitive ability, the answer to the first question is yes but for school achievement the answer appears to be no, as suggested, for example, by the results in Table 9.5. However, studies with a larger age range for school achievement measures would be necessary to answer this question more definitely. For general cognitive ability, the answer to the second question is that genetic factors largely contribute to continuity even from childhood to adulthood, although some evidence for genetic change exists, especially during the transition to school. Results appear to be similar for school achievement: Genetics appears to contribute largely to continuity with some evidence for genetic change (Bartels, Rietveld, van Baal, & Boomsma, 2002a; Byrne et al., 2007; Petrill et al., 2007), especially during the transition to school (Byrne et al., 2005). For example, longitudinal analyses of the data in Table 9.5 yielded age-to-age genetic correlations from 7 to 10 years of about .70 (Kovas et al., 2007).

A third example is multivariate genetic analysis among learning abilities and between learning abilities and general cognitive ability. Earlier in this chapter, multivariate genetic research on specific cognitive abilities was presented which suggested that most genetic effects are general. In other words, genes associated with verbal ability are also likely to be associated with spatial ability. A similar finding is emerging from multivariate genetic research on learning abilities: Genetic correlations are high between learning abilities. In a recent review of such studies, genetic correlations varied from .67 to 1.0 between reading and language (five studies), .47 to .98 between reading and mathematics (three studies), and .59 to .98 between language and mathematics (two studies) (Plomin & Kovas, 2005). The average genetic correlation across all of these studies was about .70. Genetic correlations among the measures shown in Table 9.5 are .79 on average (Kovas et al., 2007).

Could this general genetic factor that affects scores on diverse tests of school achievement be general cognitive ability? Multivariate genetic analyses between tests of school achievement and general cognitive ability suggest that genetic effects on school achievement test scores show moderate genetic correlations with general cognitive ability, but the genetic correlations are lower than they are among the school achievement measures. For example, a multivariate genetic analysis of several reading-related processes yielded an average genetic correlation of .51 with general cognitive ability, considerably lower than the average genetic correlation of .76 among the reading-related processes (Gayán

& Olson, 2003). A review of a dozen such studies reaches a similar conclusion (Plomin & Kovas, 2005), as do subsequent studies (Harlaar, Hayiou-Thomas, & Plomin, 2005; Kovas, Harlaar, Petrill, & Plomin, 2005; Kovas, Petrill, & Plomin, 2007; Wainwright et al., 2005). Also in line with this conclusion are multivariate genetic results based on the school achievement measures shown in Table 9.5. The average genetic correlation is .61 between the school achievement measures and general cognitive ability, in contrast to the average genetic correlation of .79 between the school achievement measures (Kovas et al., 2007). Model-fitting analyses of these data indicate that about a third of the genetic variance of academic performance is in common with general cognitive ability, about a third of the genetic variance is general to academic performance independent of general cognitive ability, and about a third is specific to each domain. In other words, although their genetic overlap is considerable, learning abilities are not the same thing genetically as general cognitive ability.

SUMMING UP

Although, in the field of education, school achievement has been assumed to be environmental in origin, twin studies consistently show substantial genetic influence, not just for reading but also for other subject areas such as mathematics and science. In each of these domains, the abnormal is normal; that is, learning disabilities are the low end of the same genetic and environmental influences responsible for variation in the normal distributions of learning abilities. Similar to general cognitive ability, genetic effects on learning abilities largely contribute to continuity during childhood, although some significant change is observed. Multivariate genetic analyses of diverse measures of school achievement indicate that a general genetic factor underlies these measures. Multivariate genetic analyses between tests of school achievement and general cognitive ability suggest that genetic influence on school achievement overlaps substantially with genetic influence on general cognitive ability, although some genetic influences are specific to achievement.

Identifying Genes

As mentioned in earlier chapters, research has begun to identify specific genes associated with cognitive disabilities such as dementia and reading disability (Chapter 7) and with general cognitive ability (Chapter 8). DF extremes results mentioned in the previous section suggest that QTLs associated with learning disabilities such as reading disability are also likely to be associated with learning abilities such as reading ability, that is, with normal variation throughout the distribution. This QTL hypothesis has yet to be tested in the most studied area of reading (Williams & O'Donovan, 2006).

The first QTL linkage analyses of specific cognitive abilities and school achievement in the normal range have begun to be reported. QTL linkage studies reporting weak linkages have recently been reported for a memory task (Singer, Falchi, MacGregor, Cherkas, & Spector, 2006), reading ability and spelling (Bates et al., 2007), and academic achievement (Wainwright et al., 2006). More research will be needed to determine the reliability of these reported linkages.

Genomewide association studies have also begun to be reported for specific cognitive abilities, including reading ability (Meaburn et al., 2005). Much attention was attracted by a genomewide association scan for various memory tasks that identified a significant association with a gene encoding the brain protein KIBRA, which is expressed in memory-related brain structures (Papassotiropoulos et al., 2006). As in the case of QTL linkage analysis, such results need further confirmation.

The intense molecular genetic research on cognitive disabilities such as dementia and reading disability seems likely to ignite an explosion of research that attempts to identify genes responsible for the heritability of specific cognitive abilities.

Summary

Many specific cognitive abilities show genetic influence in twin studies, although the magnitude of the genetic effect is generally lower than that for general cognitive ability. Family and twin studies suggest that the genetic contribution may be stronger for some cognitive abilities such as verbal and spatial than for other abilities, especially memory. Recent studies of twins reared apart confirm these findings. Shared environmental influence is modest.

Information-processing measures also show genetic influence in twin studies. Multivariate genetic analyses provide evidence for a top-down model in which genes primarily affect general cognitive ability, although each level of processing has unique as well as common genetic effects.

School achievement tests, and even report card grades, show substantial genetic influence, even in the early school years. Multivariate genetic research indicates that genetic influence on school achievement overlaps with genetic influence on general cognitive ability, although some genetic influences are specific to achievement.

Research is under way to identify associations between DNA markers and cognitive abilities as well as disabilities. The levels-of-processing model predicts that most genes associated with specific cognitive abilities and measures of school achievement will be associated with general cognitive ability, but some genes will be specific to each domain.

CHAPTER TEN

Schizophrenia

P sychopathology has been the most active area of behavioral genetic research in recent years, largely because of the social importance of mental illness. One out of two persons in the United States has some form of disorder during his or her lifetime, and one out of three persons suffered from a disorder within the last year (Kessler et al., 2005). The costs in terms of suffering to patients and their friends and relatives, as well as the economic costs, make psychopathology one of the most pressing problems today.

The genetics of psychopathology led the way toward the acceptance of genetic influence in psychology and psychiatry. The history of psychiatric genetics is described in Box 10.1.

This chapter and the next two provide an overview of what is known about the genetics of several major categories of psychopathology: schizophrenia, mood disorders, and anxiety disorders. Other disorders such as posttraumatic stress disorder, somatoform disorders, and eating disorders are also briefly reviewed, as are disorders usually first diagnosed in childhood: autism, attention-deficit hyperactivity, and tic disorders. Other major categories in the *Diagnostic and Statistical Manual of Mental Disorders-IV* (DSM-IV) include cognitive disorders such as dementia (Chapter 7), personality disorders (Chapter 13), and drug-related disorders (Chapter 14). The DSM-IV includes several other disorders for which no genetic research is as yet available (e.g., dissociative disorders such as amnesia and fugue states). Much has been written about the genetics of psychopathology, including recent texts (Faraone, Tsuang, & Tsuang, 2002; Jang, 2005; Kendler & Prescott, 2006) and several edited books (e.g., Andreasen, 2005; Kendler & Eaves, 2005; McGuffin, Owen, & Gottesman, 2002). Many questions remain concerning diagnosis, most notably, the extent of comorbidity and heterogeneity. Diagnoses to date depend on symptoms, and it is possible that the same symptoms have different causes, and that different symptoms could have the same causes. One of the hopes for genetic research is that it can begin to provide diagnoses based on causes rather than symptoms. We will return to this issue in Chapter 11.

This chapter focuses on schizophrenia, the most highly studied area of behavioral genetic research on psychopathology. Schizophrenia involves persistent abnormal beliefs (delusions), hallucinations (especially hearing voices), disorganized speech (odd associations and rapid changes of subject), grossly disorganized behavior, and so-called negative symptoms, such as flat affect (lack of emotional response) and avolition (lack of motivation). Although it derives from Greek words meaning "split mind," the notion of a "split personality" has nothing to do with schizophrenia. A diagnosis of schizophrenia requires that such symptoms occur for at least six months. It usually strikes in late adolescence or early adulthood. Early onset in adolescence tends to be gradual but has a worse prognosis.

More genetic research has focused on schizophrenia than on other areas of psychopathology for three reasons. It is the most severe form of psychopathology and one of the most debilitating of all disorders (Ustun et al., 1999). Second, it is so common, with a lifetime risk usually given as about 1 percent of the population, although a recent meta-analysis suggests a somewhat lower risk (Goldner, Hsu, Waraich, & Somers, 2002). Third, schizophrenia generally lasts a lifetime, although a few people recover, especially if they have had just one episode (Robinson, Woerner, McMeniman, Mendelowitz, & Bilder, 2004); there are signs that recovery rates are improving (Bellack, 2006). Unlike patients of two decades ago, most of these individuals are no longer institutionalized, because drugs can control some of their worst symptoms. Nonetheless, schizophrenics still occupy half the beds in mental hospitals, and those discharged make up about 10 percent of the homeless population (Fischer & Breakey, 1991). It has been estimated that the cost to our society of schizophrenia alone is greater than that of cancer (National Foundation for Brain Research, 1992).

Family Studies

The basic genetic results for schizophrenia were described in Chapter 3 (see Figure 3.6) to illustrate genetic influence on complex disorders (see Gottesman, 1991, for details). Forty family studies consistently show that schizophrenia is familial. A handful of studies that do not show familial resemblance are small studies, lacking the power to detect a resemblance (Kendler, 1988). In contrast to the base rate of 1 percent lifetime risk in the population, the risk for relatives increases with genetic relatedness to the schizophrenic proband: 4 percent for second-degree relatives and 9 percent for first-degree relatives.

The average risk of 9 percent for first-degree relatives differs for parents, siblings, and offspring of schizophrenics. In 14 family studies of over 8000 schizophrenics, the median risk was 6 percent for parents, 9 percent for siblings, and

BOX 10.1

The Beginnings of Psychiatric Genetics: Bethlem Royal and Maudsley Hospitals

Founded in London, in 1247, Bethlem Hospital is one of the oldest institutions in the world caring for people with mental disorders. In 1948, amalgamation with its younger sister, Maudsley Hospital, and with London University's Institute of Psychiatry confirmed its importance as a center for research and training in psychiatry and the newly emerging profession of clinical psychology. However, there have been times in Bethlem's long history when it was associated with some of the worst images of mental illness, and it gave us the origin of the word *bedlam*. Perhaps the most famous portrayal is in the final scene of Hogarth's series of paintings, *A Rake's Progress*, which shows the Rake's decline into madness at Bethlem (see figure). Hogarth's portrayal assumes that madness

A Rake's Progress.
(William Hogarth, *A Rake's Progress*, 1735. Plate 8. The British Museum.)

is the consequence of high living and therefore, it is implied, a wholly environmental affliction.

The observation that mental disorders have a tendency to run in families is ancient, but among the first efforts to record this association systematically were those at Bethlem Hospital. Records from the 1820s show that one of the routine questions that doctors had to attempt to answer about the illness of a patient they were admitting was "whether hereditary?" This, of course, predated the development of genetics as a science, and it was not until 100 years later that the first research group on psychiatric genetics was established in Munich, Germany, under the leadership of Emil Kraepelin. The

Eliot Slater

Munich department attracted many visitors and scholars, including a mathematically gifted young psychiatrist from Maudsley Hospital, Eliot Slater, who obtained a fellowship to study psychiatric genetics there. In 1935, Slater returned to London and started his own research group, which led to the creation in 1959 of the Medical Research Council's (MRC) Psychiatric Genetics Unit. The unit was housed in a no-frills, austere, prefabricated building, affectionately known to those who worked there as "the Hut." The Bethlem and Maudsley Twin Register, set up by Slater in 1948, was among the important resources that underpinned a number of influential studies, and he introduced sophisticated statistical approaches to data evaluation. The Hut became one of the key centers for training and played a major role in the career development of many overseas postdoctoral students, including Irving Gottesman, Leonard Heston, and Ming Tsuang.

In 1971, Slater published the first psychiatric genetics textbook in English, the *Genetics of Mental Disorders*, with Valerie Cowie, the MRC Unit deputy director. Later in the 1970s, following Slater's retirement, psychiatric genetics became temporarily unfashionable in the United Kingdom, but was continued as a scientific discipline in North America and mainland Europe by researchers trained by Slater or influenced by his work. Although the currently flourishing field of psychiatric genetics is dominated by new molecular and statistical technologies that were unforeseen in Slater's day, a debt is surely owed to him and his followers for the foundations that they laid.

13 percent for offspring. The low risk for parents of schizophrenics (6 percent) is probably due to the fact that schizophrenics are less likely to marry and those who do marry have relatively few children. For this reason, parents of schizophrenics are less likely than expected to be schizophrenic. When schizophrenics do become parents, the rate of schizophrenia in their offspring is high (13 percent). The risk is the same regardless of whether the mother or the father is schizophrenic. When both parents are schizophrenic, the risk for their offspring shoots up to 46 percent. Siblings provide the least biased risk estimate, and their risk (9 percent) is in between the estimates for parents and for offspring. Although the risk of 9 percent is high, nine times the population risk of 1 percent, it should be remembered that the majority of schizophrenics do not have a schizophrenic first-degree relative.

The family design provides the basis for genetic high-risk studies of the development of children whose mothers were schizophrenic. In one of the first such studies, begun in the early 1960s in Denmark, 200 such offspring were followed until their forties (Parnas et al., 1993). In the high-risk group whose mothers were schizophrenic, 16 percent were diagnosed as schizophrenic (whereas 2 percent in the low-risk group were schizophrenic), and the children who eventually became schizophrenic had mothers whose schizophrenia was more severe. These children experienced a less stable home life and more institutionalization, reminding us that family studies do not disentangle nature and nurture in the way an adoption study does. The children who became schizophrenic were more likely to have had birth complications, particularly prenatal viral infection (Cannon et al., 1993). They also showed attention problems in childhood, especially problems in "tuning out" incidental stimuli like the ticking of a clock (Hollister, Mednick, Brennan, & Cannon, 1994). Similar results were found in childhood in one of the best U.S. genetic high-risk studies, which also found more personality disorders in the offspring of schizophrenic parents when the offspring were young adults (Erlenmeyer-Kimling et al., 1995).

Twin Studies

Twin studies show that genetics contributes importantly to familial resemblance for schizophrenia. As shown in Figure 3.6, the probandwise concordance for MZ twins is 48 percent and the concordance for DZ twins is 17 percent. In a meta-analysis of 14 twin studies of schizophrenia using a liability-threshold model (see Chapter 5), these concordances suggest a heritability of liability of about 80 percent (Sullivan, Kendler, & Neale, 2003). The five newest studies from Europe and Japan have confirmed earlier findings, yielding probandwise concordances of 41–65 percent in MZ and 0–28 percent in DZ pairs (Cardno & Gottesman, 2000).

A dramatic case study involved identical quadruplets, called the Genain quadruplets, all of whom were schizophrenic, although they varied considerably

Irving Gottesman earned his Ph.D., courtesy of the G.I. Bill and the Korean War, in 1960 in child and adult clinical psychology from the University of Minnesota. He was mentored by the geneticist Sheldon C. Reed of the Dight Institute of Human Genetics, producing a dissertation on an adolescent twin study of MMPI personality traits and intelligence, within the P. E. Meehl school of thought. Now the Sherrell J. Aston Professor of Psychology Emeritus (University of Virginia) and the Bernstein Professor in Adult Psychiatry at the University of Minnesota, Gottesman in the early 1960s joined forces with James Shields as a postdoctoral fellow in Eliot Slater's MRC Psychiatric Genetics Unit in London. This collaboration led to their 1967 *PNAS* paper proposing a multifactorial polygenic threshold model for schizophrenia and 1972 monograph on schizophrenia and genetics, a twin study based on 16 years of consecutive admissions to the inpatient and outpatient units of the Maudsley and Royal Bethlem hospitals. With Danish colleagues, he has reported high risks for schizophrenia in the offspring of normal MZ co-twins of probands, and he has found appreciable heritability for criminality. He introduced the concepts of reaction range, epigenetics, and endophenotype to our field.

in severity of the disorder (DeLisi et al., 1984) (Figure 10.1). For 14 pairs of reared-apart identical twins in which at least one member of each pair became schizophrenic, 9 pairs (64 percent) were concordant (Gottesman, 1991).

Despite the strong and consistent evidence for genetic influence provided by the twin studies, it should be remembered that the average concordance for identical twins is only about 50 percent. In other words, half of the time these genetically identical pairs of individuals are discordant for schizophrenia, an outcome that provides strong evidence for the importance of nongenetic factors, despite the heritability of the hypothetical construct of liability being 80 percent.

Because differences within pairs of identical twins cannot be genetic in origin, the co-twin control method can be used to study nongenetic reasons why one identical twin is schizophrenic and the other is not. One early study of discordant identical twins found few life history differences except that the schizophrenic co-twins were more likely to have had birth complications and some neurological abnormalities (Mosher, Pollin, & Stabenau, 1971). Follow-up studies also found differences in brain structures and more frequent birth complications for the

Figure 10.1 Identical quadruplets (known under the fictitious surname Genain), each of whom developed symptoms of schizophrenia between the ages of 22 and 24 years. (Courtesy of Miss Edna Morlok.)

schizophrenic co-twin in discordant identical twin pairs (Torrey, Bowler, Taylor, & Gottesman, 1994).

An interesting finding has emerged from another use of discordant twins: studying their offspring or other first-degree relatives. Discordant identical twins provide direct proof of nongenetic influences, because the twins are identical genetically yet discordant for schizophrenia. Even though one twin in discordant pairs is spared from schizophrenia for environmental reasons, that twin still carries the same high genetic risk as the twin who is schizophrenic. That is why nearly all studies find rates of schizophrenia as high in the families of discordant as in concordant pairs (Gottesman & Bertelsen, 1989; McGuffin, Farmer, & Gottesman, 1987).

For the offspring of discordant fraternal twins, the offspring of the twin who was schizophrenic are at much greater risk than are the offspring of the non-schizophrenic twin. Members of discordant fraternal twin pairs, unlike identical twins, differ genetically as well as environmentally. However, sample sizes are small, and one such small study did not support earlier conclusions (Kringlen &

Cramer, 1989; see also, Torrey, 1990). Nonetheless, these data provide food for thought about the complex interactions between nature and nurture.

Adoption Studies

Results of adoption studies agree with those of family and twin studies in pointing to genetic influence in schizophrenia. As described in Chapter 5, the first adoption study of schizophrenia by Leonard Heston in 1966 is a classic study. The results (see Box 5.1) showed that the risk of schizophrenia in adopted-away offspring of schizophrenic biological mothers was 11 percent (5 of 47), much greater than the 0 percent risk for 50 adoptees whose birth parents had no known mental illness. The risk of 11 percent is similar to the risk for offspring reared by their schizophrenic biological parents. This finding not only indicates that family resemblance for schizophrenia is largely genetic in origin, but it also implies that growing up in a family with schizophrenics does not increase the risk for schizophrenia beyond the risk due to heredity.

Box 5.1 also mentioned that Heston's results have been confirmed and extended by other adoption studies. Two Danish studies began in the 1960s with 5500 children adopted between 1924 and 1947, and 10,000 of their 11,000 biological parents. One of the studies (Rosenthal et al., 1968; Rosenthal, Wender, Kety, & Schulsinger, 1971) used the adoptees' study method. This method is the same as that used in Heston's study, but important experimental controls were added. Because biological parents typically relinquish children for adoption when the parents are in their teenage years, but schizophrenia does not usually occur until later, the adoption agencies and the adoptive parents were not aware of the diagnosis in most cases. In addition, both schizophrenic fathers and mothers were studied to assess whether Heston's results, which involved only mothers, were influenced by prenatal maternal factors.

The first Danish study began by identifying biological parents who had been admitted to a psychiatric hospital. Biological mothers or fathers who were diagnosed as schizophrenic and whose children had been placed in adoptive homes were selected. This procedure yielded 44 birth parents (32 mothers and 12 fathers) who were diagnosed as chronic schizophrenics. Their 44 adopted-away children were matched to 67 control adoptees whose birth parents had no psychiatric history, as indicated by the records of psychiatric hospitals. The adoptees, with an average age of 33 years, were interviewed for three to five hours by an interviewer blind to the status of their birth parents.

Three (7 percent) of the 44 proband adoptees were chronic schizophrenics, whereas none of the 67 control adoptees were (Figure 10.2). Moreover, 27 percent of the probands showed schizophrenic-like symptoms, whereas 18 percent of the controls had similar symptoms. Results were similar for 69 proband adoptees whose parents were selected by using broader criteria for schizophrenia. Results were also similar, regardless of whether the mother or the father was

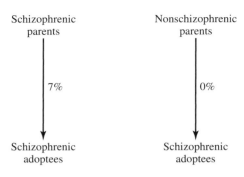

Schizophrenic
parents

7%

Schizophrenic
adoptees

Nonschizophrenic
parents

0%

Schizophrenic
adoptees

Figure 10.2 Danish adoption study of schizophrenia: adoptees' study method.

schizophrenic. The unusually high rates of psychopathology in the Danish control adoptees may have occurred because the study relied on hospital records to assess psychiatric status of the birth parents. For this reason, the study may have overlooked psychiatric problems of control parents that had not come to the attention of psychiatric hospitals. To follow up this possibility, the researchers interviewed the birth parents of the control adoptees and found that one-third fell in the schizophrenic spectrum. Thus, the researchers concluded that "our controls are a poor control group and our technique of selection has minimized the differences between the control and index groups" (Wender, Rosenthal, Kety, Schulsinger, & Weiner, 1974, p. 127). This bias is conservative in terms of demonstrating genetic influence.

An adoptees study in Finland confirmed these results (Tienari et al., 2004). About 10 percent of adoptees who had a schizophrenic biological parent showed some form of psychosis, whereas 1 percent of control adoptees had similar disorders. This study also suggested genotype-environment interaction, because adoptees whose biological parents were schizophrenic were more likely to have schizophrenia-related disorders when the adoptive families functioned poorly.

The second Danish study (Kety et al., 1994) used the adoptees' family method, focusing on 47 of the 5500 adoptees diagnosed as chronic schizophrenic. A matched control group of 47 nonschizophrenic adoptees was also selected. The biological and adoptive parents and siblings of the index and control adoptees were interviewed. For the schizophrenic adoptees, the rate of chronic schizophrenia was 5 percent (14 of 279) for their first-degree biological relatives and 0 percent (1 of 234) for the biological relatives of the control adoptees. The adoptees' family method also provides a direct test of the influence of the environmental effect of having a schizophrenic relative. If familial resemblance for schizophrenia were caused by family environment brought about by schizophrenic parents, schizophrenic adoptees should be more likely to come from adoptive families with schizophrenia, relative to the control adoptees. To the contrary, 0 percent (0 of 111) of the schizophrenic adoptees had adoptive parents or siblings who were schizophrenic—like the 0 percent incidence (0 of 117) for the control adoptees (Figure 10.3).

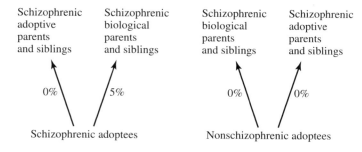

Figure 10.3 Danish adoption study of schizophrenia: adoptees' family method.

This study also included many biological half siblings of the adoptees (Kety, 1987). Such a situation arises when biological parents relinquish a child for adoption and then later have another child with a different partner. The comparison of biological half siblings who have the same father (paternal half siblings) with those who have the same mother (maternal half siblings) is particularly useful for examining the possibility that the results of adoption studies may be affected by prenatal factors, rather than by heredity. Data on paternal half siblings are less likely to be influenced by prenatal factors because they were born to different mothers. For half siblings of schizophrenic adoptees, 16 percent (16 of 101) were schizophrenic; for half siblings of control adoptees, only 3 percent (3 of 104) were schizophrenic. The results were the same for maternal and paternal half siblings, an outcome suggesting that prenatal factors are not likely to be of major importance in the origin of schizophrenia, although this does not rule out some prenatal factors such as mother's exposure to flu virus during pregnancy.

In summary, the adoption studies clearly point to genetic influence. Moreover, adoptive relatives of schizophrenic probands do not show increased risk for schizophrenia. These results imply that familial resemblance for schizophrenia is due to heredity rather than to shared family environment.

Schizophrenia or Schizophrenias?

Is schizophrenia one disorder or is it a heterogeneous collection of disorders? When the disorder was named in 1908, it was called "the schizophrenias." Multivariate genetic analysis can address this fundamental issue of heterogeneity. The classic subtypes of schizophrenia—such as catatonic (disturbance in motor behavior), paranoid (persecution delusions), and disorganized (both thought disorder and flat affect are present)—are not supported by genetic research. That is, although schizophrenia runs in families, the particular subtype does not. This result is seen most dramatically in a follow-up of the Genain quadruplets (DeLisi et al., 1984). Although they were all diagnosed as schizophrenic, their symptoms varied considerably.

There is evidence that more severe schizophrenia is more heritable than milder forms (Gottesman, 1991). Furthermore, the evidence from both early studies and more recent work, using multivariate statistical methods such as cluster analysis, suggests that the classic "disorganized" subtype of schizophrenia, even if it does not "breed true," shows an especially high rate of affected family members (Cardno et al., 1998; Farmer, McGuffin, & Gottesman, 1987). An alternative to the classic subtypes is a distinction largely based on severity (Crow, 1985). Type I schizophrenia, which has a better prognosis and a better response to drugs, involves active symptoms such as hallucinations. Type II schizophrenia, which is more severe, has a poorer prognosis and passive symptoms such as withdrawal and lack of emotion. Type II schizophrenia appears to be more heritable than type I (Dworkin & Lenzenweger, 1984).

Another approach to the problem of heterogeneity divides schizophrenia on the basis of family history (Murray, Lewis, & Reveley, 1985), although there are problems with this approach (Eaves, Kendler, & Schulz, 1986) and there is

CLOSE UP

Kenneth Kendler is Banks Distinguished Professor of Psychiatry and director (with Dr. Lindon Eaves) of the Virginia Institute for Psychiatric and Behavioral Genetics (VIPBG) at Virginia Commonwealth University (VCU). The VIPBG conducts multidisciplinary research aiming to clarify the role of genetic and environmental risk factors in psychiatric and substance abuse disorders. Kendler received a B.A. in 1972 from the University of California at Santa Cruz and an M.D. from Stanford University School of Medicine in 1977. He completed a residency in psychiatry at Yale University in 1980 and went to VCU in 1983. His interest in psychiatric genetics was first kindled by searching for empirical methods to test diagnostic problems in psychiatry. Kendler's research employs methods from both genetic epidemiology and molecular genetics and focuses on understanding the roles of genetic and environmental risk factors in the etiology of psychiatric and substance abuse disorders. Recent studies include large-scale family, linkage and association studies of schizophrenia and alcoholism in Ireland, population-based twin studies of major depression, anxiety disorders, alcoholism, and substance abuse. His Virginia twin studies were recently summarized in a book with Carol Prescott entitled *Genes, Environment, and Psychopathology: Understanding the Causes of Psychiatric and Substance Use Disorder.*

clearly no simple dichotomy (Jones & Murray, 1991). These typologies seem more likely to represent a continuum from less to more severe forms of the same disorder, rather than genetically distinct disorders (McGuffin et al., 1987).

A related strategy is the search for behavioral or biological markers of genetic liability, called endophenotypes (Gottesman & Gould, 2003), discussed in Chapter 15. An example of a behavioral endophenotype in genetic research on schizophrenia is called smooth-pursuit eye tracking. This term refers to the ability to follow a moving object smoothly with one's eyes without moving the head (Levy, Holzman, Matthysse, & Mendell, 1993). Some studies have shown that schizophrenics, whose eye tracking is jerky, tend to have more negative symptoms, and their relatives with poor eye tracking are more likely to show schizophrenic-like behaviors (Clementz, McDowell, & Zisook, 1994). However, some research does not support this hypothesis (Torrey et al., 1994). Many other cognitive (Sitskoorn, Aleman, Ebisch, Appels, & Kahn, 2004) and brain (Harrison & Weinberger, 2005) endophenotypes for schizophrenia have been investigated. The hope is that such endophenotypes will clarify the inheritance of schizophrenia and assist attempts to find specific genes responsible for schizophrenia.

Although some researchers assume that schizophrenia is heterogeneous and needs to be split into subtypes, others argue in favor of the opposite approach, lumping schizophrenia-like disorders in a broader spectrum of schizoid disorders (Farmer et al., 1987; McGue & Gottesman, 1989). It is possible that schizophrenia represents the extreme of a quantitative dimension that extends into normality.

Ultimately, these crucial issues about splitting and lumping may be resolved by molecular genetics. When genes that are associated with schizophrenia are found, the question is whether they will relate to a particular type of schizophrenia, as assumed by the "splitters." Or, at the other extreme, will these genes for schizophrenia relate to a continuum of thought disorders that extends into normal behavior, such as social withdrawal, attention problems, and magical thinking?

Identifying Genes

Before the new DNA markers were available, attempts were made to associate classic genetic markers, such as blood groups, with schizophrenia. For example, a weak association with schizophrenia marked by paranoid delusions was suggested by several early studies, with the major genes encoding human leukocyte antigens (HLAs) of the immune response, a gene cluster associated with many diseases (McGuffin & Sturt, 1986).

Although schizophrenia was one of the first behavioral domains put under the spotlight of molecular genetic analysis, it has been slow to reveal evidence for specific genes. During the euphoria of the 1980s when the new DNA markers were first being used to find genes for complex traits, some claims were made for linkage, but they could not be replicated. The first was a claim for linkage with an autosomal dominant gene on chromosome 5 for Icelandic and British families (Sherrington et al., 1988). However, combined data from five other studies in other countries failed to confirm the linkage (McGuffin et al., 1990).

More than 20 genomewide linkage scans (with more than 350 genetic markers) have been published, but none has suggested a gene of major effect for schizophrenia (Riley & Kendler, 2006). Hundreds of reports of linkage for schizophrenia in the 1990s led to a confusing picture, because few reports were replicated. However, greater clarity has emerged since 2000. For example, a meta-analysis of 20 genomewide linkage scans of schizophrenia in diverse populations indicated greater consistency of linkage results than previously recognized (Lewis et al., 2003). Significant linkage was found on the long arm of chromosome 2 ($2q$); linkage was suggested for ten other regions, including $6p$ and $8p$. It is now generally accepted, as we have seen in previous chapters on cognitive disabilities and abilities, that genetic influence on schizophrenia is caused by multiple genes of small effect, none of which is either necessary or sufficient to cause the disorder. It has been difficult detecting linkage signals because linkage analysis requires very large samples to detect small effects.

Nonetheless, there has been progress toward identifying at least two genes responsible for some of the heritability of schizophrenia (Owen, Craddock, & O'Donovan, 2005). Although the strongest linkage region on $2q$ is only now being followed up, fine mapping of one of the suggestive linkage regions ($8p22$-$p11$) led to a gene called *neuregulin 1* (Stefansson et al., 2002). Meta-analyses of *neuregulin 1* yield evidence for significant but small associations with different polymorphisms in Caucasian and Asian populations (Li, Collier, & He, 2006). *Neuregulin* is a plausible candidate gene because of its multifaceted role in the development of the nervous system (Harrison & Law, 2006). Fine mapping of another linkage region ($6p24$-21) pointed to a candidate gene called *dysbindin* at $6p22.3$ (Straub et al., 2002). Like *neuregulin*, *dysbindin* shows significant but small associations with schizophrenia (Norton, Williams, & Owen, 2006) although not as consistent as *neuregulin* (Mutsuddi et al., 2006). Although its function is not clear, *dysbindin* is expressed throughout the brain and expression is diminished in schizophrenics. Other candidate genes are also being pursued but accumulating evidence from large collaborative studies is less convincing (e.g., Jonsson, Kaiser, Brockmoller, Nimgaonkar, & Crocq, 2004; Talkowski et al., 2006).

The next exciting development will be the results soon to emerge from ongoing genomewide association scans using very large samples of schizophrenics and controls. As discussed in previous chapters, such genomewide association studies may identify genes of even smaller effect size than those identified by linkage studies.

Summary

Psychopathology is the most active area of research in behavioral genetics. For schizophrenia, lifetime risk is about 1 percent in the general population, 10 percent in first-degree relatives whether reared together or adopted apart, 17 percent for fraternal twins, and 48 percent for identical twins. This pattern of results indicates substantial genetic influence as well as nonshared family environmental influence. Genetic high-risk studies and co-twin control studies suggest that, within genetic high-risk groups, birth complications and attention problems in childhood are weak predictors of schizophrenia, which usually strikes in early adulthood. Genetic influence has been found for both the adoptees' study method, like that used in the first adoption study by Heston, and the adoptees' family method. More severe schizophrenia may be more heritable than less severe forms.

Recent meta-analyses of the many linkage studies of schizophrenia have begun to yield consistent results and have led to the identification of two genes (*neuregulin* and *dysbindin*) that have significant but small associations with schizophrenia. Overall, it seems likely that genetic liability to schizophrenia results from multiple genes of small effect.

CHAPTER ELEVEN

Other Adult Psychopathology

Although schizophrenia has been the most highly studied disorder in behavioral genetics, in recent years the spotlight has turned to mood disorders. In this chapter, we provide an overview of genetic research on mood disorders as well as other adult psychopathology. The chapter ends with a discussion of the extent to which genes that affect one disorder also affect other disorders.

Mood Disorders

Mood disorders involve severe swings in mood, not just the "blues" that all people feel on occasion. For example, the lifetime risk for suicide for people diagnosed as having mood disorders has been estimated as 19 percent (Goodwin & Jamison, 1990). There are two major categories of mood disorders: major depressive disorder, consisting of episodes of depression, and bipolar disorder, in which there are episodes of both depression and mania.

Major depressive disorder usually has a slow onset over weeks or even months. Each episode typically lasts several months and ends gradually. Characteristic features include depressed mood, loss of interest in usual activities, disturbance of appetite and sleep, loss of energy, and thoughts of death or suicide. Major depressive disorder affects an astounding number of people. In a U.S. survey, the lifetime risk is about 16 percent, with about half of these in a severe or very severe category; risk is two times greater for women than for men after adolescence (Kessler et al., 2005). Moreover, the problem is getting worse: Each successive generation born since World War II has higher rates of depression (Burke, Burke, Roe, & Regier, 1991). These temporal trends could possibly be due to changes in environmental influences, diagnostic criteria, or clinical referral rates. Major depressive disorder is sometimes called unipolar depression because it involves only depression. In contrast, bipolar disorder, also known as manic-depressive illness, is a disorder in which the mood of the affected individual alternates between the depressive pole and the other pole of mood, called mania. Mania involves euphoria, inflated self-esteem, sleeplessness, talkativeness, racing thoughts, distractibility, hyperactivity, and reckless behavior. Mania typically begins and ends suddenly, and lasts from several days to several months. Mania is sometimes difficult to diagnose and, for this reason, DSM-IV (the American Psychiatric Association's *Diagnostic and Statistical Manual of Mental Disorders-IV*) has distinguished bipolar I disorder, with a clear manic episode, from bipolar II disorder, with a less clearly defined manic episode. Bipolar disorder is much less common than major depression, with an incidence of about 3 percent of the adult population and no gender difference (Kessler et al., 2005), although this estimate is based on a broader concept than has traditionally been applied.

Family Studies

For 70 years, family studies have shown increased risk for first-degree relatives of individuals with mood disorders (Slater & Cowie, 1971). Since the 1960s, researchers have considered major depression and bipolar depression separately. In seven family studies of major depression, the family risk was 9 percent on average, whereas risk in control samples was about 3 percent (McGuffin & Katz, 1986). Age-corrected morbidity risk estimates that take into account lifetime risk (see Chapter 3) are about twice as high (Sullivan, Neale, & Kendler, 2000). A review of 18 family studies of bipolar I and II disorder yielded an average risk of 9 percent, as compared to less than 1 percent in control individuals (Smoller & Finn, 2003). (See Figure 11.1.) The risks in these studies are low relative to the frequency of the disorder mentioned earlier, because these studies focused on severe depression, often requiring hospitalization.

It has been hypothesized that the distinction between unipolar major depression and bipolar depression is primarily a matter of severity; bipolar depression may be a more severe form of mood disorder (McGuffin & Katz, 1986). The basic multivariate finding from family studies is that relatives of unipolar probands are not at increased risk for bipolar depression (less than 1 percent), but relatives of bipolar probands are at increased risk (14 percent) for unipolar depression (Smoller & Finn, 2003). If we postulate that bipolar depression is a more severe form of depression, this model would explain why familial risk is greater for bipolar depression, why bipolar probands have an excess of unipolar relatives, and why unipolar probands do not have many relatives with bipolar depression. However, a twin study discussed in the next section does not provide much support for the hypothesis that bipolar disorder is a more severe form of unipolar depression (McGuffin et al., 2003). Identifying genes associated with these disorders will provide crucial evidence for resolving such issues.

Are some forms of depression more familial? For example, there is a long history of trying to subdivide depression into reactive (triggered by an event) and endogenous (coming from within) subtypes, but family studies provide little

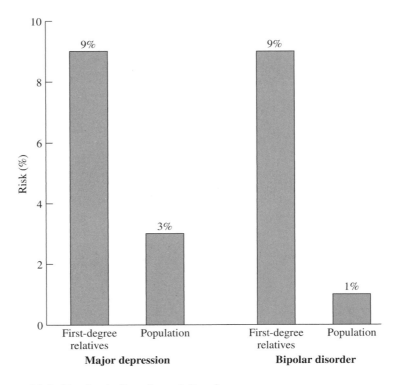

Figure 11.1 Family studies of mood disorders.

support for this distinction (Rush & Weissenburger, 1994). However, severity and especially recurrence show increased familiality for major depressive disorder (Sullivan et al., 2000). Early onset appears to increase familial risk for bipolar disorder (Smoller & Finn, 2003). Drug use and suicide attempts are also familial features of bipolar disorder (Schulze, Hedeker, Zandi, Rietschel, & McMahon, 2006). Another potentially promising direction for subdividing depression is in terms of response to drugs (Binder & Holsboer, 2006). For example, there is some evidence that the therapeutic response to specific antidepressants tends to run in families (Tsuang & Faraone, 1990). The main drug treatment for bipolar disorder is lithium; responsiveness to lithium appears to be strongly familial (Grof et al., 2002).

Twin Studies

Twin studies yield evidence for substantial genetic influence for mood disorders. For major depressive disorder, six twin studies yielded average twin probandwise concordances of .43 for MZ twins and .28 for DZ twins (Sullivan et al., 2000). Liability-threshold model fitting of these data estimated heritability of liability as .37 with no shared environmental influence. A recent Swedish twin study, the

Dorret Boomsma is head of the Department of Biological Psychology at the Free University in Amsterdam, where she has established the Netherlands Twin Register.

As a student, Boomsma spent a year at the Institute for Behavioral Genetics in Boulder, Colorado, where she received her introduction to quantitative genetics from Steven Vandenberg and John DeFries. Back in Amsterdam, her doctoral research involved the application of structural equation models to twin-family data on cardiovascular risk factors. This work led to several extensions of multivariate genetic models that had not been considered before, such as the estimation of single-subject genetic scores, and to later QTL studies.

The Netherlands Twin Register was established in 1986 and includes about 50 percent of all newborn twins in the Netherlands. These twins, the oldest of whom are now 20 years of age, participate in longitudinal studies of behavioral development and childhood psychopathology. Additionally, families of adolescent and young adult twins take part in longitudinal studies of addiction, personality, and psychopathology. Subsamples of twins participate in cognitive, information processing, imaging, and psychophysiological projects designed to study neural mechanisms that may mediate the influence of genes on behavior. Highly selected subsamples of dizygotic twins and their siblings participate in a genomewide QTL search of anxiety and depression.

largest twin study to date, yielded highly similar results: .38 heritability, and no shared environmental influence (Kendler, Gatz, Gardner, & Pedersen, 2006a). However, family studies suggest that more severe depression might be more heritable. In line with this suggestion, the only clinically ascertained major depressive disorder twin sample large enough to perform model-fitting analyses estimated heritability of liability as 70 percent (McGuffin, Katz, Watkins, & Rutherford, 1996). However, it is also possible that the higher heritability of depression in the clinical sample represents higher reliability of clinical assessment.

For bipolar depression, average twin concordances were 72 percent for MZ twins and 40 percent for DZ twins in early studies (Allen, 1976); three more recent twin studies yield average twin concordances of 65 and 7 percent, respectively (Smoller & Finn, 2003). The two most recent twin studies yield strikingly similar results: MZ and DZ twin concordances were 40 and 5 percent in a U.K. study (McGuffin et al., 2003) and 43 and 6 percent in a Finnish study (Kieseppa, Partonen, Haukka, Kaprio, & Lonnqvist, 2004). Model-fitting liability-threshold analyses suggest extremely high heritabilities of liability (.89 and .93, respectively) and no shared environmental influence. The average MZ and DZ twin concordances for the five recent studies are 55 and 7 percent, respectively. (See Figure 11.2.)

As mentioned earlier, one of the most important implications of genetic research is to provide diagnostic classifications based on etiology rather than symptoms. For example, are unipolar depression and bipolar disorder genetically distinct? One twin study investigated the model described earlier that bipolar disorder is a more extreme version of major depressive disorder (McGuffin et al., 2003). Part of the problem in addressing this issue is that conventional diagnostic rules assume that an individual either has unipolar or bipolar disorder and bipolar disorder trumps unipolar disorder. However, in this twin study, this diagnostic assumption was relaxed and a genetic correlation of .65 was found between depression and mania, which supports the model. However, 70 percent of the genetic variance on mania was independent of depression, which does not support the model. A model that explicitly tested the assumption that bipolar disorder is a more extreme form of unipolar depression was rejected, but so too was a model in which the two disorders were assumed to be genetically distinct. This lack of resolution is probably due to a lack of power: Although this was the largest clinically ascertained twin study, there were only 67 pairs with bipolar disorder and 244 pairs with unipolar depression. Resolution of this important diagnostic issue can be addressed definitively when genes are identified for the two

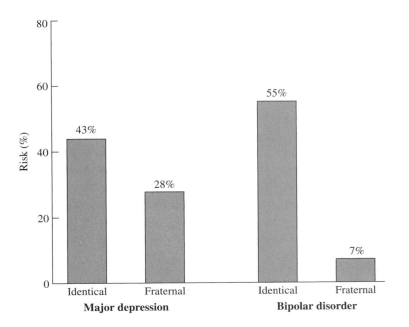

Figure 11.2 Approximate twin results for mood disorders.

disorders: To what extent will the same genes be associated with depression and mania? There is some emerging evidence from linkage studies (discussed below) for overlapping linkage regions (Farmer, Elkin, & McGuffin, 2007).

As in the research on schizophrenia, a study of offspring of identical twins discordant for bipolar depression has been reported (Bertelsen, 1985). As in the results for schizophrenia, the same 10 percent risk for mood disorder was found in the offspring of the unaffected twin and in the offspring of the affected twin. This outcome implies that the identical twin who does not succumb to bipolar depression nonetheless transmits a liability for the illness to offspring to the same extent as does the ill twin.

Adoption Studies

Results of adoption research on mood disorders are mixed. The largest study began with 71 adoptees with a broad range of mood disorders (Wender et al., 1986). Mood disorders were found in 8 percent of the 387 biological relatives of the probands, a risk only slightly greater than the risk of 5 percent for the 344 biological relatives of control adoptees. The biological relatives of the probands showed somewhat greater rates of alcoholism (5 percent versus 2 percent) and of attempted or actual suicide (7 percent versus 1 percent). Two other adoption studies relying on medical records of depression found little evidence for genetic influence (Cadoret, O'Gorman, Heywood, & Troughton, 1985; von Knorring, Cloninger, Bohman, & Sigvardsson, 1983). Although the sample size is necessarily small, 12 pairs of identical twins reared apart have been identified in which at least one member of each pair had suffered from major depression (Bertelsen, 1985). Eight of the 12 pairs (67 percent) were concordant for major depression, which is consistent with a hypothesis of at least some genetic influence on depression.

An adoption study that focused on adoptees with bipolar depression found stronger evidence for genetic influence (Mendlewicz & Rainer, 1977). The rate of bipolar disorders in the biological parents of the bipolar adoptees was 7 percent, but it was 0 percent for the parents of control adoptees. As in the family studies, biological parents of these bipolar adoptees also showed elevated rates of unipolar depression (21 percent) relative to the rate for biological parents of control adoptees (2 percent), a result suggesting that the two disorders are not distinct genetically. Adoptive parents of the bipolar and control adoptees differed little in their rates of mood disorders.

Identifying Genes

For decades, the greater risk of major depression for females led to the hypothesis that a dominant gene on the X chromosome might be involved. As explained in Chapter 3, females can inherit the gene on either of their two X chromosomes, whereas males can only inherit the gene on the X chromosome they receive from their mother. Although initially linkage was reported between depression and color blindness, which is caused by genes on the X chromosome (Chapter 3), studies of DNA markers on the X chromosome failed to confirm linkage (Baron, Freimer, Risch, Lerer, & Alexander, 1993). Father-to-son inheritance is common for both major depression and bipolar depression, which argues against X-linkage inheritance. Moreover, as mentioned earlier, bipolar depression shows little sex difference. For these reasons, X linkage seems unlikely (Hebebrand, 1992).

In 1987, researchers reported linkage between bipolar depression and markers on chromosome 11 in a genetically isolated community of Old Order Amish in Pennsylvania (Egeland et al., 1987). Unfortunately, this highly publicized finding was not replicated in other studies. The original report was withdrawn when follow-up research on the original pedigree with additional data showed that the evidence for linkage disappeared (Kelsoe et al., 1989).

These false starts led to greater caution in the search for genes for mood disorders. Linkage studies of major depressive disorder have lagged behind those for schizophrenia and bipolar disorder because, as discussed above, major depressive disorder appears to be less heritable, at least in community-based samples (McGuffin, Cohen, & Knight, 2007). Three genomewide linkage studies of major depressive disorder converge on linkage at 15q (Camp et al., 2005; Holmans et al., 2007; McGuffin et al., 2005). Follow-up fine mapping showed modestly positive evidence for linkage at 15q25-q26 (Levinson et al., 2007). These studies focused on early-onset depression and recurrent depression, because quantitative genetic results mentioned above suggest that early-onset depression and recurrent depression are more heritable.

Genomewide linkage scans of bipolar disorder led to a surprising discovery. A meta-analysis of 11 linkage studies with more than 1200 individuals diagnosed as having bipolar disorder found strong evidence for linkage at 13q and 22q (Badner & Gershon, 2002). The same study also conducted a meta-analysis of 18 linkage studies of schizophrenia and found the strongest evidence for linkage in the same two regions, 13q and 22q, in addition to other regions. Moreover, although many candidate genes have been reported to be associated with bipolar disorder (Craddock & Forty, 2006; Farmer et al., 2007), the genes that appear to replicate most are those that were identified initially for their association with schizophrenia: neuregulin and dysbindin (see Chapter 10) (Farmer et al., 2007; Kato, 2007). However, more molecular genetic research is needed to resolve this critical issue of genetic overlap between schizophrenia and bipolar disorder. For example, later meta-analyses of linkage studies of bipolar disorder do not support the linkages to 13q and 22q (McQueen et al., 2005; Segurado et al., 2003); however, these meta-analyses used different analytic techniques.

Nonetheless, the possibility of genetic overlap between schizophrenia and bipolar disorder is important because the distinction between them is fundamental to diagnostic classifications, such as described in the DSM-IV. Bipolar disorder can only be diagnosed if the individual is *not* schizophrenic, which precludes comorbidity, although a mixed category called schizoaffective disorder has been acknowledged. For this reason, relatives of schizophrenics have not shown an increased risk for bipolar disorder in family studies. However, the assumption that schizophrenia and bipolar disorder are distinct has been brought into question by the molecular genetic evidence (Craddock & Owen, 2005). Based on symptoms rather than accepting the traditional diagnostic categories, family studies are beginning to show overlap between schizophrenic symptoms and bipolar symptoms (Craddock, O'Donovan, & Owen, 2005). Similarly, a twin study has shown genetic overlap between symptoms of schizophrenia and bipolar disorder (Cardno, Rijsdijk, Sham, Murray, & McGuffin, 2002).

As is the case for schizophrenia, the next major development will come from the results of ongoing genomewide association scans of large samples of individuals with bipolar disorder and control individuals.

SUMMING UP

Family, twin, and adoption data indicate moderate genetic influence on major depressive disorder and substantial genetic influence on bipolar disorder. More severe and recurrent forms of these disorders appear to be more heritable. Bipolar disorder may be a more severe form of depression. Some convergence has begun to emerge from genomewide linkage studies of major depressive disorder and bipolar disorder. A surprising finding is that two genes that are associated with schizophrenia also appear to be associated with bipolar disorder. Genetic research has begun to call into question the assumption that schizophrenia and bipolar disorder are distinct disorders.

Anxiety Disorders

A wide range of disorders involve anxiety (panic disorder, generalized anxiety disorder, phobias) or attempts to ward off anxiety (obsessive-compulsive disorder). In panic disorder, recurrent panic attacks come on suddenly and unexpectedly, and usually last for several minutes. Panic attacks often lead to a fear of being in a situation that might bring on more panic attacks (e.g., agoraphobia, which literally means "fear of the marketplace"). Generalized anxiety refers to a more chronic state of diffuse anxiety marked by excessive and uncontrollable worrying. In phobia, the fear is attached to a specific stimulus, such as fear of heights (acrophobia) and enclosed places (claustrophobia), or fear of social situations (social phobia). In obsessive-compulsive disorder, anxiety occurs when the person does not perform some compulsive act driven by an obsession—for example, repeated hand-washing in response to an obsession with hygiene.

Anxiety disorders are usually not as crippling as schizophrenia or severe depressive disorders. However, they are the most common form of mental illness, with a lifetime prevalence of 29 percent (Kessler et al., 2005), and can lead to

CLOSE UP

Soo Hyun Rhee has been an assistant professor in the clinical and behavioral genetics programs in the Department of Psychology and a faculty fellow at the Institute for Behavioral Genetics, University of Colorado, since 2002. She received a B.A. in psychology in 1993 from Washington University in St. Louis, where she completed an honors thesis on verbal and spatial memory span with Sandra Hale and Joel Myerson. She learned about behavioral genetics for the first time from Irwin Waldman while interviewing for graduate school at Emory University and immediately became interested in the field. She conducted her master's thesis and dissertation with Dr. Waldman, examining sex differences in attention-deficit hyperactivity disorder. After receiving a Ph.D. in clinical psychology in 1999, she chose to complete a postdoctoral fellowship with John Hewitt and Thomas Crowley at the Center for Antisocial Drug Dependence (CADD) in Colorado. Working for CADD provided a unique opportunity to work with a data set combining a clinical family sample, a twin sample, and an adoption sample, and to obtain clinical training in the assessment and treatment of adolescents with severe conduct disorder and substance abuse. During her postdoctoral fellowship, she became interested in studying comorbidity, given conflicting conclusions regarding the causes of comorbidity between psychiatric disorders in the literature. She conducted studies examining the validity of methods testing alternative comorbidity models. Her current research includes the examination of comorbidity, development, and sex differences in childhood externalizing disorders and substance use disorders.

other disorders, notably depression and alcoholism. Median age of onset is much earlier for anxiety (11 years) than for mood disorders (30 years). The lifetime risks for anxiety disorders are 5 percent for panic disorder, 6 percent for generalized anxiety disorder, 13 percent for specific phobias, 12 percent for social phobia, and 2 percent for obsessive-compulsive disorder. Panic disorder, generalized anxiety disorder, and specific phobias are twice as common in women as in men.

There has been much less genetic research on anxiety disorders than on schizophrenia and mood disorders. In general, results for anxiety disorders appear to be similar to those for depression in suggesting moderate genetic influence, as compared to the more substantial genetic influence seen for schizophrenia and bipolar disorder. As discussed later, the similarity in results for anxiety and depression may be caused by genetic overlap between them. Nonetheless, we will briefly review evidence for genetic influence for panic disorder, generalized anxiety disorder, phobias, and obsessive-compulsive disorder.

For panic disorder, a review of eight family studies of panic disorder yielded an average morbidity risk of 13 percent in first-degree relatives of cases and 2 percent in controls (Shih, Belmonte, & Zandi, 2004). In an early twin study of panic disorder, the concordance rates for identical and fraternal twins were 31 and 10 percent, respectively (Torgersen, 1983). In the largest twin study with a nonclinical sample, the heritability of liability was about 40 percent with no evidence of shared environmental influence (Kendler, Gardner, & Prescott, 2001). A meta-analysis of five twin studies yielded a similar liability heritability (43 percent) with no shared environment (Hettema, Neale, & Kendler, 2001). No adoption data are available for panic disorder or any other anxiety disorders.

Generalized anxiety disorder appears to be as familial as panic disorder, but the evidence for heritability is weaker. A review of family studies indicates an average risk of about 10 percent among first-degree relatives as compared to a risk of 2 percent in controls (Eley, Collier, & McGuffin, 2002). However, two twin studies found no evidence for genetic influence (Andrews, Stewart, Allen, & Henderson, 1990; Torgersen, 1983); three other twin studies suggested modest genetic influence of about 20 percent and little shared environmental influence (Hettema, Prescott, & Kendler, 2001; Kendler, Neale, Kessler, Heath, & Eaves, 1992; Scherrer et al., 2000).

Phobias show familial resemblance: 30 percent familial risk versus 10 percent in controls for specific phobias excluding agoraphobia (Fyer, Mannuzza, Chapman, Martin, & Klein, 1995), 5 percent versus 3 percent for agoraphobia (Eley et al., 2002), and 20 percent versus 5 percent for social phobia (Stein et al., 1998). One twin study found heritability of liability of about 30 percent for these phobias (Kendler, Myers, Prescott, & Neale, 2001). Although there is little evidence of shared environmental influence, phobias are learned, even fears of evolutionary fear-relevant stimuli such as snakes and spiders. An interesting twin study of fear conditioning showed moderate genetic influence on individual differences in learning and extinguishing fears (Hettema, Annas, Neale, Kendler, & Fredrikson, 2003).

For obsessive-compulsive disorder (OCD), family studies yield a wide range of results because of differences in diagnostic criteria. However, nine family studies that used criteria from the DSM and had more than 100 cases yielded more consistent results, with an average risk of 7 percent for family members and 3 percent for controls (Shih et al., 2004). Family studies also suggest that early-onset OCD is more familial. Only three small twin studies of OCD have been reported, and two of these found no heritability (Shih et al., 2004), although twin studies of OCD-like symptoms in unselected samples of twins suggest moderate heritability (van Grootheest, Cath, Beekman, & Boomsma, 2005).

DSM-IV also includes posttraumatic stress disorder (PTSD) as an anxiety disorder, even though its diagnosis depends on a prior traumatic event that threatens death or serious injury, such as war, assault, or natural disaster. PTSD symptoms include reexperiencing the trauma (intrusive memories and nightmares) and denying the trauma (emotional numbing). One survey estimated that the lifetime risk for one PTSD episode is about 1 percent (Davidson, Hughes, Blazer, & George, 1991). The risk is much higher, of course, in those who have experienced trauma. For example, after a plane crash, as many as one-half of the survivors develop PTSD (Smith, North, McColl, & Shea, 1990). About 10 percent of U.S. veterans of the Vietnam War still suffered from PTSD many years later (Weiss et al., 1992). Response to trauma appears to show familial resemblance (Eley et al., 2002). The Vietnam War provided the possibility to conduct a twin study of PTSD, because more than 4000 twin pairs were veterans of the war. A series of studies of these twins began by dividing the sample into those who served in Southeast Asia (who were much more likely to experience trauma) and those who did not (True et al., 1993). The results were similar for both groups regardless of the type of trauma experienced: Heritabilities of 15 PTSD symptoms were all about 40 percent, and there was no evidence of shared environmental influence.

Other Disorders

As mentioned earlier, DSM-IV includes many other categories of disorders, but next to nothing is known about their genetics. Interesting results are emerging, however, from the early stages of genetic research on four of these categories of disorders: seasonal affective disorder, somatoform disorders, chronic fatigue, and eating disorders. Other disorders are discussed in later chapters: impulse-control disorders such as hyperactivity in Chapter 12, antisocial personality disorder in Chapter 13, and substance abuse disorders in Chapter 14.

Seasonal affective disorder (SAD) is a type of major depression that occurs with the seasons, typically in the fall or winter (Rosenthal et al., 1984). Family and twin studies suggest results similar to depression with modest heritability (about 30 percent) and little shared environmental influence (Sher, Goldman, Ozaki, & Rosenthal, 1999). However, the most recent twin study reported heritability twice as high (Jang, Lam, Livesley, & Vernon, 1997). It is noteworthy that this recent study was conducted in British Columbia (Canada) and yielded very high rates of SAD compared to the other studies, which suggests the possibility that the higher heritability and prevalence in the Canadian sample might be due to the northern latitude and more severe winters of Canada (Jang, 2005).

In somatoform disorders, psychological conflicts lead to physical (somatic) symptoms such as stomach pains. Somatoform disorders include somatization disorder, hypochondriasis, and conversion disorder. Somatization disorder involves multiple symptoms with no apparent physical cause. Hypochondriacs worry that a specific disease is about to appear. Conversion disorder, which was formerly called hysteria, involves a specific disability such as paralysis with no physical cause. Somatoform disorders show some genetic influence in family, twin, and adoption studies (Guze, 1993). Somatization disorder, which is much more common in women than in men, shows strong familial resemblance for women, but for men it is related to increased family risk for antisocial personality (Guze, Cloninger, Martin, & Clayton, 1986; Lilienfeld, 1992). An adoption study suggests that this link between somatization disorder in women and antisocial behavior in men may be genetic in origin (Bohman, Cloninger, von Knorring, & Sigvardsson, 1984). Biological fathers of adopted women with somatization disorder showed increased rates of antisocial behavior and alcoholism. A twin study of somatic distress symptoms in an unselected sample showed genetic as well as shared environmental influence, and also suggested that some of the genetic influence is independent of depression and phobia (Gillespie, Zhu, Heath, Hickie, & Martin, 2000).

Chronic fatigue refers to fatigue of more than six months' duration that cannot be explained by a physical or other psychiatric disorder. A family study suggests that chronic fatigue is moderately familial (Walsh, Zainal, Middleton, & Paykel, 2001). A twin study of diagnosed chronic fatigue found concordance rates of 55 percent in MZ twins and 19 percent in DZ twins (Buchwald et al., 2001). Twin studies of chronic fatigue symptoms in unselected samples found modest genetic and shared environmental influence (Sullivan, Evengard, Jacks, & Pedersen, 2005; Sullivan, Kovalenko, York, Prescott, & Kendler, 2003), even in childhood (Farmer, Scourfield, Martin, Cardno, & McGuffin, 1999).

Eating disorders include anorexia nervosa (extreme dieting and avoidance of food) and bulimia nervosa (binge eating followed by vomiting), both of which occur mostly in adolescent girls and young women. Both types of eating disorders appear to run in families (Eley et al., 2002); in twin studies both appear to be moderately heritable with little influence of shared environment (Eley et al., 2002; Kendler & Prescott, 2006). For example, the largest twin study of anorexia found a heritability of liability estimate of 56 percent and no shared environmental influence (Bulik et al., 2006). Eating disorders is an area that is especially promising for studies of the interplay between genes and environment (Bulik, 2005).

SUMMING UP

Most anxiety disorders—panic disorder, generalized anxiety disorder, phobias, obsessive-compulsive disorder, posttraumatic stress disorder—show moderate genetic influence with little evidence of shared environmental influence. There is a suggestion that panic disorder is the most heritable and obsessive-compulsive disorder is the least heritable of these anxiety disorders. As was the case for the mood disorders, early onset of the disorders is more familial and more heritable. For many other DSM-IV categories of disorders, no genetic research has as yet been reported, although evidence for genetic influence has been found for seasonal affective disorder, somatoform disorders, chronic fatigue, and eating disorders.

Co-Occurrence of Disorders

Co-occurrence or comorbidity of psychiatric disorders is striking. People with one disorder have almost a 50 percent chance of having more than one disorder during a 12-month period (Kessler, Chiu, Demler, Merikangas, & Walters, 2005). In addition, more serious disorders are much more likely to involve comorbidity. Are these really different disorders that co-occur, or does the co-occurrence call into question current diagnostic systems? Diagnostic systems are based on phenotypic descriptions of symptoms rather than on causes. Genetic research offers the hope of systems of diagnosis that take into account evidence on causation. As explained in the Appendix, multivariate genetic analysis of twin and adoption data can be used to ask whether genes that affect one trait also affect another trait.

More than a hundred genetic studies have addressed this key question of comorbidity in psychopathology. Earlier in this chapter, we considered the surprising finding of genetic overlap between major depressive disorder and bipolar disorder, and the even more surprising possibility of genetic overlap between bipolar disorder and schizophrenia. Scores of multivariate family and twin studies have examined comorbidity across the many anxiety disorders, as well as between anxiety disorders and other disorders such as depression and alcoholism. Rather than describe studies that compare two or three disorders (see, for example, Jang, 2005; McGuffin et al., 2002), we will provide an overview of multivariate genetic results that point to a surprising degree of genetic comorbidity.

For example, consider the diverse anxiety disorders. A multivariate genetic analysis of lifetime diagnoses of major anxiety disorders indicated substantial genetic overlap between generalized anxiety disorder, panic disorder, agoraphobia, and social phobia (Hettema, Prescott, Myers, Neale, & Kendler, 2005). The only specific genetic effects were found for specific phobias such as fear of animals. Differences between the disorders are largely caused by nonshared environmental factors. Results were similar for men and women despite the much greater frequency of anxiety disorders in women. Although this study did not include obsessive-compulsive disorder, other research suggests that it too is part of the general genetic anxiety factor (Nestadt et al., 2001).

Broadening this multivariate genetic approach beyond anxiety disorders to include major depression yields the most surprising finding in this area: Anxiety (especially generalized anxiety disorder) and depression are largely the same thing genetically. This finding was initially reported in a paper in 1992 for lifetime estimates (Kendler et al., 1992), with results summarized in Figure 11.3. Heritability of liability in this study was 42 percent for major depression and 69 percent for generalized anxiety disorder. There was no significant shared environmental influence; nonshared environment accounted for the remainder of the liability of the two disorders. The amazing finding was the genetic correlation of 1.0 between the two disorders, indicating that the same genes affect

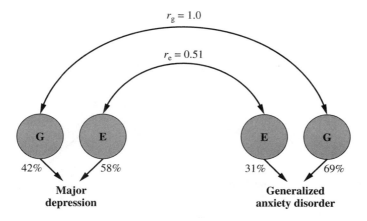

$r_g = 1.0$

$r_e = 0.51$

G   E   E   G

42%   58%   31%   69%

**Major
depression**

**Generalized
anxiety disorder**

Figure 11.3 Multivariate genetic results for major depression and generalized anxiety disorder. (Adapted from Kendler, Neale, Kessler, Heath, & Eaves, 1992 [Figure 2]. Copyright 1992 by the American Medical Association. Used with permission.)

depression and anxiety. Nonshared environmental influences correlated .51, suggesting that nonshared environmental factors differentiate the disorders to some extent. These findings for lifetime estimates of depression and anxiety were replicated using one-year prevalences obtained from follow-up interviews (Kendler, 1996a). A review of 23 twin studies and 12 family studies confirms that anxiety and depression are largely the same disorder genetically and that the disorders are differentiated by nonshared environmental factors (Middeldorp, Cath, Van Dyck, & Boomsma, 2005).

Going beyond depression and anxiety disorders to include drug abuse and antisocial behavior suggests a genetic structure of common psychiatric disorders (not including schizophrenia and bipolar disorder) that differs substantially from current diagnostic classifications based on symptoms (Kendler, Prescott, Myers, & Neale, 2003). As summarized in Figure 11.4, genetic research suggests two broad categories of disorder, called *internalizing* and *externalizing*. Internalizing disorders include depression and anxiety disorders; externalizing disorders include alcohol and other drug abuse, and antisocial behavior in adulthood (and conduct disorder in children). Internalizing disorders can be separated into an anxious/misery factor that includes depression and anxiety disorders, and a fear factor which includes phobias. Panic disorder is involved in both internalizing factors. As discussed in Chapter 13, the internalizing disorders might represent the extreme of the broad personality trait called *neuroticism*.

Externalizing disorders will be discussed in Chapters 13 and 14, but two points are noteworthy now: The disparate externalizing disorders are part of a general genetic factor, and both alcohol dependence and other drug abuse

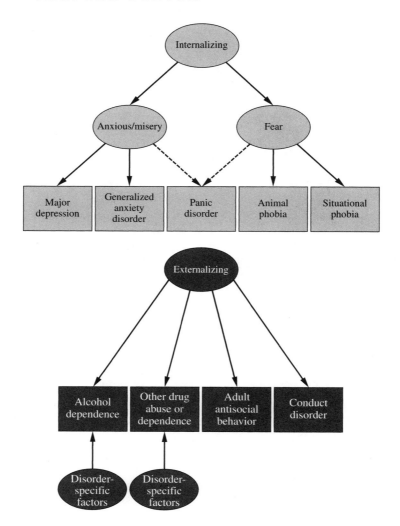

Figure 11.4 A structure for the genetic risk factors for common psychiatric and substance use disorders. Strong relationships are depicted by solid lines and, in the case of panic disorder only, weaker relationships by dotted lines. (From "The Structure of Genetic and Environmental Risk Factors for Common Psychiatric and Substance Use Disorders in Men and Women," by K. S. Kendler, C. A. Prescott, J. Myers, & M. C. Neale, 2003, *Archives of General Psychiatry, 60,* p. 936. Used with permission.)

include some disorder-specific genetic effects. In general, the genetic structure of internalizing and externalizing disorders applies equally to men and women despite the much greater risk of internalizing disorders for women and externalizing disorders for men. Because few disorders show shared environmental influence, it does not affect the structure. Nonshared environment largely contributes to heterogeneity rather than comorbidity. Thus, the phenotypic structure of comorbidity is largely driven by the genetic structure shown in Figure 11.4 (Krueger, 1999).

These multivariate genetic results predict that when genes are found that are associated with any of the internalizing disorders, the same genes are highly likely to be associated with other internalizing disorders. Similarly, genes associated with any of the externalizing disorders will be associated with the others, but not with the internalizing disorders. This result suggests that genetic influences are broad in their effect in psychopathology. It mirrors a similar finding concerning "generalist genes" in the area of cognitive abilities (see Chapters 8 and 9).

Identifying Genes

Although multivariate genetic research suggests that the genetic action lies at the level of broad categories of internalizing and externalizing disorders, molecular genetic research on anxiety disorders has focused on traditional diagnoses. Moreover, not nearly as much molecular genetic research has been conducted on these disorders as compared to the mood disorders. As a result, linkage studies have not yet converged and the "usual suspect" candidate gene studies have not yet revealed replicable results (e.g., Eley et al., 2002; Jang, 2005).

Panic disorder has been studied most, in part because it appears to be more heritable than the other anxiety disorders and in part because it can be so debilitating. Five earlier linkage studies of panic disorder have not yielded consistent results (Villafuerte & Burmeister, 2003), suggesting that genetic effects may be relatively small. However, the largest and most recent report has yielded more promising results, suggesting linkage on 15q and possibly 2q (Fyer et al., 2006). As with other complex traits, candidate gene associations have largely failed to replicate (e.g., Maron et al., 2007). The strongest case so far can be made for an association between panic disorder in females and a functional polymorphism (Val158Met) in the catechol-O-methyltransferase (*COMT*) gene (McGrath et al., 2004; Rothe et al., 2006), a polymorphism that has been reported to be associated with many other common disorders and complex traits (Craddock, Owen, & O'Donovan, 2006). Similar mixed stories are beginning to emerge for candidate gene studies of obsessive-compulsive disorders (Hemmings & Stein, 2006; Stewart et al, 2007) and for linkage and candidate gene association studies of eating disorders (Slof-op 't Landt et al., 2005).

Summary

Moderate genetic influence has been found for major depressive disorder, and substantial genetic influence has been found for bipolar disorder. More severe and recurrent forms of these mood disorders are more heritable. Bipolar disorder may be a more severe form of depression. Surprisingly, molecular genetic studies for bipolar disorder suggest similar linkages and associations found for schizophrenia.

Anxiety disorders yield quantitative genetic results that are similar to depression—moderate genetic influence with little evidence of shared environmental influence. Some evidence for genetic influence has also been found for seasonal affective disorder, somatoform disorders, chronic fatigue, and eating disorders.

Some of the most far-reaching genetic findings in psychopathology concern genetic comorbidity. Genetic research has begun to call into question the fundamental diagnostic distinction between schizophrenia and bipolar disorder, including the molecular genetic findings of similar linkages and associations for the two disorders. The most striking finding within the mood disorders is that major depressive disorder and generalized anxiety disorder are the same disorder from a genetic perspective. Multivariate genetic research suggests a genetic structure of common psychiatric disorders that includes just two broad categories, internalizing and externalizing disorders.

CHAPTER TWELVE

Developmental Psychopathology

Schizophrenia and mood disorders are typically diagnosed in adulthood. Other disorders emerge in childhood. General cognitive disability, learning disorders, and communication disorders were discussed in Chapter 7. Other DSM-IV diagnostic categories that first appear in childhood include pervasive developmental disorders (e.g., autistic disorder), attention-deficit and disruptive behavior disorders (e.g., attention-deficit hyperactivity disorder, conduct disorder), anxiety disorders, tic disorders (e.g., Tourette disorder), and elimination disorders (e.g., enuresis). It has been estimated that as many as one out of four children has a diagnosable disorder (Cohen et al., 1993) and one in five has a moderate or severe disorder (Brandenburg, Friedman, & Silver, 1990).

Only in the past two decades has genetic research begun to focus on disorders of childhood (Rutter, Silberg, O'Connor, & Simonoff, 1999). Developmental psychopathology is not limited to childhood: It considers change and continuity throughout the lifecourse, such as dementia which develops later in life (see Chapter 7). However, genetic research on childhood disorders has blossomed recently, which is reflected in this chapter. One reason to consider childhood disorders is that some adult disorders emerge in childhood. Median age of onset is much earlier for anxiety disorders (11 years) and impulse-control disorders (11 years) than for mood disorders (30 years). Half of all lifetime cases of diagnosed disorders start by age 14, which suggests that interventions aimed at prevention or early treatment need to focus on childhood and adolescence (Kessler et al., 2005). However, the main reason for the increased interest in the genetics of childhood disorders is that the two major childhood disorders—autism and attention-deficit hyperactivity disorder—have been shown to be among the most heritable of all mental disorders, as described in the following sections.

Autism

Autism was once thought to be a childhood version of schizophrenia, but it is now known to be a distinct disorder marked by abnormalities in social relationships, communication deficits, and restricted interests. As traditionally diagnosed, it is relatively uncommon, occurring in three to six individuals out of every 10,000; and it occurs about four times more often in boys than in girls. During the 1990s, there was a fivefold increase in the diagnosis of autism, probably because of heightened awareness and changing diagnostic criteria (Muhle, Trentacoste, & Rapin, 2004). The diagnosis has been broadened to *autistic spectrum disorders* (ASDs) that include autism, Asperger syndrome, and other pervasive developmental disorders. Traditionally, a diagnosis of autism was limited to children who showed impairments in all three areas (social, communication, interests) before 3 years of age. In contrast, Asperger syndrome was diagnosed if children were impaired in the social and interests domains but appeared to have normal language and cognitive development before 3 years of age. The "other" diagnosis was used for children who show severe impairment in just one or two of the domains. Most researchers now consider these three disorders as part of a single continuum or spectrum of disorder. In the early 2000s, great concern among parents was driven by media reports that the supposed increase in ASDs was caused environmentally by the measles-mumps-rubella (MMR) vaccine. However, the evidence on this putative environmental cause of ASDs has been consistently negative (Rutter, 2005a; Smeeth et al., 2004).

Family and Twin Studies

When Kanner first characterized autism in 1943, he assumed it was caused "constitutionally" (Kanner, 1943). However, in subsequent decades, autism was thought to be environmentally caused, either by cold and rejecting parents or by brain damage (Hanson & Gottesman, 1976). Genetics did not seem to be important, because there were no reported cases of an autistic child having an autistic parent and because the risk to siblings was only about 5 percent (Bailey, Phillips, & Rutter, 1996; Smalley, Asarnow, & Spence, 1988). However, this rate of 5 percent is 100 times greater than the population rate of autism as diagnosed in these studies, a difference implying strong familial resemblance. The reason why autistic children do not have autistic parents is that few autistic individuals marry and have children.

In 1977, the first systematic twin study of autism began to change the view that autism was environmental in origin (Folstein & Rutter, 1977). Four of 11 pairs of identical twins were concordant for autism, whereas none of 10 pairs of fraternal twins were. These pairwise concordance rates of 36 and 0 percent rose to 92 and 10 percent when the diagnosis was broadened to include communication and social problems. Co-twins of autistic children are more likely to have communication problems as well as social difficulties. In a follow-up of the twin sample into adult life, problems with social relationships were prominent (Le Couteur et al., 1996). These findings were replicated in other twin studies (Folstein & Rosen-Sheidley, 2001). A conservative estimate of the concordance in monozygotic pairs is 60 percent—a thousandfold increase in risk over the general population base rate. A review of four independent twin studies suggests a heritability of liability for autism greater than 90 percent (Freitag, 2007).

On the basis of these twin and family findings, views regarding autism have changed radically. Instead of being seen as an environmentally caused disorder, it is now considered to be one of the most heritable mental disorders (Freitag, 2007). One unusual aspect of genetic research on autism is that, as traditionally diagnosed, autism is so severe that it nearly always results in affected children being seen by clinical services rather than remaining undetected in the community (Thapar & Scourfield, 2002). As a result, nearly all twin studies have been based on clinical cases rather than community samples. However, recent research has considered ASDs as continua that extend well into common behavioral problems seen in undiagnosed children in the community. This trend was driven in part by the results of early family studies in which relatives of autistic individuals were found to have some communication and social difficulties (Bailey, Palferman, Heavey, & Le Couteur, 1998). Twin studies have also supported the hypothesis that the genetic and environmental causes of ASD symptoms are distributed continuously throughout the population (Constantino & Todd, 2003). This is an emerging rule in behavioral genetics, that disorders are actually the quantitative extreme of a continuum of normal variation (see Chapters 10 and 11).

In contrast to the assumption that autism involves a triad of impairments—poor social interaction, language and communication problems, and restricted range of interests and activities—a twin study of ASD symptoms in a community sample has found evidence for genetic heterogeneity, especially between social impairments (interaction and communication) and nonsocial impairments (interests and activities). In the first multivariate genetic analysis of the triad of symptoms, high heritability (about 80 percent) was found for the three types of symptoms but surprisingly low genetic correlations were found among them in the general population (Ronald, Happé, & Plomin, 2005; Ronald, Happé, Price, Baron-Cohen, & Plomin, 2006) and for children with extreme ASD symptoms (Ronald et al., 2006). These findings suggest that, although some children by chance have all three types of symptoms, the ASD triad of symptoms are different genetically. This surprising conclusion, which contradicts the traditional diagnosis of autism, is supported by other genetic data (Kolevzon, Smith, Schmeidler, Buxbaum, & Silverman, 2004) as well as cognitive and brain data (Happé, Ronald, & Plomin, 2006).

Identifying Genes

Quantitative genetic evidence suggesting substantial genetic influence on autism led to autism being the early target of affected sib-pair linkage analysis after the success of QTL linkage in the area of reading disability in 1994 (see Chapter 7). In 1998, an international collaborative linkage study reported evidence of a locus on chromosome 7 (7q31-q33) in a study of 87 affected sibling pairs (International Molecular Genetic Study of Autism Consortium, 1998). This 7q linkage was replicated in other studies, although several studies did not replicate the linkage (Trikalinos et al., 2006). No specific gene has been implicated reliably (Freitag, 2007). Many other linkage regions have been reported in 12 genomewide linkage studies but none has been replicated in more than two studies (Ma et al., 2007). Despite the sex difference in ASDs, no consistent evidence for linkage to the X chromosome has emerged.

As with other common disorders, these linkage results could be viewed as demonstrating that there are no genes of sufficiently large effect size to be detected by sib-pair linkage analyses with samples of fewer than 100 affected sibling pairs. The most straightforward way to address the issue of power to detect smaller QTL effect sizes is to increase the sample size, although it is difficult to obtain such samples because only about 5 percent of the siblings of autistic children are also autistic. Collaboration led to a recent sib-pair linkage analysis of more than 1000 families across 19 countries, involving 120 scientists from more than 50 institutions (Szatmari et al., 2007). Although previously reported linkages were not replicated, including the linkage on 7q, linkage was suggested for 11p12-q13. Linkage results appeared stronger when families with copy number variants (see Chapter 4) were removed from the analysis.

Similar to other disorders, more than 100 candidate gene associations have been reported, especially for the 7q linkage region, but no consistent associations have as yet been found (Bacchelli & Maestrini, 2006). Again similar to other disorders, we await with interest the results of ongoing genomewide association studies and other new genomic approaches (Gupta & State, 2007). Molecular genetic studies of diagnosed ASDs focus on children with impairments in all three types of ASD symptoms. However, the multivariate genetic research described above indicating genetic heterogeneity for the three types of symptoms suggests that molecular genetic studies might profit by focusing on the three types of symptoms separately rather than beginning with diagnoses of autism, which requires the presence of all three impairments.

SUMMING UP

Genetic research has begun to be applied to disorders that appear in childhood. One of the biggest surprises from behavioral genetic research involves autism, which had been thought to be environmental in origin, but yielded heritability estimates greater than 90 percent. Although the three components of traditional diagnoses of autistic spectrum disorders (ASDs)—problems with social interaction, communication deficits, and restricted interests—are all highly heritable, multivariate genetic evidence is mounting that these three components differ genetically. The high heritability of ASDs led to a dozen linkage studies and association studies of more than 100 candidate genes, but no solid linkages or associations have yet been found, perhaps because the three components required for a diagnosis of autism differ genetically.

Attention-Deficit and Disruptive Behavior Disorders

The DSM-IV grouping of attention-deficit and disruptive behavior disorders is interesting because it includes a disorder that appears to be substantially heritable, attention-deficit hyperactivity disorder, and a disorder that shows only modest genetic influence, conduct disorder, when it occurs in the absence of overactivity/inattention. Although all children have trouble learning self-control, most have

CLOSE UP

Anita Thapar is professor of child and adolescent psychiatry at the University of Wales College of Medicine, Cardiff. She qualified in medicine in Cardiff and then trained as a clinical child and adolescent psychiatrist. She was awarded a medical research training fellowship and during this time trained in psychiatric genetics with Peter McGuffin. Her doctoral research involved a twin study of psychiatric symptoms in childhood and adolescence. After she moved to the University of Manchester, she set up the Manchester Twin Register of 3000 school-aged twin pairs and continued examining the influence of genetic and environmental influences on psychiatric traits in children. Curiosity about the molecular genetic basis of attention-deficit hyperactivity disorder and related traits was stimulated by her initial twin study results and her clinical work, which involved many children with ADHD. As a result of this, she started molecular genetic studies of attention-deficit hyperactivity disorder. She returned to Cardiff in 1999, where she continues her research into the genetic basis of childhood psychiatric disorders and traits, using a variety of methods. Her current research includes genetic studies of ADHD, depression, and juvenile-onset psychosis.

made considerable progress by the time they enter school. Those who have not learned self-control are often disruptive, impulsive, and aggressive, and have problems adjusting to school.

Attention-deficit hyperactivity disorder (ADHD), as defined by DSM-IV, refers to children who are very restless, have a poor attention span, and act impulsively. Estimates of the prevalence of ADHD in North America are about 4 percent of elementary school children, with boys greatly outnumbering girls (Moldin, 1999). European psychiatrists have tended to take a more restricted approach to diagnosis, with an emphasis on hyperactivity that not only is severe and pervasive across situations but also is of early onset and unaccompanied by high anxiety (Taylor, 1995). There is continuing uncertainty about the merits of these narrower and broader approaches to diagnosis. However conceptualized, ADHD usually continues into adolescence and, in about a third of cases, continues through into adulthood (Klein & Mannuzza, 1991).

Twin Studies

ADHD runs in families, with relative risks of about 5 for first-degree relatives (Biederman et al., 1992), and with greater familial risk when ADHD persists into adulthood (Faraone, Biederman, & Monuteaux, 2000). Twin studies have consistently shown a strong genetic effect on hyperactivity regardless of whether it is measured by questionnaire (Goodman & Stevenson, 1989; Silberg et al., 1996) or by standardized and detailed interviewing (Eaves et al., 1997), regardless of whether it is rated by parents or teachers (Saudino, Ronald, & Plomin, 2005), and regardless of whether it is treated as a continuously distributed dimension (Thapar, Langley, O'Donovan, & Owen, 2006) or as a clinical diagnosis (Gillis, Gilger, Pennington, & DeFries, 1992). A review of 20 twin studies estimates heritability of ADHD as 76 percent, with no influence of shared environment (Faraone et al., 2005). Heritability is over 70 percent regardless of whether ADHD is assessed as a category or as a continuum (Jepsen & Michel, 2006). These results suggest that heritability is greater for ADHD than for other childhood disorders with the exception of autism. As is almost always the case in behavioral genetics, stability of ADHD symptoms is largely driven by genetics (Kuntsi, Rijsdijk, Ronald, Asherson, & Plomin, 2005; Larsson, Larsson, & Lichtenstein, 2004; Price et al., 2005; Rietveld, Hudziak, Bartels, Van Beijsterveldt, & Boomsma, 2004). Usually the case for psychopathology, heritability appears to be greater for persistent ADHD, in this case for ADHD that extends into adulthood (Faraone, 2004). An unusual aspect of ADHD results is that DZ correlations are lower than expected relative to MZ correlations, especially for parental ratings. This could be due to a contrast effect in which parents inflate differences between their DZ twins, but this pattern of twin results is also consistent with nonadditive genetic variance (Eaves et al., 1997; Hudziak, Derks, Althoff, Rettew, & Boomsma, 2005; Rietveld, Posthuma, Dolan, & Boomsma, 2003), as discussed in Chapter 9. Although adoption studies to date have been few and quite limited methodologically (McMahon, 1980), they lend some support to the hypothesis of genetic influence for ADHD (e.g., Cantwell, 1975).

The activity and attention components of ADHD are both highly heritable (Rietveld et al., 2004). Multivariate genetic twin analyses of the inattention and hyperactivity components of ADHD indicate substantial genetic overlap between the two components, providing genetic justification for the syndrome of ADHD (Eaves et al., 2000; Larsson, Lichtenstein, & Larsson, 2006; McLoughlin, Ronald, Kuntsi, Asherson, & Plomin, 2007; Rasmussen et al., 2004). Another multivariate issue concerns the genetic overlap between parental ratings and teacher ratings of ADHD, both of which are highly heritable. Multivariate genetic analyses suggest some genetic overlap but also some genetic effects specific to parents and teachers (Thapar et al., 2006). In other words, these results predict that to some extent different genes will be associated with ADHD viewed by parents in the home and ADHD viewed by teachers in school. In addition, pervasive ADHD that is seen both at home and in school is more heritable than ADHD specific to just one situation (Thapar et al., 2006).

Genetic studies of conduct disorder yield results quite different from those for ADHD. DSM-IV criteria for conduct disorder include aggression, destruction of property, deceitfulness or theft, and other serious violations of rules such as running away from home. Some 5 to 10 percent of children and adolescents meet these diagnostic criteria, with boys again greatly outnumbering girls (Cohen et al., 1993; Rutter et al., 1997). In contrast to ADHD, the combined data from a number of early twin studies of juvenile delinquency yield concordance rates of 87 percent for identical twins and 72 percent for fraternal twins, rates that suggest only modest genetic influence and substantial shared environmental influence (McGuffin & Gottesman, 1985). This pattern is broadly supported by the results of a twin study of self-reported teenage antisocial behavior in U.S. Army Vietnam era veterans (Lyons et al., 1995). However, some twin studies of delinquent acts and conduct disorder symptoms in normal samples of adolescents have shown genetic influence (Thapar et al., 2006).

A twin study of adolescent males used a technique called *latent class analysis*, which attempts to account for patterning among symptoms by hypothesizing underlying (latent) classes (Eaves et al., 1993). One class involves symptoms from both ADHD and conduct disorder, for which strong genetic influence was found (Nadder, Rutter, Silberg, Maes, & Eaves, 2002; Silberg et al., 1996). By sharp contrast, there was almost no significant genetic influence for a "pure" class of conduct disorder without hyperactivity, for which there was a strong shared environmental influence (Silberg et al., 1996). This finding fits with multivariate twin studies that found substantial genetic overlap between ADHD and conduct problems (Thapar et al., 2006), suggesting that what ADHD and conduct problems have in common is largely genetic and that what conduct problems do not share with ADHD is largely environmental.

Heterogeneity in antisocial behavior also contributes to some of the inconsistencies in the published research findings on conduct problems. For example, there is evidence from several twin studies that aggressive antisocial behavior is more heritable than nonaggressive antisocial behavior (e.g., Eley, Lichtenstein, & Stevenson, 1999). Moreover, different genetic factors affect aggressive and nonaggressive conduct problems (Gelhorn et al., 2006). Genetic effects are probably greatest with respect to early-onset aggressive antisocial behavior that is accompanied by hyperactivity and that shows a strong tendency to persist into adulthood as an antisocial personality disorder (DiLalla & Gottesman, 1989; Lyons et al., 1995; Moffitt, 1993; Robins & Price, 1991; Rutter et al., 1999). (See Chapter 13 for a discussion of personality disorders, including antisocial personality disorder.) In addition, antisocial behavior that is persistent across situations (home, school, laboratory) is more heritable (Arseneault et al., 2003). In contrast, environmentally mediated risks are probably strongest with respect to nonaggressive juvenile delinquency that has an onset in the adolescent years and does not persist into adult life. The development of conduct disorder and antisocial behavior is a rich vein for studies of gene-environment interplay (Moffitt, 2005), as discussed in Chapter 16.

Another aspect of genetic heterogeneity in childhood antisocial behavior is callous-unemotional personality, which involves psychopathic tendencies such as lack of empathy and guilt. In a large twin study of 7-year-old children rated by their teachers, antisocial behavior accompanied by callous-unemotional tendencies is highly heritable (80 percent) with no shared environmental influence, whereas antisocial behavior without callous-unemotional tendencies is only modestly heritable (30 percent) and shows moderate shared environmental influence (35 percent) (Viding, Blair, Moffitt, & Plomin, 2005).

Identifying Genes

As was the case for autism, the consistent evidence of a large genetic contribution to ADHD attracted the attention of molecular geneticists. However, this recognition came later for ADHD than for autism and at a time when molecular genetic studies had moved on from linkage to association studies in an attempt to identify QTLs of small effect size. Because genomewide association was not available at that time, these early studies were limited to candidate genes. Interest has centered on genes involved in the dopamine pathway because many children with ADHD improve when given psychostimulants, such as methylphenidate, which affect dopamine pathways. The dopamine transporter gene *DAT1* was an obvious candidate because methylphenidate inhibits the dopamine transporter mechanism and *DAT1* knock-out mice are hyperactive (Caron, 1996). An exciting initial finding of an association for *DAT1* (Cook et al., 1998) was replicated in three studies but failed to replicate in three other studies (Thapar & Scourfield, 2002). Somewhat stronger results were found for two other dopamine genes that code for dopamine receptors called *DRD4* and

DRD5. A recent meta-analysis found small (odds ratios of about 1.3) but significant associations for *DRD4* and *DRD5*, although no association was found for *DAT1* (Li, Sham, Owen, & He, 2006). As expected from the multivariate genetic results indicating substantial genetic overlap between ADHD symptoms, these patterns of associations are similar across symptoms (Thapar et al., 2006).

Associations have been reported for more than 30 other candidate genes but none have been consistently replicated (Bobb et al., 2006; Thapar, O'Donovan, & Owen, 2005; Waldman & Gizer, 2006). Although candidate gene association studies have dominated genetic research on ADHD, three genomewide linkage screens have been reported as well as a bivariate linkage scan for ADHD and reading disability (Gayán et al., 2005), and a follow-up fine-mapping study of nine candidate linkage regions (Ogdie et al., 2004). No consistent linkage regions have been identified.

SUMMING UP

The DSM-IV categories that encompass attention deficit and disruptive behaviors include attention-deficit hyperactivity disorder (ADHD), which is substantially heritable, and conduct disorder, which shows less genetic influence and greater influence of shared family environment. A meta-analysis of 20 twin studies of ADHD yielded a heritability estimate of about 75 percent, with no evidence of shared environmental influence. Unlike autism, the attentional and hyperactivity components of ADHD are strongly related genetically, providing support for the ADHD syndrome. Although conduct disorder, especially non-aggressive conduct disorder, is less heritable than ADHD, the overlap between conduct disorder and ADHD is largely mediated genetically. Conduct disorder is unusual in showing evidence for shared environmental influence. Similar to autism, the high heritability of ADHD attracted molecular genetic researchers, although in research on ADHD, molecular genetic studies focused on candidate genes, especially genes in the dopamine system that are involved in the treatment of ADHD with methylphenidate. Meta-analyses support small but significant associations between two dopamine receptor genes (*DRD4* and *DRD5*) and ADHD.

Anxiety Disorders

The median age of onset for anxiety disorders is 11 years, and for this reason some genetic research has considered anxiety in childhood (Rutter et al., 1999). A twin study in the United Kingdom, with more than 4500 four-year-old twin pairs rated by their mothers, examined five components of anxiety (Eley et al., 2003). Three components are comparable to adult anxiety disorders (see Chapter 11): generalized anxiety, fears, and obsessive-compulsive behaviors; and two are specific to childhood: separation anxiety and shyness/inhibition. Heritability was greatest for obsessive-compulsive behaviors (65 percent) and shyness/inhibition (75 percent), with no evidence of shared environmental influence. A study of obsessive-compulsive symptoms in the United States and the Netherlands also found high heritability (55 percent) in both countries in twins aged 7, 10, and 12 years (Hudziak et al., 2004). Heritabilities of generalized anxiety and fears were about 40 percent. For fears, there was some evidence of shared environment, which is similar to results for specific fears in adults (Chapter 11).

Separation anxiety is interesting because in addition to showing moderate heritability (about 40 percent), substantial shared environmental influence was also found (35 percent) (Feigon, Waldman, Levy, & Hay, 2001). It is noteworthy that studies of maternal attachment of young children, which is indexed in part by separation anxiety, have also found evidence for shared environmental influence (Fearon et al., 2006; O'Connor & Croft, 2001; Roisman & Fraley, 2006). However, a follow-up of the 4-year-old UK twins at 6 years of age using DSM-IV diagnoses of separation anxiety disorder found high heritability of liability (73 percent) and no shared environmental influence (Bolton et al., 2006). These results are not necessarily contradictory because the studies that found shared environmental influence and modest heritability analyzed individual differences throughout the distribution whereas the latter study focused on the diagnosable extreme of separation anxiety.

Multivariate genetic analysis of the study of 4-year-old twins indicated that the five components of anxiety were moderately correlated genetically, although obsessive-compulsive behaviors were least related genetically to the others (Eley et al., 2003). The strong genetic overlap between anxiety and depression in adulthood (Chapter 11) suggests that depressive symptoms might also be profitably studied in childhood (Thapar & Rice, 2006). One twin study found differences before and after puberty in the etiology of the association between anxiety and depression (Silberg, Rutter, & Eaves, 2001).

Other Disorders

Although schizophrenia and bipolar disorder do not generally appear until early adulthood, genetic research on possible childhood forms of these disorders has been motivated by the principle that more severe forms of disorders are likely to have an earlier onset (Nicolson & Rapoport, 1999). In relation to childhood-onset schizophrenia, relatives of affected individuals are at increased risk of schizophrenia, suggesting a link between the child and adult forms of the disorder (Nicolson et al., 2003). The only twin study of childhood schizophrenia yielded high heritability, although the sample size was small (Kallmann & Roth, 1956). Interesting results concerning links with adult schizophrenia are emerging from a program of molecular genetic research incorporating brain endophenotypes (Addington et al., 2005; Gornick et al., 2005).

Childhood bipolar disorder appears to be more likely in families with adult bipolar disorder (Pavuluri, Birmaher, & Naylor, 2005). Linkage and candidate gene studies of childhood bipolar disorder have been reported, but no consistent results have emerged (Althoff, Faraone, Rettew, Morley, & Hudziak, 2005). When genes are identified that are responsible for the high heritabilities of adult schizophrenia and bipolar disorder, one of the next research questions will be whether these genes are also associated with juvenile forms of these disorders.

Other childhood disorders for which some genetic data are available include enuresis (bedwetting) and tics. Enuresis in children after 4 years of age is common, about 7 percent for boys and 3 percent for girls. An early family study found substantial familial resemblance (Hallgren, 1957). Strong genetic influence was found in three small twin studies (Bakwin, 1971; Hallgren, 1957; McGuffin et al., 1994). A large study of adult twins reporting retrospectively on enuresis in childhood yielded substantial heritability (about 70 percent) for both males and females (Hublin, Kaprio, Partinen, Koskenvuo, 1998). However, an equally large study of 3-year-old twins found only moderate genetic influence on nocturnal bladder control as reported by parents for boys (about 30 percent) and an even smaller effect in girls (about 10 percent) (Butler, Galsworthy, Rijsdijk, & Plomin, 2001). Candidate gene studies have not yielded replicable results (von Gontard, Schaumburg, Hollmann, Eiberg, & Rittig, 2001).

Tic disorders involve involuntary twitching of certain muscles, especially of the face, that typically begin in childhood. A twin study indicated that heritability of tics in children and adolescents was modest (about 30 percent) (Ooki, 2005). The same study showed that stuttering was highly heritable (about 80 percent) but that tics and stuttering are genetically different. Genetic research has focused on the most severe form, called Tourette disorder. Tourette disorder is rare (about 0.4 percent), whereas simple tics are much more common. Although family studies show little familial resemblance for simple tics, relatives of probands with chronic, severe tics characteristic of Tourette disorder are at increased risk for tics of all kinds (Pauls, 1990), for obsessive-compulsive disorder (Pauls, Towbin, Leckman, Zahner, & Cohen, 1986), and for ADHD (Pauls, Leckman, & Cohen, 1993). A twin study of Tourette disorder found concordances of 53 percent for identical twins and 8 percent for fraternal twins (Price, Kidd, Cohen, Pauls, & Leckman, 1985). Molecular genetic studies have so far not yielded replicable results. Linkage studies of large family pedigrees have been reported (e.g., Verkerk et al., 2006), but no clear major-gene linkages have been detected. The largest genomewide QTL linkage study of Tourette disorder recently suggested linkage on chromosome 2p (The Tourette Syndrome Association International Consortium for Genetics, 2007). Of the many candidate gene associations that have been reported (Pauls, 2003), only one gene has as yet yielded consistent associations: rare variants of a gene (SLITRK1) involved in dendritic growth (Abelson et al., 2005; Grados & Walkup, 2006).

Summary

Genetic research on childhood disorders has increased dramatically during the past two decades, in part fueled by finding such high heritability for autism and for hyperactivity. A general summary of twin results for the major domains of childhood psychopathology is presented in Figure 12.1. In addition to the high heritabilities of autism and its components, and of ADHD and its components, heritability is also exceptionally high for aggressive conduct disorder, obsessive-compulsive symptoms, and shyness. Just as interesting, however, are the moderate heritabilities for nonaggressive conduct disorder, generalized anxiety, fears, and separation anxiety. Especially noteworthy is the evidence for shared environmental influence for nonaggressive conduct disorder and separation anxiety, because evidence for shared environmental influence is so rarely seen. Nearly all of these results in childhood are based on parent or teacher reports of children's behavior. A twin study of psychopathology in adolescence using interviews with the twins themselves yielded quite different results (Ehringer, Rhee, Young, Corley, & Hewitt, 2006).

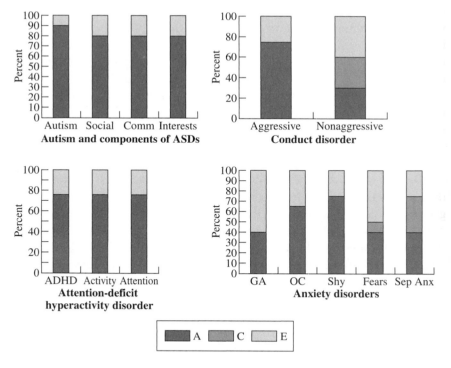

Figure 12.1. Summary of twin study estimates of genetic and environmental variances for major domains of childhood psychopathology. A = additive genetic variance; C = common (shared) environmental variance; E = nonshared environmental variance.

Two decades ago, autism was thought to be an environmental disorder. Now, twin studies suggest that it is one of the most heritable disorders, although the results of linkage studies and candidate gene studies have not as yet been successful. This lack of success might be due in part to the possibility that the components of the autistic triad—social, communication, and interests—are different genetically, even though each is highly heritable.

The DSM-IV category of attention-deficit and disruptive behavior disorders includes attention-deficit hyperactivity disorder (ADHD), which is highly heritable and shows no shared environmental influence. Multivariate genetic research suggests that its components of activity and attention overlap genetically, providing support for the construct of ADHD. Candidate gene studies of ADHD have yielded two dopamine receptor genes that show small but significant associations. This DSM-IV category also includes conduct disorder. Genetic research suggests that conduct disorder is heterogeneous, with aggressive conduct disorder showing substantial genetic influence and no shared environmental influence, in contrast to nonaggressive conduct disorder showing only modest genetic influence and moderate shared environmental influence.

Twin studies of parental ratings of anxiety in childhood suggest an interestingly diverse pattern of results. The highest heritability emerges for shyness, which is one of the most highly heritable personality traits (Chapter 13). Heritability is also very high for obsessive-compulsive symptoms, although results in adulthood are more mixed (Chapter 11). Heritability is more modest for generalized anxiety, which is comparable to results in adulthood (Chapter 11). These three aspects of anxiety show no evidence for shared environmental influence, which is similar to results for adult psychopathology but is even more surprising in childhood because children are living with their families. In contrast, fears and especially separation anxiety are notable for the evidence they show of shared environmental influence.

Some genetic influence has also been reported for childhood schizophrenia, childhood bipolar disorder, enuresis, and chronic tics, although much less genetic research has targeted these disorders.

CHAPTER THIRTEEN

Personality and Personality Disorders

I f you were asked what someone is like, you would probably describe various personality traits, especially those depicting extremes of behavior. "Jennifer is full of energy, very sociable, and unflappable." "Steve is conscientious, quiet, but quick tempered." Genetic researchers have been drawn to the study of personality because, within psychology, personality has always been the major domain for studying the normal range of individual differences, with the abnormal range being the provenance of psychopathology. A general rule emerging from behavioral genetic research is that common disorders are the quantitative extreme of the same genetic and environmental factors that contribute to the normal range of variation. In other words, some psychopathology may be the extreme of normal variation in personality. We will return to the links between personality and psychopathology near the end of this chapter, after we have described basic research on personality.

Personality traits are relatively enduring individual differences in behavior that are stable across time and across situations (Pervin & John, in press). In the 1970s, there was an academic debate about whether personality exists, a debate reminiscent of the nature-nurture debate. Some psychologists argued that behavior is more a matter of the situation than of the person, but it is now generally accepted that both are important and can interact (Kenrick & Funder, 1988; Rowe, 1987). Cognitive abilities (Chapters 8 and 9) also fit the definition of enduring individual differences, but they are usually considered separately from personality. Another definitional issue concerns temperament, personality traits that emerge early in life and, according to some researchers (e.g., Buss & Plomin, 1984), may be more heritable. However, there are many different definitions of temperament (Goldsmith et al., 1987), and the supposed distinction between temperament and personality will not be emphasized here.

Genetic research on personality is extensive and is described in several books (Cattell, 1982; Eaves, Eysenck, & Martin, 1989; Loehlin, 1992; Loehlin & Nichols, 1976) and hundreds of research papers (Krueger, Johnson, & Caspi, in press). We shall provide only an overview of this huge literature, in part because its basic message is quite simple: Genes make a major contribution to individual differences in personality whereas shared environment does not, especially when assessed by a self-report questionnaire; environmental influence on personality is almost entirely of the nonshared variety. After a brief overview of these results, we shall describe other findings from genetic research on personality and personality disorders, and recent reports of specific genes associated with personality and personality disorders.

Genetic research on animal personality has focused on traits such as fearfulness and activity level. Some of this work was described in Chapter 5. Animal research is especially useful for identifying specific genes related to personality, as described in Chapter 6, and for functional genomic studies of how genes affect behavior, as described in Chapter 15. The present chapter focuses on human personality.

Self-Report Questionnaires

The vast majority of genetic research on personality involves self-report questionnaires administered to adolescents and adults. Such questionnaires include from dozens to hundreds of items like, "I am usually shy when meeting people I don't know well" or "I am easily angered." People's responses to such questions are remarkably stable, even over several decades (Costa & McCrae, 1994).

Thirty years ago, a landmark study involving nearly 800 pairs of adolescent twins and dozens of personality traits reached two major conclusions that have stood the test of time (Loehlin & Nichols, 1976). First, nearly all personality traits show moderate heritability. This conclusion might seem surprising, because you would expect some traits to be highly heritable and other traits not to be heritable at all. Second, although environmental variance is also important, virtually all the environmental variance makes children growing up in the same family no more similar than children in different families. This category of environmental effects is called nonshared environment. The second conclusion is also surprising, because theories of personality from Freud onward assumed that parenting played a critical role in personality development. This important finding is discussed in Chapter 16.

Genetic research on personality has focused on five broad dimensions of personality, called the *Five-Factor Model (FFM)*, that encompass many aspects of personality (Goldberg, 1990). The best studied of these are extraversion and neuroticism. Extraversion includes sociability, impulsiveness, and liveliness. Neuroticism (emotional instability) involves moodiness, anxiousness, and irritability. These two traits plus the three others included in the FFM create the acronym OCEAN: openness to experience (culture), conscientiousness (conformity, will to achieve), extraversion, agreeableness (likability, friendliness), and neuroticism.

Genetic results for extraversion and neuroticism are summarized in Table 13.1(Loehlin, 1992). In five large twin studies in five different countries, with a total sample size of 24,000 pairs of twins, results indicate moderate genetic influence. Correlations are about .50 for identical twins and about .20 for fraternal twins. Studies of twins reared apart also indicate genetic influence, as do adoption studies of extraversion. For neuroticism, adoption results point to less genetic influence than do the twin studies—indeed, the sibling data indicate no genetic influence at all. Lower heritability in adoption than in twin studies could be due to nonadditive genetic variance, which makes identical twins more than twice as similar as first-degree relatives (Eaves, Heath, Neale, Hewitt, & Martin, 1998; Eaves, Heath, et al., 1999; Keller, Coventry, Heath, & Martin, 2005; Loehlin, Neiderhiser, & Reiss, 2003; Plomin, Corley, Caspi, Fulker, & DeFries, 1998). It could also be due to a special environmental effect that boosts identical twin similarity (Plomin & Caspi, 1999). Model-fitting analyses across these twin and adoption designs produce heritability estimates of about 50 percent for extraversion and about 40 percent for neuroticism (Loehlin, 1992). The fact that the heritability estimates are much less than 100 percent implies that environmental factors are important, but, as mentioned earlier, this environmental influence is almost entirely due to nonshared environmental effects, although there is some evidence that shared environmental influence may be more important at the extremes of personality (Pergadia, Madden, et al., 2006).

TABLE 13.1

Twin, Family, and Adoption Results for Extraversion and Neuroticism

	Correlation	
Type of Relative	**Extraversion**	**Neuroticism**
Identical twins reared together	.51	.46
Fraternal twins reared together	.18	.20
Identical twins reared apart	.38	.38
Fraternal twins reared apart	.05	.23
Nonadoptive parents and offspring	.16	.13
Adoptive parents and offspring	.01	.05
Nonadoptive siblings	.20	.09
Adoptive siblings	−.07	.11

SOURCE: *Loehlin (1992).*

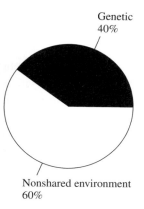

Genetic
40%

Nonshared environment
60%

Figure 13.1 Genetic results for personality traits assessed by self-report questionnaires are remarkably similar, suggesting that 30 to 50 percent of the variance is due to genetic factors. Environmental variance is also important, but hardly any environmental variance is due to shared environmental influence.

Heritabilities in the 30 to 50 percent range are typical of personality results (Figure 13.1), although much less genetic research has been done on the other three traits of the FFM. Also, openness to experience, conscientiousness, and agreeableness have been measured differently in different studies because, until recently, no standard measures were available. A model-fitting summary of family, twin, and adoption data for scales of personality thought to be related to these three traits yielded heritability estimates of 45 percent for openness to experience, 38 percent for conscientiousness, and 35 percent for agreeableness, with no evidence of shared environmental influence (Loehlin, 1992). The first genetic study to use a measure specifically designed to assess the FFM factors found similar estimates in an analysis of twins reared together and twins reared apart, except that agreeableness showed lower heritability (12 percent) (Bergeman et al., 1993). Another twin study yielded similar moderate heritabilities for all of the FFM factors (Jang, Livesley, & Vernon, 1996).

Do these broad FFM factors represent the best level of analysis for genetic research? Multivariate genetic research supports the FFM structure but also suggests a hierarchical model in which subtraits within each FFM factor show significant unique genetic variance not shared with other traits in the factor (Jang, McCrae, Angleitner, Riemann, & Livesley, 1998; Jang et al., 2006; Loehlin, 1992). For example, extraversion includes diverse traits such as sociability, impulsiveness, and liveliness, as well as activity, dominance, and sensation seeking. Each of these traits has received some attention in genetic research but not nearly as much as the more global traits of extraversion and neuroticism.

Several theories of personality development have been proposed about other ways in which personality should be sliced, and similar results have been found for the different traits highlighted in these theories (Kohnstamm, Bates, & Rothbart, 1989). For example, a neurobiologically oriented theory organizes personality into four different domains: novelty seeking, harm avoidance, reward dependence, and persistence (Cloninger, 1987). Similar twin study results have been found for these dimensions (Heiman, Stallings, Young, & Hewitt, 2004;

Stallings, Hewitt, Cloninger, Heath, & Eaves, 1996). Sensation seeking, which is related to conscientiousness as well as to extraversion (Zuckerman, 1994), is especially interesting because it is the domain of the first association reported between a specific gene and normal personality, as described later. In two large twin studies, heritabilities of about 60 percent were found for a measure of general sensation seeking (Fulker, Eysenck, & Zuckerman, 1980; Koopmans, Boomsma, Heath, & van Doornen, 1995). This evidence for substantial genetic influence is supported by results from a study of identical twins reared apart, which yielded a correlation of .54 (Tellegen et al., 1988). Sensation seeking itself can be broken down into components, such as disinhibition (seeking sensation through social activities such as parties), thrill seeking (desire to engage in physically risky activities), experience seeking (seeking novel experiences through the mind and senses), and boredom susceptibility (intolerance for repetitive experience). Each of these subscales also shows moderate heritability.

One of the most surprising findings from genetic research on personality questionnaires is that the many traits that have been studied all show moderate genetic influence and no influence of shared environment. It is also surprising that studies have not found any personality traits assessed by self-report questionnaire that consistently show low or no heritability in twin studies, in contrast to childhood psychopathology (Chapter 12), where some disorders are more heritable than others and some disorders yield more shared environmental influence than others. But can this be true for personality? One way to explore this issue is to use measures of personality other than self-report questionnaires to investigate whether this result is somehow due to self-report measures.

Other Measures of Personality

A study of adult twins in Germany and Poland compared twin results from self-report questionnaires and from ratings by peers for measures of the FFM personality factors for nearly a thousand pairs of twins (Riemann, Angleitner, & Strelau, 1997). Each twin's personality was rated by two different peers. The average correlation between the two peer ratings was .61, a result indicating substantial agreement concerning each twin's personality. The averaged peer ratings correlated .55 with the twins' self-report ratings, a result indicating moderate validity of self-report ratings. Figure 13.2 shows the results of twin analyses for self-report data and peer ratings averaged across two peers. The results for self-report ratings are similar to other studies. The exciting result is that peer ratings also show significant genetic influence, although somewhat less than self-report ratings. For two of the five traits (extraversion and agreeableness), peer ratings suggest greater influence of shared environment than self-report ratings, although these differences are not nearly statistically significant. Importantly, multivariate genetic analysis indicates that the same genetic factors are largely involved in self-report and peer ratings, a result providing strong evi-

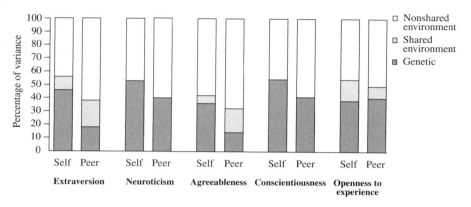

Figure 13.2 Genetic (dark), shared environment (gray), and nonshared environment (white) components of variance for self-report ratings and peer ratings for the FFM personality traits. Components of variance were calculated from identical twin (660 pairs) and same-sex fraternal twin (200 pairs) correlations presented by Riemann, Angleitner, & Strelau (1997). Heritability was estimated by doubling the difference between the identical and fraternal twin correlations, with the proviso that heritability cannot exceed the identical twin correlation. Shared environment was estimated as the difference between the identical twin correlation and heritability. The remainder of the variance was attributed to nonshared environment, which includes error of measurement. These estimates differ somewhat from the model-fitting results reported by Riemann et al. because their results were presented for best-fitting models in which parameters were dropped unless they were significant. As a result, shared environment, which the twin method has little power to detect, was not significant and thus not included in the best-fitting models presented by Riemann et al., and heritability estimates sometimes exceeded identical twin correlations. (Used with permission from Plomin & Caspi, 1999.)

dence for the genetic validity of self-report ratings. An earlier study used twin reports about each other, and it also found similar evidence for genetic influence on personality traits, whether assessed by self-report or by the co-twin (Heath, Neale, Kessler, Eaves, & Kendler, 1992).

Genetic researchers interested in personality in childhood were forced to use measures other than self-report questionnaires. For the past 30 years, this research has relied primarily on ratings by parents, but twin studies using parent ratings have yielded odd results. Correlations for identical twins are high and correlations for fraternal twins are very low, sometimes even negative. It is likely that these results are due to contrast effects in which parents of fraternal twins contrast the twins (Plomin, Chipuer, & Loehlin, 1990). For example, parents might report that one twin is the active twin and the other is the inactive twin, even though, relative to other children that age, the twins are not really very different from each other (Carey, 1986; Eaves, 1976; Neale & Stevenson, 1989).

Furthermore, adoption studies using parent ratings in childhood find little evidence for genetic influence (Loehlin, Willerman, & Horn, 1982; Plomin, Coon, Carey, DeFries, & Fulker, 1991; Scarr & Weinberg, 1981; Schmitz, 1994). A combined twin study and stepfamily study of parent ratings of adolescents found significantly greater heritability estimates for twins than for nontwins and confirmed that parent ratings are subject to contrast effects (Saudino, McGuire, Reiss, Hetherington, & Plomin, 1995). As mentioned in relation to self-report questionnaires, such findings might also be due to nonadditive genetic variance. However, the weight of evidence indicates that genetic results for parent ratings of personality are due in part to contrast effects (Saudino, Wertz, Gagne, & Chawla, 2004).

Other measures of children's personality, such as behavioral ratings by observers, show more reasonable patterns of results in both twin and adoption studies (Braungart, Plomin, DeFries, & Fulker, 1992; Cherny et al., 1994; Goldsmith & Campos, 1986; Matheny, 1980; Plomin & Foch, 1980; Plomin, Foch, & Rowe, 1981; Plomin et al., 1993; Saudino, Plomin, & DeFries, 1996; Wilson & Matheny, 1986). For example, genetic influence has been found in observational studies of young twins for a dimension of fearfulness called behavioral inhibition (Matheny, 1989; Robinson, Kagan, Reznick, & Corley, 1992), for shyness observed in the home and the laboratory (Cherny et al., 1994), and for activity level that is measured by using actometers, which record movement (Saudino & Eaton, 1991). Because evidence for genetic influence is so widespread, even for observational measures, it is interesting that observer ratings of personality in the first few days of life have found no evidence for genetic influence (Riese, 1990) and that individual differences in smiling in infancy also show no genetic influence (Plomin, 1987).

Although the focus of this overview has been on human personality, a recent study of over 10,000 behaviorally tested dogs of two breeds also found moderate heritability (about 25 percent) across 16 personality traits (Saetre et al., 2006). Moreover, a single genetic factor which the authors called *shyness-boldness* accounted for most the genetic variance of the 16 traits with the possible exception of aggressiveness.

SUMMING UP

Twin studies using self-report questionnaires of personality typically find heritabilities ranging from 30 to 50 percent, with no evidence for shared environmental influence. Adoption studies find somewhat less genetic influence, perhaps as a result of the presence of nonadditive genetic variance. Extraversion and neuroticism have been studied most and yield heritability estimates of 50 and 40 percent, respectively, across both twin and adoption studies. A twin study of peer ratings yielded similar results. Parent ratings of children's personality are affected by contrast effects that exaggerate estimates of genetic influence in twin studies. Observational measures of children's personality also show genetic influence in twin and adoption studies.

More systematic genetic research is needed that explicitly incorporates different sources of personality data (Goldsmith, 1993). Nonetheless, the results so far are encouraging in that the pervasive evidence for genetic influence on personality gleaned from self-report questionnaires can be confirmed by using other measures. Moreover, multivariate genetic analyses across multiple sources suggest genetic validity for personality assessment: Genetic factors largely account for what is in common across ratings in the home and laboratory (Cherny et al., 1994), across teacher and tester ratings (Schmitz, Saudino, Plomin, Fulker, & DeFries, 1996), and across parent and laboratory ratings (Goldsmith, Buss, & Lemery, 1997).

Other Findings

There is a renaissance of genetic research on personality, which will be accelerated by research showing the association between personality and psychopathology and by reports of specific genes associated with personality (Bouchard & Loehlin, 2001). Both of these trends will be discussed later in this chapter. As just described, another example of new directions for personality research is increasing interest in measures other than self-report questionnaires. Three other examples include research on personality in different situations, developmental change and continuity, and the role of personality in the interplay between nature and nurture.

Situations

It is interesting, in relation to the person-situation debate mentioned earlier, that some evidence suggests that genetics is involved in situational change as well as in stability of personality across situations (Phillips & Matheny, 1997). For example, in one study, observers rated the adaptability of infant twins in two laboratory settings, unstructured free play and test taking (Matheny & Dolan, 1975). Adaptability differed to some extent across these situations, but identical twins changed in more similar ways than fraternal twins did, an observation implying that genetics contributes to change as well as to continuity across situations for this personality trait, a finding confirmed in a more recent study of person-situation interaction (Borkenau, Riemann, Spinath, & Angleitner, 2006). Such results might differ for other personality traits. For example, a twin study of shyness found that genetic factors largely contribute to stability across observations in the home and in the laboratory; environmental factors account for shyness differences between these situations (Cherny et al., 1994). A twin study using a questionnaire to assess personality in different situations found that genetic factors contribute to personality changes across situations (Dworkin, 1979). Even patterns of responding across items of personality questionnaires show genetic influence (Eaves & Eysenck, 1976; Hershberger, Plomin, & Pedersen, 1995).

Development

Does heritability change during development? Unlike general cognitive ability, which shows increases in heritability throughout the life span (Chapter 8), it is more difficult to draw general conclusions concerning personality development, in part because there are so many personality traits. In general, heritability appears to increase during infancy (Goldsmith, 1983; Loehlin, 1992), starting with zero heritability for personality during the first days of life (Riese, 1990). Of course, what is assessed as personality during the first few days of life is quite different from what is assessed later in development, and the sources of individual differences might also be quite different in neonates. Throughout the rest of the life span, it is clear that twins become less similar as time goes by, but this decreasing similarity occurs for identical twins as much as for fraternal twins for most personality traits, an observation suggesting that heritability does not change in childhood (McCartney et al., 1990), adolescence (Rettew, Vink, Willemsen, Doyle, Hudziak, and Boomsma, 2006), or adulthood (Loehlin & Martin, 2001).

A second important question about development concerns the genetic contribution to either continuity or change from age to age. For cognitive ability, genetic factors largely contribute to stability from age to age rather than to change, although some evidence can be found, especially in childhood, for genetically influenced change (Chapter 8). Although less well studied than cognitive ability, developmental findings for personality appear to be similar (Bratko & Butkovic, 2007; Loehlin, 1992), even in late adulthood (Johnson, McGue, & Krueger, 2005; Read, Vogler, Pedersen, & Johansson, 2006).

Nature-Nurture Interplay

Another new direction for genetic research on personality involves the role of personality in explaining a fascinating finding: Environmental measures widely used in psychological research show genetic influence. As discussed in Chapter 16, genetic research consistently shows that family environment, peer groups, social support, and life events often show as much genetic influence as measures of personality. The finding is not as paradoxical as it might seem at first. Measures of psychological environments in part assess genetically influenced characteristics of the individual. Personality is a good candidate to explain this genetic influence, because personality can affect how people select, modify, construct, or perceive their environments. For example, in adulthood, genetic influence on personality has been reported to contribute to genetic influence on parenting in two studies (Chipuer & Plomin, 1992; Losoya, Callor, Rowe, & Goldsmith, 1997) but not in another (Vernon, Jang, Harris, & McCarthy, 1997).

Genetic influence on perceptions of life events can be entirely accounted for by the FFM personality factors (Saudino, Pedersen, Lichtenstein, McClearn, & Plomin, 1997). These findings are not limited to self-report questionnaires. For example, genetic influence found on an observational measure of home environments can be explained entirely by genetic influence on a tester-rated measure of attention called task orientation (Saudino & Plomin, 1997).

CLOSE UP

Hill Goldsmith was a biology major as an undergraduate. During his senior year, he developed interests in human genetics and the psychology of individual differences. These interests eventually led to graduate study in behavioral genetics at the University of Minnesota, where his adviser was Irving Gottesman. Goldsmith shared Gottesman's interest in psychiatric genetics but focused his early research on personality development during infancy and early childhood. He developed a theory of temperament within the context of what later became known as "affective science," the study of emotion. Goldsmith also became a full-fledged developmental psychologist, with both behavioral genetic and nongenetic aspects to his research. He continues to study the early emotional development of twins, with emphasis on laboratory-based assessment and on physiological as well as behavioral measures. Over the past 15 years, his research and teaching have begun to focus more on childhood anxiety and the autism spectrum.

Goldsmith's main advice to students who enter the field of behavioral genetics is to become a true expert in both genetic methods and some major content area of behavioral study (e.g., cognitive science, developmental neuroscience, clinical psychology). In short, "Know your phenotype!" He also emphasizes that young researchers must learn to collaborate effectively with researchers from other disciplines. Goldsmith currently works in the developmental and clinical areas of the Department of Psychology and in the Social and Emotional Processes Group at the Waisman Center for developmental disabilities, both at the University of Wisconsin–Madison.

Personality and Social Psychology

Social psychology focuses on the behavior of groups, whereas individual differences are in the spotlight for personality research. For this reason, there is not nearly as much genetic research relevant to social psychology as there is for personality. However, some areas of social psychology border on personality, and genetic research has begun at these borders. Three examples are relationships, self-esteem, and attitudes.

Relationships

Genetic research has addressed parent-offspring relationships, romantic relationships, and sexual orientation.

Parent-offspring relationships   Relationships between parents and offspring vary widely in their warmth (such as affection and support) and control (such as monitoring and organization). To what extent do genetic influences on parents and on offspring contribute to relationships? If identical twins are more similar in the qualities of their relationships than fraternal twins, this difference indicates genetic influence on relationships. For example, the first research of this sort involved adolescent twins' perceptions of their relationships with their parents. In two studies with different samples and different measures, genetic influence was found for twins' perceptions of their mothers' and fathers' warmth toward them (Rowe, 1981, 1983a). In contrast, adolescents' perceptions of their parents' control did not show genetic influence. One possible explanation is that parental warmth reflects genetically influenced characteristics of their children, but parental control does not (Lytton, 1991). Dozens of subsequent twin and adoption studies have found similar results that point to substantial genetic influences in most aspects of relationships, not just between parents and offspring, but also between siblings and friends (Plomin, 1994).

A major area of developmental research on parent-offspring relationships involves attachment between infant and caregiver, as assessed in the so-called Strange Situation, a laboratory-based assessment in which mothers briefly leave their child with an experimenter and then return (Ainsworth, Blehar, Waters, & Wall, 1978). Sibling concordance of about 60 percent has been reported for attachment classification (van Ijzendoorn et al., 2000; Ward, Vaughn, & Robb, 1988). The first systematic twin study of attachment using the Strange Situation found only modest genetic influence and substantial influence of shared environment (O'Connor & Croft, 2001). For 110 twin pairs, MZ and DZ concordances for attachment type were 70 and 64 percent, respectively; for a continuous measure of attachment security, MZ and DZ correlations were .48 and .38, respectively. Although another twin study based on observations rather than the Strange Situation found evidence for greater genetic influence (Finkel et al., 1998), three other studies using the Strange Situation also found modest heritability and substantial shared environmental influence (Bokhorst et al., 2003; Fearon et al., 2006; Roisman & Fraley, 2006). In addition, a small twin study using a different measure of attachment found similar results for infant-father attachment (Bakermans-Kranenburg, van Uzendoorn, Bokhorst, & Schuengel, 2004). As described in Chapter 12, twin studies of separation anxiety disorder, which is related to attachment, also generally show modest heritability and substantial shared environmental influence.

Another component of relationships is empathy. One twin study of infants used videotape observations of the empathic responding of infant twins following simulations of distress in the home and in the laboratory (Zahn-Waxler, Robinson, & Emde, 1992). Evidence was found for genetic influence for some aspects of the infants' empathic responses. A twin study of parent and teacher ratings of children's prosocial behavior during infancy and childhood found increasing genetic influence and decreasing shared environmental influence (Knafo & Plomin, 2006a). Twin studies of empathic emotional responses also yielded evidence for genetic influence in adolescence (Davis, Luce, & Kraus, 1994) and in adulthood (Rushton, 2004).

Romantic relationships Like parent-offspring relationships, romantic relationships differ widely in various aspects, such as closeness and passion. The first genetic study of styles of romantic love is interesting because it showed no genetic influence (Waller & Shaver, 1994). The average twin correlations for six scales (for example, companionship and passion) were .26 for identical twins and .25 for fraternal twins, results implying some shared environmental influence but no genetic influence. Similar results have been found for initial attraction in mate selection (Lykken & Tellegen, 1993). In other words, genetics plays no role in the type of romantic relationships we choose. Perhaps love is blind, at least from the DNA point of view.

Sexual orientation An early twin study of male homosexuality reported remarkable concordance rates of 100 percent for identical twins and 15 percent for fraternal twins (Kallmann, 1952). However, a later twin study found less extreme concordances of 52 and 22 percent, respectively, and concordance of 22 percent for genetically unrelated adoptive brothers (Bailey & Pillard, 1991); other twin studies found even less genetic influence and more influence of shared environment (Bailey, Dunne, & Martin, 2000; Kendler, Thornton, Gilman, & Kessler, 2000). A small twin study of lesbians also yielded evidence for moderate genetic influence (Bailey, Pillard, Neale, & Agyei, 1993). This area of research received considerable attention because of reports of linkage between homosexuality and a region at the tip of the long arm of the X chromosome (Hamer, Hu, Magnuson, Hu, & Pattatucci, 1993; Hu et al., 1995). The X chromosome has been targeted because it was thought that male homosexuality is more likely to be transmitted from the mother's side of the family, but recent studies do not find an excess of maternal transmission (Bailey et al., 1999). The

X linkage was not replicated in a subsequent study (Rice, Anderson, Risch, & Ebers, 1999). When genetic research touches on especially sensitive issues such as sexual orientation, it is important to keep in mind earlier discussions (see Chapter 5) about what it does and does not mean to show genetic influence (Pillard & Bailey, 1998).

Self-Esteem

A key variable for adjustment is self-esteem, which is also referred to as a sense of self-worth. Research on the etiology of individual differences in self-esteem has focused on the family environment (Harter, 1983). It is surprising that the possibility of genetic influence had not been considered previously, because it seems likely that genetic influence on personality and psychopathology (especially depression, for which low self-esteem is a core feature) could also affect self-esteem. Twin and adoption studies of self-esteem have been reported for teacher and parent ratings in middle childhood (Neiderhiser & McGuire, 1994), for self-ratings in adolescence (Kamakura, Ando, & Ono, 2007; Neiss, Sedikides, & Stevenson, 2006), for teacher, parent, and self-ratings in adolescence (McGuire, Neiderhiser, Reiss, Hetherington, & Plomin, 1994), and for self-ratings in adulthood (Roy, Neale, & Kendler, 1995). These studies point to modest genetic influence on self-esteem, but no influence of shared family environment.

Attitudes and Interests

Social psychologists have long been interested in the impact of group processes on change and continuity in attitudes and beliefs. Although it is recognized that social factors are not solely responsible for attitudes, it has been a surprise to find that genetics makes a major contribution to individual differences in attitudes. A core dimension of attitudes is traditionalism, which involves conservative versus liberal views on a wide range of issues. A measure of this attitudinal dimension was included in an adoption study of personality as a control variable, in the sense that it was not expected to be heritable (Scarr & Weinberg, 1981). However, the results indicated that this measure was as heritable as the personality measures. In several twin studies (Eaves et al., 1989), including a study of twins reared apart (McCourt, Bouchard, Lykken, Tellegen, & Keyes, 1999; Tellegen et al., 1988), identical twin correlations are typically about .65 and fraternal twin correlations are about .50.

   This pattern of twin correlations suggests heritability of about 30 percent and shared environmental influence of about 35 percent. However, assortative mating is higher for traditionalism than for any other psychological trait, with spouse correlations of about .50, unlike personality, which shows little assortative mating. Assortative mating inflates the fraternal twin correlation for interests, thereby lowering estimates of heritability and raising estimates of shared environment (Chapter 8). When assortative mating is taken into account, heritability is estimated to be about 50 percent and shared environmental influence is about 15 percent (Eaves et al., 1989; Olson et al., 2001). A twin-family analysis confirmed heritabilities of about 50 percent for traditionalism as well as showing similarly high heritabilities for sexual and religious attitudes but lower heritabilities for attitudes about taxes, the military, and politics (15 to 30 percent) (Eaves, Heath, et al., 1999). Religious attitudes were the focus of a special issue of the journal *Twin Research* (Eaves, D'Onofrio, & Russell, 1999). A recent study suggests that the heritability of religiousness increases from adolescence to adulthood (Koenig, McGue, Krueger, & Bouchard, 2005).

Sometimes these results are held up for ridicule—how can attitudes about royalty or nudist camps be heritable? We hope that by now you can answer this question (see Chapter 5), but it has been put particularly well in the context of social attitudes:

> We may view this as a kind of cafeteria model of the acquisition of social attitudes. The individual does not inherit his ideas about fluoridation, royalty, women judges, and nudist camps; he learns them from his culture. But his genes may influence which ones he elects to put on his tray. Different cultural institutions—family, church, school, books, television—like different cafeterias, serve up somewhat different menus, and the choices a person makes will reflect those offered him as well as his own biases. (Loehlin, 1997, p. 48)

This theme of nature operating via nurture will be picked up again in Chapter 16.

Social psychology traditionally uses the experimental approach rather than investigating naturally occurring variation (Chapter 5). There is a need to bring together these two research traditions. For example, Tesser (1993), a social psychologist, separated attitudes into those that were more heritable (such as attitudes about the death penalty) and those that were less heritable (coeducation and the truth of the Bible). In standard social psychology experimental situations, the more heritable items were found to be less susceptible to social influence and more important in interpersonal attraction (Tesser, Whitaker, Martin, & Ward, 1998).

A related area is vocational interests, which involve personality-like dimensions such as realistic, intellectual, social, enterprising, conventional, and artistic. Results from twin studies for vocational interests are similar to results for personality questionnaires, with identical twin correlations of about .50 and fraternal twin correlations of about .25 (Roberts & Johansson, 1974). Moderate genetic influence also emerged in an adoption study of vocational interests (Scarr & Weinberg, 1978a). A combined twin and adoption study indicated about 35 percent heritability for most vocational interests (Betsworth et al., 1994). A multivariate genetic analysis between vocational interests and personality found some common genetic basis between them (Harris, Vernon, Johnson, & Jang, 2006).

Evidence for genetic influence was also found in twin studies of work values (Keller, Bouchard, Segal, & Dawes, 1992) and job satisfaction (Arvey,

Bouchard, Segal, & Abraham, 1989). Some evidence for slight (about 10 percent) shared environmental influence is found in vocational interests and job satisfaction (Gottfredson, 1999).

SUMMING UP

Genetic research on personality across situations and across time suggests that genetics is largely responsible for continuity and that change is largely due to environmental factors. Some research indicates that heritability increases during development. New directions for research include the use of measures other than self-report questionnaires and the role of personality in explaining genetic influence on measures of the environment. Another new direction is the interface between personality and social psychology. Recent research has found evidence for genetic influence on social relationships (parent-offspring relationships and sexual orientation, but not romantic relationships), self-esteem, attitudes, and vocational interests.

CLOSE UP

Matt McGue is a professor of psychology and a member of the Institute of Human Genetics at the University of Minnesota. His interest in human behavioral genetics dates from the late 1970s, when he was fortunate to be a graduate student at the University of Minnesota. At that time, Minnesota was one of the leaders in the nascent field of behavioral genetics: Irv Gottesman was undertaking his important research on the genetics of schizophrenia, Tom Bouchard was initiating his landmark study of reared-apart twins, and David Lykken was establishing the Minnesota Twin Registry. McGue's current research focuses on the behavioral genetics of substance abuse disorders and the normal aging process. He codirects the Minnesota Center for Twin and Family Research, a coordinated series of twin and adoption studies that aim to investigate the influence of genetic and environmental factors on the transition from late adolescence to early adulthood. He also collaborates with colleagues from the Danish Twin Register on a series of twin, sibling, and molecular genetic studies aimed at identifying and characterizing the combined effect of genetic and environmental factors on late-life cognitive, physical, and emotional functioning.

Personality Disorders

To what extent is psychopathology the extreme manifestation of normal dimensions of personality? It has long been suggested that this is the case for some psychiatric disorders (e.g., Cloninger, 2002; Eysenck, 1952; Livesley, Jang, & Vernon, 1998). An important general lesson from behavioral genetic research on psychopathology (Chapter 11 and 12) as well as cognitive disabilities (Chapter 7) is that common disorders are the quantitative extreme of the same genetic and environmental factors that contribute to the normal range of variation. With cognitive disabilities such as reading disability, it is easy to see what normal variation is—variation in reading ability is normally distributed and reading disability is the low end of that distribution. However, what are the dimensions of normal variation associated with depression or other types of psychopathology?

Chapter 11 ended with a multivariate genetic model that proposes two broad categories of psychopathology. The internalizing category includes depression and the anxiety disorders, and the externalizing category includes antisocial behavior and drug abuse. One of the most important findings from genetic research on personality is the extent of genetic overlap between the internalizing category of psychopathology and the personality factor of neuroticism. As mentioned earlier, neuroticism does not mean *neurotic* in the sense of being nervous; neuroticism refers to a general dimension of emotional instability, which includes moodiness, anxiousness, and irritability. Recent twin studies found that genetic factors shared between neuroticism and internalizing disorders accounted for between one-third and one-half of the genetic risk (Hettema, Neale, Myers, Prescott, & Kendler, 2006; Mackintosh, Gatz, Wetherell, & Pedersen, 2006). Another study reported genetic correlations of about .50 between neuroticism and major depression (Kendler, Gatz, Gardner, & Pedersen, 2006b). Similar findings have emerged from earlier multivariate genetic studies (Eaves et al., 1989).

In summary, the internalizing category of psychopathology is similar genetically to the personality factor of neuroticism. What about the externalizing category of psychopathology? Although it would be wonderfully symmetrical if extraversion predicted externalizing psychopathology, this is not the case (Khan, Jacobson, Gardner, Prescott, & Kendler, 2005). However, several studies have shown that aspects of extraversion—especially novelty seeking, impulsivity, and disinhibition—predict externalizing psychopathology (Krueger, Caspi, Moffitt, Silva, & McGee, 1996). A twin study that addressed the causes of overlap between a disinhibitory dimension of personality and externalizing psychopathology found that some of the overlap is genetic in origin, although most of the genetic influence on disinhibitory personality is independent of externalizing psychopathology (Kreuger et al., 2002).

Genetic research on the overlap between personality and psychopathology has focused on an area of psychopathology called *personality disorders*. Unlike psychopathology, described in Chapters 10 and 11, personality disorders are personality traits that cause significant impairment or distress. People with personality disorders regard their disorder as part of who they are, their personality, rather than as a condition that can be treated. That is, they do not feel that they were once well and are now ill. For this reason, DSM-IV separates personality disorders from clinical syndromes. This category of disorders (called Axis II), which also includes general cognitive disability (called *mental retardation* in DSM-IV), refers to long-term disorders that date from childhood. Although the reliability, validity, and utility of personality disorders have long been questioned, research has addressed the genetics of personality disorders and their links to normal personality and to other psychopathology (Jang, 2005; Nigg & Goldsmith, 1994). Increasingly, personality disorders are being considered as dimensions rather than categories, and this will also increase their use in genetic research (Widiger & Trull, 2007).

DSM-IV recognizes ten personality disorders, but only three have been investigated systematically in genetic research: schizotypal, obsessive-compulsive, and antisocial personality disorders. Most genetic research has targeted antisocial personality disorder because of its relevance to criminal behavior. For this reason, antisocial personality disorder is discussed in a separate section that follows a brief presentation of the other two personality disorders.

Schizotypal and Obsessive-Compulsive Personality Disorders

Schizotypal personality disorder involves less intense schizophrenic-like symptoms and, like schizophrenia, clearly runs in families (e.g., Baron, Gruen, Asnis, & Lord, 1985; Siever et al., 1990). The results of a small twin study suggested genetic influence, yielding 33 percent concordance for identical twins and 4 percent for fraternal twins (Torgersen et al., 2000). Twin studies using dimensional measures of schizotypal symptoms in unselected samples of twins also found evidence for genetic influence (Claridge & Hewitt, 1987; Kendler, Czajkowski, et al., 2006; Kendler & Hewitt, 1992).

Genetic research on schizotypal personality disorder focuses on its relationship to schizophrenia and has consistently found an excess of the disorder among first-degree relatives of schizophrenic probands. A summary of such studies found that the risks of schizotypal personality disorder are 11 percent for the first-degree relatives of schizophrenic probands and 2 percent in control families (Nigg & Goldsmith, 1994). Adoption studies have played an important role in showing that the disorder is part of the genetic spectrum of schizophrenia. For example, in the Danish adoption study (see Chapter 10), the rate of schizophrenia in the biological first-degree relatives of schizophrenic adoptees was 5 percent, but 0 percent in their adoptive relatives and relatives of control adoptees (Kety et al., 1994). When schizotypal personality disorder was included in the diagnosis, the rates rose to 24 and 3 percent, respectively, implying greater genetic influence for the spectrum of schizophrenia that includes schizotypal personality disorder (Kendler, Gruenberg, & Kinney, 1994). Twin studies also suggest that schizotypal personality disorder is related genetically to schizo-phrenia (Farmer et al., 1987), especially for the negative (anhedonia) rather than the positive (delusions) aspects of *schizotypy* (Torgersen et al., 2002). Community samples of twins suggest that the negative and positive aspects of schizotypy dif-fer genetically (Linney et al., 2003) and that schizotypy is genetically related to the schizophrenia spectrum (Jang, Woodward, Lang, Honer, & Livesley, 2005).

Obsessive-compulsive personality disorder sounds as if it is a milder version of the obsessive-compulsive type of anxiety disorder (OCD, described in Chap-ter 11); family studies provide some empirical support for this. However, the diagnostic criteria for these two disorders are quite different. The compulsion of OCD is a single sequence of bizarre behaviors, whereas the personality dis-order is more pervasive, involving a general preoccupation with trivial details that leads to difficulties in making decisions and getting anything accomplished. Only one small twin study of diagnosed obsessive-compulsive personality dis-order has been reported, and it found substantial genetic influence (Torgersen et al., 2000). However, twin studies of obsessional symptoms in unselected sam-ples of twins suggest modest heritability (Torgersen, & Psychol, 1980; Young, Fenton, & Lader, 1971). Family studies indicate that obsessional traits are more common (about 15 percent) in relatives of probands with obsessive-compulsive disorder than in controls (5 percent) (Rasmussen & Tsuang, 1984). This find-ing implies that obsessive-compulsive personality disorder might be part of the spectrum of the obsessive-compulsive type of anxiety disorder.

In summary, schizotypal and obsessive-compulsive personality disorders show results typical of personality traits: moderate heritability and no evidence for shared environmental influence. A dimensional approach to personality dis-orders yields 18 traits, all of which show this same pattern of results (Jang, Livesley, Vernon, & Jackson, 1996). Multivariate genetic analysis among four higher-order factors derived from these 18 traits (emotional dysregulation, dis-social behavior, inhibitedness, and compulsivity) indicate substantial genetic overlap but also some independent genetic influence (Livesley et al. 1998). Moreover, a multivariate genetic analysis comparing symptoms of personality disorders and major dimensions of personality finds substantial genetic correla-tions, especially with neuroticism (Jang & Livesley, 1999).

Antisocial Personality Disorder and Criminal Behavior

Much more genetic research has focused on *antisocial personality disorder (ASP)* than on other personality disorders. Lying, cheating, and stealing are examples of antisocial behavior. Antisocial personality disorder is at the extreme of antisocial behavior, with chronic indifference to and violation of the rights of others.

DSM-IV criteria for antisocial personality disorder include a history of illegal or socially disapproved activity beginning before age 15 and continuing into adult-hood—irresponsibility, irritability, aggressiveness, recklessness, and disregard for truth. Although antisocial personality disorder shows early roots, the vast majority of juvenile delinquents and children with conduct disorders do not develop anti-social personality disorder (Robins, 1978). For this reason, there is a need to distinguish conduct disorder that is limited to adolescence from antisocial behavior that persists throughout the life span (Caspi & Moffitt, 1995). As diagnosed by DSM-IV criteria, antisocial personality disorder affects about 1 percent of females and 4 percent of males from 13 to 30 years of age (Kessler et al., 1994).

Family studies show that ASP runs in families (Nigg & Goldsmith, 1994); an adoption study found that familial resemblance is largely due to genetic rather than to shared environmental factors (Schulsinger, 1972). The risk for ASP is increased fivefold for first-degree relatives of ASP males, whether living together or adopted apart. For relatives of ASP females, risk is increased tenfold, a result suggesting that, to be affected for this disproportionately male disorder, females need a greater genetic loading. Although no twin studies of diagnosed ASP are available, a personality questionnaire assessing symptoms of antisocial behavior yielded average correlations of .50 for identical twins and .22 for fraternal twins in three studies of unselected samples (Nigg & Goldsmith, 1994). A meta-analysis of 51 twin and adoption studies of antisocial behavior found evidence for significant shared environmental influences (15 percent) as well as significant genetic effects (30 percent) (Rhee & Waldman, 2002). The magnitude of both shared environment and heritability was lower in parent-offspring designs than in twin and sibling designs, which could signal developmental changes between childhood (offspring) and adulthood (parents), in contrast to twins, who are exactly the same age. This hypothesis is supported by the results of a twin study of more than 3000 pairs of adult male twins that assessed current ASP symptoms as well as a retrospective report of adolescent ASP symptoms (Lyons et al., 1995). For adult ASP symptoms, results similar to other studies were found—correlations were .47 for identical twins and .27 for fraternal twins, suggesting moderate genetic influence and modest shared environmental influence. In contrast, for adolescent ASP symptoms, results indicated little genetic influence and substantial shared environmental influence (correlations of .39 for identical twins and .33 for fraternal twins). These developmental changes have been confirmed in another twin study (Jacobson, Prescott, & Kendler, 2002). It is now generally accepted that, from adolescence to adulthood, genetic influence increases and shared environmental influence decreases for ASP symptoms (Figure 13.3) (Jacobson, 2005). In addition, as is typically found in longitudinal genetic analyses, genetics largely contributes to stability and nonshared environment contributes to change during development (Blonigen, Hicks, Krueger, Patrick, & Iacono, 2006; Burt, McGue, Carter, & Iacono, 2007).

A type of antisocial personality disorder called psychopathy has recently become the target of genetic research because of its prediction of violent crime and recidivism. Although there is no precise equivalent in DSM-IV, psychopathic personality disorder involves a lack of empathy, callousness, irresponsibility, and manipulativeness (Hare, 1993; Viding, 2004). As mentioned in Chapter

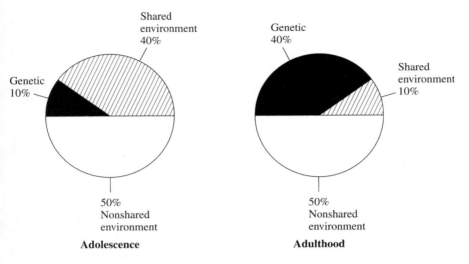

Figure 13.3 The causes of antisocial symptoms change from adolescence to adulthood, with genetics becoming more important and shared environment becoming less important. (Based on Lyons et al., 1995.)

12, psychopathic tendencies appear to be highly heritable in childhood, with no influence of shared environment (Viding et al., 2005). Similar results have been found in late adolescence (Larsson, Andershed, & Lichtenstein, 2006). A follow-up report also indicated that the overlap between psychopathic personality and antisocial behavior is largely genetic in origin (Larsson et al., 2007).

ASP shows interesting genetic relations with other disorders; of particular interest is the relation between ASP and criminal behavior. For example, two adoption studies of biological parents with criminal records found increased rates of ASP in their adopted-away offspring (Cadoret & Stewart, 1991; Crowe, 1974), suggesting that genetics contributes to the relationship between criminal behavior and ASP. Most genetic research in this area has focused on criminal behavior itself rather than on ASP, because crime can be assessed objectively by using criminal records. However, criminal behavior, although important in its own right, is only moderately associated with ASP. About 40 percent of male criminals and 8 percent of female criminals qualify for a diagnosis of ASP (Robins & Regier, 1991). Clearly, breaking the law cannot be equated with psychopathology (Rutter, 1996b).

The best twin study of criminal behavior included male twins born in Denmark from 1881 to 1910 (Christiansen, 1977). Evidence from more than a thousand twin pairs was found for genetic influence for criminal convictions, with an overall concordance of 51 percent for male identical twins and 30 percent for male-male fraternal twins. In 13 twin studies of adult criminality, identical twins are consistently more similar than fraternal twins (Raine, 1993). The average concordances for identical and fraternal twins are 52 and 21 percent, respectively.

A more recent twin study in the United States of self-reported arrests and criminal behavior involved more than 3000 male twin pairs in which both members served in the Vietnam War (Lyons, 1996). Genetics contributed to self-reported arrests and criminal behavior. However, self-reported criminal behavior before age 15 showed negligible genetic influence. Shared environment made a major contribution to arrests and criminal behavior before age 15, but not later. These results before age 15 are similar to results discussed earlier for ASP symptoms in adolescence and adulthood (Lyons et al., 1995) and for conduct disorder in adolescence (Chapter 12).

Adoption studies are consistent with the hypothesis of significant genetic influence on adult criminality, although adoption studies point to less genetic influence than twin studies do. It has been hypothesized that twin studies overestimate genetic effects because identical twins are more likely to be partners in crime (Carey, 1992). Adoption studies include both the adoptees' study method (Cloninger, Sigvardsson, Bohman, & von Knorring, 1982; Crowe, 1972) and the adoptees' family method (Cadoret et al., 1985). One of the best studies used the adoptees' study method, beginning with more than 14,000 adoptions in Denmark between 1924 and 1947 (Mednick, Gabrielli, & Hutchings, 1984). Using court convictions as an index of criminal behavior, the researchers found evidence for genetic influence and for genotype-environment interaction, as shown in Figure 13.4. Adoptees were at greater risk for criminal behavior when their biological parents had criminal convictions, a finding implying genetic influence. Unlike the twin study just described, this adoption study (and others) found genetic influence for crimes against property but not for violent crimes (Bohman, Cloninger, Sigvardsson, & von Knorring, 1982; Brennan, Mednick, & Jacobsen, 1996). Genotype-environment interaction is also found. Adoptive parents with criminal convictions had no effect on the criminal behavior of adoptees unless the adoptees' biological parents also had criminal convictions.

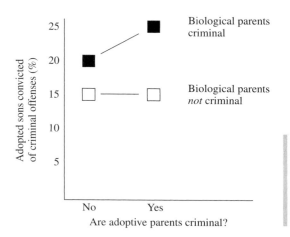

Figure 13.4 Evidence for genetic influence and genotype-environment interaction for criminal behavior in a Danish adoption study. (Adapted from Mednick, Gabrielli, & Hutchings, 1984.)

In other words, the highest rate of criminal behavior was found for adoptees who had both biological parents and adoptive parents with criminal records.

A Swedish adoption study of criminality using the adoptees' family method found similar evidence for genotype-environment interaction as well as interesting interactions with alcohol abuse, which greatly increases the likelihood of violent crimes (Bohman, 1996; Bohman et al., 1982). When adoptees' crimes did not involve alcohol abuse, their biological fathers were found to be at increased risk for nonviolent crimes. In contrast, when adoptees' crimes involved alcohol abuse, their biological fathers were not at increased risk for crime. These findings suggest that genetics contributes to criminal behavior but not to alcohol-related crimes, which are likely to be more violent crimes.

Research on the genetics of crime has been controversial. For example, a conference on the topic in the United States was postponed and then disrupted (Roush, 1995), although a similar conference in the United Kingdom created little stir (Bock & Goode, 1996). Especially for such sensitive topics, it is important to keep in mind the discussion in Chapter 5 concerning the interpretation of genetic effects (see also Raine, 1993; Rutter, 1996a). These issues will become increasingly important as specific genes that contribute to genetic risk are identified (Dinwiddie, Hoop, & Gershon, 2004).

SUMMING UP

Genetic research suggests that, at least for some conditions, there may be a continuum of individual differences from normal personality to personality disorders to psychopathology. For example, genetic variation in neuroticism largely accounts for genetic variation in anxiety and depression. Genetic overlap has also been reported between personality disorders and psychopathology: for example, between schizotypal personality disorder and schizophrenia, and between obsessive-compulsive personality disorder and obsessive-compulsive anxiety disorder. The relationship between antisocial personality disorder and criminal behavior may also be due in part to genetic influences. For symptoms of antisocial personality disorder and for criminal records, genetic influence increases and shared environmental influence decreases from adolescence to adulthood. Some evidence for genotype-environment interaction is found for criminal behavior in that the highest rate of criminal behavior was found for adoptees who had both biological and adoptive parents with criminal records.

Identifying Genes

In contrast to molecular genetic research on psychopathology, molecular genetic research on personality has just begun (Benjamin, Ebstein, & Belmaker, 2002; Hamer & Copeland, 1998). The field began in 1996 with reports from two studies of an association between a DNA marker for a certain neuroreceptor gene (*DRD4*, dopamine *D4* receptor) and the personality trait of novelty seeking in unselected samples (Benjamin et al., 1996; Ebstein et al., 1996). *DRD4* is the gene mentioned in Chapter 12 that shows an association with attention-deficit hyperactivity disorder (ADHD). Novelty seeking is one of the four traits included in a theory of temperament developed by Cloninger (Cloninger, Svrakic, & Przybeck, 1993). Novelty seeking is very similar to the impulsive sensation-seeking dimension studied by Zuckerman (1994), and it is this impulsiveness that creates the genetic link with the impulsive component of ADHD. Individuals high in novelty seeking are characterized as impulsive, exploratory, fickle, excitable, quick-tempered, and extravagant. Cloninger's theory predicts that novelty seeking involves genetic differences in dopamine transmission.

The DNA marker consists of seven alleles involving 2, 3, 4, 5, 6, 7, or 8 repeats of a 48-base pair sequence in a gene on chromosome 11 that codes for the *D4* receptor of dopamine and is expressed primarily in the brain limbic system. The number of repeats changes the receptor's structure, which has been shown to affect the receptor's efficiency in vitro. The shorter alleles (2, 3, 4, or 5 repeats) code for receptors that are more efficient in binding dopamine than are the receptors coded for by the larger alleles (6, 7, or 8 repeats). The theory is that individuals with the long-repeat *DRD4* allele are dopamine deficient and seek novelty to increase dopamine release. For this reason, the *DRD4* alleles are usually grouped as *short* (about 85 percent of alleles) or *long* (15 percent of alleles).

In both studies, individuals with the longer *DRD4* alleles had significantly higher novelty-seeking scores than did individuals with the shorter alleles. Figure 13.5 shows the distributions of novelty-seeking scores for individuals with

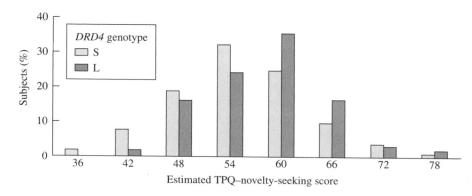

Figure 13.5 The longer allele for the *DRD4* gene has been reported to be associated with increased novelty seeking. The overlap in novelty-seeking scores for those with the shorter allele (S) and those with the longer allele (L) shows that the effect is modest, accounting for about 4 percent of the variance in novelty seeking. (From Benjamin et al., 1996, used with permission.)

the short and the long *DRD4* alleles. The overlap in scores shows that the effect is small, accounting for about 4 percent of the variance in this trait. As would be expected for an association of small effect, many studies have failed to replicate the association with novelty seeking, in part because studies have not been large enough to detect the small effect, which is probably much smaller than the 4 percent effect found in the original reports (Jang, 2005). A meta-analysis of 17 studies of extraversion rather than the narrow trait of novel seeking found a weak association (Munafo et al., 2003). *DRD4* has also recently been reported to be associated with novelty seeking in the vervet monkey (Bailey, Breidenthal, Jorgensen, McCracken, and Fairbanks, 2007).

The *DRD4* association with ADHD is also in the expected direction— longer *DRD4* repeat alleles are associated with greater risk for ADHD. Moreover, an association between *DRD4* long repeat alleles and heroin addiction has been found in three studies (Ebstein & Belmaker, 1997; Kotler et al., 1997; Li et al., 1997; see summary by Ebstein & Kotler, 2002), a finding that is interesting because there is an extensive literature relating sensation seeking to drug abuse, including opiate abuse (Zuckerman, 1994).

Neuroticism has been another focus of candidate gene studies of personality, in part because of the force of an early report of association with a polymorphism in a serotonin transporter gene (*5-HTTLPR*; Lesch et al., 1996) quickly followed by reports of its association with depression (Collier et al., 1996). A meta-analysis of 22 studies found some evidence for an association (Munafo et al., 2003; see also Munafo, Clark, & Flint, 2005), although a subsequent study of more than 100,000 individuals found no association with either neuroticism or depression (Willis-Owen et al., 2005). Nonetheless, serotonin genes have been reported to show associations in studies of brain measures related to emotional responding (Chapter 15) and in interaction with an adverse childhood environment (Chapter 16) (Ebstein, 2006; Serretti, Calati, Mandelli, & De Ronchi, 2006).

As noted in Chapter 7, early reports of an association between XYY males and violence were overblown, although there seems to be some increase in hyperactivity and perhaps conduct problems (Ratcliffe, 1994). In a four-generation study of a Dutch family, a deficiency in a gene on the X chromosome that codes for an enzyme (monoamine oxidase A, MAOA),which is involved in the breakdown of several neurotransmitters, was associated with impulsive aggression and borderline mental retardation in males (Brunner, 1996; Brunner, Nelen, Breakefield, Ropers, & van Oost, 1993), although this genetic effect has not yet been found in any other families. However, as described in Chapter 16, the MAOA gene has been strongly associated with antisocial behavior in individuals who suffered severe childhood maltreatment, a genotype-environment interaction (Caspi et al., 2002). This finding has held up in a meta-analysis (Kim-Cohen et al., 2006). Two linkage studies of neuroticism suggested several, but different, linkages, and none included 17*q*, where the serotonin transporter gene is located (Fullerton et al., 2003; Nash et al., 2004).

No genomewide association studies of personality have as yet been reported. More powerful methods for identifying such QTLs for personality are available in research on nonhuman animals, as described in Chapter 6 (see also Flint, 2004; Willis-Owen & Flint, 2007). For example, several QTLs for fearfulness have been identified in mice, as assessed in open-field activity (Flint et al., 1995; Henderson, Turri, DeFries, & Flint, 2004; Talbot et al., 1999). Also, transgenic knock-out gene studies in mice often find personality effects, such as increased aggression when one or the other of two genes were knocked out, either the gene for a receptor for an important neurotransmitter (serotonin; Saudou et al., 1994) or the gene for an enzyme (neuronal nitric oxide synthase) that plays a basic role in neurotransmission (Nelson et al., 1995).

Summary

More twin data are available from self-report personality questionnaires than from any other domain of psychology, and they consistently yield evidence for moderate genetic influence for dozens of personality dimensions. Most well studied are extraversion and neuroticism, which yield heritability estimates of about 50 percent for extraversion and about 40 percent for neuroticism across twin and adoption studies. Other personality traits assessed by personality questionnaire also show heritabilities ranging from 30 to 50 percent. There is no replicated example of zero heritability for any specific personality trait. Environmental influence is almost entirely due to nonshared environmental factors. These surprising findings are not limited to self-report questionnaires. For example, a twin study using peer ratings yielded similar results. Although the degree of genetic influence suggested by twin studies using parent ratings of their children's personalities appears to be inflated by contrast effects, more objective measures, such as behavioral ratings by observers, indicate genetic influence in twin and adoption studies.

New directions for genetic research include looking at personality continuity and change across situations and across time. Results so far indicate that genetics is largely responsible for continuity and that change is largely due to environmental factors. Other new findings include the central role that personality plays in producing genetic influence on measures of the environment. Another new direction for research lies at the border with social psychology. For example, genetic influence has been found for relationships, such as parent-offspring relationships and sexual orientation, but not romantic relationships. Other examples include evidence for genetic influence on self-esteem, attitudes, and vocational interests.

A major new direction for genetic research on personality is to consider its role in psychopathology. For example, depression and other internalizing psychopathology are, to a large extent, the genetic extreme of normal variation in the major personality dimension of neuroticism. Personality disorders, which are at the border between personality and psychopathology, are another growth area for genetic research in personality. It is likely that some personality disorders are part of the genetic continuum of psychopathology: schizotypal personality disorder and schizophrenia, and obsessive-compulsive personality disorder and obsessive-compulsive anxiety disorder. Most genetic research on personality disorders has focused on antisocial personality disorder and its relationship to criminal behavior. From adolescence to adulthood, genetic influence increases and shared environmental influence decreases for symptoms of antisocial personality disorder, including juvenile delinquency and adult criminal behavior.

QTL associations have been reported for several candidate genes and personality traits; for example, a dopamine receptor gene may be associated with the personality trait of novelty seeking. However, similar to research on psychopathology, replication of associations has been difficult in part because effects sizes are much smaller than originally anticipated. Identifying genes associated with personality is likely to be helped by employing powerful strategies using mouse models.

CHAPTER FOURTEEN

Health Psychology and Aging

G enetic research in psychology has focused on cognitive disabilities and abilities (Chapters 7–9), psychopathology (Chapters 10–12), and personality (Chapter 13). The reason for this focus is that these are the areas of psychology that have had the longest history of research on individual differences. Much less is known about the genetics of other major domains of psychology that have not traditionally emphasized individual differences, such as perception, learning, and language.

The purpose of this chapter is to provide an overview of genetic research in two new areas of the behavioral sciences. One of the newest areas is health psychology, sometimes called psychological or behavioral medicine because it lies at the intersection between psychology and medicine. Research in this area focuses on the role of behavior in promoting health, and in preventing and treating disease. Although genetic research has just begun in this area, some conclusions can be drawn about relevant topics such as responses to stress, body weight, and addictive behaviors.

The second area is aging. Although behavioral genetic research has neglected the last half of the life span, new research has produced interesting results, especially about issues unique to aging, such as quality of life in the later years. The explosion of molecular genetic research on cognitive decline and dementia in the elderly has added momentum to genetic research on behavioral aging.

Health Psychology

Most of the central issues about the role of behavior in promoting health and in preventing and treating disease have not yet been addressed in genetic research (Baum & Posluszny, 1999). For example, the first book on genetics and health psychology was not published until 1995 (Turner, Cardon, & Hewitt,

┤ CLOSE **UP** ├

John K. Hewitt is director of the Institute for Behavioral Genetics at the University of Colorado, Boulder. He was educated in England, first at the University of Birmingham, where he studied psychology and genetics, and then at the Institute of Psychiatry, receiving his doctorate in 1978. He was drawn to behavioral genetics because it offered the possibility of applying rigorous scientific and hypothesis-testing methods to the complex questions involved in understanding individual behavior, because it provided an integrative framework for thinking about a wide range of questions, and also because the field seemed to be populated by some of the brightest and most creative people working in psychology and psychiatry. He notes that the great thing about doing research in behavioral genetics is that there really never is a dull moment. The common theme of Hewitt's research has been the application of biometrical genetics to the elucidation of the development of individual differences in human and animal behavior. His current research emphasizes cross-sectional and longitudinal studies of twins and families, together with DNA collection and genotyping, to understand the development of behavior problems, vulnerability to drug use, abuse, and dependence, and genetics and health. He also has a keen interest in statistical genetics and the application of linkage and association methods in the study of behavioral traits.

1995). Nonetheless, conclusions can be drawn about the genetics of two areas relevant to health psychology: body weight and obesity, and addictions, including alcoholism, smoking, and other drug abuses.

Body Weight and Obesity

Obesity is a major health risk for several medical disorders, especially diabetes and heart disease. Although it is often assumed that individual differences in weight are largely due to factors, such as eating habits and exercise, that are often mistakenly assumed to be environmental, twin and adoption studies consistently lead to the conclusion that genetics accounts for the majority of the variance for weight (Grilo & Pogue-Geile, 1991). For example, as illustrated in Figure 14.1, twin correlations for weight based on thousands of pairs of twins are .80 for identical twins and .43 for fraternal twins. Identical twins reared apart correlate .72. Biological parents and their adopted-away offspring are almost as similar in weight (.23) as are nonadoptive parents and their offspring (.26), who share

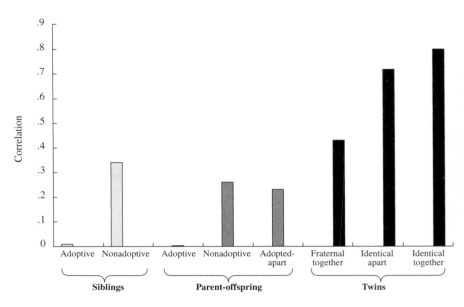

Figure 14.1 Family, adoption, and twin results for body weight. (Derived from Grilo & Pogue-Geile, 1991.)

both nature and nurture. Adoptive parents and their offspring, and adoptive siblings, who share nurture but not nature, do not resemble each other at all for weight.

Together, the results in Figure 14.1 imply a heritability of about 70 percent. Similar results have been found across eight European countries despite average differences in weight, with some suggestion of greater shared environmental influence for women (Schousboe et al., 2003). Similar results are also found for body mass index, which corrects weight on the basis of height, and for skinfold thickness, which is an index of fatness (Grilo & Pogue-Geile, 1991; Maes, Neale, & Eaves, 1997). There are few genetic studies of overweight or obesity, in part because weight shows a continuous distribution, a situation rendering diagnostic criteria somewhat arbitrary (Bray, 1986). Nonetheless, using an obesity cutoff based on body mass index, twin studies have indicated similarly high heritabilities for obesity in childhood (Koeppen-Schomerus, Wardle, & Plomin, 2001) and adulthood (Stunkard, Foch, & Hrubec, 1986). A parent-offspring family study indicates that, although assortative mating (see Chapter 8) for body weight is modest (a correlation of about .20), the risk of obesity in adult offspring is 20 percent if both parents are obese, 8 percent if only one parent is obese, and only 1 percent if neither parent is obese (Jacobson, Torgerson, Sjostrom, & Bouchard, 2007).

The dramatic increase in obesity throughout the world is sometimes thought to deny a role for genetics but, as discussed in Chapter 5, the causes of population means and variance are not necessarily related. That is, the mean population increase in weight is probably due to the increased availability and reduced costs of energy-dense food and a reduction in physical activity (Abelson & Kennedy, 2004). However, despite our increasingly "obesogenic" environments, a wide range of variation in weight remains—many people are still thin. Obesogenic environments could shift the entire distribution upwards while the causes of individual differences, including genetic causes, could remain unchanged.

As also emphasized in Chapter 5, finding genetic influence does not mean that the environment is unimportant. Anyone can lose weight if they stop eating. The issue is not what *can* happen but rather what *does* happen. That is, to what extent are the obvious differences in weight among people due to genetic and environmental differences that exist in a particular population at a particular time? The answer provided by the research summarized in Figure 14.1 is that genetic differences largely account for individual differences in weight. If everyone ate the same amount and exercised the same amount, people would still differ in weight for genetic reasons.

This conclusion was illustrated dramatically in an interesting study of dietary intervention in 12 pairs of identical twins (Bouchard, Lykken et al., 1990). For three months, the twins were given excess calories and kept in a controlled sedentary environment. Individuals differed greatly in how much weight they gained, but members of identical twin pairs correlated .50 in weight gain. Similar twin studies show that the effects on weight of physical activity and exercise are also influenced by genetic factors (Fagard, Bielen, & Amery, 1991; Heitmann et al., 1997).

Such studies do not indicate the mechanisms by which genetic effects occur. For example, even though genetic differences occur when calories and exercise are controlled, in the world outside the laboratory, genetic contributions to individual differences might be mediated by individual differences in proximal processes such as food intake and exercise. In other words, individual differences in eating habits and in the tendency to exercise, although typically assumed to be environmental factors responsible for body weight, might be influenced by genetic factors. Twin studies suggest that genetic factors do affect many aspects of eating, such as the number, timing, and composition of meals as well as degree of hunger and sense of fullness after eating (de Castro, 1999), eating styles such as emotional eating and uncontrolled eating (Tholin, Rasmussen, Tynelius, & Karlsson, 2005), and food preferences in general (Breen, Plomin, & Wardle, 2006).

Previous chapters have indicated that environmental variance is of the non-shared variety for most areas of behavioral research. This is also the case for body weight. As noted in relation to Figure 14.1, adoptive parents and their adopted children and adoptive siblings do not resemble each other at all for weight. This finding is surprising, because theories of weight and obesity have largely focused on weight control by means of dieting; yet individuals growing up in the same families do not resemble each other for environmental reasons (Grilo & Pogue-Geile, 1991). Attitudes toward eating and weight also show substantial heritability and no influence of shared family environment (Ruther-ford, McGuffin, Katz, & Murray, 1993). In other words, environmental factors that affect individual differences in weight are factors that do not make children growing up in the same family similar. The next step in this research is to iden-tify environmental factors that differ for children growing up in the same fam-ily. For example, although it is reasonable to assume that children in the same family share similar diets, this may not be the case.

Genetic factors that affect body weight begin to have their effects in early childhood (Meyer, 1995). Longitudinal genetic studies are especially informative. The first longitudinal twin study from birth through adolescence found no heri-tability for birth weight, increasing heritability during the first year of life, and stable heritabilities of 60 to 70 percent thereafter (Figure 14.2; see Matheny, 1990). These results have been replicated in other twin studies in chil-dren (Estourgie-van Burk, Bartels, van Beijsterveldt, Delemarre-van de Waal, & Boomsma, 2006; Pietilainen et al., 1999) and in a parent-offspring adoption study, which also suggested that there is substantial genetic continuity from childhood to adulthood (Cardon, 1994a). Twin studies in adulthood also report heritabilities of 60 to 80 percent, and longitudinal twin studies indicate that genetic influences are largely stable but show some significant change during adulthood (Romeis, Grant, Knopik, Pedersen, & Heath, 2004; Stunkard et al., 1986).

Obesity has become the target of intense molecular genetic research in part because of the so-called obese gene in mice. In the 1950s, a recessive mutation that caused obesity in the homozygous condition was discovered in mice. When these obese mice were given blood from a normal mouse, they lost weight, a

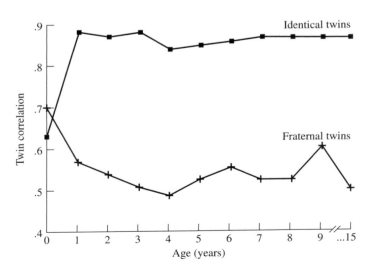

Figure 14.2 Identical and fraternal twin correlations for weight from birth to 15 years of age. (Derived from Matheny, 1990.)

result suggesting that the obese mice were missing some factor important in control of weight. The gene was cloned and was found to be similar to a human gene (Zhang et al., 1994). The gene's product, a hormone called leptin, was shown to reduce weight in mice by decreasing appetite and increasing energy use (Halaas et al., 1995). However, with rare exceptions (Montague et al., 1997), obese humans do not appear to have defects in the leptin gene. The gene that codes for the leptin receptor in the brain has also been cloned from another mouse mutant (Chua et al., 1996). Mutations in this gene might contribute to genetic risk for obesity. Animal models of obesity, including mutational screening and targeted mutations (see Chapter 6), continue to be the forefront of genetic research on obesity (Speakman, Hambly, Mitchell, & Krol, 2007).

Like most complex traits, major single-gene effects on human obesity involve rare and often severe disorders. In addition, 244 genes in mice have been shown to affect body weight when mutated or otherwise altered (Rankinen et al., 2006). However, QTLs, multiple genes of various effect sizes, are likely to be responsible for the substantial genetic contribution to common overweight and obesity (Bell, Walley, & Froguel, 2005). More than 60 genomewide linkage scans have reported more than 250 linkages, and more than 50 regions have been supported by two or more studies (Rankinen et al., 2006). Candidate gene studies have also produced a welter of results, with 426 reports of positive associations involving 127 candidate genes; encouragingly, 22 of these genes have been supported in at least five studies (Rankinen et al., 2006). The first genomewide association scan reported one highly publicized association that replicated in several samples (Herbert et al., 2006), but several subsequent studies failed to replicate the finding (Herbert et al., 2007). Another genomewide association scan of type 2 diabetes identified a SNP whose association with diabetes appears to be mediated by body weight (Frayling et al., 2007). This SNP, in a gene of unknown function on chromosome 16 called *FTO* (a member of a family of genes identified on the basis of a mouse mutation that causes fused toes), yielded a small but significant association with body weight in 13 samples with a total of nearly 40,000 individuals. As predicted by quantitative genetic research, the SNP is associated with body weight throughout the distribution, not just with the obese end of the distribution. Also as predicted by quantitative genetic research, the SNP is not associated with birth weight but shows correlations with body weight beginning at 7 years of age.

SUMMING UP

Body weight shows high heritabilities, about 70 percent, and little influence of shared environment. Genetic effects on weight are largely stable after infancy, although there is some evidence for genetic change. Obesity is the target of much molecular genetic research, with some consistent linkages and associations beginning to emerge.

Addictions

Alcohol abuse, smoking, and abuse of other drugs are major health-related behaviors. As discussed in Chapters 11 and 13, drug abuse is part of a general genetic factor of externalizing disorders, but alcohol and other drugs include significant disorder-specific genetic effects (Kendler et al., 2003). Most behavioral genetic research in this area has focused on alcoholism.

Alcoholism   Clearly there are many steps in the path to alcoholism. For example, there is choice in whether or not to drink alcohol at all, in the amount one drinks, in the way that one drinks, and in the development of tolerance and dependence. Each of these steps might involve different genetic mechanisms. For this reason, alcoholism is likely to be highly heterogeneous. Nonetheless, over 30 family studies have shown that alcoholism runs in families, although the studies vary widely in the size of the effect and in diagnostic criteria (Cotton, 1979). For males, alcoholism in a first-degree relative is by far the single best predictor of alcoholism. For example, a family study of 300 alcohol-abusing probands found an average risk of about 40 percent in first-degree male relatives and 20 percent in female relatives. The risk rates in the general population are about 20 percent for males and 5 percent for females, using the same diagnostic procedures for assessment (Reich & Cloninger, 1990). Assortative mating for alcohol use is substantial (correlation of .38), which a twin analysis indicated is caused by initial selection of the spouse rather than the effect of living with the spouse (Agrawal et al., 2006). Assortative mating of this magnitude could inflate estimates of shared environment. Assortative mating could also create a genotype-environment correlation in which children are more likely to experience both genetic and environmental risks. (See Chapter 8 for more discussion of assortative mating.)

Results of twin and adoption studies of alcoholism vary greatly, but taken together they indicate moderate heritability for males and modest heritability for females (Legrand, McGue, & Iacono, 1999). For example, a major twin study of male alcoholics in Sweden reported concordances of about 50 percent for identical twins and about 35 percent for fraternal twins; these results contrast with a prevalence of about 15 percent in Swedish birth cohorts from 1902 to 1949 (Kendler, Prescott, Neale, & Pedersen, 1997). Using the liability-threshold model that assumes an underlying quantitative continuum, heritability estimates were 60 percent. Eight adoption studies of alcoholism yield compatible results (Ball & Collier, 2002). The largest adoption study of alcoholism, which included more than 600 reared-away offspring of alcoholic biological parents from Stockholm, also found evidence for genetic influence (Cloninger, Bohman, & Sigvardsson, 1981). The rates of alcoholism were 23 percent for adopted males and 15 percent for control males. As is often found in psychopathology (Chapters 10 and 11), earlier onset and more severe alcoholism appears to be more heritable. An interesting developmental finding is that shared environment appears to be related to the initial use of alcohol in adolescence and young adulthood, but not to later alcohol abuse (Pagan et al., 2006).

Research on alcoholism has also provided examples of genotype-environment interaction. Heritability has been reported to be lower for married individuals (Heath, Jardine, & Martin, 1989), for individuals with a religious upbringing (Koopmans, Slutske, van Baal, & Boomsma, 1999) and stricter and closer families (Miles, Silberg, Pickens, & Eaves, 2005), and in regions with lower alcohol sales (Dick, Rose, Viken, Kaprio, & Koskenvuo, 2001). These findings suggest that genetic risk for alcoholism is greater in more permissive environments (unmarried, nonreligious upbringing, greater alcohol availability). Two adoption studies suggest a different kind of genotype-environment interaction: Adoptees who had both genetic risk (an alcoholic biological parent) and environmental risk (an alcoholic adoptive parent) were most likely to abuse alcohol (Sigvardsson, Bohman, & Cloninger, 1996). (See Chapter 16 for more discussion of genotype-environment interaction.)

In contrast to these studies on male alcoholics, early twin and adoption studies of female alcoholics did not yield consistent results (McGue, 2000). For example, in the Stockholm adoption study (Cloninger et al.,1981), the rates of alcoholism were 5 percent for adopted females with alcoholic biological parents and 3 percent for control females. A review of nine twin studies of alcoholism concluded that heritability estimates are about 50 percent for males and 25 percent for females (Ball & Collier, 2002). However, two of the largest and most recent twin studies of alcoholism in women report heritabilities comparable to those found for twin studies of males (Heath et al., 1997); recent reviews conclude that heritabilities are similar for men and women (Dick & Bierut, 2006).

Alcoholism is interesting because it shows some evidence for shared environment that is shared by siblings but not by parents and offspring. For example, in an adoption study of alcohol use and misuse among adolescents, the correlation between problem drinking in parents and adolescent alcohol use was .30 for biological offspring but only .04 for adoptive offspring (McGue, Sharma, & Benson, 1996). Despite the lack of resemblance between adoptive parents and their adoptive offspring, adoptive sibling pairs who were not genetically related correlated .24. Moreover, the adoptive sibling correlation was significantly greater for like-sexed siblings ($r = .45$) than for opposite-sexed siblings ($r = .01$). These results suggest the reasonable hypothesis that sibling effects (or perhaps peer effects) may be more important than parent effects in the use of alcohol in adolescence. However, as mentioned earlier, assortative mating might be responsible for apparent shared environmental influences (Agrawal et al., 2006).

Although depression often co-occurs with alcoholism, multivariate genetic research indicates that alcoholism and depression are largely due to different genes (McGuffin et al., 1994; Merikangas, 1990). Other possible mediators of genetic influence on alcoholism have also been explored, such as personality, alcohol sensitivity, and cognitive factors (McGue, 1993). An influential classification based on the adoption study mentioned earlier (Cloninger et al., 1981) suggests that early-onset alcoholism in males associated with alcohol-related aggression, called type II alcoholism, is especially heritable. Another direction for genetic research is toward understanding mechanisms by which genetic influence affects vulnerability to alcohol use and abuse. For example, a twin study found substantial genetic influence on changes in EEG following alcohol; these changes are a brain indicator of acute tolerance and sensitivity to the effects of alcohol (O'Connor, Sorbel, Morxorati, Li, & Christian, 1999).

One of the strongest areas of behavioral genetic research in rodents is called *psychopharmacogenetics*, that is, genetic effects on behavioral responses to drugs. The larger field of *pharmacogenetics* (Roses, 2000), often called *pharmacogenomics* in recognition of the ability to examine genetic effects on a genomewide basis, focuses on genetic differences in positive and negative effects of drugs in order to individualize and optimize drug therapy (Evans & Relling, 2004; Goldstein, Tate, & Sisodiya, 2003). Most research in psychopharmacogenetics involves alcohol (Bloom & Kupfer, 1995; Broadhurst, 1978; Crabbe & Harris, 1991). In 1959, it was shown that inbred strains of mice differ markedly in their preference for drinking alcohol, an observation that implies genetic influence (McClearn & Rodgers, 1959). Inbred strain differences have subsequently been found for many behavioral responses to alcohol (Phillips & Crabbe, 1991).

Selection studies provide especially powerful demonstrations of genetic influence. For example, one study successfully selected for sensitivity to the effects of alcohol (McClearn, 1976). When mice are injected with the mouse equivalent of several drinks, they will "sleep it off" for various lengths of time. "Sleep time" in response to alcohol injections was measured by the time it took mice to right themselves after being placed on their backs in a cradle (Figure 14.3). Selection

Figure 14.3 The "sleep cradle" for measuring loss of righting response after alcohol injections in mice. In cradle 2, a long-sleep mouse is still on its back, sleeping off the alcohol injection. In cradle 3, a short-sleep mouse has just begun to right itself. (Courtesy of E. A. Thomas.)

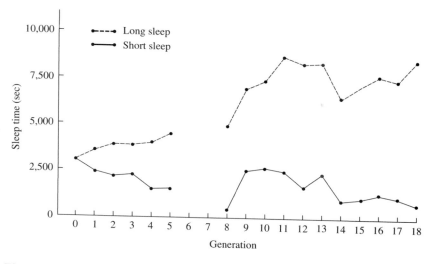

Figure 14.4 Results of alcohol sleep-time selection study. Selection was suspended during generations 6 through 8. (From McClearn, unpublished.)

for this measure of alcohol sensitivity was successful, an outcome providing a powerful demonstration of the importance of genetic factors (Figure 14.4). After 18 generations of selective breeding, the long-sleep (LS) animals "slept" for an average of two hours. Many of the short-sleep (SS) mice were not even knocked out, and their average "sleep time" was only about ten minutes. By generation 15, there was no overlap between the LS and SS lines (Figure 14.5). That is, every mouse in the LS line slept longer than any mouse in the SS line.

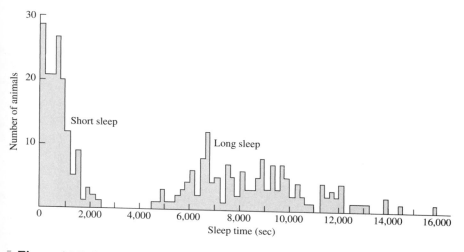

Figure 14.5 Distributions of alcohol sleep time after 15 generations of selection. (From McClearn, unpublished.)

The steady divergence of the lines over 18 generations indicates that many genes affect this measure. If just one or two genes were involved, the lines would completely diverge in a few generations. Selected lines provide important animal models for additional research on pathways between genes and behavior. For example, the LS and SS lines have been extensively used as mouse models of alcohol sensitivity (Collins, 1981). Other selection studies include successful selection in mice for susceptibility to seizures during withdrawal from alcohol dependence and for voluntary alcohol consumption in rats (Crabbe, Kosobud, Young, Tam, & McSwigan, 1985). These are powerful genetic effects. For example, mice in the line selected for susceptibility to seizures are so sensitive to withdrawal that they show symptoms after a single injection of alcohol. Mouse genetic models continue to be widely used for behavioral genetic research on alcohol-related traits as well as for molecular genetic research (Bennett et al., 2006).

Early psychopharmacogenomic studies of mice were used to identify QTLs associated with drug-related behavior (Crabbe, Phillips et al., 1999). For example, several groups mapped QTLs for alcohol preference drinking in mice to the middle of mouse chromosome 9 (Phillips, Belknap et al., 1998), a region that includes the gene coding for the dopamine *D2* receptor subtype. Studies with *D2* receptor knock-out mice revealed that they showed reduced alcohol preference drinking. Each QTL conferred a difference in "sleep time" of about 20 minutes. That is, an average individual mouse possessing the *LS* allele at one of these loci would be sedated for about 20 minutes longer than an individual with the *SS* allele. However, if an individual possessed all five *LS* alleles, its genotype could account for 130 minutes of the total of 170 minutes in sleep-time difference between the LS and SS mice. Such differences in brain sensitivity to ethanol in human populations could be responsible for the lethal consequences of binge drinking in some individuals (Heath et al., 2003).

Many QTLs have been mapped for alcohol-related responses such as alcohol drinking, alcohol-induced loss of righting reflex, and acute alcohol withdrawal, as well as cocaine seizures, morphine preference, and analgesia (Figure 14.6) (Crabbe, Phillips et al., 1999; Lovinger & Crabbe, 2005). Mouse QTL research is especially exciting because it can nominate candidate QTLs that can then be tested in human QTL research (Lovinger & Crabbe, 2005). Knock-out studies in mice also demonstrate the effects of specific genes on behavioral responses to drugs. For example, knocking out a serotonin receptor gene in mice leads to increased alcohol consumption (Crabbe et al., 1996) and to increased vulnerability to cocaine (Rocha et al., 1998). Another study found supersensitivity to alcohol, cocaine, and methamphetamine in mice whose dopamine *D4* receptor gene was knocked out (Rubinstein et al., 1997). In addition, antisense DNA "knockdowns" that block drug effects by preventing the synthesis of receptor molecules in specific brain regions (see Chapter 6) have been shown to affect behavioral responses for dozens of drugs (Buck, Crabbe, & Belknap, 2000).

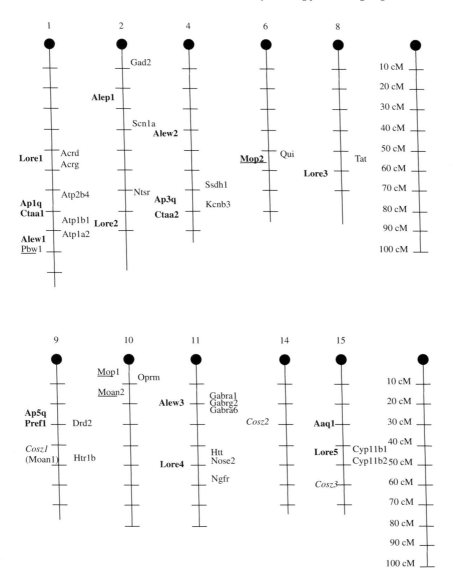

Figure 14.6 Drug-related QTLs and candidate genes. QTLs identified on mouse chromosomes 1, 2, 4, 6, 8, 9, 10, 11, 14, and 15 are indicated in their most likely location. Alcohol-related QTLs are indicated in bold, cocaine-related QTLs in italics, pentobarbitol-related QTLs with an underline, and morphine-related QTLs in parentheses. Plausible candidate genes mapped near these QTLs are indicated to the right of each chromosome. (Adapted with permission from Crabbe, Phillips et al., 1999, p. 175.)

Alcoholism in humans has also long been a target for molecular genetic studies (Reich, Hinrichs, Culverhouse, & Bierut, 1999) because much is known about genes involved in the metabolism of alcohol (Lovinger & Crabbe, 2005). An early impetus for molecular genetic research involved an aldehyde dehydrogenase gene (*ALDH2*). An *ALDH2* allele (*ALDH2*2*) that leads to inactivity of a key enzyme in the metabolism of alcohol occurs in 25 percent of Chinese and 40 percent of Japanese but is hardly ever found in Caucasians. The resulting buildup of aldehyde leads to unpleasant symptoms such as flushing and nausea when alcohol is consumed. This is an example of a mutant allele that protects against development of alcoholism. This genetic variant results in reduced alcohol consumption and has been implicated as the reason why rates of alcoholism are much lower in Asian populations than in Caucasian populations (Hodgkinson, Mullan, & Murray, 1991). The same symptoms are produced by the drug disulfiram (Antabuse), which is the basis for an alcoholism therapy used to deter drinking.

As in many other areas of psychopathology, large-scale studies scanning the genome for linkages have been reported but show little convergence (e.g., Hill et al., 2004; Long et al., 1998; Prescott, Sullivan et al., 2006; Reich et al., 1998). Unlike other areas, alcoholism has a solid candidate gene finding for *ALDH2* that in part explains ethnic differences in alcoholism. However, this gene and other genes related to alcohol metabolism have not shown consistent associations with alcoholism within ethnic groups (Ball & Collier, 2002). Associations for other candidate genes have been reported, especially genes that code for GABA (γ-aminobutyric acid) and dopamine (Dick & Bierut, 2006; Edenberg & Foroud, 2006), although replication has been limited (Higuchi, Matsushita, & Kashima, 2006). The *GABA* genes, which code for a major inhibitory system in the human nervous system, became a target for genetic research because linkage studies pointed to chromosome 4*p* where several *GABA* genes reside. The strongest evidence to date has emerged for one of these GABA receptor genes (*GABRA2*), which also mirrors the genotype-environment interaction mentioned above: The high-risk *GABRA2* genotype conveyed less risk for married individuals (Dick, Agrawal et al., 2006). The first genome scan for association with alcoholism reported many associations with small effect sizes but no large effects (Johnson et al., 2006).

SUMMING UP

Results of twin and adoption studies of alcoholism vary greatly, but all together they suggest moderate heritability and little evidence for shared environment. Several examples of genotype-environment interaction have been reported in which genetic risk for alcoholism is greater in more permissive environments. Psychopharmacogenetics has been a very active area of research using mouse models of drug use and abuse, especially for alcohol. For example, selection studies have documented genetic influence on many behavioral responses to drugs. Many QTLs for alcohol-related behavior in mice have been identified. In human populations, an aldehyde dehyrogenase gene (*ALDH2*) accounts for some ethnic differences in alcoholism. Linkage and association studies have not yet yielded consistent results, although the strongest available evidence involves linkage and association results for a GABA receptor gene (*GABRA2*).

Smoking About a third of American adults currently smoke, and most are dependent on nicotine. About a third of these adults try to quit each year, but only about 3 percent succeed—nicotine is more addictive than most illicit drugs. It is also one of the most lethal drugs—tobacco use is associated with the death of hundreds of thousands of people each year in the United States alone (Peto, Lopez, Boreham, Thun, & Heath, 1992). Although nicotine is an environmental agent, individual differences in susceptibility to its addictive properties are influenced by genetic factors. Five twin studies with more than a thousand twin pairs each from

CLOSE UP

John Crabbe has been at the VA Medical Center and in Behavioral Neuroscience at the Oregon Health & Science University in Portland since 1979. He is director of the NIH Portland Alcohol Research Center. He entered graduate school at the University of Colorado in 1968 to obtain a Ph.D. in social psychology.

Fortuitously, Crabbe was sidetracked into studying behavioral neuroscience at the fledgling Institute for Behavioral Genetics in Boulder. He has been surrounded by mice ever since. His interest is in understanding individual differences in behavioral susceptibility to drugs of abuse and their neurobiological bases. He uses inbred strains, selectively bred lines, specialized populations for genetic mapping, and mice with null mutations for various genes. He has recently been developing improved mouse behavioral assays that can be used to dissect genetic contributions to different aspects of related responses. For example, he is studying how it is that some genes can affect one measure of motor coordination but not other tasks that seem to the experimenter virtually identical. He is developing new measures of alcohol withdrawal. Most recently, he has begun to selectively breed mouse lines that voluntarily drink alcohol until they become intoxicated, that is, a mouse model of college students.

four countries all point to genetic influence on smoking (Heath & Madden, 1995). For example, the largest study includes 12,000 pairs from Sweden, of whom half smoked (Medlund, Cederlof, Floderus-Myrhed, Friberg, & Sorenson, 1977). If one twin currently smoked, the probability that the co-twin smoked was 75 percent for identical twins and 63 percent for fraternal twins. Across these twin studies, analyses based on the liability-threshold model (see Chapter 3) suggest heritabilities of liability to smoking of about 60 percent and some shared environmental influence (Heath & Madden, 1995). More recent twin studies, such as a population study in Finland, continue to yield similar results (Broms, Silventoinen, Madden, Heath, & Kaprio, 2006), with even higher heritability for studies using direct measures of nicotine dependence (Prescott, Madden, & Stallings, 2006). One study suggests that a key index of a genetic factor of nicotine dependence is how soon after waking a smoker smokes (Haberstick et al., 2007). These results refer to smoking cigarettes; an interesting study found that smoking tobacco in pipes and cigars showed no genetic influence and substantial shared environmental influence (Schmitt, Prescott, Gardner, Neale, & Kendler, 2005).

The reasons why people start to smoke appear to differ from the reasons why people persist in smoking and in the amount they smoke (Heath & Martin, 1993). For example, shared environment, probably due to peers rather than parents, plays a larger role in smoking initiation than in smoking persistence (Hamilton et al., 2006; Rowe & Linver, 1995), which is similar to the developmental result mentioned above in relation to alcohol use. An extended twin design that included parents found that shared environment is shared by twins but not by parents and offspring (Maes et al., 2006). Multivariate studies suggest that genetic effects on initiation differ from genetic effects on persistent smoking (Broms et al., 2006). Another interesting multivariate result is that the genetics of persistent smoking appear to be mediated by genetic vulnerability to nicotine withdrawal (Pergadia, Heath, Martin, & Madden, 2006).

Researchers are now attempting to identify specific genes responsible for these genetic effects. Molecular genetic studies of smoking have primarily investigated genes related to nicotine receptors (e.g., *CHRNB2*) and nicotine metabolism (*CYP* genes), as well as dopamine and serotonin genes involved in brain reward pathways, although no consistent findings have as yet emerged (Prescott, Madden, & Stallings, 2006). Four linkage studies have been conducted in four different countries; no consistent linkage regions have been found, perhaps because studies have focused on smoking rather than nicotine dependence (Prescott, Madden, & Stallings, 2006).

Other drugs   Inbred strain and selection studies in mice have documented genetic influence on sensitivity to almost all drugs subject to abuse (Crabbe & Harris, 1991). Human studies are difficult to conduct because drugs such as amphetamines, heroine, and cocaine are illegal and exposure to these drugs changes over time (Seale, 1991). Family studies have shown about an eightfold increased risk of drug abuse in relatives of probands with drug abuse for a wide range of drugs such as cannabis, sedatives, opioids, and cocaine (Ball & Collier, 2002; Merikangas et al., 1998). Two major twin studies of a broad range of drug abuse have been conducted in the U.S., one involving veterans of the U.S. war in Vietnam (Tsuang, Bar, Harley, & Lyons, 2001) and the other involving twins in Virginia (Kendler, Karkowski, Neale, & Prescott, 2000). Both studies yielded evidence of substantial heritabilities of liability (about 30 percent to 70 percent) and little evidence of shared environmental influence across various drugs of abuse. Similar results have been found in a more recent population twin study in Norway (Kendler, Aggen, Tambs, & Reichborn-Kjennerud, 2006). A focus for recent research has been on developmental issues (Zucker, 2006). For example, as found for alcohol and smoking, shared family environmental factors play a role in initiation, but genetic factors are largely responsible for subsequent use and abuse (Kendler & Prescott, 1998; Rhee et al., 2003). Multivariate genetic analyses indicate that the same genes largely mediate vulnerability across different drugs (Jang, 2005), but shared environmental influence in adolescence is more drug specific (Young, Rhee, Stallings, Corley, & Hewitt, 2006). An interesting finding is that exposure to drugs shows genetic influence, a type of genotype-environment correlation (Kendler, 2001). That is, significant heritability was found, not just for use of marijuana, stimulants, sedatives, cocaine, opiates, and psychedelics, but also for exposure to each drug. Exposure to drugs is usually, and reasonably, thought to be an environmental risk factor. However, results such as these raise the possibility that genetic factors contribute to experience, a topic discussed in Chapter 16.

The molecular genetics of drug-related behaviors has been examined in mice, especially for transgenic models of responses to opiates, cocaine, and amphetamine. More than three dozen transgenic mouse models have been established for responses to these drugs (Ball & Collier, 2002). QTL research in mice has also been active (Crabbe, Phillips et al., 1999), including genes involved in reward mechanisms as well as drug preference and response (Goldman, Oroszi, & Ducci, 2005). There have been few linkage studies of drugs other than alcohol. A QTL linkage study in adolescence suggested two linkage regions for vulnerability to substance abuse (Stallings et al., 2003); these two regions also show linkage to general antisocial behavior (Stallings et al., 2005). Candidate gene studies have not yet yielded consistent findings (Ball & Collier, 2002). As was the case for alcoholism, the first genome scan for addiction to drugs other than alcohol reported many associations with small effect sizes but no large effects (Liu et al., 2005).

SUMMING UP

As was the case for alcoholism, moderate genetic influence and little shared environmental influence have been found for smoking and illicit drug abuse, although shared environmental influence plays a larger role for initiation of smoking. Multivariate genetic research suggests that the same genes largely mediate vulnerability across different drugs, although shared environmental influences are more drug specific. Exposure to drugs is also influenced by genetic factors, a genotype-environment correlation. Far fewer linkage and candidate gene studies have been reported for use and abuse of these drugs as compared to alcohol; no consistent results have as yet emerged.

Psychology and Aging

Aging is another example of a new area in the behavioral sciences that is being introduced to genetic research. Like health psychology, aging is an area of great social significance. The average age of most societies is increasing, primarily as a result of improvements in health care. For example, in the United States, the number of people age 65 and older will double from 10 to 20 percent during the next 30 years (U.S. Bureau of the Census, 1995). The fastest growing group of adults is those over age 85. Worldwide, this group is growing nearly twice as fast as the population as a whole (Chawla, 1993). Although obvious changes occur later in life, it is not possible to lump these older individuals into a category of "the elderly" because older adults differ greatly biologically and psychologically. The question for genetics is the extent to which genetic factors contribute to individual differences in functioning later in life.

Surprisingly little genetic research in the behavioral sciences has been directed toward the last half of the life span. Chapter 7 described genetic research on dementia, for which moderate genetic influence has been found. Dementia is a focal area for molecular genetic research. Several genes have been identified that account for most cases of a rare form of dementia that occurs in middle adulthood. The best example of a QTL in behavioral genetics is the association between apolipoprotein E and typical late-onset dementia.

Another interesting finding about genetics and cognitive aging was described in Chapter 8: The heritability of general cognitive ability increases throughout the life span. In later life, heritability estimates reach 80 percent, one of the highest heritabilities reported for behavioral traits, although in the very oldest individuals, heritability may decline again (Figure 14.7). Not enough research has been conducted on specific cognitive abilities throughout the life span to be able to conclude whether heritabilities of specific cognitive abilities also increase during development. However, this conclusion seems likely, at least as a general rule, because genetic influence on specific cognitive abilities largely overlaps with genetic influence on general cognitive ability (Chapter 9). Not mentioned in this discussion of multivariate genetic analysis in Chapter 9 is a distinction made in the field of cognition and aging between "fluid" abilities, such as spatial ability, that decline with age and "crystallized" abilities, such as vocabulary, that increase with age (Baltes, 1993). Although it has been assumed that fluid abilities are more

Figure 14.7 Ninety-three-year-old MZ twins participating in a twin study of cognitive functioning late in life, and photos of them going back to childhood (McClearn et al., 1997). Not only do MZ twins continue to look physically similar late in life, they also continue to perform similarly on measures of cognitive ability. (Reproduced with permission from *Science*, June 6, 1997. Copyright 1997 American Association for the Advancement of Science.)

biologically based and crystallized abilities more culturally based, genetic research so far has found that fluid and crystallized abilities are equally heritable (Pedersen, 1996). Research has begun to identify genes associated with normal cognitive aging separate from the study of pathological dementia (Deary, Wright, Harris, Whalley, & Starr, 2004; Zubenko, Hughes, Zubenko, & Maher, 2007).

For psychopathology and personality, the few genetic studies in later life yield results similar to those described in Chapters 10–13 for research earlier in life (Bergeman, 1997). For example, for depression in later life, twin studies indicate modest heritabilities similar to those found earlier in life (Gatz, Pedersen, Plomin, Nesselroade, & McClearn, 1992; Johnson, McGue, Gaist, Vaupel, & Christensen, 2002). For personality, Type A behavior—hard-driving and competitive behavior that is of special interest because of its reputed link with heart attacks—shows moderate heritability typical of other personality measures in older twins (Pedersen, Lichtenstein et al., 1989). Another interesting personality domain is locus of control, which refers to the extent that outcomes are believed to be due to one's own behavior or chance. For some older individuals, this sense of control declines, and the decline is linked to declines in psychological functioning and poor health. A twin study later in life found moderate genetic influence for two aspects of locus of control, sense of responsibility and life direction (Pedersen, Gatz, Plomin, Nesselroade, & McClearn, 1989). However, the key variable of the perceived role of luck in determining life's outcomes showed no genetic influence and substantial shared environmental influence. This finding, although in need of replication, stands out from the usual finding in personality research of moderate genetic influence and no shared environmental influence. The high stability of personality in later life is largely mediated by genetic factors (Johnson, McGue, & Krueger, 2005; Read et al., 2006).

The famous U.S. Supreme Court Justice Oliver Wendell Holmes quipped that "those wishing long lives should advertise for a couple of parents, both belonging to long-lived families" (Cohen, 1964, p. 133). Genetic research, however, indicates only modest genetic influence on longevity, with heritabilities of about 25 percent (Bergeman, 1997), although there is some suggestion that genetic influence on longevity increases at the most advanced ages (Hjelmborg et al., 2006). *APOE* is the only gene that has been shown to be associated with individual differences in human longevity, probably because of its links with cardiovascular disease rather than dementia (Christensen, Johnson, & Vaupel, 2006). Much genetic research in nonhuman species—especially mice, fruit flies, and nematode worms—is attempting to identify genes associated with longevity (Finch & Ruvkun, 2001; Shmookler Reis, Kang, & Ayyadevara, 2006). For example, seven genetic mouse models have been reported to show increased life span (Liang et al., 2003). In the fruit fly *Drosophila melanogaster*, selective breeding, QTL analysis, and mutational analysis have identified more than 40 genes related to the aging process (Poirier & Seroude, 2005). In the nematode worm (*C. elegans*), more than 70 genes have been found to influence life span (Braeckman & Vanfleteren, 2007).

Psychologists are especially interested in how well we live, the quality of life, not just how long we live. Health and functioning in daily life show moderate genetic influence later in life, as does the relationship between health and psychological well-being (Harris, Pedersen, Stacey, McClearn, & Nesselroade, 1992) and life satisfaction (Plomin & McClearn, 1990). Another aspect of quality of life is self-perceived competence. One study of older twins found that six dimensions of self-perceived competence—including interpersonal skills, intellectual abilities, and domestic skills—show heritabilities of about 50 percent (McGue, Hirsch, & Lykken, 1993). For some of us, hair loss in later life contributes to reduced quality of life, and a recent twin study suggests that individual differences in baldness in men over 70 years is 80 percent heritable (Rexbye et al., 2005).

SUMMING UP

Surprisingly few behavioral genetic studies have been directed toward the last half of the life span. Nonetheless, dementia is one of the most intense areas of molecular genetic research. The best example of a QTL in behavioral genetics is the association between apolipoprotein E and late-onset dementia. Longevity shows only modest genetic influence, although much research on nonhuman animals has attempted to identify genes that extend the life span. The few twin and adoption studies of behavioral traits in the later years yield results that are generally similar to those found earlier in the life span. Quality of life indicators later in life also show some genetic influence.

Summary

Two new areas of psychology from which interesting genetic results are emerging are health psychology and aging. One example of genetic research on health psychology concerns body weight and obesity. Although most theories of weight gain are environmental, genetic research consistently shows substantial genetic influence on individual differences in body weight, with heritabilities of about 70 percent. Also interesting in light of environmental theories is the consistent finding that shared family environment does not affect weight. Longitudinal genetic studies indicate that genetic influences on weight are surprisingly stable after infancy, although there is some evidence for genetic change even during adulthood. Body weight and obesity are the target of much molecular genetic research in mice and humans with some success.

Another example from health psychology concerns addictions. Alcoholism shows moderate genetic influence, with stronger genetic influence for alcoholism that is early in onset, severe, and associated with aggression. Several examples of genotype-environment interaction have been reported in which more permissive environments yield greater heritability of alcoholism. Selection studies of alcohol-related behaviors in mice demonstrate genetic influence, provide animal models for research, and yield QTLs. Persistence and quantity of smoking also show moderate genetic influence; initiation of smoking shows a larger role for shared environment, probably due to peers rather than to parents. Genetics also affects risk for use of illicit drugs such as amphetamines, heroin, and cocaine; the same genes largely mediate vulnerability across these drugs. An active area of behavioral genetic research is pharmacogenetic studies of drug-related responses in mice. Molecular genetic studies have also begun to yield some QTLs for genetic vulnerability to human alcoholism and to a lesser extent for other drugs.

Genetic research has only recently addressed the last half of the life span. Dementia and cognitive decline in later life are intense areas of molecular genetic research. For general cognitive ability, twin and adoption studies indicate that heritability increases during adulthood. Psychopathology and personality generally show results similar to those for younger ages, moderate heritability, and no shared family environment. Although longevity shows only modest genetic influence, several quality of life measures show moderate genetic influence in studies of elderly individuals.

CHAPTER FIFTEEN

Pathways between Genes and Behavior

As indicated in previous chapters, quantitative genetic research consistently shows that genetics contributes importantly to individual differences in nearly all behaviors such as learning abilities and disabilities, psychopathology, and personality. We have also seen in these chapters that quantitative genetics and molecular genetics are coming together in the study of complex traits and common disorders. Molecular genetic research, which attempts to identify the specific genes (QTLs) responsible for the heritability of these behaviors, has only begun to identify such genes. The most recent research using genomewide association scans with large samples suggests that the heritabilities of complex traits and common disorders are due to many genes of small effect. Nonetheless, the bottom line for behavioral genetics is this: Heritability means that DNA variation creates behavioral variation, and we need to find these DNA sequences to understand the mechanisms by which genes affect behavior.

The goal is not only finding genes associated with behavior but also understanding the pathways between genes and behavior, the mechanisms by which genes affect behavior, sometimes called *functional genomics* (Figure 15.1). This

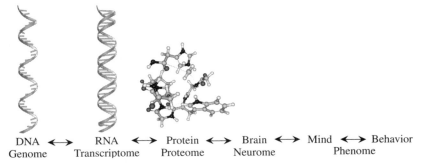

DNA ⟷ RNA ⟷ Protein ⟷ Brain ⟷ Mind ⟷ Behavior
Genome   Transcriptome   Proteome   Neurome   Phenome

Figure 15.1 Functional genomics includes all levels of analysis from genome to behavior. This chapter focuses on the transcriptome, proteome, and the brain.

BOX 15.1

Levels of Analysis

The relationship between brain and "mind" (mental constructs) has been a central issue in philosophy for four centuries since Descartes advocated a mind-body dualism in which the mind was nonphysical. Because this dualism between mind and body is now generally rejected (see Bolton & Hill, 2004; Kendler, 2005), we will simply assert the view that all behavior is biological in the general sense that behavior depends on physical processes. Does this mean that behavior can be reduced to biology (Bickle, 2003)? Because all behavior is biological, it would seem that the answer must logically be "yes." However, the sense in which all behavior is biological is similar to saying that all behavior is genetic (because without DNA there can be no behavior) or that all behavior is environmental (because without the environment there can be no behavior).

Behavioral genetics' way out of this philosophical conundrum is to focus empirically on individual differences in behavior and to investigate the extent to which genetic and environmental differences can account for these differences in behavior (see Chapter 5). The point of this chapter is to consider some of the levels of analysis that lie between genes and behavior. The ultimate goal of behavioral genetics is to understand the links between genes and behavior at all levels of analysis.

Different levels of analysis are more or less useful to address different questions, such as questions about causes and questions about cures (Bolton & Hill,

---

chapter considers ways in which researchers are attempting to connect the dots between genes and behavior. (See Box 15.1 for a discussion of some relevant philosophical issues.) We begin with gene expression, not just expression of one gene at a time but all the genes in the genome, called the *transcriptome*. The next step along the pathways from genes to behavior is all the proteins coded by the transcriptome, called the *proteome*. Next is the brain, which, continuing the –*omics* theme, has been referred to as the *neurome*. This chapter stops at the brain level of analysis because the mind (cognition and emotion) and behavior—sometimes called the *phenome*—have been the focus of Chapters 7 to 14.

It should be noted that this chapter is not about the transcriptome, the proteome, or the brain per se, three of the most active areas of research in all of the life sciences. Instead, the chapter is about the more focused issue of the role of gene expression, proteins, and the brain in mediating genetic effects on individual differences in behavior. The environment plays a crucial role at each step in the pathways between genes and behavior; it is the topic of Chapter 16.

2004). *Functional genomics* generally assumes a bottom-up approach that begins at the level of cells and molecular biology. The phrase *behavioral genomics* has been proposed as an antidote that emphasizes the value of a top-down approach that attempts to understand how genes work at the level of the behavior of the whole organism (Plomin & Crabbe, 2000). Behavioral genomics may be more fruitful than other levels of analysis in terms of prediction, diagnosis, intervention, and prevention of behavioral disorders.

Finally, relationships between levels of analysis should be considered correlational until proven causal, which is why the connections between levels in Figure 15.1 are double-headed arrows. For example, associations between brain differences and behavioral differences are not necessarily caused by the brain differences: Behavior can cause changes in brain structure and function. A striking example is that the posterior hippocampus, a part of the brain that stores spatial representations of the environment, is significantly larger in London taxi drivers (Maguire et al., 2000); the size is correlated with the number of years of driving a taxi (Maguire, Woollett, & Spiers, 2006). Similarly, correlations between gene expression and behavior are not necessarily causal because behavior can change gene expression. A crucial point is that the only exception to this rule is DNA: Correlations between differences in DNA sequence and differences in behavior are causal in the sense that behavior does not change the nucleotide sequence of DNA. In this sense, DNA is in a causal class of its own.

The Transcriptome: Gene Expression throughout the Genome

Gene expression is the first step on any pathway from genes to behavior: A polymorphism in DNA can only have an effect when the gene is expressed. Some genes, called housekeeping genes, are expressed at a steady rate in most of our cells. Other genes are expressed as their product is needed in response to the environment.

As explained in Chapter 4, gene expression is a multistep process that begins when DNA is transcribed into RNA. The focus of gene expression has been on genes that code for proteins. These DNA sequences are transcribed into messenger RNA and then translated into amino acid sequences that constitute proteins. For protein-coding genes, expression is most affected by altering the rate of transcription initiation, but other factors that affect expression include alteration of the RNA transcript, passage of the messenger RNA through the nuclear membrane, protection or degradation of the RNA transcript in the cytoplasm, the rate of translation, and posttranslational modification of the protein. (An animation of these processes is available at www.maxanim.com/genetics/Gene%20expression/Gene%20expression.htm).

It has become clear in recent years that much DNA is transcribed into RNA, but the RNA is not translated into proteins and is thus called *non-coding RNA* (see Chapter 4). Non-coding RNA transcripts can regulate the expression of other genes without being translated into proteins. This regulation of gene expression by non-coding RNA is primarily affected by altering the rate of transcription, but other factors include changes in the RNA transcript itself and the way the RNA transcript interacts with its regulatory targets, which are often messenger RNA transcripts.

Gene Expression Profiles: RNA Microarrays

For both protein-coding and non-protein-coding DNA, gene expression can be indexed by the number of RNA transcripts, which is the end result of the various processes mentioned, not just the initial transcription process. In contrast to DNA, which faithfully preserves the genetic code in all cells, at all ages, and at all times, RNA degrades quickly and is tissue specific, age specific, and state specific, as noted in Chapter 4. An exciting development is the ability of microarrays to assess the expression of all genes in the genome simultaneously, called *gene expression profiling*. Gene expression (RNA) microarrays are the same as the SNP microarrays described in Box 6.2 except that the probes in RNA microarrays detect a particular sequence of DNA that corresponds to an RNA fragment, rather than identifying a particular SNP allele. In addition, the goal of RNA microarrays is to detect the quantity of each of the RNA fragments; for this reason, each probe is represented with millions of copies. In contrast, SNP probes detect the presence or absence of SNP alleles; multiple probes for each allele are used only in order to increase the accuracy of genotyping. RNA microarrays had been limited to probes for exons which assessed transcription of the 2 percent of the genome that involves protein-coding genes. In 2005, *tiling microarrays* became available that have millions of probes that span the genome at regular intervals in order to detect transcription systematically throughout the genome, thus including non-coding RNA as well as protein-coding RNA (Royce et al., 2005). Tiling microarrays confirmed the suspicion that much of the genome is transcribed into RNA, the so-called dark matter of the genome (Johnson, Edwards, Shoemaker, & Schadt, 2005). Other advances in understanding the complexity of gene expression are developing rapidly (Carninci, 2006).

RNA microarrays make it possible to take snapshots of gene expression throughout the genome at different times (for example, during development, or before and after interventions) and in different tissues (for example, in different brain regions). There are scores of studies that have investigated changes in gene expression profiling in response to drugs (Yuferov, Nielsen, Butelman, & Kreek, 2005) and between groups such as psychiatric cases and controls (Konradi, 2005). Although a review in 2004 of 5000 early gene expression studies found problems with reliability or validity (Miklos & Maleszka, 2004), improvements have been made in recent years in relation to gene expression arrays and their analysis (e.g., Allison, Cui, Page, & Sabripour, 2006; Cobb et al., 2005). Nonetheless, limitations remain, such as difficulties in detecting RNA transcripts when there are very few copies (Draghici, Khatri, Eklund, & Szallasi, 2006; Wang et al., 2006).

Gene expression profiling of the brain is like structural genetic neuroimaging in that it can create an atlas of localized patterns of gene expression throughout the brain. Because genetic neuroimaging requires brain tissue, its use in the human species is limited to post-mortem brains and tissue samples removed during surgery, such as tumors (Yamasaki et al., 2005), which raises questions about lack of control concerning gene expression at the time of death (Konradi, 2005). For this reason, structural genetic neuroimaging research has primarily been investigated in mice rather than the human species. For example, a comprehensive atlas of expression profiles of 20,000 genes in the adult mouse brain is publicly accessible online (Lein et al., 2007; www.brain-map.org).

Structural brain maps of gene expression are fundamental because genes can only function if they are expressed. The next goal is functional genetic neuroimaging—studying changes in gene expression in the brain across time, for example, during development, or following interventions such as drugs or cognitive tasks. Such research on brain function needs to use animal models. For example, research on mice is under way that aims to create an atlas of profiles of gene expression throughout the brain during learning and memory tasks in the Genes to Cognition research consortium (Grant, 2003; www.genes2cognition.org).

CLOSE UP

Seth Grant graduated in Science, Medicine, and Surgery from the University of Sydney, Australia. After clinical training, he started in basic molecular research at Cold Spring Harbor Laboratory in New York and worked on transgenic mice in cancer and diabetes. He then worked with Eric Kandel (Nobel laureate 2000) at Columbia University and New York State Psychiatric Institute. In 1994, he joined the University of Edinburgh, where he was professor of Molecular Neuroscience and director of the Centre for Neuroscience. He is currently working at the Wellcome Trust Sanger Institute as a principal scientist and directs the Genes to Cognition research consortium.

Because of the practical and scientific limitations of using post-mortem brain tissue, RNA microarrays will be much more widely applicable to human research if easily available tissue such as blood can be used for gene expression profiling. Some similarities between expression in blood and brain have been reported (e.g., Gladkevich, Kaufman, & Korf, 2004; Glatt et al., 2005; Nicholson, Unger, Mangalathu, Ojaniemi, & Vernon, 2004; Pahl, 2005; Sharp et al., 2006; Sullivan, Fan, & Perou, 2006). Although gene expression profiling in the blood cannot be used to localize patterns of gene expression in the brain, blood could be used to address some important questions, most notably, gene expression profile differences as a function of development or interventions.

Rather than studying the expression of each gene in isolation, RNA microarrays make it possible to study profiles of gene expression across the transcriptome, which leads to understanding the coordination of gene expression throughout the genome (Ghazalpour et al., 2006; Schadt, 2006).

Genetical Genomics

So far, we have discussed gene expression from a normative perspective rather than considering individual differences. The field of gene expression has only recently turned to individual differences, and their causes and consequences (Cobb et al., 2005; Rockman & Kruglyak, 2006). Much recent research has been directed toward treating gene expression as a phenotypic trait and finding QTLs (called *expression QTLs* or *eQTLs*) associated with gene expression in mice (Schadt, 2006; Williams, 2006) and humans (Morley et al., 2004). This field has been called *genetical genomics* to emphasize the links between the genome and the transcriptome (Jansen & Nap, 2001; Li & Burmeister, 2005; Petretto et al., 2006). These links have become explicit because recent research using DNA microarrays (see Chapter 6) can be used to scan the genome for associations with genomewide gene expression assessed on RNA microarrays.

Research on genomewide gene expression in rodents has profited from the availability of inbred lines and especially recombinant inbred lines, which facilitate both quantitative genetic and molecular genetic research (Chesler et al., 2005; Letwin et al., 2006; Peirce et al., 2006) and provide access to brain tissue. However, for rodent research as well as human research, many eQTL associations have been reported but few have been replicated. This is a repeat of the story told in Chapter 6 in which genetic effects on complex traits, including individual differences in gene expression, appear to be caused by many QTLs of small effect size. As a result, very large samples will be needed to attain adequate statistical power to detect reliable associations with gene expression traits.

Gene Expression as a Biological Basis for Environmental Influence

Genetical genomics attempts to identify the QTLs responsible for the genetic contribution to individual differences in gene expression, but to what extent are these individual differences genetic in origin? It cannot be assumed that individual differences in gene expression are highly heritable because gene expression has evolved to be responsive to intra-cellular and extra-cellular environmental variation. Indeed, the few small quantitative genetic studies of human RNA transcript levels suggest that heritabilities appear to be modest on average across the genome, which implies that most of the variability in transcript levels is due to environmental factors (Cheung et al., 2003; Correa & Cheung, 2004; McRae et al., 2007; Monks et al., 2004; Sharma et al., 2005). Members of identical twin pairs become increasingly different in gene expression profiles throughout the life span (Fraga et al., 2005; Petronis, 2006).

Environmental factors involved in gene expression are part of a rapidly expanding area of research called *epigenesis*, which involves long-term changes in gene expression that continue across generations of cells, including genomic imprinting (see Chapter 3) and changes in chromosome structure (Jaenisch & Bird, 2003). More than 1600 papers on epigenesis were published in 2006 alone. A new field called *epigenomics* considers the entire transcriptome (Callinan & Feinberg, 2006). Epigenesis is sometimes mistakenly discussed as if it were an alternative to genetics (Rakyan & Beck, 2006), in part because epigenetic research tends to focus on environmental causes of gene expression, as seen in the example of differences in gene expression profiles within pairs of identical twins. Also, the long-term changes in gene expression across generations of cells are often incorrectly referred to as "inherited." It should be reiterated that gene expression is a phenotype; individual differences in expression itself or in epigenetic processes that lead to individual differences in expression may be due to genetic differences (Richards, 2006) or environmental differences.

RNA microarrays could lead to a paradigm shift in studying environmental influences on complex behavioral disorders by considering gene expression as the fundamental biological basis of environmental influence. This perspective could provide a biological foundation upon which to build an understanding of more complex levels of environmental analysis typically studied in behavioral research. It could also have far-reaching impact on translational research by providing biomarkers for differential diagnosis and providing a biological basis for monitoring environmental interventions such as drugs and other therapies. For example, an extremely active area of research is pharmacogenomics, the study of changes in gene expression profiles in response to drugs (Yuferov, Nielsen, Butelman, & Kreek, 2005). Other environmental interventions could be investigated similarly, even for complex environments such as parenting or stress.

SUMMING UP

The first systematic step in the pathways between genes and behavior is gene expression. Genetic effects on behavior can take place only to the extent that genes are expressed, a process that begins with transcription of DNA to RNA.

The transcriptome refers to the expression of all the genes in the genome, which can now be assessed using RNA microarrays, both for coding and non-coding RNA. Individual differences in gene expression should be considered as a phenotypic trait that can be caused by genetic and environmental factors. The field of genetical genomics looks for eQTLs that account for heritable variation in gene expression, and the field of epigenetics looks for environmental mechanisms in gene expression. Quantitative genetic studies of gene expression as a phenotype suggest that individual differences in expression may not be highly heritable. The transcriptome could provide a biological foundation for studies of the environment.

As noted at the outset of this chapter, we cannot hope to provide a review of all that is known about gene expression. Of special interest in terms of pathways between genes and behavior is the extent to which DNA associations with behavior are mediated by individual differences in gene expression. In the following section, we will continue along the pathways between genes and behavior by considering the next level of analysis, the proteome.

The Proteome: Proteins Coded throughout the Transcriptome

The proteome, which refers to the entire complement of proteins, is studied even more than the transcriptome, reaching 5000 publications per year in 2006. The increase in complexity from the genome to the transcriptome is multiplied many times as we move from the transcriptome to the proteome for three reasons. First, there are many more proteins than genes, in part because alternative splicing of genes can produce different messenger RNA transcripts (Brett et al., 2002). Second, after amino acid sequences are translated from messenger RNA, they undergo modifications, called *posttranslational modifications*, which change their structure and thus change their function. Third, proteins do not work in isolation; their function is affected by their interactions with other proteins as they form protein complexes.

The proteome can be identified using gels in an electrical field (electrophoresis) to separate proteins in one dimension on the basis of their charge and in a second dimension on their molecular weight, called *two-dimensional gel electrophoresis*. The precision of identifying proteins has been greatly improved by the use of mass spectrometry, which analyzes mass and charge at an atomic level (Aebersold & Mann, 2003). Using these techniques, a proteome atlas of nearly 5000 proteins and 5000 protein complexes is available in the fruit fly (Giot et al., 2003); similar resources are available for the hippocampus of the mouse (Pollak, John, Hoeger, & Lubec, 2006) and the hippocampus of the rat (Fountoulakis, Tsangaris, Maris, & Lubec, 2005).

The relative quantity of each protein can also be estimated from two-dimensional gel electrophoresis. Individual differences in the quantity of a protein in a particular tissue represent a protein trait which is analogous to the RNA transcript traits discussed in the previous section. As with the transcriptome, the proteome needs to be considered as a phenotype that can be attributed to genetic and environmental factors. Such protein traits can be related to individual differences in behavior. For example, human studies using cerebrospinal fluid have reported about 300 differences in protein levels and protein modifications in psychiatric disorders (Fountoulakis & Kossida, 2006).

Although the transcriptome is the target of much recent genetic research, far less genetic research has considered the proteome. One reason is that the measurement of the proteome is more difficult than the transcriptome, which can be measured easily using RNA microarrays. Microarrays are being developed for protein analysis, but the problem is that, unlike the simple nucleotide chains assessed by RNA or DNA microarrays, proteins come in diverse sizes, shapes, and chemistry (MacBeath, 2002). The earliest and still most widely used array consists of antibodies specific to each protein. Antibodies are proteins used by the immune system to identify and neutralize foreign organisms like bacteria and viruses. In protein microarrays, the antibodies are designed to detect specific proteins that have been fluorescently labeled. However, the largest arrays can only measure a few hundred proteins. Many new techniques are being developed to create protein microarrays that use DNA rather than antibodies to detect proteins, which will complete the bridge between the transcriptome and the proteome (Eisenstein, 2006).

Similar to research on the transcriptome, the mouse has been the focus of proteomic work because of the availability of brain tissue. A pioneering study examined 8767 proteins from the mouse brain as well as other tissues and found that 1324 of these proteins showed reliable differences in a large backcross (see Chapter 2) (Klose et al., 2002). Of these proteins, 466 were mapped to chromosomal locations. Although such linkages need to be replicated, the genetic results are interesting for two reasons: Most proteins showed linkage to several regions, and the chromosomal positions often differed from those of the genes that code for the proteins. These results suggest that multiple genes affect protein traits. A more recent study, yielding similar results, focused on protein expression in the hippocampus (Pollak, John, Schneider, Hoeger, & Lubec, 2006).

SUMMING UP

The second step in the pathways between genes and behavior is proteomics, all the proteins created by the transcriptome. The proteome is considerably more complex than the transcriptome and more difficult to measure. Nonetheless, some preliminary proteomic atlases are available and genetic research has begun, relating the proteome to the genome.

Much remains to be learned about individual differences in the proteome and their genetic and environmental origins. The goal of behavioral genetics is to relate the proteome back to the transcriptome and the genome, and forward to the brain and behavior.

The Brain

Each step along the pathways from genome to transcriptome to proteome involves huge increases in complexity, but these increases in complexity pale in comparison to the complexity of the brain. The brain has trillions of junctions between neurons (*synapses*) instead of billions of DNA base pairs, and hundreds of neurotransmitters, not just the four bases of DNA. Although the three-dimensional structure of proteins and their interaction in protein complexes contribute to the complexity of the proteome, this complexity is nothing as compared to the complexity of the three-dimensional structure and interactions among neurons in the brain.

Neuroscience, the study of brain structure and function, is another extremely active area of research. This section provides an overview of neurogenetics as it relates to behavior. Because the brain is so central in the pathways between genes and behavior, brain phenotypes are sometimes referred to as *endophenotypes*, as discussed in Box 15.2. In this chapter, we will refer to areas of the brain depicted in Figure 15.2 and to the structure of the neuron shown in Figure 15.3.

As discussed earlier in this chapter, research on the transcriptome and proteome has begun to build bridges to the brain by creating atlases of gene and protein expression throughout the brain. Most of this research involves animal models because of the access to brain tissue in nonhuman animals. A huge

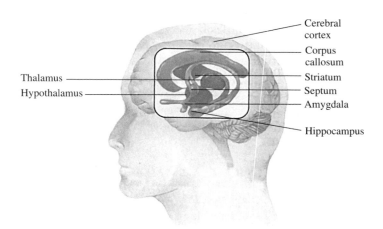

Figure 15.2 Basic structures of the part of the human brain called the forebrain.

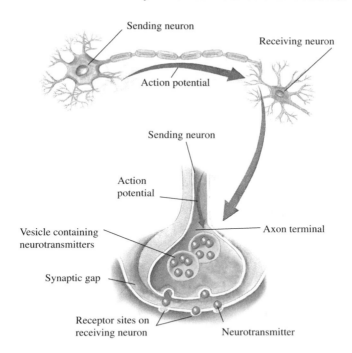

Figure 15.3 The neuron. Electrical impulses (action potentials) travel from one neuron to another across a gap at the end of the axon known as a synapse. Action potentials release neurotransmitters into the synaptic gap. The neurotransmitters bind to receptor sites on the receiving neuron.

advantage for neurogenetic research in the human species is the availability of neuroimaging, which, as discussed later, makes it possible to assess the structure and function of the human brain. We begin, however, with two major areas of neurogenetic research on behavior that focus on animal models, particularly the fruit fly *Drosophila* and the mouse: circadian rhythms, and learning and memory. The advantage of neurogenetic research with animal models is the ability to use both natural and induced genetic mutations to dissect pathways between neurons and behavior. The third example of neurogenetic research involves emotion in the human species, using neuroimaging.

Circadian Rhythms

The neurogenetics of the daily sleep-wake cycles, called *circadian* (from the Latin, meaning "about a day") *rhythms,* have been studied primarily in *Drosophila* and mice (Rosato, Tauber, & Kyriacou, 2006). In 1971, the *period* mutation was isolated in *Drosophila;* this mutation substantially alters circadian period length (Konopka & Benzer, 1971). Much effort has been devoted to understanding how this mutation and others act as pacemakers in the brain to affect circadian rhythms.

BOX 15.2

Endophenotypes

The goal of behavioral genetics is to understand pathways between genes and behavior at all levels of analysis. In addition, each level of analysis warrants attention in its own right. (See Box 15.1.) Using the brain level of analysis as an example, there is much to learn about the brain itself regardless of the brain's relationship to genes or to behavior. However, the focus of behavioral genetics, and this chapter, is on the brain as a pathway between genes and behavior.

Levels of analysis lower than behavior itself are sometimes called endophenotypes where *endo* means "inside." It has been suggested that these lower levels of analysis, such as the brain, might be more amenable to genetic analysis than behavior (Bearden & Freimer, 2006; Gottesman & Gould, 2003). In addition, lower-level processes, such as neurotransmitter levels in the brain, can be modeled more closely in animals and humans than can behavior itself (Gould & Gottesman, 2006). Specifically, it is hoped that genes will have larger effects on lower levels of analysis and will thus be easier to identify. Recent genetic research on brain neuroimaging phenotypes supports this hypothesis (see text), as does research on alcoholism (Dick, Jones et al., 2006). However, caution is warranted until these DNA associations are replicated, because genetic influences are likely to be pleiotropic and polygenic for brain traits as well as behavioral traits (Kovas & Plomin, 2006). Moreover, a meta-analysis of genetic associations reported for endophenotypes concluded that genetic effect sizes are no greater for endophenotypes than for other phenotypes (Flint & Munafo, 2007).

Although less complex than behavioral traits, brain traits are nonetheless very complex, and complex traits are generally influenced by many genes of small effect (see Chapter 6). Indeed, the most basic level of analysis, gene expression, appears to be influenced by many genes of small effect, as well as by substantial environmental influence. One might think that lower levels of analysis are more heritable, but this does not seem to be the case. Using gene expression again as an example because it is the most basic level of analysis, individual differences in transcript levels across the genome do not appear to be highly heritable.

Another issue is that the goal of behavioral genetics is to understand pathways between genes, brain, and behavior. Genes found to be associated with brain phenotypes are important in terms of the brain level of analysis, but their usefulness for behavioral genetics depends on their relationship with behavior. In other words, when genes are found to be associated with brain traits, the extent to which the genes are associated with behavioral traits needs to be assessed rather than assumed.

It is now known that *period* and other genes constitute a core unit of the clock mechanism that operates in a certain part of the hypothalamus called the *suprachiasmatic nucleus (SCN)* (Moore, 1999). Destruction of the SCN obliterates the normal sleep-wake cycle. The proteins coded by the clock genes regulate the sleep-wake cycle by means of a cascade of changes in levels of hormones, melatonin, and body temperature (Figure 15.4). The proteins begin the process by interacting with the neural cell membrane and synchronizing the SCN neurons into a system called the SCN pacemaker. The 24-hour cycle is caused in part by a self-regulatory feedback loop in which the genes are normally transcribed, but as the proteins build up, they inhibit transcription of the

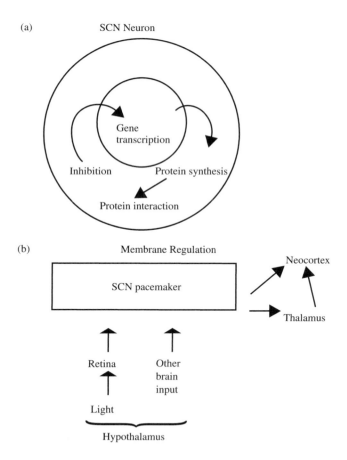

Figure 15.4 The circadian rhythm is driven by a feedback loop of gene transcription in neurons in the suprachiasmatic nucleus (SCN) in the hypothalamus (a). The same proteins involved in this feedback loop also regulate the neuron's membrane, which synchronizes the SCN neurons into a pacemaker. The SCN pacemaker receives information about light from the retina and information about other stimuli from other brain areas (b).

genes that code for them. As the proteins degrade during the next 24 hours, the genes are again transcribed.

In addition, one of the clock genes, *timeless*, encodes a protein that degrades rapidly in response to light; as a result, the transcription cycle of the *timeless* gene has a cycle shorter than 24 hours in the presence of light. This property allows the circadian system to be responsive to light cycles and may be responsible for the fact that people sleep as little as three hours a night in the far north during the summer months, when there is no darkness.

Clock genes have been found in many other species, suggesting that they serve an evolutionarily ancient function (Ko & Takahashi, 2006). However, the rule of pleiotropy applies—these genes have many other effects. As one of many examples of pleiotropy, the *period* gene has been found to play an important role in long-term memory (Sakai, Tamura, Kitamoto, & Kidokoro, 2004). Clock genes also affect another cyclical aspect of behavior in *Drosophila*, the interpulse intervals in the male courtship "song," which is generated by wing vibration, as well as female mating activity (Sakai & Ishida, 2001). One of the first dramatic demonstrations of transgenic research involved the transfer of the *period* gene for one species (*Drosophila simulans*) to another species (*Drosophila melanogaster*). The species-specific song cycle and other courtship behaviors were transferred along with the gene (Wheeler et al., 1991).

The first mammalian circadian mutation (*tau*) appeared spontaneously in a laboratory stock of hamsters (Ralph & Menaker, 1988). This mutation inspired the search for a similar one in mice, and a mutation that lengthens the circadian period by one hour in heterozygotes and four hours in homozygotes was eventually found (Vitaterna et al., 1994). The mouse gene, called *clock*, was cloned in 1997 (King et al., 1997). Its crucial role in circadian rhythms was proved by inserting the normal *clock* gene into mutant mouse embryos and demonstrating that these mice with "rescued" mutations had normal circadian rhythms (Antoch et al., 1997). Further studies have shown that certain neurons isolated from *clock* mutant mice have arrhythmic firing patterns. This finding suggests that the *clock* gene uses these neurons to synchronize neuronal firing patterns, which ultimately regulate the daily activity cycles of the mouse (Herzog, Takahashi, & Block, 1998).

Most of the thousands of papers on circadian rhythms and genetics use mutations, either natural or induced, to disassemble neural machinery. However, few studies have considered normal variation in these neural processes. One commonly experienced source of variation is the environmental disruption of circadian rhythms called jet lag. In addition, circadian rhythms are disrupted by the aging process (Hofman & Swaab, 2006), and natural variations in clock genes have been found to be associated with latitude (Costa & Kyriacou, 1998). Although circadian rhythms are ancient processes honed by evolution, trait variation in the human species is apparent, for example, in the way we classify people as "larks" (morning people) and "owls" (evening people) (Merrow,

Spoelstra, & Roenneberg, 2005). Dozens of SNPs have been identified in ten circadian clock genes in the human species and are being studied in relation to behaviors such as bipolar disorder (Nievergelt et al., 2006).

Learning and Memory

Even more neurogenetic research has considered learning and memory, one of the most obvious functions of the brain. Much of this research involves the fruit fly *Drosophila*. *Drosophila* can indeed learn and remember, which has been studied primarily in relation to spatial learning and olfactory learning (Skoulakis & Grammenoudi, 2006). Learning and memory in *Drosophila* is becoming the first area to connect the dots between genes, brain, and behavior (Margulies, Tully, & Dubnau, 2005; McGuire, Deshazer, & Davis, 2005). For example, in studies of chemically created mutations in *Drosophila melanogaster*, investigators have identified dozens of genes that, when mutated, disrupt learning (Waddell & Quinn, 2001). A model of memory has been built by using these mutations to dissect memory processes. Beginning with dozens of mutations that affect overall learning and memory, investigators found, on closer examination, that some mutations (such as *dunce* and *rutabaga*) disrupt early memory processing, called short-term memory (STM). In humans, this is the memory storage system you use when you want to remember a telephone number temporarily. Although STM is diminished in these mutant flies, later phases of memory consolidation, such as long-term memory (LTM), are normal. Other mutations affect LTM but do not affect STM. Genes identified as necessary for learning in *Drosophila* also appear to be important in mammals (Davis, 2005).

Neurogenetic research is now attempting to identify the brain mechanisms by which these genes have their effect. Several of the mutations from mutational screening were found to affect a fundamental signaling pathway in the cell involving cyclic AMP (cAMP). *Dunce*, for example, blocks an early step in the learning process by degrading cAMP prematurely. Normally, cAMP stimulates a cascade of neuronal changes including production of a protein kinase that regulates a gene called *cAMP-responsive element* (*CRE*). CRE is thought to be involved in stabilizing memory by changing the expression of a system of genes that can alter the strength of the synaptic connection between neurons, called *synaptic plasticity*, which has been the focus of research in mice (see below). In terms of brain regions, a major target for research in *Drosophila* has been a type of neuron called *mushroom body neuron* that appears to be the major site of olfactory learning in insects (Heisenberg, 2003), although many other neurons are also involved (Davis, 2004). Pairing shock with olfactory cues triggers a complex series of signals that results in a cascade of expression of different genes. These changes in gene expression produce long-lasting functional and structural changes in the synapse (Liu & Davis, 2006). Because neuronal function is not easily investigated in flies, much work along these lines has been done with *Aplysia*, a marine invertebrate with a simple nervous system amenable to electrophysiological recording

(Hawkins, Kandel, & Bailey, 2006), and with mice (Grant, Marshall, Page, Cumiskey, & Armstrong, 2005).

Learning and memory constitute an intense area of research activity in the mouse. However, rather than relying on randomly created mutations, neuro-genetic research on learning and memory in the mouse uses targeted mutations. It also focuses on one area of the brain called the hippocampus (see Figure 15.2), which has been shown in studies of human brain damage to be crucially involved in memory. In 1992, one of the first gene targeting experiments for behavior was reported (Silva, Paylor, Wehner, & Tonegawa, 1992). Investigators knocked out a gene (α-*CaMKII*) that normally codes for the protein α-Ca^{2+}-calmodulin kinase II, which is expressed postnatally in the hippocampus and other forebrain areas critical for learning and memory. Mutant mice homozygous for the knock-out gene learned a spatial task significantly more poorly than control mice did, although otherwise their behavior seemed normal. A spatial memory task used in most of the research of this type is a water maze. In studies using this task, various mutant and control mice are trained to escape from a large pool of opaque water by finding a platform hidden just beneath the water's surface (Figure 15.5).

In the 1990s, there was an explosion of research using targeted mutations in the mouse to study learning and memory (Mayford & Kandel, 1999), with 22 knock-out mutations shown to affect learning and memory in mice (Wahlsten, 1999). Many of these targeted mutations involve changes in the strength of con-

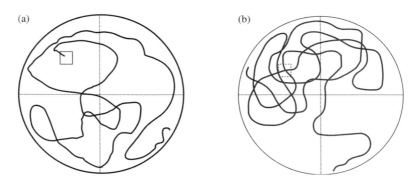

Figure 15.5 The Morris water maze is frequently used in neurogenetic research on spatial memory. A mouse escapes the water by using spatial cues to find a submerged platform. Shown in these diagrams are swim paths to a platform (upper left quadrant) in the Morris water maze. The mouse is trained to know the location of a submerged invisible platform. The animal usually navigates by using distal room clues such as doors and posters on the walls, but it can also be given more proximal cues to control for orientation. (a) The trained animal is tested on its efficiency in finding the platform (time, path length, erroneous entries into the wrong quadrants). (b) The submerged platform is removed, and the time the trained animal spends searching in the correct quadrant is assessed.

nections across the synapse and have been the topic of more than 10,000 papers, with 500 papers focused on the genetics of synaptic plasticity. Memories are made of long-term synaptic changes, called *long-term potentiation* (Lynch, 2004). The idea that information is stored in neural circuits by changing synaptic links between neurons was first proposed in 1949 (Hebb, 1949).

Although genes drive long-term potentiation, understanding how this occurs is not going to be easy because each synapse is affected by more than a thousand protein components. The α-*CaMKII* gene, mentioned earlier in relation to the first reported knock-out study of learning and memory, activates *CRE-encoded* expression of a protein called CRE-binding protein (CREB), which affects long-term but not short-term memory (Silva, Kogan, Frankland, & Kida, 1998). CREB expression is a critical step in cellular changes in the synapse in the mouse, as it is in *Drosophila*. In *Drosophila*, another gene that activates CREB was the target of a *conditional* knock-out that can be turned on and off as a function of temperature. These changes in CREB expression were shown to correspond to changes in long-term memory (Yin, Vecchio, Zhou, & Tully, 1995). A complete knock-out of *CREB* in mice is lethal, but deletions that substantially reduce CREB have also been shown to impair long-term memory (Mayford & Kandel, 1999).

A receptor involved in neurotransmission via the basic excitatory neurotransmitter glutamate plays an important role in long-term potentiation and other behaviors in mice as well as humans (Newcomer & Krystal, 2001). The N-methyl-D-aspartate (NMDA) receptor serves as a switch for memory formation by detecting coincident firing of different neurons; it affects the cAMP system among others. Overexpressing one particular *NMDA* gene (*NMDA receptor 2B*) enhanced learning and memory in various tasks in mice (Tang et al., 1999). A conditional knock-out was used to limit the mutation to a particular area of the brain—in this case, the forebrain. Normally, expression of this gene has slowed down by adulthood; this pattern of expression may contribute to decreased memory in adults. In this research, the gene was altered so that it continued to be expressed in adulthood, resulting in enhanced learning and memory. However, this particular *NMDA* gene is part of a protein complex (N-methyl-D-aspartate receptor complex) that involves 185 proteins; mutations in many of the genes responsible for this protein complex are associated with behavior in mice and humans (Grant et al., 2005).

Targeted mutations indicate the complexity of brain systems for learning and memory. For example, none of the genes and signaling molecules in flies and mice found to be involved in learning and memory are specific to learning processes. They are involved in many basic cell functions, a finding that raises the question of whether they merely modulate the cellular background in which memories are encoded (Mayford & Kandel, 1999). It seems likely that learning involves a network of interacting brain systems. Another example of complexity can be seen in work on the gene for the *dunce* mutant in *Drosophila*. When it was altered by disabling various combinations of its five DNA start sites for transcription, the investigators found that each combination has different effects on learning and memory processes (Dubnau & Tully, 1998).

Chemical-induced and targeted mutations in *Drosophila* and mouse have shown that long-term potentiation of the synapse is a necessary facet of learning and memory, although other processes are also important (Mayford & Kandel, 1999). The number of papers using mutations to study learning and memory has declined since 2001, in part due to the problems with gene targeting described in Chapter 6 and in part due to the increased use in research of pharmacological and neural interventions rather than genetic interventions. Similar to circadian rhythms, relatively little neurogenetic research has as yet been conducted on normal variation in learning and memory.

Emotion

In the human species, the structure and function of brain regions can be assessed using noninvasive neuroimaging techniques. There are many ways to scan the brain, each with a different pattern of strengths and weaknesses. For example, brain structures can be seen clearly using magnetic resonance imaging (MRI) (Figure 15.6). Functional MRI (fMRI) is able to visualize changing blood flow in the brain, which is associated with neural activity. The spatial resolution of fMRI is good, about two millimeters, but its temporal resolution is limited to events that take place over several seconds. Electroencephalography (EEG) measures voltage differences across the brain that index electrical activity, using electrodes placed on the scalp. It provides excellent temporal resolution (less than one millisecond) but its spatial resolution is very poor, because it averages activity across adjacent regions on the brain's surface. It is possible to combine the spatial strength of fMRI and the temporal strength of EEG (Debener, Ullsperger,

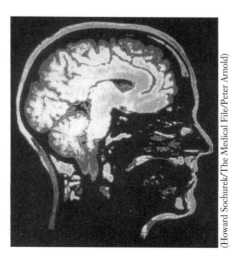

(Howard Sochurek/The Medical File/Peter Arnold)

Figure 15.6. Magnetic resonance imaging (MRI) scan of the human brain.

Siegel, & Engel, 2006), which can also accomplished using a different technology, *magnetoencephalography* (MEG; Ioannides, 2006).

Neuroimaging is just beginning to be used in genetic research. For example, in several twin studies, structural neuroimaging has shown that individual differences in the volume of many brain regions are highly heritable and correlated with general cognitive ability (Posthuma et al., 2002; Thompson et al., 2001; Wallace et al., 2006). Candidate gene studies have also begun to report associations with several types of brain function (Mattay & Goldberg, 2004; Winterer, Hariri, Goldman, & Weinberger, 2005). Although much neuroimaging research investigates human learning and memory, recent attention in neurogenetics has turned to emotion (LeDoux, 2000) and especially the role of the amygdala (see Figure 15.2; Phelps & LeDoux, 2005). For example, a highly cited paper reported that a serotonin transporter gene polymorphism (*5-HTTLPR*) is associated with amygdala neuronal activity in response to threat-related stimuli (looking at angry and fearful faces) as assessed by fMRI (Hariri et al., 2002).

CLOSE ⬛ UP

Ahmad Hariri's interest in identifying the biological mechanisms that give rise to individual differences in behavior is rooted in his undergraduate studies of evolutionary biology. While a doctoral student working with Susan Bookheimer at UCLA, he pursued the unique opportunity to study the neurobiological basis of complex human behaviors with fMRI. His early work led to important models of how the dynamic interactions between the amygdala and prefrontal cortex mediate emotional arousal and regulation. He continued his research on the neurobiology of emotional behaviors as a postdoctoral fellow with Daniel Weinberger at the U.S. National Institute of Mental Health. There he realized the powerful synergy of molecular genetics and neuroimaging. Through such "imaging genetics" research, he and his collaborators have found specific neurobiological pathways through which common genetic polymorphisms impact normal variability in behavior, as well as risk for neuropsychiatric disease. He has recently added positron emission tomography (PET) to his arsenal of imaging genetics tools, revealing molecular substrates that mediate the effects of genetic polymorphisms on the neural circuitry supporting emotional behavior. He is extremely enthusiastic about the potential of imaging genetics to reveal the complex interplay of genes, brain, and behavior.

Importantly, this finding has been replicated in several studies and has also received support from mouse knock-out research (Hariri & Holmes, 2006); it may have more general behavioral implications in terms of how we react to environmental stress (Hariri et al., 2005).

SUMMING UP

Three examples of the complexities of the brain level of analysis include circadian rhythms, learning and memory, and emotion. Natural and induced mutations have been used in *Drosophila* to dissect neural machinery involved in circadian rhythms, and learning and memory. Neurogenetic research on learning and memory in mice has relied on targeted mutations of specific genes. In the human species, neurogeneticists have begun to use neuroimaging to investigate brain structure and function, and their relation to cognition and emotion.

Summary

As genes associated with behavior are identified, genetic research will switch from finding genes to using genes to understand the pathways from genes to behavior, that is, the mechanisms by which genes affect behavior. Three general levels of analysis between genes and behavior are the transcriptome (gene expression throughout the genome), the proteome (protein expression throughout the transcriptome), and the brain. The RNA microarray is an exciting development that makes it possible to study the expression of all genes in the genome across the brain, across development, across states, and across individuals. Although there are as yet no microarrays to assess the proteome, this level of analysis has much to offer in terms of building bridges from the transcriptome to the brain. All pathways between genes and behavior travel through the brain, which is by far the most complicated organ on earth, as can be glimpsed in neurogenetic research on circadian rhythms, learning and memory, and emotion. Two general genetic themes emerge repeatedly and with greater force with increasingly complex levels of analysis: pleiotropy (each gene influences many traits) and polygenicity (each trait is influenced by many genes). Much more work needs to be done at all levels of analysis of the pathways from genes to behavior in relation to individual differences.

CHAPTER SIXTEEN

The Interplay between Genes and Environment

The previous chapter considered three segments of pathways between genes and behavior: the transcriptome (gene expression throughout the genome), the proteome (protein expression throughout the transcriptome), and the brain. The environment plays a crucial role at each step in these pathways. The most exciting direction for environmental research in relation to these pathways is the transcriptome. As indicated in the previous chapter, gene expression evolved to be responsive to intracellular and extracellular environments. Individual differences in gene expression appear to be only moderately heritable, which implies that most of the variance in gene expression is due to environmental factors. It was suggested that the transcriptome could lead to a paradigm shift in studying environmental influences on behavior: Gene expression can be considered as a biological index of environmental influence. In other words, environmental influence could be assessed in terms of its change in gene expression profiles across the genome. The effects of drugs in changing gene expression profiles have begun to be investigated, but we predict that this approach will be used much more widely to study short-term and long-term environmental influences.

Although much remains to be learned about the specific mechanisms involved in the pathways between genes and behavior, we know much more about genes than we do about the environment. We know that genes are located on chromosomes in the nucleus of cells, how their information is stored in the four nucleotide bases of DNA, and how they are transcribed and then translated using the triplet code. In contrast, where in the brain are environmental influences expressed, how do they change in development, and how do they cause individual differences in behavior? Given these differences in levels of understanding, genetic influences on behavior may be construed as being easier to study than environmental influences.

One thing we know for sure about the environment is that it is important. Quantitative genetic research reviewed in Chapters 7 to 14 provides the best available evidence that the environment is an important source of individual differences throughout the domain of behavior. Moreover, quantitative genetic research is changing the way we think about the environment. Three of the most important discoveries from genetic research in the behavioral sciences are about nurture rather than nature. The first discovery is that environmental influences tend to make children growing up in the same family no more similar than children growing up in different families. These environmental influences are called *nonshared environment*. The second discovery is equally surprising: Many environmental measures widely used in the behavioral sciences show genetic influence. This research suggests that people create their own experiences, in part for genetic reasons. This topic has been called the *nature of nurture*, although in genetics it is known as *genotype-environment correlation* because it refers to experiences that are correlated with genetic propensities. The third discovery at the interface between nature and nurture is that the effects of the environment can depend on genetics, and that the effects of genetics can depend on the environment. This topic is called *genotype-environment interaction*, genetic sensitivity to environments.

Nonshared environment, genotype-environment correlation, and genotype-environment interaction are the topics of this chapter. The goal of this chapter is to show that some of the most important questions in genetic research involve the environment, and some of the most important questions for environmental research involve genetics. Genetic research will profit if it includes sophisticated measures of the environment, environmental research will benefit from the use of genetic designs, and behavioral science will be advanced by collaboration between geneticists and environmentalists. These are ways in which some behavioral scientists are putting the nature-nurture controversy behind them, and bringing nature and nurture together in the study of development in their attempt to understand the processes by which genotypes eventuate in phenotypes (Moffitt, 2005; Rutter, 2006, 2007; Rutter, Moffitt, & Caspi, 2006).

Three reminders about the environment are warranted. First, genetic research provides the best available evidence for the importance of environmental factors. The surprise from genetic research has been the discovery that genetic factors are so important throughout the behavioral sciences, sometimes accounting for as much as half of the variance. However, the excitement about this discovery should not overshadow the fact that environmental factors are at least as important. Heritability rarely exceeds 50 percent and thus "environmentality" is rarely less than 50 percent.

Second, in quantitative genetic theory, the word *environment* includes all influences other than inheritance, a much broader use of the word than is usual in the behavioral sciences. By this definition, environment includes, for instance, prenatal events and biological events such as nutrition and illness, not just family socialization factors.

Third, as explained in Chapter 5, genetic research describes *what is* rather than predicts *what could be*. For example, high heritability for height means that height differences among individuals are largely due to genetic differences, given the genetic and environmental influences that exist in a particular population at a particular time (*what is*). Even for a highly heritable trait such as height, an environmental intervention such as improving children's diet or preventing illness could affect height (*what could be*). Such environmental factors are thought to be responsible for the average increase in height across generations, for example, even though individual differences in height are highly heritable in each generation.

Nonshared Environment

From Freud onward, most theories about how the environment works in behavioral development implicitly assume that offspring resemble their parents because parents provide the family environment for their offspring, and that siblings resemble each other because they share that family environment. Twin and adoption research during the past two decades has dramatically altered this view. In fact, genetic designs, such as twin and adoption methods, were devised specifically to address the possibility that some of this widespread familial resemblance may be due to shared heredity rather than to shared environment. The surprise is that genetic research consistently shows that for many behavioral traits, family resemblance is almost entirely due to shared heredity rather than to shared environment (Plomin & Daniels, 1987). As indicated in Chapters 10–13, shared environment plays a negligible role in much of psychopathology and personality. Chapter 14 showed that shared environment also has little effect on alcoholism and, most surprisingly, on body weight. Only a few possible exceptions to this rule have been found, such as conduct disorder in adolescence. Results for cognitive abilities and disabilities are more complex (Chapters 7–9). The evidence is clear that, in childhood, about a quarter of the variance of general cognitive ability is due to shared environment. However, after adolescence, the role of shared environment is negligible. Although twin studies suggest some influence of shared environment for specific cognitive abilities and especially for school achievement, studies of adoptive relatives are needed to provide direct tests of the importance of shared environment in this context.

Environmental influence is important, accounting for at least half of the variance for most behavioral domains, but it is generally not shared environment that causes family members to resemble each other. The salient environmental influences are not shared by family members (Figure 16.1). This remarkable finding means that environmental influences that affect development operate to make children growing up in the same family no more similar than children

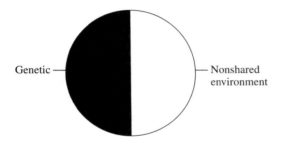

Figure 16.1 Phenotypic variance for most behavioral traits is caused by environmental variance as well as genetic variance, but the environmental variance is largely of the nonshared variety.

growing up in different families. Shared and nonshared environments are not limited to family environments. Experiences outside the family can also be shared or not shared by siblings, such as peer groups, life events, and educational and occupational experiences.

The nonshared environment component of variance refers to variance not explained by heredity or by shared environment, and it includes error of measurement. For example, for self-report personality questionnaires (Chapter 13), genetics typically accounts for about 40 percent of the variance, shared environment for 0 percent, and nonshared environment for 60 percent. Such questionnaires are usually at least 80 percent reliable, which means that about 20 percent of the variance is due to error of measurement. In other words, systematic nonshared environmental variance excluding error of measurement accounts for about 40 percent of the variance (i.e., 60 percent minus 20 percent).

Quantitative genetic designs provide an essential starting point in the quantification of the net effect of genetic and environmental influences in the populations studied. If the net effect of genetic factors is substantial, there may be value in seeking to identify the specific genes responsible for that genetic effect. Similarly, if environmental influences are largely nonshared rather than shared, this finding should deter researchers from relying solely on family-wide risk factors that pay no attention to the ways in which these influences impinge differentially on different children in the same family. Current research is trying to identify specific sources of nonshared environment and to investigate associations between nonshared environment and behavioral traits, as discussed later.

Estimating Nonshared Environment

How do genetic designs estimate the net effect of nonshared environment? Chapter 5 focused on heritability, which is estimated, for example, by comparing identical and fraternal twin resemblance or by using adoption designs. In quantitative genetics, environmental variance is variance not explained by genetics. Shared environment is estimated as family resemblance not explained by genetics. Nonshared environment is the rest of the variance: variance not explained by genetics or by shared environment. The conclusion that environmental variance is largely nonshared refers to this residual component of variance, usually estimated by model-fitting analyses. However, more direct tests of shared and nonshared environments make it easier to understand how they can be estimated.

A direct test of shared environment is resemblance among adoptive relatives. Why do genetically unrelated adoptive "siblings" correlate about .25 for general cognitive ability in childhood? The answer must be shared environment because adoptive siblings are unrelated genetically. This result fits with the conclusion in Chapter 8 that about one-quarter of the variance of general cognitive ability in childhood is due to shared environment. By adolescence, the correlation for adoptive siblings plummets to zero and is the basis for the conclusion that shared environment has negligible impact in the long run. For personality and much psychopathology, adoptive siblings correlate near zero, a value implying that shared environment is unimportant and that environmental influences, which are substantial, are of the nonshared variety.

Just as genetically unrelated adoptive siblings provide a direct test of shared environment, identical twins reared together provide a direct test of nonshared environment. Because they are identical genetically, differences within pairs of identical twins can only be due to nonshared environment. For example, for self-report personality questionnaires, identical twins typically correlate about .45. This value means that about 55 percent of the variance is due to nonshared environment plus error of measurement. Identical twin resemblance is also only moderate for most mental disorders, an observation implying that nonshared environmental influences play a major role.

Differences within pairs of identical twins provide a conservative estimate of nonshared environment, because twins often share special environments that increase their resemblance but do not contribute to similarity among "normal" siblings. For example, for general cognitive ability, identical twins correlate about .85, a result that does not seem to leave much room for nonshared environment (i.e., $1 - .85 = .15$). However, fraternal twins correlate about .60 and nontwin siblings correlate about .40, implying that twins have a special shared twin environment that accounts for as much as 20 percent of the variance (Koeppen-Schomerus et al., 2003). For this reason, the identical twin correlation of .85 may be inflated by .20 because of this special shared twin environment. In other words, about a third of the variance of general cognitive ability may be due to nonshared environment, that is, $1 - (.85 - .20) = .35$.

Identifying Specific Nonshared Environment

The next step in research on nonshared environment is to identify specific factors that make children growing up in the same family so different. To identify nonshared environmental factors, it is necessary to begin by assessing aspects of the environment specific to each child, rather than aspects shared by siblings. Many measures of the environment used in studies of behavioral development are general to a family rather than specific to a child. For example, whether or not their parents have been divorced is the same for two children in the family.

Assessed in this family-general way, divorce cannot be a source of differences in siblings' outcomes, because it does not differ for two children in the same family. However, research on divorce has shown that divorce affects children in a family differently (Hetherington & Clingempeel, 1992). If the divorce is assessed in a child-specific way (e.g., by assessing the children's perceptions about the stress caused by the divorce, which may, in fact, differ among siblings), divorce could well be a source of differential sibling outcome.

Even when environmental measures are specific to a child, they can be shared by two children in a family. Research on siblings' experiences is needed to assess the extent to which aspects of the environment are shared. For example, to what extent are maternal vocalizing and maternal affection toward the children shared by siblings in the same family? Observational research on maternal interactions with siblings assessed when each child was 1 and 2 years old indicates that mothers' spontaneous vocalizing correlates substantially across the siblings (Chipuer & Plomin, 1992). This research implies that maternal vocalizing is an experience shared by siblings. In contrast, mothers' affection yields negligible correlations across siblings, a result indicating that maternal affection is not shared and is thus a better candidate for nonshared influence.

Some family structure variables, such as birth order and sibling age spacing, are, by definition, nonshared environmental factors. However, these factors have generally been found to account for only a small portion of variance in behavioral outcomes. Research on more dynamic aspects of nonshared environment has found that children growing up in the same family lead surprisingly separate lives (Dunn & Plomin, 1990). Siblings perceive their parents' treatment of themselves and the other siblings as quite different, although parents report that they treat all their children similarly. Observational studies tend to back up the children's perspective.

Table 16.1 shows sibling correlations for measures of family environment in a study focused on these issues, called the Nonshared Environment and Adolescent Development (NEAD) project (Reiss et al., 2000). During two 2-hour visits to 720 families with two siblings ranging in age from 10 to 18 years, a large battery of questionnaire and interview measures of the family environment was administered to both parents and offspring. Parent-child interactions were videotaped during a session when problems in family relationships were discussed. Sibling correlations for children's reports of their family interactions (e.g., children's reports of their parents' negativity) were modest; they were also modest for observational ratings of child-to-parent interactions and parent-to-child interactions. This finding suggests that these experiences are largely nonshared. In contrast, parent reports yielded high sibling correlations, for example, when parents reported on their own negativity toward each of the children. Although this may be due to a "rater" effect, in that the parent rates both children, the high sibling correlations indicate that parent reports of children's environments are not good sources of candidate variables for assessing nonshared environmental factors.

TABLE 16.1	

Sibling Correlations for Measures of Family Environment

Type of Data	Sibling Correlation
Child reports	
Parenting	.25
Sibling relationship	.40
Parent reports	
Parenting	.70
Sibling relationship	.80
Observational data	
Child to parent	.20
Parent to child	.30

SOURCE: *Adapted from Reiss et al. (2000).*

Nonshared environment is not limited to measures of the family environment. Indeed, experiences outside the family, as siblings make their own way in the world, are even more likely candidates for nonshared environmental influence (Harris, 1998). For example, how similarly do siblings experience peers, social support, and life events? The answer is "only to a limited extent"; correlations across siblings for these experiences range from about .10 to .40 (Plomin, 1994). It is also possible that nonsystematic factors, such as accidents and illnesses, initiate differences between siblings. Compounded over time, small differences in experience might lead to large differences in outcome. We return to the topic of chance at the end of this section.

Identifying Specific Nonshared Environment That Predicts Behavioral Outcomes

Once child-specific factors are identified, the next question is whether these nonshared experiences relate to behavioral outcomes. For example, to what extent do differences in parental treatment account for the nonshared environmental variance known to be important for personality and psychopathology? Although research in this area has only just begun, some success has been achieved in predicting differences in adjustment from sibling differences in their experiences. The NEAD project mentioned earlier provides an example in that negative parental behavior directed specifically to one adolescent sibling (controlling for parental treatment of the other sibling) relates strongly to that child's antisocial behavior and, to a lesser extent, to that child's depression (Reiss et al., 2000). Most of these associations involve negative aspects of parenting, such as conflict, and negative outcomes, such as antisocial behavior. Associations are generally weaker for positive parenting, such as affection.

A meta-analysis of 43 papers that addressed associations between nonshared experiences and siblings' differential outcomes concluded that "measured non-shared environmental variables do not account for a substantial portion of the nonshared variability" (Turkheimer & Waldron, 2000, p. 78). Looking at the same studies, however, an optimist could conclude that this research is off to a good start (Plomin, Asbury, & Dunn, 2001). The proportion of total variance accounted for in adjustment, personality, and cognitive outcomes was 0.01 for family constellation (e.g., birth order), 0.02 for differential parental behavior, 0.02 for differential sibling interaction, and 0.05 for differential peer or teacher interaction. Moreover, these effects are largely independent, because their effects add up in predicting the outcomes—incorporating all of these measures of differential environment accounts for about 13 percent of the total variance of the outcome measures.

When associations are found between nonshared environment and outcome, the question of direction of effects is raised. That is, is differential parental negativity the cause or the effect of sibling differences in antisocial behavior? Genetic research is beginning to suggest that most differential parental treatment of siblings is in fact the effect rather than the cause of sibling differences. One of the reasons why siblings differ is genetics. Siblings are 50 percent similar genetically, but this statement implies that siblings are also 50 percent different. Research on nonshared environment needs to be embedded in genetically sensitive designs in order to distinguish true nonshared environmental effects from sibling differences due to genetics. For this reason, the NEAD project included identical and fraternal twins, full siblings, half siblings, and genetically unrelated siblings. Multivariate genetic analysis of associations between parental negativity and adolescent adjustment yielded an unexpected finding: Most of these associations were mediated by genetic factors, although some nonshared environmental influence was also found (Pike, McGuire, Hetherington, Reiss, & Plomin, 1996). This finding and similar research (Reiss et al., 2000) implies that differential parental treatment of siblings to a substantial extent reflects genetically influenced differences between the siblings, such as differences in personality. The role of genetics in environmental influences is given detailed consideration in the next section.

Differences within pairs of MZ twins provide a simple and direct test of nonshared experience that controls for the role of genetics. That is, because MZ twins are identical genetically, nonshared environmental influence is implicated if MZ differences in experience correlate with MZ differences in outcome. In the NEAD project, analyses of MZ differences confirmed the results of the full multivariate genetic analysis mentioned above (Pike, McGuire et al., 1996) in showing that MZ differences in experiences of parental negativity correlated modestly with MZ differences in adjustment outcomes (Pike, Reiss, Hetherington, & Plomin, 1996). Other MZ differences studies have also identified nonshared environmental factors free of genetic confound (Asbury, Dunn, Pike,

& Plomin, 2003; Caspi et al., 2004). A longitudinal MZ differences study from infancy to middle childhood found that MZ differences in birth weight and family environment during infancy related to their differences in behavior problems and academic achievement as assessed by their teachers at age 7 (Asbury, Dunn, & Plomin, 2006b). Another longitudinal MZ differences study suggested a pernicious downward spiral of interplay of nonshared environmental influence between negative parenting and children's behavior problems (Burt, McGue, Krueger, & Iacono, 2006).

Because such studies have only been able to identify specific nonshared environmental factors that account for a small portion of nonshared environment, the MZ difference method has been used to search for other sources of nonshared environment (Asbury, Dunn, & Plomin, 2006a). From a sample of 1590 MZ pairs rated by their teachers for anxiety at the age of 7, the most discordant pairs were selected and interviewed with their mothers to explore reasons why the twins may have become so different in their level of anxiety. Some of the top reasons reported by the mothers were negative school experiences, peer rejection, illness and accidents, and perinatal life events such as birth weight. Perinatal factors are receiving increased attention as a source of nonshared environmental influence later in life (Stromswold, 2006).

No matter how difficult it may be to find specific nonshared environmental factors within the family, it should be emphasized that nonshared environment is generally the way the environment works in the behavioral sciences. It seems reasonable that experiences outside the family, for example, experiences with peers or life events, might be richer sources of nonshared environment (Harris, 1998). It is also possible that chance contributes to nonshared environment in the sense of random noise, idiosyncratic experiences, or the subtle interplay of a concatenation of events (Dunn & Plomin, 1990). Francis Galton, the founder of behavioral genetics, suggested that nonshared environment is largely due to chance: "The whimsical effects of chance in producing stable results are common enough. Tangled strings variously twitched, soon get themselves into tight knots" (Galton, 1889, p. 195).

Support for the hypothesis that chance plays an important role in nonshared environment comes from longitudinal genetic analyses of age-to-age change and continuity. Longitudinal genetic research indicates that nonshared environmental influences are age-specific for psychopathology (Kendler et al., 1993a; Van den Oord & Rowe, 1997), personality (Loehlin, Horn, & Willerman, 1990; McGue, Bacon, & Lykken, 1993; Pogue-Geile & Rose, 1985), and cognitive abilities (Cherny, Fulker, & Hewitt, 1997). That is, nonshared environmental influences at one age are largely different from nonshared environmental influences at another age. It is difficult to imagine environmental processes, other than chance, that could explain these results. Nonetheless, our view is that chance is the null hypothesis—systematic sources of nonshared environment need to be thoroughly examined before we conclude that chance factors are responsible for nonshared environment. Chance might only be a label for our current inability to identify the environmental processes by which children growing up in the same family—even pairs of identical twins—come to be so different.

SUMMING UP

Environmental influences largely operate in a nonshared manner, making children growing up in the same family no more similar than children in different families. Differences within pairs of identical twins provide a direct test of nonshared environment. Resemblance for adoptive siblings directly tests the importance of shared environment. Attempts to identify specific sources of nonshared environment indicate that many sibling experiences differ. Some of these sibling differences in experience relate to behavioral outcomes. However, the associations between sibling differences in experience and sibling differences in outcome are in part mediated genetically. Some systematic nonshared environmental effects have been identified in multivariate genetic analyses of associations between environmental measures and behavioral outcomes, and in analyses of MZ environmental and outcome differences. Chance experiences may also contribute to nonshared environment.

Implications

The discovery of the major importance of nonshared environment has far-reaching implications for understanding how the environment works in behavioral development. Whatever the salient factors might be, they operate to make two children growing up in the same family no more alike than children growing up in different families. Environmental influences that affect behavioral development do not operate on a family-by-family basis but rather on an individual-by-individual basis. That is, their effects are relatively specific to each child rather than general for all children in a family.

Theories of socialization and much behavioral research focus on environmental factors at a level of analysis that does not consider differences between children growing up in the same family. For example, when viewed in this way, parental education, parental attitudes about child-rearing, and parents' marital relationships are shared by siblings. Shared environmental factors that do not differ between children growing up in the same family cannot explain why children growing up in the same family are different. However, the effects of such factors might not be shared, as mentioned earlier. For example, the effects of variables such as parental divorce may be nonshared because the events affect children in the family differently. The message is not that family experiences are unimportant but that the effects of environmental influences are specific to each child, not general to an entire family.

The critical question for understanding how the environment influences behavioral development is why children in the same family are so different. To address this question, it is obviously necessary to study more than one child per family in an attempt to identify sibling differences in experience, and to investigate the relationship between these different experiences and differences in their behavioral outcomes. Answers to the question of why children in the same family are so different pertain not only to sibling differences. These answers provide a key to unlock the environmental origins of behavioral development for all children.

Genotype-Environment Correlation

In addition to showing that environmental influences in the behavioral sciences are largely of the nonshared variety, genetic research is also changing the way we think about the environment by showing that we create our experiences in part for genetic reasons. That is, genetic propensities are correlated with individual differences in experiences, an example of a phenomenon known as genotype-environment correlation. In other words, what seem to be environmental effects can reflect genetic influence because these experiences are influenced by genetic differences among individuals. This genetic influence is just what genetic research during the past decade has found: When environmental measures are used as outcome measures in twin and adoption studies, the results consistently point to some genetic influence, as discussed later. For this reason, genotype-environment correlation has been described as genetic control of exposure to the environment (Kendler & Eaves, 1986).

Genotype-environment correlation adds to phenotypic variance for a trait (see Appendix), but it is difficult to detect the overall extent to which phenotypic variance is due to the correlation between genetic and environmental effects (Plomin, DeFries, & Loehlin, 1977b). For this reason, these discussions focus on detection of specific genotype-environment correlations rather than on estimating their overall contribution to phenotypic variation.

The Nature of Nurture

Even though the first research on this topic was published nearly two decades ago, several dozen studies using various genetic designs and measures have converged on the conclusion that measures of the environment show genetic influence (Plomin & Bergeman, 1991). After providing some examples of this research, we will consider how it is possible for measures of the environment to show genetic influence.

A widely used measure of the home environment that combines observations and interviews is the Home Observation for Measurement of the Environment (HOME) (Caldwell & Bradley, 1978). HOME assesses aspects of the home environment such as parental responsivity, encouraging developmental advance, and provision of toys. In an adoption study of HOME, correlations for nonadoptive and adoptive siblings were compared when each child was 1 year old and again when each child was 2 years old (Braungart, Fulker, & Plomin, 1992). HOME scores are more similar for nonadoptive siblings than for adoptive siblings at both one and two years (.58 versus .35 at one year and .57 versus .40 at two years), results suggesting genetic influence on HOME. Genetic factors were estimated to account for about 40 percent of the variance of HOME scores.

Other observational studies of mother-infant interaction in infancy, using the adoption design (Dunn & Plomin, 1986) and the twin design (Lytton, 1977, 1980), show genetic influence. The NEAD project, mentioned earlier, included videotaped observations for each parent interacting with each adolescent child when the parent-child dyad was engaged in ten-minute discussions around problems and conflict relevant to the dyad. Significant heritability was found for all measures (O'Connor, Hetherington, Reiss, & Plomin, 1995).

These observational studies suggest that genetic effects on family interactions are not solely in the eye of the beholder. Most genetic research on the nature of nurture has used questionnaires rather than observations. Questionnaires add another source of possible genetic influence: the subjective processes involved in perceptions of the family environment. The pioneering research in this area was two twin studies of adolescents' perceptions of their family environment (Rowe, 1981, 1983b). Both studies found substantial genetic influence on adolescents' perceptions of their parents' acceptance and no genetic influence on perceptions of parents' control.

The NEAD project was designed in part to investigate genetic contributions to diverse measures of family environment. As shown in Table 16.2,

TABLE 16.2

Model-Fitting Heritability Estimates for Questionnaire Assessments of Parenting

Rater	Ratee	Measure	Heritability
Adolescent	Mother	Positivity	.30
		Negativity	.40
Adolescent	Father	Positivity	.56
		Negativity	.23
Mother	Mother	Positivity	.38
		Negativity	.53
Father	Father	Positivity	.22
		Negativity	.30

SOURCE: *Plomin, Reiss et al. (1994).*

significant genetic influence was found for adolescents' ratings of composite variables of their parents' positivity and negativity (Plomin, Reiss, Hetherington, & Howe, 1994). The highest heritability of the 12 scales that contributed to these composites was for a measure of closeness (e.g., intimacy, supportiveness), which yielded heritabilities of about 50 percent for both mothers' closeness and fathers' closeness as rated by the adolescents. As found in Rowe's original studies and in several other studies (Bulik, Sullivan, Wade, & Kendler, 2000), measures of parental control showed lower heritability than measures of closeness (Kendler & Baker, 2006). The NEAD project also assessed parents' perceptions of their parenting behavior toward the adolescents (lower half of Table 16.2). Parents' ratings of their own behavior yielded heritability estimates similar to those for the adolescents' ratings of their parents' behavior. Because the twins were children in these studies, genetic influence on parenting comes from parents' response to genetically influenced characteristics of their children. In contrast, when the twins are parents, genetic influence on parenting can come from other sources such as the parents' personality. Nonetheless, studies of twins as parents have generally yielded similar results that show widespread genetic influence (Neiderhiser et al., 2004).

More than a dozen other studies of twins and adoptees have reported genetic influence on family environment (Plomin, 1994). For example, for 3-year-olds, observations and ratings of parent-child mutuality (shared positive affect and responsiveness) showed genetic influence in both a twin study and an adoption study (Deater-Deckard & O'Connor, 2000). A longitudinal twin study from 11 to 17 years found significant genetic influence at both ages but greater genetic influence at 17 years (Elkins, McGue, & Iacono, 1997), a finding replicated in another study (McGue, Elkins, Walden, & Iacono, 2005). Multivariate genetic research suggests that genetic influence on perceptions of family environment is mediated by personality (Kendler, 1996b; Krueger, Markon, & Bouchard, 2003).

Genetic influence on environmental measures also extends beyond the family environment. For example, several studies have found genetic influence on measures of life events and stress, especially for life events over which we have some control, such as problems with relationships and financial disruptions (Bolinskey, Neale, Jacobson, Prescott, & Kendler, 2004; Federenko et al., 2006; Kendler, Neale, Kessler, Heath, & Eaves, 1993b; McGuffin, Katz, & Rutherford, 1991; Middeldorp, Cath, Vink, & Boomsma, 2005; Plomin et al., 1990; Thapar & McGuffin, 1996). As in the case of genetic influence on perceptions of family environment, genetic influence on life events and stress is also mediated in part by personality (Kendler, Gardner, & Prescott, 2003; Saudino et al., 1997).

Genetic influence has also been found for characteristics of children's friends and peer groups (e.g., Bullock, Deater-Deckard, & Leve, 2006; Guo, 2006;

Iervolino et al., 2002; Manke, McGuire, Reiss, Hetherington, & Plomin, 1995) as well as adults' friends (Rushton & Bons, 2005), with genetic influence increasing during adolescence and young adulthood as children leave their homes and create their own social worlds (Kendler et al., 2007). Other environmental measures that have shown genetic influence include television viewing (Plomin et al., 1990), classroom environments (Jacobson & Rowe, 1999; Jang, 1993; Walker & Plomin, 2006), work environments (Hershberger, Lichtenstein, & Knox, 1994), social support (Agrawal, Jacobson, Prescott, & Kendler, 2002; Bergeman, Plomin, Pedersen, McClearn, & Nesselroade, 1990; Kessler, Kendler, Heath, Neale, & Eaves, 1992), accidents in childhood (Phillips & Matheny, 1995), the propensity to marry (Johnson, McGue, Krueger, & Bouchard, 2004), marital quality (Spotts, Prescott, & Kendler, 2006), divorce (McGue & Lykken, 1992), exposure to drugs (Tsuang et al., 1992), and exposure to trauma (Lyons et al., 1993). In fact, there are few measures of experience examined in genetically sensitive designs that do *not* show genetic influence. It has been suggested that other fields, such as demography, also need to consider the impact of genotype-environment correlation (Hobcraft, 2006).

In summary, diverse genetic designs and measures converge on the conclusion that genetic factors contribute to experience. A recent review of 55 independent genetic studies using environmental measures found an average heritability of .27 across 35 different environmental measures (Kendler & Baker, 2006). A major direction for research on the interplay between genes and environment is to investigate the causes and consequences of genetic influence on measures of the environment.

SUMMING UP

Measures of the environment widely used in behavioral research show genetic influence. Most research has used questionnaires about the family environment, but evidence for genetic influence has also been found using observational methods and measures of the environment both inside and outside the family.

Three Types of Genotype-Environment Correlation

What are the processes by which genetic factors contribute to variations in environments that we experience? For example, to what extent are behavioral traits, such as cognitive abilities, personality, and psychopathology, mediators of this genetic contribution? What is even more important, does genetic influence on environmental measures contribute to the prediction of behavioral outcomes from environmental measures? The latter question can be viewed as a question about genotype-environment correlation, although there are other ways to view this question and to address it (Rutter, 2005b, 2006).

TABLE 16.3

Three Types of Genotype-Environment Correlation

Type	Description	Source of Environmental Influence
Passive	Children receive genotypes correlated with their family environment	Parents and siblings
Evocative	Individuals are reacted to on the basis of their genetic propensities	Anybody
Active	Individuals seek or create environments correlated with their genetic proclivities	Anybody or anything

SOURCE: *Plomin, DeFries, & Loehlin (1977b).*

There are three types of genotype-environment correlation: passive, evocative, and active (Plomin et al., 1977b). The passive type occurs when children passively inherit from their parents family environments that are correlated with their genetic propensities. The evocative, or reactive, type occurs when individuals, on the basis of their genetic propensities, evoke reactions from other people on the basis of their genetic propensities. The active type occurs when individuals select, modify, construct, or reconstruct experiences that are correlated with their genetic propensities (Table 16.3).

For example, consider musical ability. If musical ability is heritable, musically gifted children are likely to have musically gifted parents who provide them with both genes and an environment conducive to the development of musical ability (passive genotype-environment correlation). Musically talented children might also be picked out at school and given special opportunities (evocative). Even if no one does anything about their musical talent, gifted children might seek out their own musical environments by selecting musical friends or otherwise creating musical experiences (active).

Passive genotype-environment correlation requires interactions between genetically related individuals. The evocative type can be induced by anyone who reacts to individuals on the basis of their genetic proclivities. The active type can involve anybody or anything in the environment. We tend to think of positive genotype-environment correlation, such as providing a musical environment, as being positively correlated with children's musical propensities, but genotype-environment correlation can also be negative. As an example of negative genotype-environment correlation, slow learners might be given special attention to boost their performance.

Three Methods to Detect Genotype-Environment Correlation

Three methods are available to investigate the contribution of genetic factors to the correlation between an environmental measure and a behavioral trait. These methods differ in the type of genotype-environment correlation they can detect. The first method is limited to detecting the passive type. The second method detects the evocative and active types. The third method detects all three types. All three methods can also provide evidence for environmental influence free of genotype-environment correlation.

The first method compares correlations between environmental measures and traits in nonadoptive and adoptive families (Figure 16.2). In nonadoptive families, a correlation between a measure of family environment and a behavioral trait of children could be environmental in origin, as is usually assumed. However, genetic factors might also contribute to the correlation. Genetic mediation would occur if genetically influenced traits of parents are correlated with the environmental measure and with the children's trait. For example, a correlation between the Home Observation for Measurement of the Environment (HOME) and children's cognitive abilities could be mediated by genetic factors that affect both the cognitive abilities of parents and their scores on HOME. In contrast, in adoptive families, this indirect genetic path between family environment and children's traits is not present, because adoptive parents are not genetically related to their adopted children. For this reason, a genetic contribution to the covariation between family environment and children's traits is implied if the correlation is greater in nonadoptive families than in adoptive families. The genetic contribution reflects passive genotype-environment correlation, because children in nonadoptive families passively inherit from their parents both genes and environment that are correlated with the trait. In both nonadoptive and adoptive families, the environmental measure might be the consequence rather

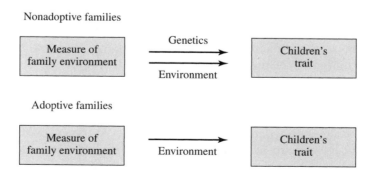

Figure 16.2 Passive genotype-environment correlation can be detected by comparing correlations between family environment and children's traits in nonadoptive and adoptive families.

than the cause of the children's traits, which could involve genetic influence of the evocative or active type of genotype-environment correlation. However, this source of genetic influence would contribute equally to environment-outcome correlations in nonadoptive and adoptive families. Increased correlations in nonadoptive families would occur only in the presence of passive genotype-environment correlation.

This method has uncovered significant genetic contributions to associations between family environment and children's behavioral development in the Colorado Adoption Project. For example, the correlation between HOME and cognitive development of 2-year-olds is higher in nonadoptive families than in adoptive families (Plomin, Loehlin, & DeFries, 1985). The same pattern of results was found for correlations between HOME and language development. The children-of-twins (COT) method can be used to address similar questions (D'Onofrio et al., 2003). In one application of the COT method, the relationship between parental divorce and early drug use of offspring was shown to be genetically mediated, whereas the relationship between parental divorce and emotional difficulties of offspring appeared to be a direct environmental effect (D'Onofrio et al., 2006). Another COT analysis suggested that harsh physical punishment has a true environmental effect on children's behavior problems (Lynch et al., 2006), although a twin study of children found that corporal punishment was genetically influenced but more severe physical maltreatment was not (Jaffee et al., 2004).

Evocative and active genotype-environment correlations are assumed to affect both adopted and nonadopted children and would not be detected using this first method. The second method for finding specific genotype-environment correlations involves correlations between biological parents' traits and adoptive families' environment (Figure 16.3). This method addresses the other two types of genotype-environment correlation, evocative and active. Traits of biological parents can be used as an index of adopted children's genotype, and this index can be correlated with any measure of the adopted children's environment. Although biological parents' traits are a weak index of adopted-away children's genotype, a finding that biological parents' traits correlate with the environment of their adopted-away children suggests that the environmental measure reflects genetically influenced characteristics of the adopted children. That is, adopted children's genetic propensities evoke reactions from adoptive parents. Attempts

Figure 16.3 Evocative and active genotype-environment correlation can be detected by the correlation between biological parents' traits (as an index of adopted children's genotype) and the environment of adoptive families.

to use this method in the Colorado Adoption Project yielded only meager evidence for evocative and active genotype-environment correlation. For example, biological mothers' general cognitive ability did not correlate significantly with HOME scores in the adoptive families of their adopted-away children (Plomin, 1994).

A developmental theory of genetics and experience predicts that the evocative and active forms of genotype-environment correlation become more important as children experience environments outside the family and begin to play a more active role in the selection and construction of their experiences (Scarr & McCartney, 1983). For example, an adoption study found evidence for an evocative genotype-environment correlation for antisocial behavior in adolescence (Ge et al., 1996). Genetic risk for the adoptees was indexed by antisocial personality disorder or drug abuse in their biological parents. Adoptees at genetic risk had adoptive parents who were more negative in their parenting than adoptive parents of control adoptees. Moreover, this effect was shown to be mediated by the adolescent adoptees' own antisocial behavior, an observation suggesting evocative genotype-environment correlation. These results were replicated in another adoption study (O'Connor, Deater-Deckard, Fulker, Rutter, & Plomin, 1998).

The third method to detect genotype-environment correlation involves multivariate genetic analysis of the correlation between an environmental measure and a trait (Figure 16.4). This method is the most general in the sense that it detects genotype-environment correlation of any kind—passive, evocative, or active. As explained in the Appendix, multivariate genetic analysis estimates the extent to which genetic effects on one measure overlap with genetic effects on another measure. In this case, genotype-environment correlation is implied if genetic effects on an environmental measure overlap with genetic effects on a trait measure.

Multivariate genetic analysis can be used with any genetic design and with any type of environmental measure, not just measures of the family environment. However, because all genetic analyses are analyses of individual differences, the environmental measure must be specific to each individual. For example, an environmental measure that is the same for family members such as socioeconomic status of the family could not be used in these analyses. However, a child-specific measure such as children's perceptions of their family's

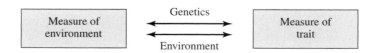

Figure 16.4 Passive, evocative, and active genotype-environment correlation can be detected by using multivariate genetic analysis of the correlation between environmental measures and traits.

socioeconomic status could be analyzed in this way. One of the first studies of this type used the sibling adoption design to compare cross-correlations between one sibling's HOME score (a child-specific rather than family-general measure of the environment) and the other sibling's general cognitive ability for nonadoptive and adoptive siblings at 2 years of age in the Colorado Adoption Project (Braungart, Fulker, & Plomin, 1992). Multivariate genetic model fitting indicated that about half of the phenotypic correlation between HOME and children's cognitive ability is mediated genetically. A twin study in childhood found that the association between parental negativity and children's prosocial behavior is largely mediated genetically (Knafo & Plomin, 2006b). In adolescence, multivariate genetic analyses also found substantial genetic mediation of correlations between measures of family environment and adolescents' depression and antisocial behavior in the NEAD project mentioned earlier (Reiss et al., 2000) and in other studies (Burt, Kreuger, McGue, & Iacono, 2003; Jacobson & Rowe, 1999; Silberg et al., 1999; Thapar, Harold, & McGuffin, 1998). For each of these correlations, more than half of the correlation is mediated genetically.

In adulthood, genetic influence on personality has also been reported to contribute to genetic influence on parenting in two studies (Chipuer, Plomin, Pedersen, McClearn, & Nesselroade, 1993; Losoya et al., 1997) but not in a third (Vernon et al., 1997). In one study, genetic effects on personality traits completely explain genetic influences on life events in a sample of older women (Saudino et al., 1997). Evidence for genetic mediation has also been found in adulthood in correlations between stressful life events and depression (Kendler & Karkowski-Shuman, 1997), between social support and depression (Bergeman, Plomin, Pedersen, & McClearn, 1991; Kessler et al., 1992; Spotts et al., 2005), between socioeconomic status and health (Lichtenstein, Harris, Pedersen, & McClearn, 1992), between socioeconomic status and general cognitive ability (Lichtenstein, Pedersen, & McClearn, 1992; Rowe, Vesterdal, & Rodgers, 1999; Tambs, Sundet, Magnus, & Berg, 1989; Taubman, 1976), between education and occupational status (Saudino et al., 1997), and between education and cognitive functioning in elderly individuals (Carmelli, Swan, & Cardon, 1995).

Multivariate genetic analysis can be combined with longitudinal analysis to disentangle cause and effect in the relationship between environmental measures and behavioral measures. For example, if negative parenting at one age is related to children's antisocial behavior at a later age, it would seem reasonable to assume that the negative parenting caused the children's antisocial behavior. However, the first twin study of this type found that this pathway is primarily mediated genetically (Neiderhiser, Reiss, Hetherington, & Plomin, 1999). Similar results were found in another study (Burt, McGue, Krueger, & Iacono, 2005).

Research on the interplay between genes and environment will be greatly facilitated by identifying some of the genes responsible for the heritability of behavior (Jaffee & Price, 2007). The conclusion from research reviewed in this section is that we ought to be able to identify genes associated with environmental measures because these are heritable. Of course, environments per se are not inherited; genetic influence comes into the picture because these environmental measures involve behavior. For example, many life events and stressors are not things that happen to us passively—to some extent, we contribute to these experiences. The first study to consider the association between DNA and environmental measures used a set of five SNPs associated with general cognitive ability in 7-year-old children (Butcher, Meaburn, Dale et al., 2005; Butcher Meaburn, Knight et al., 2005). In a sample of more than 4000 children, this "SNP set" was found to be associated with early proximal measures of the family environment (chaos and discipline) but not with distal measures (maternal education and father's occupational status), suggesting evocative rather than passive genotype-environment correlation (Harlaar, Butcher et al., 2005). Two other studies have reported associations between genes and marital status (Dick, Agrawal et al., 2006) and adults' retrospective reports of how they were parented (Lucht et al., 2006).

SUMMING UP

We create our experiences in part for genetic reasons. Three types of genotype-environment correlation are involved: passive, evocative, and active. Results of three methods to detect genotype-environment correlation suggest that the passive type is most important in childhood. A developmental theory predicts that evocative and active forms of genotype-environment correlation become more important later in development.

Implications

Research using diverse genetic designs and measures leads to the conclusion that genetic factors often contribute substantially to measures of the environment, especially the family environment. The most important implication of finding genetic contributions to measures of the environment is that the correlation between an environmental measure and a behavioral trait does not necessarily imply exclusively environmental causation. Genetic research often shows that genetic factors are importantly involved in correlations between environmental measures and behavioral traits. In other words, what appears to be an environmental risk might actually reflect genetic factors. Conversely, of course, what appears to be a genetic risk might actually reflect environmental factors.

This research does not mean that experience is entirely driven by genes. Widely used environmental measures show some significant genetic influence, but most of the variance in these measures is not genetic. Nonetheless, environmental measures cannot be assumed to be entirely environmental just because they are called environmental. Indeed, research to date suggests that it is safer to assume that measures of the environment include some genetic effects. Especially in families of genetically related individuals, associations between measures of the family environment and children's developmental outcomes cannot be assumed to be purely environmental in origin. Taking this argument to the extreme, two books have concluded that socialization research is fundamentally flawed because it has not considered the role of genetics (Harris, 1998; Rowe, 1994).

These findings support a current shift from thinking about passive models of how the environment affects individuals toward models that recognize the active role we play in selecting, modifying, and creating our own environments. Progress in this field depends on developing measures of the environment that reflect the active role we play in constructing our experience.

Genotype-Environment Interaction

The previous section focused on correlations between genotype and environment. Genotype-environment correlation refers to the role of genetics in exposure to environments. Genotype-environment interaction involves genetic sensitivity, or susceptibility, to environments. There are many ways of thinking about genotype-environment interaction (Rutter, 2005b, 2006), but in quantitative genetics the term generally means that the effect of the environment on a phenotype depends on genotype, or conversely that the effect of the genotype on a phenotype depends on the environment (Kendler & Eaves, 1986; Plomin, DeFries, & Loehlin, 1977a). As discussed in Chapter 5, this is quite different from saying that genetic and environmental effects cannot be disentangled because they "interact." As illustrated in Figure 5.13, when considering the variance of a phenotype, genes can affect the phenotype independent of environmental effects, and environments can affect the phenotype independent of genetic effects. In addition, genes and environments can interact to affect the phenotype beyond the independent prediction of genes and environments.

This point can be seen in Figure 16.5, in which scores on a trait are plotted against low- versus high-risk genotypes for individuals reared in low- versus high-risk environments. Genetic risks can be assessed using animal models, adoption designs, or DNA, as discussed below. The figure shows examples in which (a) genes have an effect with no environmental effect, (b) environment has an effect with no genetic effect, (c) both genes and environment have effects, and (d) both genes and environment have effects *and* there is also an interaction between genetics and environment. In this case, the interaction involves a greater effect of genetic risk in a high-risk environment. In psychiatric genetics, this type of interaction is called the *diathesis-stress* model (Gottesman, 1991; Paris, 1999). That is, individuals at genetic risk for psychopathology (diathesis, or predisposition) are especially sensitive to the effects of stressful environments. Although

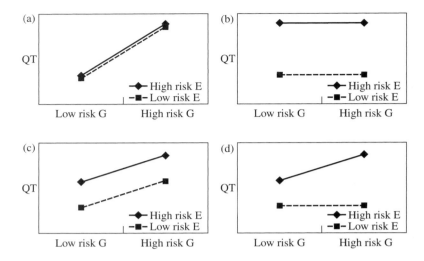

Figure 16.5 Genetic (G) and environmental (E) effects and their interaction. QT refers to a phenotypic quantitative trait. (a) G can have an effect without an effect of E, (b) E can have an effect without G, (c) both G and E can have an effect, and (d) both G and E can have an effect *and* there can also be an interaction between G and E.

there is evidence for genotype-environment interactions of this sort, some studies show greater genetic influence in permissive, low-risk environments (Kendler, 2001).

As was the case for genotype-environment correlation, genotype-environment interaction adds to phenotypic variance for a trait (see Appendix), but it is difficult to detect the overall extent to which phenotypic variance is due to the interaction between genetic and environmental effects (Jinks & Fulker, 1970; Plomin et al., 1977b; van der Sluis et al., 2006). For this reason, the following discussion focuses on detection of specific genotype-environment interactions rather than on estimating their overall contribution to phenotypic variation.

Animal Models

Genotype-environment interaction is easier to study in animals in the laboratory because both genotype and environment can be manipulated. Chapter 8 described one of the best-known examples of genotype-environment interaction. Maze-bright and maze-dull selected lines of rats responded differently to "enriched" and "restricted" rearing environments (Cooper & Zubek, 1958). The enriched condition had no effect on the maze-bright selected line, but it improved the maze-running performance of the maze-dull rats. The restricted environment was detrimental to the performance of the maze-bright rats but had little effect on the maze-dull rats. This result is an interaction in that the effect of restricted versus enriched environments depends on the genotype of the animals. Other examples from animal research in which environmental effects on behavior differ as a function of genotype have been found (Erlenmeyer-Kimling, 1972; Fuller & Thompson, 1978; Mather & Jinks, 1982), although a series of learning studies in mice failed to find replicable genotype-environment interactions (Henderson, 1972).

As mentioned in Chapter 5, an influential paper reported genotype-environment interaction in which genotype was assessed using inbred strains of mice, and environment was indexed by different laboratories (Crabbe, Wahlsten, & Dudek, 1999). However, subsequent studies found much less evidence for genotype-environment interaction of this particular type (Valdar, Solberg, Gauguier, Burnett et al., 2006; Wahlsten et al., 2003, 2006). Despite the power of animal model research to manipulate genotype and environment, there is surprisingly little systematic research on genotype-environment interaction. (Animal model research in the laboratory is less suited to the study of genotype-environment correlation because such research requires that animals be free to select and modify their environment, which rarely happens in laboratory experiments.)

Adoption Studies

Although in the human species genes and environment cannot be manipulated experimentally as in animal model research, the adoption design can explore genotype-environment interaction as illustrated in Figure 16.5. Chapter 13 described an example of genotype-environment interaction for criminal behavior found in two adoption studies (Bohman, 1996; Brennan, Mednick, & Jacobsen, 1996). Adoptees whose biological parents had criminal convictions had an increased risk of criminal behavior, suggesting genetic influence; adoptees whose adoptive parents had critimal convictions also had an increased risk of criminal behavior, suggesting environmental influence. However, genotype-environment interaction was also indicated because criminal convictions of adoptive parents led to increased criminal convictions of their adopted children mainly when the adoptees' biological parents also had criminal convictions.

Another example of a similar type of genotype-environment interaction has been reported for adolescent conduct disorder (Cadoret, Yates, Troughton, & Woodworth, 1995). Genetic risk was indexed by biological parents' antisocial personality diagnosis or drug abuse, and environmental risk was assessed by marital, legal, or psychiatric problems in the adoptive family. Adoptees at genetic risk were more sensitive to the environmental effects of stress in the adoptive family. Adoptees at low genetic risk were unaffected by stress in the adoptive family. This result confirms previous research that also showed interactions between genetic risk and family environment in the development of adolescent antisocial behavior (Cadoret, Cain, & Crowe, 1983; Crowe, 1974).

However, there are many examples in which genotype-environment interaction could not be found. For example, using data from the classic adoption study of Skodak and Skeels (1949), general cognitive ability scores were compared for adopted children whose biological parents were high or low in level of education (as an index of genotype) and whose adoptive parents were high or low in level of education (as an index of environment) (Plomin et al., 1977b). Although the level of education of the biological parents showed a significant effect on the adopted children's general cognitive ability, no environmental effect was found for adoptive parents' education and no genotype-environment interaction was found. A similar adoption analysis using more extreme groups found both genetic and environmental effects but, again, no evidence for genotype-environment interaction (Capron & Duyme, 1989, 1996; Duyme et al., 1999). Other attempts to find genotype-environment interaction for cognitive ability in infancy and childhood have not been successful in adoption analyses (Plomin, DeFries, & Fulker, 1988).

Twin Studies

The twin method has also been used to identify genotype-environment interaction. One twin's phenotype can be used as an index of the co-twin's genetic risk in an attempt to explore interactions with measured environments. Using this method, the effect of stressful life events on depression was greater for individuals at genetic risk for depression (Kendler et al., 1995). Another study found that the effect of physical maltreatment on conduct problems was greater for children with high genetic risk (Jaffee et al., 2005). The approach is stronger when twins reared apart are studied, an approach that has also yielded some evidence for genotype-environment interaction (Bergeman, Plomin, McClearn, Pedersen, & Friberg, 1988).

The most common use of the twin method in studying genotype-environment interaction simply asks whether heritability differs in two environments. Large samples are needed to detect this type of genotype-environment interaction. For example, about 1000 pairs of each type of twin are needed to detect a heritability difference of 60 percent versus 40 percent. For example, Chapter 14 mentioned several examples in which the heritability of alcohol use and abuse is greater in more permissive environments. Analyses of differences in heritability as a function of the environment can treat the environment as a continuous variable rather than dichotomizing it (Purcell, 2002).

Another analysis of this type showed that heritability of general cognitive ability is significantly greater in families with more highly educated parents (74 percent) than in families with less well educated parents (26 percent) (Rowe, Jacobson, & van den Oord, 1999), a finding replicated in three other studies for parental education and socioeconomic status (Harden, Turkheimer, & Loehlin, 2007; Kremen et al., 2005; Turkheimer, Haley, Waldron, D'Onofrio, & Gottesman, 2003), although opposite results were found in a fourth study (Asbury, Wachs, & Plomin, 2005). Higher heritability was also found for adolescent antisocial behavior in more economically advantaged families (Tuvblad, Grann, & Lichtenstein, 2006).

Caspi and Moffitt

CLOSE **UP**

Terrie E. Moffitt grew up in the American South and trained in clinical psychology and neuropsychology, completing her Ph.D. (1984) at the University of Southern California and her postdoctoral training at UCLA. **Avshalom Caspi** grew up in Israel and trained in developmental psychology, completing his Ph.D. (1986) at Cornell University.

They met in 1987 and discovered many common interests, including developmental psychopathology. In 1997, they moved to London to join the MRC Social, Genetic, and Developmental Psychiatry Centre at the Institute of Psychiatry and to learn about psychiatric genetics. As they entered this new field, they noticed that reports of connections between genes and diseases often failed to replicate, and wondered if that failure was because the connection was conditional upon some unmeasured, unobserved factor that contributed to disease causation. It was during one of their trips to a malarial zone that they hit upon the idea that the missing factor may be the environment: Almost everyone is bitten by mosquitoes, but not everyone develops malaria. They then turned their attention to the study of gene × environment interactions.

For diseases that are common in the population and that already have known nongenetic environmental causes, a person's risk of developing the disease can depend on whether a genetic susceptibility in their family is met with the environmental cause. Their work provided the first evidence for interactions between specific genes and environments in the behavioral sciences and thereby proof of the principle that (a) some susceptibility genes influence the brain's response to environmental pathogens, and (b) the effects of certain genes on psychiatric disorders may be stronger than heretofore known, within environmentally vulnerable groups of people. Their research recommends the following: To find genes for diseases with known environmental causes, ascertain environmental exposure and use it as a magnifying research tool to reveal the connection between gene and disease outcome. This can be done using both observational and experimental studies.

DNA

Quantitative genetic results showing only modest evidence for genotype-environment interaction do not restrict the possibility that molecular genetic research will find interactions between specific genes and specific environments. Indeed, DNA studies of gene-environment interaction have yielded exciting

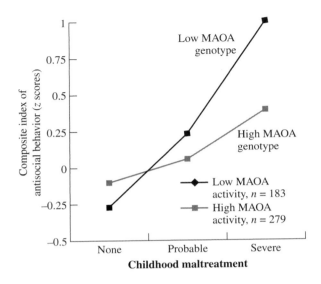

Figure 16.6 Gene-environment interaction: The effect of a polymorphism in the *MAOA* gene on antisocial behavior depends on childhood maltreatment. (From Caspi et al., 2002. Reprinted with permission from AAAS.)

results in two of the most highly cited papers in behavioral genetics. The first study involved adult antisocial behavior, childhood maltreatment, and a functional polymorphism in the monoamine oxidase A (*MAOA*) gene, which is widely involved in metabolizing a broad range of neurotransmitters (Caspi et al., 2002). As shown in Figure 16.6, childhood maltreatment was associated with adult antisocial behavior, as has been known for decades. *MAOA* was not related to antisocial behavior for most individuals who experienced no childhood maltreatment—that is, there was no difference in antisocial behavior between children with low and high *MAOA* genotypes. However, *MAOA* was strongly associated with antisocial behavior in individuals who suffered severe childhood maltreatment, which suggests a genotype-environment interaction of the diathesis-stress type. The rarer form of the gene which lowers *MAOA* levels made individuals especially vulnerable to the effects of childhood maltreatment. Although attempts to replicate this finding have been mixed, it is supported by a meta-analysis of all extant studies (Kim-Cohen et al., 2006).

The second study involved depression, stressful life events, and a functional polymorphism in the promoter region of the serotonin transporter gene (*5-HTT*) (Caspi et al., 2003). As shown in Figure 16.7, there was no association between the gene and depressive symptoms in individuals reporting few stressful life events. An association appeared with increasing number of life events, which is another example of the diathesis-stress model of genotype-environment interaction. This interaction has been replicated in several studies (e.g., Kendler, Kuhn, Vittum, Prescott, & Riley, 2005; Zalsman et al., 2006) but not all (e.g.,

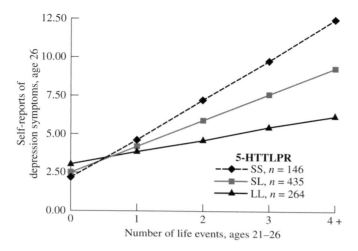

Figure 16.7 Gene-environment interaction: The effect of a polymorphism in the 5-HTTLPR gene depends on the number of life events. (From Caspi et al., 2003. Reprinted with permission from AAAS.)

Gillespie, Whitfield, Williams, Heath, & Martin, 2005). This finding has also received support from mouse research, mentioned in Chapter 15, which showed that the serotonin transporter gene was involved in emotional reactions to environmental threats (Hariri & Holmes, 2006).

Another example of genotype-environment interaction was reported by the same group: Cannabis use was associated with later psychotic symptoms such as hallucinations and delusions only in individuals with a particular allele of the catechol-o-methyltransferase (*COMT*) gene (Caspi et al., 2005). More genotype-environment interactions are likely to be found, as measures of the environment are included in genetic research, as researchers examine the extremes of genotypes and environments, and especially as specific genes whose effects can be studied in interaction with experience are identified (Moffitt, Caspi, & Rutter, 2005).

Genomewide association approaches have also begun to be applied in the search for genotype-environment interaction (Plomin & Davis, 2006). For example, the set of five SNPs associated with general cognitive ability (*g*) that was mentioned in the previous section yielded significant genotype-environment interaction (Harlaar, Butcher et al., 2005). One significant genotype-environment interaction was in line with the quantitative genetic analyses mentioned above: Genetic effects on *g* are stronger for children in families of higher socioeconomic status. Two other significant interactions showed greater associations between the SNP set and general cognitive ability (*g*) at both the low and high ends of the environment. That is, children with a genetic propensity towards high *g* profit disproportionately from a good environment, and children with a genetic propensity towards low *g* suffer disproportionately from a bad environment.

SUMMING UP

Animal studies have provided examples in which environmental effects on behavior differ as a function of genotype: examples of a genetic sensitivity to environment known as genotype-environment interaction. It is more difficult to identify genotype-environment interaction for human behavior. In quantitative genetic research, a few examples have been found in which stressful environments primarily affect individuals who are at genetic risk. Molecular genetic research using specific genes has yielded replicable results showing genetic vulnerability to environmental risks.

Summary

Genetic research can tell us as much about nurture as it can about nature. Three of the most important findings from genetic research in the behavioral sciences involve the environment. First, genetic research has shown that environmental influences work in a nonshared manner, making children growing up in the same family no more similar than children growing up in different families. Second, genetic factors often contribute to measures of the environment that are widely used in behavioral research and are responsible in part for the correlation between environmental measures and behavioral traits. Third, the effect of environments on behavior can depend on genetics, and the effect of genetics on behavior can depend on the environment.

In addition to providing the best available evidence for the importance of the environment for behavioral traits, genetic research shows that environmental effects tend not to be shared by family members. For example, resemblance between adoptive siblings is negligible for many traits, an observation indicating that shared environment is not important. Differences within pairs of identical twins also suggest the importance of nonshared environment. Attempts to identify specific sources of nonshared environment have found that family environments are experienced differently by children growing up in the same family. Experiences outside the family, such as those with peers, are likely to be even more important sources of nonshared environment. Nonshared experiences are related to behavioral outcomes, especially for negative aspects of parenting and negative outcomes. However, cause and effect in these associations is not clear. Recent research suggests that genetic factors largely mediate the association between nonshared experiences and differences in siblings' outcomes.

This suggestion leads to the second finding at the interface between nature and nurture: Our experiences are influenced in part by genetic factors. This finding is the topic of genotype-environment correlation. Dozens of studies using various genetic designs and measures of the environment converge on the conclusion that genetic factors contribute to the variance of measures of the environment. Genotype-environment correlations are of three types: passive, evocative, and active. Three methods are available to assess specific genotype-environment correlations between behavioral traits and measures of the environment. These three methods have identified several examples of genotype-environment correlation.

A third aspect of the interface between nature and nurture is genotype-environment interaction. Animal studies, in which both genotype and environment can be controlled, have yielded examples in which environmental effects on behavior differ as a function of genotype. Although it is more difficult to identify genotype-environment interaction for human behavior, some examples have been found, especially in molecular genetic studies using functional polymorphisms in candidate genes. The general form of these interactions is that stressful environments primarily have their effect on individuals who are genetically at risk, a diathesis-stress type of genotype-environment interaction.

The recognition through behavioral genetic research of genotype-environment correlations and interactions emphasizes the power of genetic research to elucidate environmental risk mechanisms. Understanding how nature and nurture correlate and interact will be greatly facilitated as more genes are identified that are associated with behavior and with experience.

CHAPTER SEVENTEEN

Evolution and Behavior

E volution is the environment writ large, but it is written in the genes. Although its roots lie firmly with Darwin's ideas of more than a century ago, evolutionary thinking has only recently established itself in the behavioral sciences. This chapter offers an overview of evolutionary theory and two related fields. Population genetics provides a quantitative basis for investigating forces, especially evolutionary forces, that change gene and genotype frequencies. The second related field is evolutionary psychology, which considers behavioral adaptations on an evolutionary time scale.

Charles Darwin

One of the most influential books ever written is Charles Darwin's 1859 *On the Origin of Species* (Figure 17.1). Darwin's famous 1831–1836 voyage around the world on the *Beagle* led him to observe the remarkable adaptation of species to their environments. For example, he made particularly compelling observations about 14 species of finches found in a small area on the Galápagos Islands. The principal differences among these finches were in their beaks, and each beak was exactly appropriate for the particular eating habits of the species (Figure 17.2).

Theology of the time proposed an "argument from design," which viewed the adaptation of animals and plants to the circumstances of their lives as evidence of the Creator's wisdom. Such exquisite design, so the argument went, implied a "Designer." Darwin was asked to serve as naturalist on the surveying voyage of the *Beagle* in order to provide more examples for the "argument from design." However, during his voyage, Darwin began to realize that species, such as the Galápagos finches, were not designed once and for all. This realization led to his heretical theory that species evolve one from another: "Seeing this

Figure 17.1 Charles Darwin as a young man. (Courtesy of Trustees of the British Museum of Natural History.)

gradation and diversity of structure in one small, intimately related group of birds, one might really fancy that from an original paucity of birds in this archipelago, one species had been taken and modified for different ends" (Darwin, 1896, p. 380). For over 20 years after his voyage, Darwin gradually and systematically marshaled evidence for his theory of evolution.

Darwin's theory of evolution begins with variation within a population. Variation exists among individuals in a population due, at least in part, to heredity. If the likelihood of surviving to maturity and reproducing is influenced even to a slight degree by a particular trait, offspring of the survivors will show more of the trait than their parents' generation. In this way, generation after generation, the characteristics of a population can gradually change. Over a sufficiently long period, the cumulative changes can be so great that populations become different species, no longer capable of interbreeding successfully.

For example, the different species of finches that Darwin saw on the Galápagos Islands may have evolved because individuals in a progenitor species differed slightly in the size and shape of their beaks. Certain individuals with slightly more powerful beaks may have been more able to break open hard seeds. Such individuals could survive and reproduce when seeds were the main source of food. The beaks of other individuals may have been better at catching insects, and this shape gave those individuals a selective advantage at certain times. Generation after generation, these slight differences led to other differences, such as different habitats. For example, seed eaters made their living on

Figure 17.2 The 14 species of finches in the Galápagos Islands and Cocos Island. (a) A woodpecker-like finch that uses a twig or cactus spine instead of its tongue to dislodge insects from tree-bark crevices. (b–e) Insect eaters. (f, g) Vegetarians. (h) The Cocos Island finch. (i–n) The birds on the ground eat seeds. Note the powerful beak of (i), which lives on hard seeds. (From "Darwin's finches" by D. Lack. ©1953 by Scientific American, Inc. All rights reserved.)

the ground and insect eaters lived in the trees. Eventually, the differences became so great that offspring of the seed eaters and insect eaters rarely interbred. Different species were born. A Pulitzer prize–winning account of 25 years of repeated observations of Darwin's finches, *The Beak of the Finch* (Weiner, 1994), shows natural selection in action.

Although this is the way the story is usually told, another possibility is that behavioral differences in habitat preference led the way to evolution of beaks rather than the other way around. That is, heritable individual differences in habitat preference may have existed that led some finches to prefer life on the ground and others to prefer life in the trees. The other differences such as beak size and shape may have been secondary to these habitat differences. Although this proposal may seem to be splitting hairs, this alternative story makes two points. First, it is difficult to know the mechanisms driving evolutionary change. Second, although behavior is not as well preserved as physical characteristics, it is likely that behavior was often at the cutting edge of natural selection. Artificial selection studies (Chapter 5) show that behavior can be changed through selection, as seen in the dramatic behavioral differences between breeds of dogs (see Figure 5.1), and that form often follows function.

Darwin's most notable contribution to the theory of evolution was his principle of *natural selection:*

> Owing to this struggle [for life], variations, however slight and from whatever cause proceeding, if they be in any degree profitable to the individuals of a species, in their infinitely complex relations to other organic beings and to their physical conditions of life, will tend to the preservation of such individuals, and will generally be inherited by the offspring. The offspring, also, will thus have a better chance of surviving, for, of the many individuals of any species which are periodically born, but a small number can survive. (Darwin, 1859, pp. 51–52)

Although Darwin used the phrase "survival of the fittest" to characterize this principle of natural selection, it could more appropriately be called reproduction of the fittest. Mere survival is necessary, but it is not sufficient. The key to the spread of alleles in a population is the relative number of surviving and reproducing offspring.

Darwin convinced the scientific world that species evolved by means of natural selection. *Origin of Species* is at the top of most scientists' lists of books of the millennium—his theory has changed how we think about all the life sciences. Nonetheless, outside science, controversy continues (Pinker, 2002). For instance, in the United States, the boards of education in several states have attempted to curtail teaching evolution as a result of pressure from creationists who believe in a literal biblical interpretation of creation. However, most people, but not everyone—see, for example, Dawkins (2006) versus Collins

(2006a)—accept the notion that science and religion are distinctly different realms, with science operating in the realm of verifiable facts and religion focused on purpose, meaning, and values. "Respectful noninterference" between science and religion is needed (Gould, 1999).

Scientifically, Darwin's theory of evolution had serious gaps, mainly because the mechanism for heredity, the gene, was not yet understood. Gregor Mendel's work was not published until seven years after the publication of the *Origin of Species*, and even then it was ignored until the turn of the century. Mendel provided the answer to the riddle of inheritance, which led to an understanding of how variability arises through mutations and how genetic variability is maintained generation after generation (Chapter 2). A recent rewrite of the *Origin of Species* is interesting in pointing out how evolutionary theory and research have changed since Darwin, as well as showing how prescient Darwin was (Jones, 1999).

Darwin considered behavioral traits to be just as subject to natural selection as physical traits were. In *Origin of Species*, an entire chapter is devoted to instinctive behavior patterns. In a later book, *The Descent of Man and Selection in Relation to Sex*, Darwin discussed intellectual and moral traits in animals and humans, concluding that the difference between the mind of a human being and the mind of an animal "is certainly one of degree and not of kind" (1871, p. 101).

Inclusive Fitness

Darwin's theory of individual fitness has been extended to consider a measure called *inclusive fitness*, which is defined as the fitness of an individual plus part of the fitness of kin that is genetically shared by the individual (Hamilton, 1964). Inclusive fitness and kin selection explain altruistic acts that do not directly benefit the individual. If the net result of an altruistic act helps more of that individual's genes to survive and to be transmitted to future generations, the act is adaptive even if it results in the death of the individual.

For example, the founder of quantitative genetics, R. A. Fisher, long ago suggested an example of kin selection and inclusive fitness that involves distastefulness of some butterfly larvae (Fisher, 1930). A bird will learn that certain larvae taste bad, but the lesson costs the larva its life. However, sibling eggs are laid in a cluster, and inclusive fitness is served by the sacrifice of one larva if two siblings (the genetic equivalent of the sacrificed larva) are saved. Inclusive fitness switches the focus from the individual to the gene, which explains the title of a classic book in this area, *The Selfish Gene* (Dawkins, 1976). Acts that appear to be altruistic can be interpreted in terms of "selfish" genes that are maximizing their reproduction through inclusive fitness.

Sociobiology

Inclusive fitness was popularized by a book in 1975 called *Sociobiology: The New Synthesis*, which promoted evolutionary thinking as a unifying theme for all of the life sciences, including the behavioral sciences (Wilson, 1975). Sociobiology has offered novel and interesting hypotheses that stem from the simple principle of inclusive fitness and kin selection.

One general theory is called parental investment (Trivers, 1972, 1985). For example, why do mothers provide most of the care of offspring in the vast majority of mammalian species, including humans? Unless a species is completely monogamous (as eagles are, for example), males have less invested in their offspring. Males can have many offspring by many females, but each female must devote large amounts of energy to each pregnancy and, in mammals, provide sustenance after birth. In terms of inclusive fitness, the fitness of females is better served by increased care of each offspring, because females must make a substantial investment in each one of them. In many cases, however, the male's investment is little more than copulation, and he can maximize his inclusive fitness by having more offspring by different females.

A related reason for the relative investments of mothers and fathers in the care of their offspring is that females can always be sure that they share half of their genes with their young. Males, however, cannot be sure that offspring are theirs. The theory of parental investment led to two predictions that have received considerable support: (1) The sex that invests more in offspring (typically, but not always, the female) will be more discriminating about mating, and (2) the sex that invests less (typically, but not always, the male) will compete more for sexual access (Platek & Shackelford, 2006; Trivers, 1985).

For these reasons, in most species, males court and females choose, and dads are often cads (Miller, 2000; Symons, 1979). The greater altruism of mothers toward their offspring is no less selfish from the point of view of genes than that of fathers (Hrdy, 1999). Although there are many hypotheses of this sort in which "selfish altruism" of genes evolved through kin selection, it is also likely that some positive social behaviors evolved through the less convoluted mechanism of individual selection (de Waal, 1996).

A twist on parental investment theory is parental competition. As mentioned in Chapter 3, some genes are "imprinted" in the sense that their expression depends on whether the allele came from the mother or father. Imprinted genes may be the result of parental competition for control of the embryo (Moore & Haig, 1991; Wilkins & Haig, 2003). Female lifetime reproductive success is maximized by becoming pregnant again as soon as possible, but male reproductive success is best served by prolonged care of his progeny before and after birth. This may account for paternal imprinted genes that protect embryos (Li et al., 1999). The larger issue is that differential parental investment affects not only mating but also parental care and even investment in gametes (Wedell, Kvarnemo, Lessels, & Tregenza, 2006).

Evolutionary thinking is making major inroads in the behavioral sciences, a field called *evolutionary psychology* (Buss, 2005a, 2007; Gangestad & Simpson, 2007). Before turning to evolutionary psychology, an overview of the quantitative foundation of evolution, the field of population genetics, is in order.

SUMMING UP

Darwin proposed that species evolve one from another. Hereditary variation among individuals results in differences in reproductive fitness. This process of natural selection changes species and can lead to new species that rarely interbreed. Gaps in the theory of evolution occurred because the mechanism of heredity, the gene, was not understood in Darwin's time. Natural selection affects behavior (function) just as much as it affects anatomy (form). Sociobiology is an extension of evolutionary theory that focuses on inclusive fitness and kin selection.

Population Genetics

Darwin's evidence for the evolution of species, such as the beaks of the Galápagos finches, relied on qualitative descriptions. Population genetics provides evolution with a quantitative basis. Its unique contribution is to describe allelic and genotypic frequencies in populations and to study the forces that change these frequencies, such as natural selection. Increasingly, population genetics involves analysis of DNA rather than genotypes inferred from phenotypes.

In the absence of opposing forces, the frequencies of alleles and genotypes remain the same, generation after generation. As explained in Box 2.2, this stability is called Hardy-Weinberg equilibrium. Population geneticists investigate the forces that change this equilibrium (Hartl, 2004; Hartl & Clark, 2006). For example, selection against a rare recessive allele is very slow, and it is for this reason that most deleterious alleles are recessive. Suppose that a recessive allele were lethal in the homozygous condition, when two such alleles are inherited. Further suppose that the frequency of the allele were 2 percent in the original population. If no homozygous recessive individuals were to reproduce for 50 generations, the frequency of this undesirable allele would only change from 2 to 1 percent. In contrast, complete selection against a dominant allele would wipe out the allele in a single generation. As mentioned in Chapter 2, the dominant allele responsible for Huntington's disease persists because its lethal effect is not expressed until after the reproductive years.

Natural selection is often discussed in terms of *directional selection* of this sort, a process in which a deleterious allele is selected against. If successful, directional selection would remove genetic variability. Another type of selection maintains different alleles rather than favoring one allele over another, a process that is especially interesting because genetic variability within a species is the focus of behavioral genetics. In contrast to directional selection, this type of selection is called *stabilizing selection* because it leads to a *balanced polymorphism*. Suppose that selection operated against both dominant and recessive homozygotes. In this process, heterozygotes would reproduce relatively more than the two homozygous genotypes. However, heterozygotes always produce homozygotes as well as heterozygotes (see Box 2.2). Genetic variability is thus maintained.

Sickle-cell anemia in humans is a specific example of this kind of balanced polymorphism. This single-gene autosomal recessive disease is caused by a single nucleotide mutation which damages red blood cell membranes, so that they are unable to fulfill their function of transporting oxygen. As a result, reproductive fitness of individuals with sickle-cell anemia (recessive homozygotes) is lowered. However, there is a puzzle: The allele is maintained in relatively high frequency in some African populations and among African Americans. The high frequency of this debilitating recessive allele is due to the higher relative fitness of heterozygotes (carriers). Heterozygotes are more resistant than normal homozygotes to a form of malaria, prevalent in certain parts of Africa, which infects between 300 and 500 million people each year. It causes between one and three million deaths annually, mostly among young children in sub-Saharan Africa. In other words, the decreased relative fitness of homozygotes with sickle-cell anemia is balanced by the increased relative fitness of heterozygote carrriers.

Another sort of stabilizing selection involves environmental diversity. As noted in relation to Darwin's finches, if environments encountered by a species are diverse, selection pressures can differ and foster genetic variability. A balanced polymorphism can also occur if selection depends on the frequency of a genotype. For example, selection that favors rare alleles produces genetic variability. Individuals with a rare genotype might use resources that are not used by other members of the species and thus gain a selective edge. Predator-prey relationships can also be frequency dependent. For example, predatory birds and mammals tend to attack more common types of prey.

Another type of frequency-dependent selection involves mate selection in which rare genotypes have an edge. For example, in fruit flies, females are more likely to mate with a rare male (Ehrman, 1972; Knoppien, 1985). Like the other types of stabilizing selection, frequency-dependent sexual selection maintains genetic variability in a species.

What evolutionary forces maintain harmful genetic variants, such as those for schizophrenia, that lower reproductive fitness? As mentioned in Chapter 10, balanced selection is a possibility. However, a new theory called *polygenic mutation-selection balance* may have general relevance to highly polygenic behavioral traits (Keller & Miller, 2006). If behavioral disorders like schizophrenia are influenced by hundreds or thousands of genes, natural selection would have a very difficult time screening out new deleterious mutations. More generally, selection may stabilize genetic variation because complex traits entail trade-offs of different fitness benefits and costs (Nettle, 2006).

In addition to considering forces that change allelic frequency, population genetics also investigates systems of mating—inbreeding and assortative mating—that change genotypic frequencies without changing allelic frequencies. Inbreeding involves matings between genetically related individuals. If inbreeding occurs, offspring are more likely than average to have the same alleles at any locus; therefore, recessive traits are more likely to be expressed. Inbreeding reduces heterozygosity and increases homozygosity. In relation to the derivation of inbred strains mentioned in Chapter 5, population genetics shows that, after 20 generations of brother-sister matings, at least 98 percent of all loci are homozygous. Inbreeding often leads to a reduction in viability and fertility, called *inbreeding depression.*

Inbreeding depression is caused by the increase in homozygosity for deleterious recessive alleles. Although inbreeding reduces genetic variability, its overall effect on genetic variability in natural populations is negligible because it is relatively rare. The other side of the coin is *hybrid vigor,* or *heterosis.* These terms refer to an increase in viability and performance when different inbred strains are crossed. Outbreeding reintroduces heterozygosity and masks the effects of deleterious recessive alleles. *Assortative mating*, phenotypic similarity between mates, is another system of mating that changes genotypic, not allelic, frequency. As discussed in Chapter 8, assortative mating for a particular trait increases genotypic variance for that trait in a population. Although assortative mating is modest for most behavioral traits, increases in genetic variance due to assortative mating accumulate over generations. In other words, even a small amount of assortative mating can greatly increase genetic variability after many generations.

SUMMING UP

Population genetics is the study of allelic and genotypic frequencies in populations and of the forces that change these frequencies, such as natural selection. Stabilizing selection, such as frequency-dependent selection, increases genetic variability in a population. Inbreeding has little overall effect on genotypic variability because it is relatively rare at a population level. Assortative mating increases genotypic variability cumulatively over generations.

Evolutionary Psychology

Thinking about behavior from an evolutionary perspective has brought new insights to the behavioral sciences, as Darwin predicted it would in *On the Origin of Species* (Buss, 1995, 2007). Evolutionary thinking is essential to the behavioral sciences because it paints in the portrait of our species the broad strokes that show the similarities to and differences from other species. For example, the fact that we are mammals, defined in terms of the mammary gland, means that we have evolved a system in which mothers care for their young after birth. The fact that we are primates has many evolutionary implications, such as extremely slow postnatal development, which requires long-term care by par-

CLOSE UP

David M. Buss is an evolutionary psychologist at the University of Texas at Austin.

The first intellectual ideas that gripped him pertained to the origins of things—cosmological theories of the universe and evolutionary theories of human nature. It was not until after completing his Ph.D. at the University of California at Berkeley and taking a job at Harvard, however, that Buss started using evolutionary theorizing in his psychological research. The area that caught his attention was human mating—how we select, attract, and get rid of romantic partners. Mating seemed to be at the nexus of so many important phenomena, affecting everything from personal happiness to the genetic structure of the next generation. But psychology lacked a sound theory of human mating.

Buss's first study of what men and women want mushroomed into a study of 10,047 individuals in 37 cultures from around the world. His past 15 years of research have focused on exploring the mysteries of human mating (see *The Evolution of Desire: Strategies of Human Mating*). His recent books include *The Handbook of Evolutionary Psychology* (Ed.), *The Dangerous Passion: Why Jealousy Is as Necessary as Love and Sex,* and *The Murderer Next Door: Why the Mind Is Designed to Kill*. His current work focuses on hidden sides of human mating strategies, the psychology of sexual victimization and sexual predators, and status, prestige, and social reputation.

ents. Also fundamental for understanding our species are facts such as these: Our species uses language naturally, walks upright on two feet, and has eyes in the front of the head that permit depth perception.

Evolutionary psychology seeks to understand the adaptive value of species-wide aspects of human behavior such as our natural use of language, our similar facial expressions for basic emotions, and similarities in mating strategies across cultures. It also addresses differences between groups within a species, most notably differences between the sexes. This is a different level of analysis from most behavioral genetics research, which typically focuses on differences among individuals within a species rather than species-wide aspects of behavior. However, the definition of behavioral genetics as the genetic analysis of behavior includes behavior at all levels of analysis.

It is important to remember that the causes of the typical behavior of a species are not necessarily related to the causes of individual differences within a species. For example, young mammals typically bond to their caregivers. This behavior is an adaptation that has evolved presumably to protect them while they continue to develop after birth. But the evolution of bonding does not mean that individual differences in attachment are due to genetic factors. In fact, as noted in Chapter 13, attachment is one of the few traits that shows little evidence of genetic influence. The role of genetic factors in the origins of individual differences in behavior is an empirical issue that requires quantitative genetic analysis. It is much more difficult to investigate genetic mechanisms in an evolutionary time frame and to pin down genes responsible for particular adaptations if the genes do not vary within a species. Knock-out technology in mice (see Chapter 6) is one approach to studying the role of genes that do not vary. Nonetheless, it is difficult to glean information about evolution from knock-outs. There is a need to build bridges between evolutionary psychology and behavioral genetic theory and research on individual differences (Buss & Greiling, 1999; Nettle, 2006; Segal & MacDonald, 1998).

Instincts

Evolutionary psychologists have brought back the word *instinct*, which had been effectively banned from psychology. For example, an influential book on the evolution of language is called *The Language Instinct* (Pinker, 1994). Instincts, which refer to evolved behavioral adaptations, were accepted by psychologists early in this century. William James (1890), the founder of American psychology, presented a long list of instincts that begin at birth and increase in length with development. However, instincts were largely rejected as psychology moved more toward environmental explanations of behavior. For half a century, the only instinct discussed was a general ability to learn. During this period, cultural anthropologists focused on differences between cultures that are presumably learned, rather than on their similarities which might be due to evolution.

For example, the ease with which all members of our species learn a language suggests that language is innate, an instinct. Although the dictionary defines *innate* as "inborn," in evolutionary psychology the word refers to the ease with which certain things but not others are learned. That is, the word *innate* refers to evolved capacities and constraints rather than rigid hard-wiring that is impervious to experience. *Instinct* means an innate behavioral tendency, not an inflexible pattern of behavior. Although language is now generally accepted as innate, debate continues about what exactly is innate, whether it is a general predisposition to learn language, or to learn specific "modules" such as grammatical structures (Pinker, 1994).

Another example involves instinctive fears, such as fear of spiders and snakes, which protects against receiving poisonous bites, and fear of heights, which makes us wary of situations in which we might fall. Such instinctive fears are defined as a normal emotional response to realistic danger. Phobias, on the other hand, are fears that are out of proportion to realistic danger and overgeneralized. What is interesting from an evolutionary perspective is that phobias are not random—they typically involve overblown adaptive responses, such as fears of snakes and spiders, heights, and crowded places. Fear of snakes and spiders was adaptive in our evolutionary past, even though automobiles and guns are far more likely to harm us nowadays (Marks & Nesse, 1994; Ohman & Mineka, 2001). Darwin predicted this when he suggested: "May we not suspect that the fears of children, which are quite independent of experience, are the inherited effects of real dangers during ancient savage time?" (Darwin, 1877, p. 290). Moreover, such fears and phobias emerge in development when they are needed. For example, fear of heights and fear of strangers emerge at about six months when infants begin to crawl. A field called Darwinian or evolutionary psychiatry considers psychopathology as adaptive responses of a stone-age brain to modern times (McGuire & Troisi, 1998; Stevens & Price, 2004).

Empirical Evidence

It is easy to make up stories about how anything might be adaptive, called "just-so stories" after Kipling's book for children that includes whimsical parables like why the elephant got its trunk. The danger is starting with a known phenomenon and working backward to propose an explanation rather than making a prediction whose answer is unknown until it is tested (de Waal, 2002). For example, we know that most mammalian fathers are less involved in rearing than mothers. As mentioned earlier, this behavioral difference can be explained in terms of differential parental investment—offspring cost mothers more. But if fathers had happened to be equally invested in the care of their offspring, it could have been argued that this evolved because supportive fathers perpetuate their genes. Evolutionary psychologists try to make testable predictions that tease apart evolutionary and cultural explanations (Buss, 2005a; Gangestad, Haselton, & Buss, 2006), although it is difficult to rule out cultural explanations completely (Buller, 2005a, 2005b; Fehr & Fischbacher, 2003; Neher, 2006; Wood & Eagly, 2002). Some of the most interesting predictions are those in which behaviors that we think are pathological reflect adaptations, such as fears, men's aggressive violence, and overeating in a world of fast food.

Quantitative genetic methods, such as twin and adoption designs for addressing the origins of individual differences, are not available to evolutionary analyses. (Remember that showing genetic influence on individual differences in behavior does not imply that species-typical behavior is due to genetic adaptation.) The revolution in DNA analysis has transformed population genetics and other areas of behavioral genetics. DNA can also be used to peer into our species' distant past, for example, showing that Neanderthals could not be ancestors to modern human beings (Poinar, 1999). DNA can also be used to shed light on migrations of ancient peoples and patterns of human genetic diversity (Garrigan & Hammer, 2006), such as tracing the Saxon, Viking, and Celt origins of the British (McKie, 2007; Oppenheimer, 2006; Sykes, 2007). However, the DNA revolution has not yet come to the field of evolutionary psychology. One interesting area in which DNA has been used involves sexual attractiveness of mates. (See Box 17.1.)

Evolutionary psychology has had to rely on evidence that is less direct. For example, one strand of research involves comparisons between species, because differences between species can be assumed to have evolved. For example, as mentioned earlier, the theory of differential parental investment (that mothers invest more in their offspring than do fathers) led to the hypothesis that the parent who invests more should be choosier about selecting a mate. If this is an

BOX 17.1

Mate Preference and the Major Histocompatibility Complex

The major histocompatibility complex (MHC), which is the most gene-dense region of the mammalian genome, plays an important role in the immune system. In humans, the MHC is on chromosome 6 and includes 3.6 million base pairs and 140 highly polymorphic genes—some genes have hundreds of alleles. Why is this gene region so highly polymorphic? The answer is balanced selection: Greater diversity of the MHC makes the immune system more adaptable in its response to invaders.

Balanced selection for the MHC is driven by frequency-dependent selection. We prefer mates who are most different from us in MHC alleles. We do this on the basis of smell: Specific olfaction neurons in the nose function to detect body odors caused by the MHC locus (Boehm & Zufall, 2006). In a study known as the "T-shirt" experiment, female and male college students were genotyped for several MHC genes. Male students wore a T-shirt for two consecutive nights. The next day, each female student rated the pleasantness of the odors of the men's T-shirts. T-shirts were rated as more pleasant smelling by a woman when the man's MHC genotype was most different from hers (Wedekind, Seebeck, Bettens, & Paepke, 1995).

Choosing mates who differ from us in MHC genotypes is a type of frequency-dependent selection that leads to greater diversity of MHC genes between parents and thus produces stronger immune systems in offspring. A recent study also suggests that MHC incompatibility predicts sexual compatibility of a couple. Greater MHC differences within a couple predict greater sexual responsivity of the woman and less attraction to other men (Garver-Apgar, Gangestad, Thornhill, Miller, & Olp, 2006). This association is strongest when the women are fertile during the middle of the menstrual cycle. An *ovulatory-shift hypothesis* proposes that mate preferences of women are generally accentuated when they are fertile (Gangestad, Thornhill, & Garver-Apgar, 2005).

evolved adaptation, we would expect that in most species females will be more discriminating than males in choice of mate, because most mothers invest more in offspring than do fathers. Confirming this prediction, females are choosier than males in selecting mates in most species (Buss, 1994b). Moreover, in the few species in which males invest more than females, males are choosier. For example, in the pipefish seahorse, the male receives eggs from the female and nurtures them in a kangaroolike pouch. Male pipefish seahorses are choosier than females in selecting mates. Comparisons of this sort across species support the hypothesis that behavioral differences in maternal and paternal investment are evolved adaptations.

In addition to comparisons across species, evolutionary psychologists also consider comparisons between groups within the human species. The most influential data are from different cultures. For example, females were found to be choosier than males in selecting mates in a study of more than 10,000 people in 37 cultures (Buss, 1994a). An early example that began with Darwin involved basic facial expressions of emotions that can be recognized in all cultures, such as expressions of happiness, anger, grief, disgust, and surprise (Eckman, 1973).

Differences between sexes are often examined. For instance, the hypothesis of differential parental investment predicts that men have evolved adaptations that increase their chances of paternity. Support for this hypothesis comes from data showing that across many cultures, men are much more likely than women to be jealous about signs of sexual infidelity (Buss, 2003). A plausible adaptive explanation is that natural selection shaped sexual jealousy in men as a mechanism to prevent cuckoldry because women always know their child is theirs but men cannot be sure. Describing such sex differences and their plausible adaptive value supports but does not prove that these sex differences are caused by evolved genetic adaptations that differ between the sexes (Harris, 2003).

Consider beauty. Although it is obvious that men prefer attractive women, what is it about women that men find attractive? What do women value in men? Why? Evolutionary thinking has provided some new insights into such issues, as discussed in Box 17.2.

Another example involves murder. Most murders are caused by family members, which seems to violate the principle of inclusive fitness. Evolutionary psychologists predicted on the basis of inclusive fitness that family murders would primarily involve genetically unrelated stepfathers who harm their stepchildren rather than biological fathers and their own offspring, a hypothesis that has been confirmed (Daly & Wilson, 1999; Tooley, Karakis, Stokes, & Ozanne-Smith, 2006), although cultural explanations cannot be completely excluded (Burgess & Drais, 1999).

As a final example, consider morning sickness, which during the first three months of pregnancy includes food aversions in addition to nausea. Rather than

BOX 17.2

Selection for Beauty

Everyone knows that most men prefer attractive women, but why? In 1992, a popular book called *The Beauty Myth* argued that beauty is a mere cultural construction (Wolf, 1992). In 1999, a book called *Survival of the Prettiest* caused a stir by arguing that beauty goes much deeper, all the way down to our genes (Etcoff, 1999). The hypothesis is that a woman's appearance provides clues to her fertility.

A hip-to-waist ratio of .70 is seen as most attractive across all cultures. Although models during our time have become thinner, their hip-to-waist ratio has remained at .70. A computer "morph" program has been used to thicken each model's waist from left to right. (R. Henss, "Waist-to-hip ratio and female attractiveness. Evidence from photographic stimuli and methodological considerations," in *Personality and Individual Differences*, 2000, p. 504.)

thinking about morning sickness as something bad, evolutionists have suggested that morning sickness may be an adaptation that prevents mothers from consuming toxins damaging to the developing fetus (Nesse & Williams, 1996). Food aversions during pregnancy typically involve foods that contain toxins, such as alcohol, coffee, and meat, but hardly ever do they include foods that do not contain toxins, such as bread or cereals (Flaxman & Sherman, 2000). Moreover, the food aversions usually disappear after the first three months of pregnancy, which is the most sensitive period of fetal organ development. One piece of evidence in support of this hypothesis is that women who do not have morning sickness during the first trimester are three times more likely to experience a spontaneous abortion than women who do have morning sickness (Profet, 1992).

For example, many cultures share similar concepts of beauty that advertise health and fertility, such as shiny hair and clear skin. Also, babies as young as three months look longer at faces that adults find attractive (Langlois, Ritter, Roggman, & Vaughn, 1991). Moreover, mothers of more attractive infants are more affectionate and playful with their infants than are mothers of less attractive infants (Langlois, Ritter, Casey, & Sawin, 1995). A meta-analysis of research on facial attractiveness across cultures suggests that averageness, symmetry, and sexual dimorphism set standards of beauty (Rhodes, 2006). It's not just a pretty face: Across many cultures, a small waist circumference relative to hip circumference, an hourglass figure, is seen as most beautiful (Furnham, Moutafi, & Baguma, 2002; Singh, 1993, 2004). The adaptive value of the hourglass figure may be that it predicts health in women (Schutzwohl, 2006; Weeden & Sabini, 2005).

With makeup and plastic surgery, women create what is called a supernormal stimulus that exaggerates what is desirable evolutionarily in young women, such as color in lips and cheeks, full lips and breasts, and the shapeliness of the leg, which is exaggerated by high heels. *The Beauty Myth* focuses on beauty in women rather than in men, because men are thought to be sought less for their looks than for their ability to provide resources for offspring. In nearly all cultures, males prefer younger attractive females and females prefer older males with high social status, behaviors that support the hypothesis that the mating game focuses on fertility in women and power in men (Gangestad & Scheyd, 2005). However, recent research suggests that, like the hourglass figure, physical attractiveness in men is also a general marker of "good genes" and, for this reason, women also prefer good-looking mates, especially in relation to short-term mates and when they are fertile (Gangestad, Garver-Apgar, Simpson, & Cousins, 2007).

Evolutionary psychologists have studied many other behavioral adaptations such as parent-offspring conflict, preference for particular habitats, and cooperation and conflict in groups. As indicated in recent books, evolutionary psychologists tackle big issues such as why we kill (Buss, 2005b), why we lie (Smith, 2004), why we think (Gardenfors, 2006; Geary, 2005), why we suffer posttraumatic stress disorder (Cantor, 2005), why we believe in gods (Atran, 2005; Dawkins, 2006; Dennett, 2006; Wolpert, 2007), and public policy implications of evolutionary psychology (Crawford & Salmon, 2004). Evolutionary psychology books also address entertaining issues such as why we shop (Saad, 2007), why we like art (Pinker, 2002), and why we like literature (Barash & Barash, 2005; Gottschall & Wilson, 2005). Books by critics of evolutionary psychology have also appeared (Buller, 2005a; Fisher, 2004).

The impact of evolutionary psychology is beginning to be felt in cognitive, social, personality, developmental, clinical, and even cultural psychology (Buss, 2005a; Gangestad & Simpson, 2007; Goetz & Shackelford, 2006). It should be emphasized that even if particular behaviors are adaptive in an evolutionary sense, by no means does this imply that such behaviors are morally acceptable or desirable. Finding evolved adaptations does not imply genetic determinism or that behavior is unmodifiable. Indeed, an argument could be made that one purpose of society is to make it possible for our stone-age brains to adjust to a modern world.

SUMMING UP

Evolutionary psychologists focus on species-wide themes, such as the natural use of language, greater investment by mothers than fathers in offspring, and facial expressions that are seen in all cultures. This level of analysis that considers species-typical behavioral adaptations (instincts) in an evolutionary time scale is different from that of most behavioral genetic research, which focuses on the genetic and environmental origins of individual differences within species in current populations. There is increasing interest in integrating these perspectives.

Summary

Charles Darwin's 1859 book on the origin of species convinced the scientific world that species evolved one from the other rather than being created once and for all. Reproductive fitness is the key to natural selection. Gaps in Darwin's theory of evolution occurred because the mechanism for heredity, the gene, was not understood at that time. Darwin's theory has been extended to consider inclusive fitness and kin selection, studies that go beyond Darwin's focus on individual reproductive fitness and lead to hypotheses such as differences in parental investment for mothers and fathers. Although Darwin noted that natural selection affected behavior as much as bones, evolutionary thinking has only entered the mainstream of the behavioral sciences in recent years.

Population genetics investigates forces that change allelic and genotypic frequencies. Because behavioral genetics focuses on genetic variability within a species, types of natural selection that increase genetic variation in a population are especially interesting, such as balanced polymorphisms due to heterozygote advantage or frequency-dependent selection. Inbreeding and assortative mating change genotypic but not allelic frequencies. Although inbreeding can reduce genetic variability, its rarity in the human species makes its effects on populations negligible. Assortative mating, on the other hand, increases genotypic variability for many behavioral traits.

Evolutionary thinking is becoming increasingly influential in psychology. Most evolutionary psychology considers average differences between species on an evolutionary time scale. This level of analysis is different from that of most behavioral genetics, which focuses on contemporary individual differences. Although it is more difficult to test evolutionary genetic hypotheses rigorously, support has been found for hypotheses about behavioral adaptations (instincts), such as fears and phobias, and different mating strategies for males and females.

CHAPTER EIGHTEEN

The Future of Behavioral Genetics

Predicting the future of behavioral genetics is not a matter of crystal ball gazing because the momentum of recent developments makes the field certain to thrive, especially as behavioral genetics continues to flow into the mainstream of research. This momentum is propelled by new findings, methods, and projects, both in quantitative genetics and in molecular genetics. An excellent statement from the American Society of Human Genetics on the bright future for behavioral genetics is available in the society's journal (Sherman et al., 1997) and online (http://www.faseb.org/genetics/ashg/policy/pol-28.htm).

Another reason for optimism about the continued growth of genetics in the behavioral sciences is that leading researchers have incorporated genetic strategies in their research (Plomin, 1993). This trend has grown much stronger now that the price of admission to genetic research is just a cheek swab from which DNA is extracted, not difficult-to-obtain samples of twins or adoptees. This trend is important because the best behavioral genetic research is likely to be done by behavioral scientists who are not primarily geneticists. Experts from behavioral domains will focus on traits and theories that are pivotal to those domains and will interpret their research findings in ways that will achieve the most important advances. For this reason, a major motivation for writing this book is to enlist the aid of, and point out the opportunities to, the next generation of behavioral scientists.

Quantitative Genetics

The future will no doubt witness the application of twin and adoptee strategies, as well as genetic research using nonhuman animal models, to other behavioral traits. Behavioral genetics has only scratched the surface of possible applications, even within the domains of cognitive disabilities (Chapter 7), cognitive abilities (Chapters 8 and 9), psychopathology (Chapters 10–12), and personality (Chapter 13). For example, for cognitive abilities, most research has focused on general cognitive ability and major group factors of specific cognitive abilities. The future of quantitative genetic research in this area lies in more fine-grained analyses of cognitive abilities and in the use of information-processing, cognitive psychology and brain imaging approaches to cognition. For psychopathology, genetic research has just begun to consider disorders other than schizophrenia and the major mood disorders. Much remains to be learned about disorders in childhood, for example. Personality is so complex that it can keep researchers busy for decades, especially as they go beyond self-report questionnaires to other measures such as observations. A rich territory for future exploration is the link between psychopathology and personality.

Cognitive disabilities and abilities, psychopathology, and personality have been the targets for the vast majority of genetic research in the behavioral sciences because these areas have traditionally considered individual differences. Three new areas that are beginning to be explored genetically were described in Chapters 14 and 17: health psychology, aging, and evolutionary psychology. Some of the oldest areas of psychology—perception, learning, and language, for example—have not emphasized individual differences and as a result have yet to be explored systematically from a genetic perspective. Entire disciplines within the social and behavioral sciences, such as economics, education, and sociology, are still essentially untouched by genetic research.

Genetic research in the behavioral sciences will continue to move beyond simply demonstrating that genetic factors are important or estimating heritabilities. The questions *whether* and *how much* genetic factors affect behavioral dimensions and disorders represent important first steps in understanding the origins of individual differences. But these are only first steps. The next steps involve the question *how*, the mechanisms by which genes have their effect. How do genetic effects unfold developmentally? What are the biological pathways between genes and behavior? How do nature and nurture interact and correlate? Examples of these three directions for genetic research in psychology—developmental genetics, multivariate genetics, and "environmental" genetics—have been seen throughout the preceding chapters. The future will see more research of this type as behavioral genetics continues to move beyond merely documenting genetic influence.

Developmental genetic analysis considers change as well as continuity during development throughout the human life span. Two types of developmental questions can be asked. First, do genetic and environmental components of variance change during development? The most striking example to date involves general cognitive ability (Chapter 8). Genetic effects become increasingly important throughout the life span, at least until old age. Shared family environment is important in childhood, but its influence becomes negligible after adolescence. The second question concerns the role of genetic and environmental factors in age-to-age change and continuity during development. Using general cognitive ability again as an example, we find a surprising degree of genetic continuity from childhood to adulthood. However, some evidence has been found for genetic change as well, for example, during the transition from early to middle childhood, when formal schooling begins. Interesting developmental discoveries are not likely to be limited to cognitive development—it just so happens that most developmental genetic research so far has focused on cognitive development.

Multivariate genetic research addresses the covariance between traits rather than the variance of each trait considered by itself. A surprising finding in relation to specific cognitive abilities is that many genetic factors affect most cognitive abilities and disabilities (Chapter 9). For psychopathology, a key question is why so many disorders co-occur. Multivariate genetic research suggests that genetic overlap between disorders may be responsible for this comorbidity (Chapter 11). Another basic question in psychopathology is heterogeneity. Are there subtypes of disorders that are genetically distinct? Multivariate genetic research is critical for investigating the causes of comorbidity and heterogeneity, and for identifying the most heritable constellations (comorbidity) and components (heterogeneity) of psychopathology.

Two other general directions for multivariate genetic research are links between the normal and the abnormal, and between behavior and biology. A fundamental question is the extent to which genetic and environmental effects on disorders are merely the quantitative extremes of the same genetic and environmental factors that affect the rest of the distribution. Or are disorders different in kind, not just in quantity, from the normal range of behavior? Multivariate genetic analysis also can be used to investigate the mechanisms by which genetic factors influence behavior by identifying genetic correlations between behavior and biological processes, such as neurotransmitter systems. It cannot be assumed that the nexus of associations between biology and behavior is necessarily genetic in origin. Multivariate genetic analysis is needed to investigate the extent to which genetic factors mediate these associations.

"Environmental" genetics will continue to explore the interface between nature and nurture. As described in Chapter 16, genetic research has made some of the most important discoveries about the environment in recent decades, especially nonshared environment and the role of genetics in experience. More discoveries about environmental mechanisms can be predicted, as the environment continues to be investigated in the context of genetically sensitive designs. Much remains to be learned about interactions and correlations between nature and nurture.

In summary, no crystal ball is needed to predict that behavioral genetic research will continue to flourish as it turns to other areas of behavior and, especially, as it goes beyond the rudimentary questions of *whether* and *how much* to ask the question *how*. Such research will become increasingly important as it guides molecular genetic research to the most heritable components and constellations throughout the human life span, as they interact and correlate with the environment. In return, developmental, multivariate, and "environmental" genetics will be transformed by molecular genetics.

Molecular Genetics

The behavioral sciences are at the dawn of a new era in which molecular genetic techniques will revolutionize genetic research by identifying specific genes that contribute to genetic variance for complex dimensions and disorders. The quest is to find not the gene for a trait, but the multiple genes (quantitative trait loci, QTLs) that are associated with the trait in a probabilistic rather than a predetermined manner. The breathtaking pace of molecular genetics (Chapter 6) leads us to predict that behavioral scientists will routinely use DNA markers as a tool in their research to identify some of the relevant genetic differences among individuals. DNA markers could be profitably incorporated into any research design that considers individual differences. This is a safe prediction, because it is already happening in research on dementia and cognitive decline in the elderly. It is now standard practice for research in this area to take advantage of the genetic risk information provided by the DNA marker for apolipoprotein E (Chapter 7), even when researchers are interested primarily in psychosocial risk mechanisms. Aiding this prediction that behavioral scientists will routinely use DNA in their research is the fact that DNA is inexpensive to obtain and microarrays make genotyping increasingly inexpensive even for complex traits for which hundreds or thousands of DNA markers are genotyped.

To answer questions about how genes influence behavior, nothing can be more important than identifying specific genes responsible for the widespread genetic influence on behavior. As specific genes are found, more precise questions can be asked, using measured genotypes. Do the effects of the genes change during development? Do the genes correlate with some aspects of a trait but not others (heterogeneity) or do their effects extend across several traits (comorbidity)? Are genes for disorders also associated with normal dimensions and vice versa? Do the genetic effects interact or correlate with the environment?

The behavioral sciences will have an important role to play in understanding the pathways between genes and behavior (Chapter 15). In contrast to bottom-up functional genomics research on cells, top-down behavioral genomics research is likely to pay off more quickly in terms of prediction, diagnosis, intervention, and prevention of behavioral disorders. Behavioral genomics represents the long-term future of behavioral genetics, when we will be awash with specific genes that account for some of the ubiquitous genetic influence on behavioral dimensions and disorders. Behavioral genomics can make important contributions toward understanding the functions of genes, and DNA opens up new scientific horizons for understanding behavior. Bottom-up functional genomics will eventually meet top-down behavioral genomics in the brain. The grandest implication for science is that DNA will serve as a common denominator integrating diverse disciplines.

We also predict that DNA analysis will be used routinely in clinics. One of the great strengths of DNA analysis is that it can be used to predict risk long before a disorder appears. This predictive ability will allow the use of interventions that prevent the disorder rather than trying to reverse a disorder once it appears and has already caused collateral damage. Molecular genetics may also eventually lead to individualized behavioral medicine, using gene-based diagnoses and treatment programs.

For these reasons, it is crucial that behavioral scientists be prepared to take advantage of the exciting developments in molecular genetics. In the same way that we now assume that computer literacy is an essential goal to be achieved during elementary and secondary education, students in the behavioral sciences must be taught about genetics in order to prepare them for this future. Otherwise, this opportunity for behavioral scientists will slip away by default to geneticists, and genetics is much too important a topic to be left to geneticists! Clinicians use the acronym "DNA" to note that a client "did not attend"—it is critical to the future of the behavioral sciences that DNA mean deoxyribonucleic acid rather than "did not attend."

Nature and Nurture

The controversy that swirled around behavioral genetic research during the 1970s (Chapter 8) has largely faded. One of many signs of the increasing acceptance of genetics is that behavioral genetics was identified by the American Psychological Association at its centennial celebration in 1992 as one of the two themes that best represent the future of psychological research (Plomin & McClearn, 1993a). This is one of the most dramatic shifts in the modern history of psychology. Indeed, the wave of acceptance of genetic influence in the behavioral sciences is growing into a tidal wave that threatens to engulf the second message coming from behavioral genetic research. The first message is that genes play a surprisingly important role across all behavioral traits. The second message is just as important: Individual differences in complex behavioral traits are due at least as much to environmental influences as they are to genetic influences.

The first message will become more prominent during the next decade as more genes are identified that are responsible for the widespread influence of genetics in the behavioral sciences. As explained in Chapter 5, it should be emphasized that genetic effects on complex traits describe *what is*. Such findings do not predict *what could be* or prescribe *what should be*. Genes are not destiny. Genetic effects represent probabilistic propensities, not predetermined programming. See Box 6.4 about the 1997 science fiction film *GATTACA* in which individuals are selected for education and employment on the basis of their DNA. (See also a discussion of the medical and societal consequences of the

Human Genome Project by its director [Collins, Green, Guttmacher, & Guyer, 2003; Collins & McKusick, 2001].)

A related point is that, for complex traits such as behavioral traits, QTL effects refer to average effects in a population, not to a particular individual. For example, one of the strongest DNA associations with a complex behavioral disorder is the association between allele 4 of the gene encoding apolipoprotein E and late-onset dementia (Chapter 7). Unlike simple single-gene disorders, this QTL association does not mean that allele 4 is necessary or sufficient for the development of dementia. Many people with dementia do not have the allele and many people with the allele do not have dementia. A particular gene may be associated with a large average increase in risk for a disorder, but it is likely to be a weak predictor at an individual level. The importance of this point concerns the dangers of labeling individuals on the basis of population averages.

Genetic influence does not mean that the environment is unimportant. To the contrary, genetic research provides the best available evidence for the importance of environmental influences and has produced some of the most important findings about the environment and how it affects development (Chapter 16).

The relationship between genetics and equality, an issue that lurks in the shadows causing a sense of unease about genetics, was discussed in Chapter 5. The main point is that finding genetic differences among individuals does not compromise the value of social equality. The essence of a democracy is that all people should have legal equality *despite* their genetic differences. Knowledge alone by no means accounts for societal and political decisions. Values are just as important as knowledge in the decision-making process. Decisions, both good and bad, can be made with or without knowledge. Nonetheless, scientific findings are often misused, and scientists, like the rest of the population, need to be concerned to diminish misuse. We believe firmly, however, that better decisions can be made with knowledge than without. There is nothing to be gained by sticking our heads in the sand and pretending that genetic differences do not exist.

Finding widespread genetic influence creates new problems to consider. For example, could evidence for genetic influence be used to justify the status quo? Will people at genetic risk be labeled and discriminated against? When genes are found for behavioral traits, will parents use them prenatally to select "designer" children? New knowledge also provides new opportunities. For example, finding genes associated with a particular disorder makes it more likely that environmental preventions and interventions that are especially effective for the disorder can be found. Knowing that certain children have increased genetic risk for a disorder could make it possible to prevent or ameliorate the disorder before it appears, rather than trying to treat the disorder after it appears and causes other problems. Moreover, it should not be assumed that once a gene associated with some disorder is found, the logical next step is to get rid of it. For example, genes that persist in the population may be the result of stabilizing selection (Chapter 17), which might mean that the genes have good as well as bad outcomes.

Two other points should be made in this regard. First, most powerful scientific advances create new problems. For example, consider prenatal screening for genetic defects. This advance has obvious benefits in terms of detecting chromosomal and genetic disorders before birth. Combined with abortion, prenatal screening can relieve parents and society of the tremendous burden of severe birth defects. However, it also raises ethical problems concerning abortion and creates the possibility of abuses, such as compulsory screening and mandatory abortion. Despite the problems created by advances in science, we would not want to cut off the flow of knowledge and its benefits in order to avoid having to confront such problems.

The second point is that it is wrong to assume that environmental explanations are good and that genetic explanations are dangerous. Tremendous harm was done by the environmentalism that prevailed until the 1960s, when the pendulum swung back to a more balanced view that recognized genetic as well as environmental influences. For example, environmentalism led to blaming children's problems on what their parents did to them in the first few years of life. Imagine that, in the 1950s, you were among the 1 percent of parents who had a child who became schizophrenic in late adolescence. You faced a lifetime of concern. And then you were told that the schizophrenia was caused by what you did to the child in the first few years. The sense of guilt would be overwhelming. Worst of all, such parent blaming was not correct. There is no evidence that early parental treatment causes schizophrenia. Although the environment is important, whatever the salient environmental factors might be, they are not shared family environmental factors. Most important, we now know that schizophrenia is substantially influenced by genetic factors.

Our hope for the future is that the next generation of behavioral scientists will wonder what the nature-nurture fuss was all about. We hope they will say, "Of course, we need to consider nature and nurture in understanding behavior." The conjunction between nature and nurture is truly *and*, not *versus*.

The basic message of behavioral genetics is that each of us is an individual. Recognition of, and respect for, individual differences is essential to the ethic of individual worth. Proper attention to individual needs, including provision of the environmental circumstances that will optimize the development of each person, is a utopian ideal and no more attainable than other utopias. Nevertheless, we can approach this ideal more closely if we recognize, rather than ignore, individuality. Acquiring the requisite knowledge warrants a high priority because human individuality is the fundamental natural resource of our species.

Statistical Methods in Behavioral Genetics

Shaun Purcell

1 Introduction

Quantitative genetics offers a powerful theory and various methods for investigating the genetic and environmental etiology of any characteristic that can be measured, including both continuous and discrete traits. As discussed in Chapter 6, quantitative genetics and molecular genetics are coming together in the study of complex quantitative traits. In both fields, powerful statistical and epidemiological methods have been developed to address a series of related questions:

- Do genes influence this outcome?
- What types of genetic effects are at work?
- Can genetic effects explain the relationships between this and other outcomes?
- Where are the genes located?
- What specific form(s) of the genes cause certain outcomes?
- Do genetic effects operate similarly across different populations and environments?

This Appendix introduces some of the methods behind these research questions, in a manner designed to provide the rationale behind the methods as well as an appreciation of the directions in which the field is developing, including molecular genetics. Both quantitative genetics (with an emphasis on the components of variance model-fitting approach to complex traits) and molecular genetics (with an emphasis on linkage and association approaches to gene mapping) are covered.

We begin with a brief overview of some of the statistical tools that are commonly used in behavioral genetic research: variance, covariance, correlation, regression, and matrices. Although one need not be a fully trained statistician to use most behavioral genetic methods, understanding the main statistical concepts that underlie quantitative genetic research enables one to appreciate the ideas, assumptions, and limitations behind the methods.

Next, the classical quantitative genetic model is introduced, which relates the properties of a single gene to variation in a quantitative phenotype. This relatively simple model forms the basis for the majority of quantitative genetic methods. We then examine how the analysis of familial correlations can be used to infer the underlying

BOX A.1

Behavioral Genetic Interactive Modules

The *Behavioral Genetic Interactive Modules* are a series of freely available interactive computer programs with accompanying textual guides designed to convey a sense of the methods of modern behavioral genetic analysis to students and researchers new to the field. Currently, 11 modules covering the material in this Appendix can be accessed from the Web site at http://statgen.iop.kcl.ac.uk/bgim. Taken together, these modules listed below lead from the basic statistical foundations of quantitative genetic analysis to an introduction to some of the more advanced analytical techniques.

Variance is designed to introduce the concept of variance: what it represents, how it is calculated, and how it can be used to assess individual differences in any quantitative trait. Standardized scores are also introduced.

Covariance demonstrates how the covariance statistic can be used to represent association between two measures.

Correlation & Regression is an exploration of the relationship among variance, covariance, correlation, and regression coefficients.

Matrices provides a simple matrix calculator.

Single Gene Model introduces the basic biometrical model used to describe the effects of individual genes, in terms of additive genetic values and dominance deviations.

Variance Components: ACE illustrates the partitioning of variance into additive genetic, shared environmental, and nonshared environmental components in the context of MZ and DZ twins.

Families demonstrates the relationship between additive and dominance genetic variance, shared and nonshared environmental variance, and expected familial correlations for different types of relatives.

Model Fitting 1 defines a simple path diagram to model the covariance between observed variables and allows the user to manually adjust path coefficients to find the best-fitting model; it includes a twin ACE model and nested models that can be compared with the full ACE model.

Model Fitting 2 performs a maximum-likelihood analysis of univariate twin data and presents the parameter estimates for nested submodels.

Multivariate Analysis models the genetic and environmental etiology of two traits.

Extremes Analysis illustrates DF extremes analysis as well as individual differences analysis, in order to explore how these two methods can inform us about links between normal variation and extreme scores.

For individuals wishing to take their study of statistical analysis further, a guide is provided to help you get started on analyzing your own data as well as simulated data sets that can be used to explore these methods further. Behavioral genetic analyses using widely available statistics packages such as *Stata* are described, as well as an introduction to *Mx*, a powerful, freely available model-fitting package by Mike Neale.

CLOSE UP

Shaun Purcell develops statistical and computational tools for the design of genetic studies, the detection of gene variants influencing complex human traits, and the dissection of these effects within the larger context of other genetic and environmental factors. He is currently an assistant professor at Harvard Medical School, based at the Psychiatric and Neurodevelopmental Genetics Unit, Massachusetts General Hospital. He is also an associate member of the Broad Institute of Harvard and MIT, and its Psychiatric Disease Initiative. As an undergraduate from 1992 to 1995, he studied experimental psychology at Oxford University; in 1996, he had the opportunity to develop an interest in statistical methods while working toward a master's of science degree at University College London. In 1997, he joined the Social, Genetic and Developmental Psychiatry (SGDP) Research Centre at the Institute of Psychiatry in London, to embark on a Ph.D. with Pak Sham and Robert Plomin, working on a project designed to map quantitative trait loci for anxiety and depression. His current work involves whole genome association studies of bipolar disorder and schizophrenia, and the development of tools for whole genome studies.

etiological nature of a trait, given our knowledge of the way genes work. The basic model partitions the variance of a single trait into portions attributable to additive genetic effects, shared environmental effects, and nonshared environmental effects. The tools of model fitting and path analysis are introduced in this context. Extensions to the basic model are also considered: multivariate analysis, analysis of extremes, and interactions between genes and environments, for example.

Finally, we see how molecular genetic information on specific loci can be incorporated. In this way, the chromosomal positions of genes can be mapped. This work leads the way to the study of gene function at a molecular level—the vital next step if we really want to know *how* our genes make us what we are.

1.1 Variation and covariation: statistical descriptions of individual differences

Behavioral genetics is concerned with the study of individual differences: detecting the factors that make individuals in a population different from one another. As a first step, it is concerned with gauging the relative importance of genetic and environmental factors that cause individual differences. To assess the importance of these factors, we need to be able to *measure* individual differences. This task requires some elementary statistical theory.

A population is defined as the complete set of all individuals in a group under study. Examples of populations would include sets such as all humans, all female Americans aged 20 to 25 in the year 2000, or all the stars in a galaxy. We might measure a characteristic, such as talkativeness, intelligence, weight, or temperature, for each of the individuals in a population. We are concerned with assessing how these characteristics vary both *within* populations (e.g., among 2-year-old males) and *between* them (e.g., male versus female infants).

If all the individuals in a set are studied, population statistics such as the average or the variance can be calculated exactly. However, it is usually not practical to measure every individual in the population, so we resort to *sampling* individuals from the population. A key concept in sampling is that, ideally, it should be conducted at *random*. A nonrandom sample, such as only the tallest 20 percent of 11-year-old girls, would give an inflated (biased) estimate of the average height of 11-year-old girls. An estimate of the average height in the population gathered from a random sample would not, *on average*, be biased. However, it is important to recognize that an estimate of the population mean made on the basis of a random sample will vary somewhat from the population mean. The amount of this variation will depend on the sample size and on chance. We need to know how much we expect this variation to be so that we know how accurate estimates of the population parameters are. This assessment of accuracy is critical when we want to compare populations.

Once we have defined a population, various *parameters* such as the mean, range, and variance can be described for the trait that we wish to study. Similarly, when we have a sample of the population, we can calculate *statistics* from the sample that correspond to the parameters of the population. It is not always the case that the measure of the sample statistic is the best estimate of the corresponding population parameter. This discrepancy distinguishes descriptive from inferential statistics. Descriptive statistics simply describe the sample; inferential statistics are used to get estimates of the parameters of the entire population.

1.1.1 The mean   The arithmetic mean is one of the simplest and most useful statistics. It is a measure of the center of a distribution and is the familiar average statistic used in everyday speech. It is very simple to compute, being the sum of all observations' values divided by the number of observations in the sample:

$$\mu = \Sigma x / N$$

where Σx is the sum of all observations in the set of size N. Strictly speaking, the mean is only labeled μ (pronounced "mu") if it is computed from the entire population. Usually, the mean will be calculated from a sample, and the mean of a variable, say x, is written as $\bar{x}$ (read as "x bar").

The mean is especially useful when comparing groups. Given an estimate of how accurate the means are, it becomes possible to compare means for two or more groups. Examples might include whether women are more verbally skilled than men, whether albino mice are less active than other mice, or whether light moves faster than sound.

Some physical measures, such as the number of inches of rainfall per year, are obviously well ordered, such that the difference between 15 and 16 inches is the same as the difference between 21 and 22 inches—namely, 1 inch. Many physical measurements have the same scale throughout the distribution, which is called an *interval* scale. In behavioral research, however, it is often difficult to get measures that are on an interval scale. Some measures are *binary*, consisting simply of the presence or absence of a disease or symptom. The mean of a binary variable scored 0 for absent and 1 for present indicates the proportion of the sample that has the symptom or disease present, so again the mean is a useful summary. However, not all measures can be effectively summarized as means. The trouble starts when there are several ordered categories, such as "Not at all/Sometimes/Quite often/Always." Even if these items are scored 0, 1, 2, and 3, the mean tells us little about the frequencies in each category. This problem is even greater when the categories cannot be ordered, such as religious affiliation. Here the mean would be of no use at all.

1.1.2 Variance   Variance is a statistic that tells us how spread out scores are. This is a measure of individual differences in the population, the focus of most behavioral genetic analyses. Variances are also important when assessing differences between group means. Behavioral genetic analyses are typically less focused on group differences, although such analyses are central to most quantitative sciences. For example, a researcher may wish to ask whether a control group significantly differs from an experimental one on a measure, or whether boys and girls differ in the amount that they eat. Testing differences between means is often carried out with a statistical method called the analysis of variance (ANOVA). In fact, individual differences are treated as the "error" term in ANOVA.

The usual approach to calculating the variance, established by R. A. Fisher (1922), one of the founders of quantitative genetics, is to take the average of the squared deviations from the mean. Fisher showed that the squared deviations from the mean had more desirable statistical properties than other measures of variance that might be considered, such as the average absolute difference. In particular, the average squared deviation is the most accurate statistic.

Calculating the variance (often written as s^2) is straightforward:

1. Calculate the mean.
2. Express the scores as deviations from the mean.
3. Square the deviations and sum them.
4. Divide the sum by the number of observations minus 1.

Or, written as a formula:

$$s^2 = \frac{\sum (x - \bar{x})^2}{N - 1}$$

A second commonly used approach involves computing the contribution of each observation to the variance and correcting for the mean at the end. This alternative method produces the same answer, but it can be more efficient for computers to use. Note that $N - 1$ instead of N is used to calculate the average squared deviation in order to produce unbiased estimates of variance—for technical, statistical reasons.

Variances range from zero upward: There is no such thing as a negative variance. A variance of zero would indicate no variation in the sample (i.e., all individuals would have to have exactly the same score). The greater the spread in scores, the greater the variance.

With binary "yes/no" or "affected/unaffected" traits, measuring the variance is difficult. We may imagine that a binary trait is observed because there is an underlying normal distribution of *liability* to the trait, caused by the additive effects of a large number of factors, each of small effect. The binary trait that we observe arises because there is a threshold, and only those with liability above threshold express the trait. We cannot directly observe the underlying liability, so typically we assume that it has variance of unity (1). If the variance of the underlying distribution were increased, it would simply change the proportion of subjects that are above threshold. That is, changing the variance is equivalent to changing the threshold. Distinguishing between mean changes and variance changes is not generally possible with binary data, but it is possible if the data are ordinal, with at least three ordered categories.

Having measured the variance, quantitative genetic analysis aims to partition it—that is, to divide the total variance into parts attributable to genetic and environmental components. This task requires the introduction of another statistical concept, covariance. Before turning to covariance, we will take a brief digression to consider another way of expressing scores that facilitates comparisons of means and variances.

1.1.3 Standardized scores Different types of measures have different scales, which can cause problems when making comparisons between them. For example, differences in height could be expressed in either metric or common (imperial) terms. In a population, the absolute value of variance in height will depend on the scale used to measure it—the unit of variance will be either centimeters squared or inches squared. If we take the square root of variance, we obtain a measure of spread that has the same unit of the observed trait, called the *standard deviation (s)*. The standard deviation also has several convenient statistical properties. If a trait is normally distributed (bell-shaped curve), then 95 percent of all observations will lie within two standard deviations on either side of the mean.

The example of measuring height demonstrates the difficulties that may be encountered when we wish to compare differently scaled measures. In the case of metric and common measurements of height, which both measure the same thing, the problem of scale can be easily overcome by using standard conversion formulas. In psychology, however, measurements often will have no fixed scale. A questionnaire that measures extraversion might have a scale from 0 to 12, from 1 to 100, or from -4 to $+4$. If scale is arbitrary, it makes sense to make all measures have the same, standardized scale.

Suppose we have data on two reliable questionnaire measures of extraversion, A and B, each from a different population. Say measure A has a range of 0 to 12 and a mean score of 6.4, whereas measure B has a range of 0 to 50 with a mean of 24. If we were to assess two individuals, one scoring 8 on measure A and the other scoring 30 on measure B, how could we tell which person is the more extraverted? The most commonly used technique is to *standardize* our measures. The formula for calculating a standardized score z from a raw score x is

$$z = \frac{x - \bar{x}}{\sqrt{s_x^2}}$$

where s_x^2 is the variance of x. That is, we reexpress the scores in standard deviation units. For example, if we calculate that measure A has a variance of 4, then the standard deviation is $\sqrt{4} = 2$. If we express scores as the number of standard deviations away from

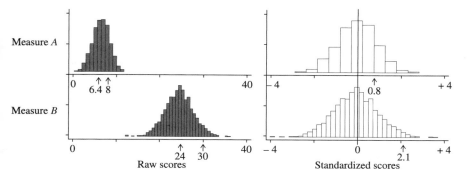

Measure A

Measure B

Raw scores

Standardized scores

Figure A.1 Standardized scores. Raw scores on the two measures cannot be directly equated. Standardizing both measures to have a mean of 0 and a standard deviation of 1 facilitates the comparison of measures A and B.

the mean, then a score of 2 raw-score units above the mean on measure A is $+1$ standard deviation units. Raw scores equaling the raw-score mean will become 0 in standard deviation units. A raw score of 2 will become $(2 - 6.4)/2 = -2.2$. Therefore, a score of 8 on measure A corresponds to a standardized score of $(8 - 6.4)/2 = 0.8$ standard deviation units above the mean.

We can also do the same for measure B, to be able to make scale-independent comparisons between our two measures of extraversion. If measure B is found to have a variance of 8 (and therefore a standard deviation of $\sqrt{8}$), then a raw score of 30 corresponds to a standardized score of $(30 - 24)/\sqrt{8} = 2.1$. We can therefore conclude that individual B is more extraverted than individual A (i.e., $2.1 > 0.8$) (Figure A.1). Converting the measures into standardized scores also allows statistical tests of the significance of such differences (the z-test).

Standardized scores are said to have *zero sum* property (they will always have a mean of 0) and unit standard deviation (i.e., a standard deviation of 1). As we have seen, standardizing is useful when comparing different measures of the same thing. Indeed, standardizing can be used to compare different measures of different things (e.g., whether a particular individual is more extreme in height or in extraversion).

However, there are some situations in which standardized scores can be misleading. Standardizing within groups (i.e., using the estimates of the mean and standard deviation from that group) will destroy between-group differences. All groups will end up with means of zero, which will hide any true between-group variation. Note that it was implicit in the example above that measures A and B are both reliable, and that the two populations are equivalent with respect to the distribution of "true" extraversion.

1.1.4 Covariance Another fundamental statistic that underlies behavioral genetic theory is *covariance*. Covariance is a statistic that informs us about the relationship between two characteristics (e.g., height and weight). Such a statistic is called a *bivariate* statistic, in contrast to the mean and variance, which are both *univariate* statistics. If two variables are associated (i.e., they *covary* together), we may have reason to believe that this covariation occurs because one characteristic influences the other. Alternatively, we might suspect that both characteristics have a common cause. Covariance, by itself, however, cannot tell us *why* two variables are associated: It is only a measure of the magnitude of association. Figure A.2 shows four possible relationships between two variables, X and Y, each of which could result in a similar covariance between the two variables. For example, it is clearly wrong to think of an individual's weight as *causing* his or her height, whereas it is fair to say that an individual's height does, in part, determine that person's weight—it should be noted that care is needed in the interpretation of *all* statistics. The methods of path analysis (as reviewed later) do offer an opportu-

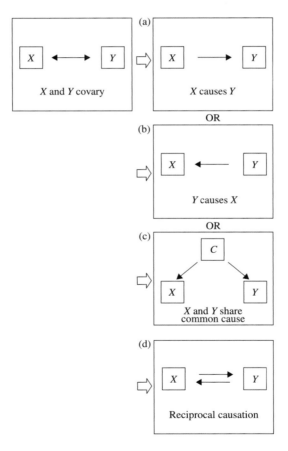

Figure A.2 Causes of covariation. Two variables can covary for a number of reasons: (a, b) One variable might cause the other, or (c) both variables might be influenced by a third variable (C), or (d) both variables might influence the other. The covariance statistic cannot by itself discriminate among these alternatives.

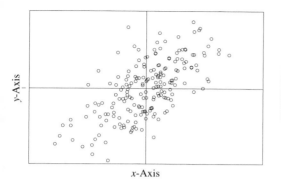

Figure A.3 Scatterplot representing 200 observations measured on two variables, X and Y. As can be seen, X and Y are not independent, because observations with higher values for X also tend to have higher values for Y.

nity to begin to "tease apart" causation from "mere" correlation, especially when applied to data sets that differ in genetic or environmental factors.

A sensible first step when investigating the relationship between two continuous variables is to begin with a *scatterplot*. The scatterplot shown in Figure A.3 represents 200 observations. In this example, it is apparent that the two measures are not independent. As X increases (the scale for X increases toward the right), we see that the scores on Y also tend to increase. Covariance is a measure that attempts to quantify this kind of relationship (as do *correlation* and *regression coefficients*, introduced later).

Calculating the covariance proceeds in much the same way as calculating the variance. However, instead of squaring the deviations from the mean, we calculate the cross-product of the deviations of the first variable with those of the second. To compute the covariance, we would

1. Calculate the mean of X.
2. Calculate the mean of Y.
3. Express the scores as deviations from the means.
4. Calculate the product of the deviations for each data pair and sum them.
5. Divide by $N - 1$ to obtain an estimate of the covariance.

Written as a formula, the covariance is

$$\mathrm{Cov}_{XY} = \frac{\sum (X - \bar{X})(Y - \bar{Y})}{N - 1}$$

Covariance values can range between plus and minus infinity. Negative values imply that high scores on one measure tend to be associated with low scores on the other measure. A covariance of 0 implies that there is no *linear* relationship between the two measures.

That covariance measures only linear association is an important issue: Consider the two scatterplots in Figure A.4. Neither of these two bivariate data sets displays any *linear* association between the two variables, so both have a covariance of zero. However, there is a clear difference between the two data sets: in one, the observations are truly *independent*, whereas it is clear that the variables in the other are related but not in a linear way.

(a)   (b)

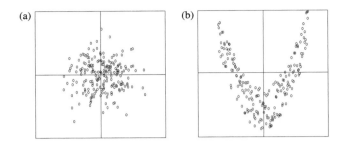

Figure A.4 Covariance and independence. The covariance statistic represents linear association. Both scatterplots represent data sets with a covariance of zero. (a) The two variables in this data set are truly independent; that is, the average value of one variable is independent of the value of the other. (b) The variables in this data set are not linearly related, but they are clearly not independent.

   A key to understanding covariance is to understand what the formula for its calculation is really doing. Figure A.5 represents the four quadrants of a scatterplot. The lines intersecting in the middle represent the mean value for each variable. When the scores are expressed as deviations from the mean, all those to the left of the vertical line (or below the horizontal line) will become negative; all values to the right of the vertical line (or above the horizontal line) will become positive. As we have seen, covariance is calculated by summing the products of these deviations. Therefore, because both the product of two positive numbers and the product of two negative numbers are positive whereas the product of one positive and one negative number is always negative, the contribution each observation makes to the covariance will depend on which quadrant it falls in. Observations in the top-right and bottom-left quadrants (both numbers above the mean and both numbers below the mean, respectively) will make a positive contribution to the covariance. The farther away from the origin (the bivariate point where the two means intersect), the larger this contribution will be. Observations in the other two quadrants will tend to decrease the covariance. If all bivariate data points were evenly distributed across this space, the positive contributions to the covariance would tend to be canceled out by an equal number of negative contributions, resulting in a near zero covariance statistic. A large positive covariance would imply that the bulk of data points fall in the bottom-left and top-right quadrants; a large negative covariance would imply that the bulk of data points fall in the top-left and bottom-right quadrants.

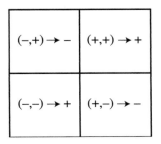

Figure A.5 Calculating covariance. The contribution each observation makes to the covariance will depend on the quadrant in which it falls.

1.1.5 *Variance of a sum*   Covariance is also important for calculating the variance of a sum of two variables. This statistic is relevant to our later discussion of the basic quantitative genetic model. Say you have variables X and Y and you know their variances and the covariance between them. What would the variance of $(X + Y)$ be? If all the data are available, you may decide to calculate a new variable that is the sum of the two variables and then calculate its variance in the ordinary manner. Alternatively, if you know only the summary statistics, you can use the formula:

$$\text{Var}(X + Y) = \text{Var}(X) + \text{Var}(Y) + 2\text{Cov}(X,Y)$$

In other words, the variance of a sum is the sum of the two variances plus twice the covariance between the two measures. If two variables are uncorrelated, then the covariance term will be zero and the variance of the sum is simply the sum of the variances. As will be seen later, the mathematics of variance is critical in the formulation of the genetic model for describing complex traits.

1.1.6 *Correlation and regression*   We have seen how using standardized scores can help when working with measures that have different scales. When creating a standardized score, we use information about the variance of a measure to rescale the raw data. As mentioned earlier, the covariance between two measures is dependent on the scales of the raw data and can range from plus to minus infinity. We can use information about the variance of two measures to standardize their covariance statistic, in a manner analogous to creating standardized scores. A covariance statistic standardized in this way is called a *correlation*.

The correlation is calculated by dividing the covariance by the square root of the product of the two variances for each measure. Therefore, the correlation between X and Y (r_{XY}) is

$$r_{XY} = \frac{\text{Cov}_{XY}}{\sqrt{s_X^2 s_Y^2}}$$

where Cov_{XY} is the covariance and s_X^2 and s_Y^2 are the variances. If both X and Y are standardized variables (i.e., s_X and s_Y, and therefore also s_X^2 and s_Y^2, both equal 1), then the correlation will be the same as the covariance (as can be seen in the formula above).

Correlations (typically labeled r) always range from $+1$ to -1. A correlation of $+1$ indicates a perfect positive *linear* relationship between two variables. A correlation of -1 represents a perfect negative linear relationship. A correlation of 0 implies no linear relationship between the two variables (in the same way that a covariance of 0 implies no linear relationship). The kind of correlations we might expect to observe in the real world are likely to fall somewhere between 0 and $+1$. How exactly do we interpret correlations of intermediate values? Does, for example, a correlation of .4 mean that the two measures are the same 40 percent of the time? In short, no. What it reflects, as seen in the equation above, is the proportion of variance that is shared by the two measures. (The square of a correlation, r^2, is a commonly used statistic that indicates the proportion of variance in one variable that can be predicted by the other. For correlations between relatives, the unsquared correlation, representing the proportion of variance common to both family members, is more useful.)

Regression is related to correlation in that it also examines the relationship between two variables. Regression is concerned with *prediction* in that it asks whether knowing the value of one variable for an individual helps us to guess what the value on another variable will be.

Regression coefficients (often called b) can be calculated by using a method similar to that used to calculate correlation coefficients. The regression coefficient of "y on x" (i.e., given X, what is our best guess for the value of Y) divides the covariance between X and Y by the variance of the variable (X) from which we are making the prediction (rather than standardizing the covariance by dividing by the product of the standard deviations of X and Y):

$$b = \frac{\text{Cov}_{XY}}{s_X^2}$$

Given this regression coefficient, an equation relating X and Y can be written:

$$\hat{Y} = bX + c$$

where c is called the regression constant. As plotted in Figure A.6, this equation describes a straight line (the *least squares regression line*) that can be drawn through the observed

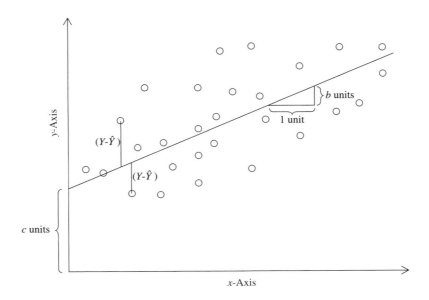

Figure A.6 Linear regression. The regression of a line of best fit between X and Y is represented by the equation $\hat{Y} = bX + c$. For each unit increase in X, we expect Y to increase b units. The two vertical lines represent the deviations between the expected and actual values of Y. The sum of the deviations squared is used to calculate the residual variance of Y, that is, the variance in Y not accounted for by X. The regression constant c represents the value of Y when X is zero (at the origin).

points and represents the best prediction of Y given information on X ($\hat{Y}$, pronounced "y hat"). The equation implies that for an increase of one unit in X, Y will increase an average of b units.

Regression equations can also be used to analyze more complicated, nonlinear relationships between two variables. For example, the variable Y might be a function of the square of X as well as of X itself. We would therefore include this higher-order term in the equation to describe the relationship between X and Y:

$$\hat{Y} = b_1 X^2 + b_2 X + c$$

This equation describes a nonlinear least squares regression line (i.e., a parabolic curve if b_1 doesn't equal zero).

It is possible to calculate the discrepancy, or error, between the predicted values of Y given X and the actual values of Y observed in the sample $(Y - \hat{Y})$. These discrepancies are called the *residuals*, and it is often useful to calculate the variance of the residuals. From the first regression equation given above, if X and Y were totally unrelated, then b would be estimated near zero and c would be the mean of Y (because this value represents the *best guess* of Y if you don't have any other information). In this case, residual error variance would be the same as the variance of Y. To the extent that knowing X actually does help you guess Y, the regression coefficient will become significantly nonzero and the error term will decrease.

We can partition the variance in a variable, Y, into the part that is associated with another variable X and the part that is independent of X. In terms of a regression of Y on X, this partitioning is reflected in the variance of the predicted Y values (the variance of $\hat{Y}$) as opposed to the variance of the residuals (the variance of $(Y - \hat{Y})$). The correlation between the two variables can actually be used to estimate these values in a straightforward way:

$$s_{\hat{Y}}^2 = r^2 s_Y \quad \text{and} \quad s_{Y-\hat{Y}}^2 = (1 - r^2) s_Y^2.$$

A common regression-based technique can be used to "regress out" or "adjust for" the effects of one variable on another. For example, we may wish to study the relationship between verbal ability and gender in children. However, we also know that verbal ability is age related, and we do not want the effects of age to confound this analysis. We can calculate an age-adjusted measure of verbal ability by performing a regression of verbal ability on age. For every individual, we subtract their predicted value (given their age) from their observed value to create a new variable that reflects verbal ability without the effects of age-related variation: The new variable will not correlate with age. If there were any mean differences in age between boys and girls in the sample, then the effects of these on verbal ability have been effectively removed.

1.1.7 Matrices   Reading behavioral genetic journal articles and books, one is likely to come across *matrices* sooner or later: "In QTL linkage the variance-covariance *matrix* for the sibship is modeled in terms of alleles shared identical-by-descent" or "The *matrix* of genotypic means can be observed. . . ." What are matrices and why do we use them? This section presents a brief introduction to matrices that will place such sentences in context.

Matrices are commonly used in behavioral genetics to represent information in a concise and easy-to-manipulate manner. A matrix is simply a block of *elements* organized in rows and columns. For example,

$$\begin{bmatrix} 34 & 23 \\ 56 & 17 \\ 65 & 38 \end{bmatrix}$$

is a matrix with three rows and two columns. Typically, a matrix will be organized such that each row and column has an associated meaning. In this example, the matrix might reflect scores for three students (each row representing one student) on English and French exams (the first column representing the score for English, the second for French). Elements are often indexed by their row and column: s_{ij} refers to the ith student's score on the jth test.

The matrix above represents raw data. In a similar way, the spreadsheet of values in statistical programs such as SPSS can be thought of as one large matrix. Perhaps the most commonly encountered form of matrix is the *correlation matrix*, which is used to represent descriptive statistics of raw data (correlations) in an orderly fashion. In a correlation matrix, the element in the ith row and jth column represents the pair-wise correlation between the ith and jth variables.

Here is a correlation matrix between three different variables:

$$\begin{bmatrix} 1.00 & 0.73 & 0.14 \\ 0.73 & 1.00 & 0.37 \\ 0.14 & 0.37 & 1.00 \end{bmatrix}$$

Correlation matrices have several easily recognizable properties. First, a correlation matrix will always be *square*—having the same number of rows as columns. For n variables, the correlation matrix will be an $n \times n$ matrix. The *diagonal* of a square matrix is the set of elements for which the row number equals the column number, so in terms of correlations, these elements represent the correlation of a variable with itself, which will always be 1. Additionally, correlation matrices will always be *symmetric* about the diagonal—that is, element r_{ij} equals r_{ji}. This symmetry represents the simple fact that the correlation between A and B is the same as the correlation between B and A. It is common practice not to write the redundant upper off-diagonal elements if a matrix is known to be symmetric. Our correlation matrix would be written

$$\begin{bmatrix} 1.00 & & \\ 0.73 & 1.00 & \\ 0.14 & 0.37 & 1.00 \end{bmatrix}$$

Correlation matrices are often presented in journal articles in tabular form to summarize correlational analyses.

A closely related type of matrix that occurs more often in behavioral genetic analysis is the *variance-covariance* matrix. In place of correlations, the elements of an $n \times n$ variance-covariance matrix are n variances along the diagonal and $(n-1)n/2$ covariances in the lower off-diagonal. A correlation matrix is a *standardized* variance-covariance matrix, just as a correlation is a standardized covariance. The variance-covariance matrix for the three variables in the correlation matrix above might be

$$\begin{bmatrix} 2.32 & & \\ 1.43 & 1.64 & \\ 0.43 & 0.98 & 4.21 \end{bmatrix}$$

A variance-covariance matrix can be transformed into a correlation matrix: $r_{ij} = v_{ij}/\sqrt{v_{ii}v_{jj}}$, where r_{ij} are the new elements of the correlation matrix and v_{ij} are the elements of the variance-covariance matrix. (This is essentially a reformulation of the equation for calculating correlations given above in matrix notation.) Note that information is lost about the relative magnitude of variances among the different variables in a correlation matrix (because they are all standardized to 1). As mentioned earlier, because correlations are not scale dependent, however, they are easier to interpret than covariances and therefore better for descriptive purposes.

Matrices can be added to or subtracted from each other as long as both matrices have the same number of rows and the same number of columns:

$$\begin{bmatrix} 4 & -5 \\ 1 & 2 \end{bmatrix} + \begin{bmatrix} -2 & x \\ 0 & y \end{bmatrix} = \begin{bmatrix} 2 & x-5 \\ 1 & y+2 \end{bmatrix}$$

Note that here the elements of the sum matrix are not simple numerical terms—elements of matrices can be as complicated as you want. The beauty of matrix notation is that we can label matrices so that we can refer to many elements with a simple letter, say, **A**. (Matrices are generally written in bold type.)

$$\mathbf{A} = \begin{bmatrix} 4 & -5 \\ 1 & 2 \end{bmatrix}$$

$$\mathbf{B} = \begin{bmatrix} -2 & x \\ 0 & y \end{bmatrix}$$

$$\mathbf{A} - \mathbf{B} = \begin{bmatrix} 6 & -5-x \\ 1 & 2-y \end{bmatrix}$$

The other common matrix algebra operations are multiplication, inversion, and transposition. Matrix multiplication does not work in the same way as matrix addition (that kind of element-by-element multiplication is actually called a *Kronecker product*). Unlike normal multiplication, where $ab = ba$, in matrix multiplication $\mathbf{AB} \neq \mathbf{BA}$. For **A** to be multiplied by **B**, matrix **A** must have the same number of columns as **B** has rows. The resulting matrix has as many rows as **A** and as many columns as **B**. Each element is the sum of products across each row of **A** and each column of **B**. Following are two examples:

$$\begin{bmatrix} a & c & e \\ b & d & f \end{bmatrix} \begin{bmatrix} g & h & i \\ j & k & l \\ m & n & o \end{bmatrix} = \begin{bmatrix} ag+cj+em & ah+ck+en & ai+cl+eo \\ bg+dj+fm & bh+dk+fn & bi+dl+fo \end{bmatrix}$$

$$\begin{bmatrix} 3 & 3 & 0 \\ 1 & -2 & 5 \end{bmatrix} \begin{bmatrix} 7 & 2 \\ 3 & 2 \\ 8 & 4 \end{bmatrix} = \begin{bmatrix} 30 & 12 \\ 41 & 18 \end{bmatrix}$$

The equivalent to division is called matrix inversion and is complex to calculate, especially for large matrices. Only square matrices have an inverse, written $\mathbf{A}^{-1}$. Matrix inversion plays a central role in solving model-fitting problems.

Finally, the transpose of a matrix, $\mathbf{A}'$, is matrix $\mathbf{A}$ but with rows and columns swapped. Therefore, if $\mathbf{A}$ were a 3×2 matrix, then $\mathbf{A}'$ will be a 2×3 matrix (note that rows are given first):

$$\begin{bmatrix} 2 & 3 \\ 0 & -1 \\ -2 & 1 \end{bmatrix}' = \begin{bmatrix} 2 & 0 & -2 \\ 3 & -1 & 1 \end{bmatrix}$$

There is a great deal more to matrix algebra than the simple examples presented here. Basic familiarization with the types of matrices and matrix operations is useful, however, if only to realize that when behavioral genetic articles and books refer to matrices they are not necessarily talking about anything particularly complicated. The main utility of matrices is their convenience of presentation—it is the actual meaning of the elements that is important.

2 Quantitative Genetics

2.1 The biometric model

When we say that a trait is *heritable* or *genetic*, we are implying that at least one gene has a measurable effect on that trait. Although most behavioral traits appear to depend on many genes, it is still important to review the properties of a single gene because the more complex models are built upon these foundations. We will begin by examining the basic quantitative genetic model that mathematically describes the genetic and environmental underpinnings of a trait.

2.1.1 Allele and genotype

The pair of alleles that an individual carries at a particular locus constitutes what we call the *genotype* at that locus. Imagine that, at a particular locus, two forms of a gene, labeled A_1 and A_2 (this would be called a *biallelic* locus), exist in the population. Because individuals have two copies of every gene (one from their father, one from their mother), individuals will possess one of three genotypes: They may have either two A_1 alleles or two A_2, in which case they are said to be *homozygous* for that particular allele. Alternatively, they may carry one copy of each allele, in which case they are said to be *heterozygous* at that locus. We would write the three genotypes as A_1A_1, A_1A_2, and A_2A_2 (or, using different notation, *AA*, *Aa*, and *aa*).

For biallelic loci, the two alleles will occur in the population at particular frequencies. If we counted all the alleles in a population and three-fourths were A_1, then we say that A_1 has an allelic frequency of .75. Because these frequencies must sum to 1, we know that the A_2 allele has a frequency of .25. It is common practice to denote the allelic frequencies of a biallelic locus as p and q (so here, p is the allelic frequency of the A_1 allele, .75, and q is the frequency of A_2, .25). Given these, we can predict the genotypic frequencies. Formally, if the two alleles A_1 and A_2 have allelic frequencies p and q, then, with random mating, we would expect to observe the three genotypes A_1A_1, A_1A_2, and A_2A_2 at frequencies p^2, $2pq$, and q^2, respectively. (See Box 2.2 and Chapter 17.)

2.1.2 Genotypic values Next, we need a way to describe any effects of the alleles at a locus on whatever trait we are interested in. A locus is said to be *associated with* a trait if some of its alleles are associated with different mean levels of that trait in the population. For qualitative diseases (i.e., diseases that are either present or not present), a single allele may be necessary and sufficient to develop the disease. In this case, the disease-predisposing allele acts in either a dominant or a recessive manner. Carrying a dominant allele will result in the disease irrespective of the other allele at that locus; conversely, if the disease-predisposing allele is recessive, then the disease will only develop in individuals homozygous for that allele.

For a quantitative trait, however, we need some way of specifying *how much* an allele affects the trait. Considering only a locus with two alleles, A_1 and A_2, we define the average value of one of the homozygotes (say, A_1A_1) as a and the average value of the other homozygote (A_2A_2) as $-a$. The value of the heterozygote (A_1A_2) is labeled d and is dependent on the mode of gene action. If there is no dominance, d will be zero (i.e., the midpoint of the two homozygotes' scores). If the A_1 allele is dominant to A_2, then d will be greater than zero. If dominance is complete (i.e., if the observed value for A_1A_2 equals that of A_1A_1), then $d = +a$.

2.1.3 Additive effects Observed genotypic values for a single locus can be defined in terms of an *additive genetic value* and a *dominance deviation*. The additive genetic value of a locus relates to the average effect of an allele. As illustrated in Figure A.7, the additive

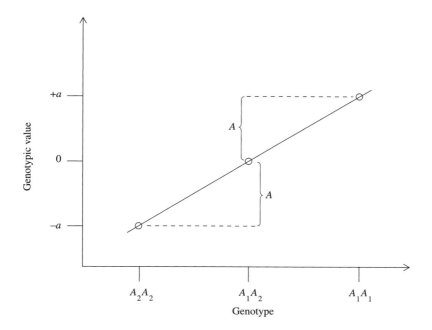

Figure A.7 Additive genetic values. The number of A_1 alleles predicts additive genetic values. Because there is no dominance (and assuming equal allelic frequency), the additive genetic values equal the genotypic values. (A, value added by each A_1 allele.)

genetic value is the genotypic value expected from the number of a particular allele (say, A_1) at that locus, either 0, 1, or 2 (each A_1 allele increases an individual's score by A units).

Additive genetic values are important in behavioral genetics because they represent the extent to which genotypes "breed true" from parents to offspring. If a parent has one copy of a certain allele, say, A_1, then each offspring has a 50 percent chance of receiving an A_1 allele. If an offspring receives an A_1 allele, then its additive effect will contribute to the phenotype to exactly the same extent as it did to the parent's phenotype. That is, it will lead to increased parent-offspring resemblance on the phenotype, irrespective of other alleles at that locus or at other loci.

2.1.4 Dominance deviation

Dominance is the extent to which the effects of alleles at a locus do not simply "add up" to produce genotypic values. The *dominance deviation* is the difference between actual genotypic values and what would be expected under a strictly additive model. Figure A.8 represents the deviations (labeled D) of the expected (or additive) genotypic values from the actual genotypic values that occur if there is an effect of dominance at the locus.

Dominance genetic variance represents genetic influence that does not "breed true." Saying that the effect of a locus involves dominance is equivalent to saying that an individual's genotypic value results from the *combination* of alleles at that particular locus. However, offspring receive only one allele from each parent, not a combination of two

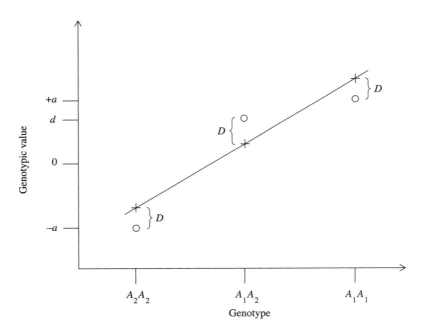

Figure A.8 Dominance deviations. The genotypic values (circles) deviate from the expected values under an additive model (crosses) when there is dominance (i.e., $d \neq 0$). (D, deviation from expected attributed to dominance.)

alleles. Genetic influence due to dominance will not be transmitted from parent to off-spring, therefore. In this way, additive and dominance genetic values are defined so as to be independent of each other.

2.1.5 Polygenic model Not only can we consider the additive and nonadditive effects at a single locus, we can also sum these effects across loci. This concept is the essence of the polygenic extension of the single-gene model. Just as additive genetic values are the summation of the average effects of two alleles at a single locus, they can also be summed across the many loci that may influence a particular phenotypic character. Similarly, dominance deviations from additive genetic values can also be summed for all the loci influencing a character. Thus, it is relatively easy to generalize the single-gene model to a polygenic one with many loci, each with its own additive and nonadditive effects. Under an *additive* polygenic model, the genetic effect G on the phenotype represents the sum of effects from different loci.

$$G = G_1 + G_2 + \ldots + G_N$$

This expression implies that the effects of different alleles simply add up—that is, there is no interaction between alleles where the effect of one allele, say, G_1 is modified by the presence of the allele with effect G_2. The polygenic model needs to consider the possibility that the effects of different loci do not add up independently but interact with each other—an interaction called *epistasis*. For example, imagine two loci, each with an allele that increases an individual's score by one point on a particular trait. If there were no epistasis, having a risk allele at both loci would increase the score by two points. If there were epistasis, however, having risk alleles at both loci might possibly lead to a ten-point increase. Epistasis therefore complicates analysis, but there is evidence that such phenomena might be quite prevalent for certain complex traits. In other words, dominance is *intralocus* interaction between alleles, whereas epistasis is *interlocus* interaction, that is, between loci.

The total genetic contribution to a phenotype is G, which is the sum of all additive genetic effects A, all dominance deviations D, and all epistatic interaction effects I:

$$G = A + D + I$$

2.1.6 Phenotypic values and variance components model Quantitative genetic theory states that every individual's phenotype is made up of genetic and environmental contributions. No behavioral phenotype will be entirely determined by genetic effects, so we should always expect an environmental effect, E, which also includes measurement error, on the phenotype P. In algebraic terms,

$$P = G + E$$

where, for convenience, we assume that P represents an individual's deviation from the population mean rather than an absolute score. In any case, behavioral genetics is not primarily interested in the score of any one individual. Rather, the focus is on explaining the causes of phenotypic differences in a population—why some individuals are more extraverted than others, or why some individuals are alcoholic, for example.

In fact, there is often no direct way of determining the relative magnitude of genetic and environmental deviations for any one individual, certainly if one has not obtained DNA from individuals. However, in a sample of individuals, especially of genetically related individuals, it is possible to estimate the variances of the terms *P*, *G*, and *E*. This approach is called the *variance components approach*, and it relies on the equation that showed us how to calculate the variance of a sum.

Recall that

$$\text{Var}(X + Y) = \text{Var}(X) + \text{Var}(Y) + 2\text{Cov}(X,Y)$$

Turning this expression around, it gives us a method for partitioning the variance of a variable that is a composite of constituent parts. That is, our goal is to "decompose the variance" of a trait into the constituent parts of genetic and environmental sources of variation.

For simplicity, we will assume no epistasis, so $P = G + E = A + D + E$. The variance of *P* is equal to the sum of the variance of the separate components, *A*, *D*, and *E*, plus twice the covariance between them:

$$\begin{aligned}\text{Var}(P) &= \text{Var}(A + D + E) \\ &= \text{Var}(A) + \text{Var}(D) + \text{Var}(E) + 2\text{Cov}(A, D) + 2\text{Cov}(A, E) + 2\text{Cov}(D, E)\end{aligned}$$

which begins to look unmanageable until we realize that we can use some theoretical assumptions of our model to constrain this equation. By definition, additive genetic influences are independent of dominance deviations. That is, Cov(*A*, *D*) will necessarily equal zero, so this term can be dropped from the model. Another assumption that we may wish to make (but one that is not necessarily true) is that genetic and environmental influences are uncorrelated. This is equivalent to saying that Cov(*A*, *E*) and Cov(*D*, *E*) equal zero and can be dropped from the model. We will see later that there are detailed reasons why this assumption might not hold (what is called a *gene-environment correlation*) (see also Chapter 16). For the time being, however, our simplified model reads:

$$\text{Var}(P) = \text{Var}(A) + \text{Var}(D) + \text{Var}(E)$$

A note on notation: Variances are often written in other ways. For example, above we have denoted additive genetic variance as Var(*A*). As we will see, this term is often written differently, depending on the context (mainly for historical reasons). In formal model fitting, the lowercase Greek letter sigma squared with a subscript (σ_A^2) might be used. A similar value calculated in the context of comparing familial correlations (narrow-sense heritability, introduced later) is typically labeled h^2, whereas it is written as a^2 in the context of path analysis (also introduced later). Under most circumstances, however, these all refer to roughly the same thing.

In conclusion, it might not seem that we have achieved very much in simply considering variances instead of values. However, as will be discussed, quantitative genetic methods can use these models to estimate the relative contribution of genetic and environmental influences to phenotypic variance.

2.1.7 Environmental variation   Because the nature of environmental effects is more varied and changeable than the underlying nature of genetic influence, it is not possible to decompose this term into constituent parts in a straightforward way. That is, if we detect genetic influence, then we know that this effect must result from at least one gene—and we know something about the properties of genes.

However, if we detect environmental influence on a trait, we cannot assume any one mechanism. But behavioral genetics is able to investigate environmental influences in two main ways. As we will see later, family-based studies using twins or adoptive relatives allow environmental influences to be partitioned into those shared between relatives (i.e., those that make relatives resemble each other) and those that are nonshared (i.e., those that do not make relatives resemble each other). This type of analysis is not at the level of specific, measured environmental variables.

A second approach is to actually measure a specific aspect of the environment (e.g., parental socioeconomic status, or nutritional content of diet) and incorporate it into genetic analysis. For example, we may wish to partition out the variation in the trait due to a measured environmental source if we consider it to represent a cause of *nuisance* or *noise* variance in the trait (i.e., to treat it as a covariate). Alternatively, we may believe that an environment is important in the expression of genetic influence. For example, we might suspect that stress might bring out genetic vulnerabilities toward depression. Therefore, depression might be expected to show greater genetic influence for individuals experiencing stress. In this case, we would not want to adjust for the effects of the environmental variable. Such a circumstance is named a gene-environment interaction ($G \times E$ interaction). In terms of the quantitative genetic model,

$$P = G + E + (G \times E)$$

where $(G \times E)$ does not necessarily represent a multiplication effect but rather any interactive effect of genes and environment that is independent of their main effects.

2.2 Estimating variance components

In the previous section we outlined a simple biometric model, describing the variation in observed phenotypes in terms of various genetic and environmental sources of variation. In this section, we consider how we can use family data to estimate some of the key parameters of such models, with a focus on heritability as estimated from the classical twin study, introducing maximum likelihood estimation and model fitting.

2.2.1 Genes and families   Until now, we have built a general genetic model of the etiology of variation in a trait among individuals. A major step in quantitative genetics is to incorporate knowledge of basic laws of heredity to allow us to extend our model to include the covariance between relatives. Conceptually, most behavioral genetic analysis contrasts phenotypic similarity between related individuals (which is measured) with their genetic similarity (which is known from genetics). If individuals who are more closely related genetically also tend to be more similar on a measured trait, then this tendency is evidence for that trait being heritable—that is, the trait is at least partially influenced by genes.

When we study families, we are not only interested in the variance of a trait—the main focus is on the covariance between relatives. Earlier we saw how we can study two variables, such as height and weight, and ask whether they are associated with each other. In a similar way, covariances and correlations can also be used to ask whether a single variable is associated between family members. For example, do brothers and sisters tend to be similar in height or not? If we measured height in sibling pairs, we could calculate the covariance between an individual's height and the sibling's height. If the covariance equaled zero, this would imply that brothers and sisters are no more likely to have similar heights than any two unrelated individuals picked at random from the population. If the covariance is greater than zero, this would imply that taller individuals tend to have taller brothers and sisters. Quantitative genetic analysis attempts to determine the factors that can make relatives similar—their shared nature or their shared nurture.

2.2.2 *Genetic relatedness in families*   An individual has two copies of every gene, one paternally inherited and one maternally inherited. When an individual passes one copy of each gene to its offspring, there is an equal chance that either the paternally inherited gene or the maternally inherited gene will be transmitted. From these two simple facts, we can calculate the expected proportion of gene sharing between individuals of different genetic relatedness. Siblings who share both biological parents will share either zero, one, or two alleles at each locus. For autosomal loci, there is a 50 percent chance that siblings will share the same paternal allele (two ways of sharing, two ways of not sharing, all with equal probability) and, correspondingly, a 50 percent chance of sharing the same maternal allele. Therefore, siblings stand a $0.50 \times 0.50 = 0.25$ (25 percent) chance of sharing both paternal and maternal alleles; a $(1.00 - 0.50) \times (1.00 - 0.50) = 0.25$ (25 percent) chance of sharing no alleles; a $1.00 - 0.25 - 0.25 = 0.50$ (50 percent) chance of sharing one allele. The average, or expected, alleles shared is therefore $(0 \times 0.25) + (1 \times 0.5) + (2 \times 0.25) = 1$. Therefore, in the average case, siblings will share half of the additive genetic variation that could potentially contribute to phenotypic variation because they share one out of two alleles. Because siblings stand only a 25 percent chance of sharing *both* alleles, in the average case, siblings will share a quarter of the dominance genetic variation that could potentially contribute to phenotypic variation.

For other types of relatives, we can work out their expected genetic relatedness in terms of genetic components of variance. Parent-offspring pairs always share precisely one allele: They will share half of the additive genetic effects that contribute to variation in the population but none of the dominance genetic effects. Half siblings, who have only one parent in common, share only a quarter of additive genetic variance but no dominance variance (because they can never inherit two alleles at the same locus from the same parent).

The majority of behavioral genetic studies focus on twins. Genetically, full sibling pairs and DZ twin pairs are equivalent. So, whereas DZ twins will only share half the additive genetic variance and one-fourth of the dominance variance, MZ twins share all their genetic makeup, so additive and dominance genetic variance components will be completely shared.

These coefficients of genetic relatedness are summarized in Table A.1. Sharing additive and dominance genetic variance contributes to the phenotypic correlation between relatives. As mentioned earlier, correlations between relatives directly estimate the proportion of variance shared between them. So we can think of the familial correlation as the sum of all the shared components of variance between two relatives.

Not only genes are shared between most relatives, however. Individuals that are genetically related are more likely to experience similar environments than unrelated individuals are. If an environmental factor influences a variable, then sharing this environment will also contribute to the phenotypic correlation between relatives. As explained in Chapter 16, behavioral genetics conceptually divides environmental influences into two

TABLE A.1

Coefficients of Genetic Relatedness

Related Pair	Proportion of Additive Genetic Variation Shared	Proportion of Dominance Genetic Variation Shared
Parent and offspring (PO)	$\frac{1}{2}$	0
Half siblings (HS)	$\frac{1}{4}$	0
Full siblings (FS)	$\frac{1}{2}$	$\frac{1}{4}$
Nonidentical twins (DZ)	$\frac{1}{2}$	$\frac{1}{4}$
Identical twins (MZ)	1	1

distinct types with regard to their impact on families. Environments that are shared by family members *and* that tend to make members more similar on a particular trait are called *shared environmental* influences. In contrast, *nonshared environmental* influences do not result in family members becoming more alike for a given trait.

Most behavioral genetic analysis focuses on three components of variance: additive genetic, shared environmental, and nonshared environmental. As we will see, this tripartite approach underlies the estimation of heritability by comparing twin correlations and is the basic model used in more sophisticated model-fitting analysis. This model is often referred to as the ACE model. (A stands for additive genetic effects, C for common (shared) environment, and E for nonshared environment.)

2.2.3 Heritability As explained in Chapter 5, heritability is the proportion of phenotypic variance that is attributable to genotypic variance. There are two types of heritability: *broad-sense heritability* refers to all sources of genetic variance, whether the genes operate in an additive manner or not. *Narrow-sense heritability* refers only to the proportion of phenotypic variance explained by additive genetic effects. Narrow-sense heritability therefore gives an indication of the extent to which a trait will "breed true"—that is, the degree of parent-offspring similarity that is expected. Broad-sense heritability, on the other hand, gives an indication of the extent to which genetic factors of any kind are responsible for trait variation in the population.

We are able to estimate the heritability of a trait by comparing correlations between certain types of family members. For simplicity, we will assume that the only influences on a trait are additive genetic effects and environmental effects that are either shared or nonshared between family members. We can describe the correlation we observe between different types of relatives in terms of the components of variance they share. For example, we expect the correlation between full siblings to represent half the additive genetic variance and, by definition, all the shared environmental variance but none of the nonshared environmental variance. As mentioned earlier, additive genetic variance is typically labeled h^2 in this context (representing narrow-sense heritability). The shared environmental variance is labeled c^2 (nonshared environment is e^2). Therefore,

$$r_{FS} = \frac{h^2}{2} + c^2$$

Suppose we observed for full siblings a correlation of .45 for a trait. We would not be able to work out what h^2 and c^2 are from this information alone because, as reflected in the equation above, nature and nurture are shared by siblings. However, by comparing sets of correlations between certain different types of relatives, we are able to estimate the relative balance of genetic and environmental effects. The most common study design uses MZ and DZ twin pairs. The correlations expressed in terms of shared variance components are therefore

$$r_{MZ} = h^2 + c^2$$

$$r_{DZ} = \frac{h^2}{2} + c^2$$

Subtracting the second equation from the first gives

$$r_{MZ} - r_{DZ} = h^2 - \frac{h^2}{2} + c^2 - c^2$$

$$= \frac{h^2}{2}$$

$$h^2 = 2(r_{MZ} - r_{DZ})$$

That is, narrow-sense heritability is calculated as twice the difference between the correlations observed for MZ and DZ twin pairs. The proportion of variance attributable to shared environmental effects can easily be estimated as the difference between the MZ correlation and the heritability ($c^2 = r_{MZ} - h^2$). Because we have estimated these two variance components from correlations, which are standardized, h^2 and c^2 represent *proportions* of variance. The final component of variance we are interested in is nonshared environmental variance, e^2. This statistic does not appear in the equations describing the correlations between relatives, of course. However, we know that h^2, c^2, and e^2 must sum to 1 if they represent proportions, so

$$h^2 + c^2 + e^2 = 1$$

$$[2(r_{MZ} - r_{DZ})] + [r_{MZ} - 2(r_{MZ} - r_{DZ})] + e^2 = 1$$

$$\therefore r_{MZ} + e^2 = 1$$

$$\therefore e^2 = 1 - r_{MZ}$$

This conclusion is intuitive: Because MZ twins are genetically identical, any variance that is not shared between them (i.e., the extent to which the MZ twin correlation is not 1) must be due to nonshared environmental sources of variance.

Let's consider an example: Suppose we observe a correlation of .64 in MZ twins and .44 in DZ twins. Taking twice the difference between the correlations, we can conclude that the trait has a heritability of 0.4 [= 2 × (.64 − .44). That is, 40 percent of variation in the population from which we sampled is attributable to the additive effects of genes. The shared family environment therefore accounts for 24 percent ($c^2 = 0.64 - 0.4 = 0.24$) of the variance; the nonshared environment accounts for 36 percent ($e^2 = 1 - 0.64 = 0.36$).

A pattern of results such as those just described would suggest that genes play a significant role in individual differences for this trait, differences between people being roughly half due to nature, half due to nurture. We have made several assumptions, however, in order to arrive at this conclusion. These assumptions will be considered more fully in the context of model fitting, but we will mention two immediate assumptions. First, we have assumed that dominance is not important for this trait (not to mention other more complex interactions such as epistasis). We have assumed that all genetic effects are additive (which is why h^2 represents narrow-sense heritability). If this assumption were not true, the heritability estimate would be biased. Second, we have assumed that MZ and DZ twins only differ in terms of the genetic relatedness. That is, the same shared environment term, c^2, appears in both MZ and DZ equations. If parents treat identical twins more similarly than they treat nonidentical twins, this assumption could result in higher MZ correlations relative to DZ correlations. This assumption, which is in theory testable, is called the *equal environments assumption* (see Chapter 5). Violations of this assumption would overestimate the importance of genetic effects.

Other types of relatives can be studied to calculate heritability; for example, we could compare correlations for full siblings and half siblings. Not all comparisons will be informative, however. Comparing the correlation for full siblings and the correlation for parent and offspring will not help to estimate heritability (because these relatives do not differ in terms of shared additive genetic variance). It is preferable to study twins for several reasons. It can be shown that for statistical reasons, twins afford greater accuracy in determining heritability because larger proportions of variance are shared by MZ twins. Additionally, twins are more closely matched for age, familial, and social influences than are half siblings or parents and offspring. The interpretation of the shared environment is much less clear for parents and offspring.

Quantitative genetic studies can also contrast family members who are genetically similar but have not shared any environmental influences. This comparison is the basis of the adoption study. The simplest form of adoption study is that of MZ twins reared apart. Because MZ twins reared apart are genetically identical but do not share any environmental influences, the correlation directly estimates heritability. That is, if there has been no selective placement, any tendency for MZ twins reared apart to be similar must be attributable to the influences of shared genes.

2.2.4 Model fitting and the classical twin design   Simple comparisons between twin correlations can indicate whether genetic influences are important for a trait. This is the important first question that any quantitative genetic analysis must ask. Here we will examine some of the more formal statistical techniques that can be used to analyze genetically informative data and to ask other, more involved questions.

Model-fitting approaches involve constructing a model that describes some observed data. In the quantitative genetic studies, the observed data that we model are typically the variance-covariance matrices for family members. The model will then consist of a variance-covariance matrix formulated in terms of various *parameters*. These will typically be the variance components (additive genetic and so on) we encountered earlier. Various combinations of different values for the model parameters will generate different expected variance-covariance matrices. The goal of model fitting is twofold: (1) to select the model with the smallest number of parameters that (2) generates expectations that match the observed data as closely as possible. As we will see, there is a payoff

CLOSE **UP**

Michael C. Neale, Ph.D., is a professor in the Departments of Psychiatry and Human Genetics, Virginia Commonwealth University, and professor of Statistical Methods for Behavioral Genetics at the Free University of Amsterdam. He is co-director of the Virginia Institute for Psychiatric and Behavioral Genetics, and director of its training program. Born and educated in England, Dr. Neale trained in behavioral genetics as a graduate student with Dr. David Fulker and as a postdoctoral trainee with Dr. Lindon Eaves. In 1990, Neale began the development of the statistical modeling software *Mx*, which has been extensively used in a wide variety of genetic studies to estimate heritability, genotype × environment interaction, genetic linkage, and association. His motivation to develop software was partly due to frustration with the tools of the time and partly the desire to simplify complex data analysis tasks for the average user, while providing a framework with which more advanced analyses could be undertaken. Much of his effort is spent on model development, where the objective is to establish (a) whether a hypothesis can be tested with available data, and (b) if so, how to do so most efficiently. Recent methodological work has focused on the assessment of phenotypes, particularly in the areas of psychiatric disorders and substance use.

between the number of parameters in a model and the accuracy with which it can model the observed data.

If we were to fit the ACE model to observed MZ and DZ twin data, the three parameter estimates selected to match the expected variance-covariance matrices with the observed ones would correspond directly to the estimates of heritability, and of shared and nonshared environmental influences that we calculated earlier in a relatively straightforward manner. Why would we ever want to perform more complicated model fitting? There are several good reasons: First, these calculations are only valid *if* the ACE model is a true reflection of reality. Model fitting allows different types of models to be explicitly tested and compared. Model fitting also facilitates the calculation of confidence intervals around the parameter estimates. It is common to read something such as "$h^2 = .35$ (0.28 − 0.42)," which means that the heritability was estimated at 35 percent, but there is a 95 percent chance that, even if it is not exactly 35 percent, it at least lies within the range of 28 to 42 percent. Model fitting can also incorporate many different types of family structures, model multivariate data, and include any *measured* genetic or environmental information we may have, in order to improve our estimates and explore potential interactions of genetic and environmental effects, or to test whether specific loci are associated with the trait or not.

Let's start from basics. Imagine that we have measured a trait in a population of twins. We have not measured any DNA, nor have we measured any other environmental factors that might influence the trait. We summarize our data as two variance-covariance matrices, one for MZ twin pairs and one for DZ twins pairs; so our "observed data" are six unique statistics:

$$\begin{bmatrix} \text{Var}_1^{MZ} & \\ \text{Cov}_{12}^{MZ} & \text{Var}_2^{MZ} \end{bmatrix}$$

$$\begin{bmatrix} \text{Var}_1^{DZ} & \\ \text{Cov}_{12}^{DZ} & \text{Var}_2^{DZ} \end{bmatrix}$$

Using our knowledge of the quantitative genetic model as outlined earlier, we can begin to construct a model that describes the two variance-covariance matrices for the twins. That is, we assume that observed trait variation is due to a certain mixture of additive genetic, dominance genetic, shared environmental, and nonshared environmental effects (we will ignore epistasis and other interactions).

Model fitting begins by creating an explicit model for the variance-covariance matrix for families, in terms of genetic and environmental variance components. Returning to the basic genetic model, phenotype, P, is a function of additive, A, and dominance, D, genetic effects. Additionally, we include environmental effects, which are either shared, C, or nonshared, E. (*Note:* The basic model did not make this distinction because it is primarily formulated to describe variation in a population of *unrelated* individuals, i.e., E referred to *all* environmental effects.)

$$P = A + D + C + E$$

In terms of variances, therefore, remembering all the assumptions outlined under the single-gene model that apply at this step (no gene-environment correlation, for example), we obtain

$$\sigma_P^2 = \sigma_A^2 + \sigma_D^2 + \sigma_C^2 + \sigma_E^2$$

where, using the model-fitting notation, $\sigma_{A/D/C/E}^2$ (pronounced "sigma") stands for the components of variance associated with the four types of effect and σ_P^2 is the phenotypic variance.

To construct our twin model, we need to explicitly write out every element of the variance-covariance matrices in terms of the parameters of the model. We have already defined the trait variance in terms of the variance components:

$$\sigma_A^2 + \sigma_D^2 + \sigma_C^2 + \sigma_E^2$$

We will write this term for all four variance elements in the model. Note that we are modeling variances and covariances instead of correlations; this is often done in model fitting because it captures more information (the variance and covariance) than a correlation does. The σ_A^2 parameter will not directly estimate narrow-sense heritability—we need to divide the additive genetic variance component by the total variance:

$$\sigma_A^2 / (\sigma_A^2 + \sigma_D^2 + \sigma_C^2 + \sigma_E^2)$$

We make the assumption that components of variance are identical for all individuals. That is, we write the same expression for all four variance elements. This assumption implies that the effects of genes and environments on an individual are not altered by that individual being a member of an MZ or DZ twin pair. Additionally, it assumes that individuals were not assigned a Twin 1 or Twin 2 label in a way that might make Twin 1's variance differ from Twin 2's variance. For example, if the first-born twin was always coded as Twin 1, then, depending on the nature of the trait, this assumption might not be warranted. (This problem is sometimes avoided by "double-entering" twin pairs so that each individual is entered twice, once as Twin 1 and once as Twin 2, when calculating the observed variance-covariance matrices. This method will, of course, ensure that Twin 1 and Twin 2 have equal variances.)

The covariance term between twins is also a function of the components of variance, in terms of the extent to which they are shared between twins, as stated earlier. All additive and dominance genetic variance, as well as shared environmental variance, is shared by MZ twins. These components contribute to the covariance between MZ twins fully. DZ twins share one-half the additive genetic variance, one-fourth the dominance genetic variance, all the shared environmental variance, and none of the nonshared environmental variance. The contributions of these components to the DZ covariance are in proportion to these coefficients of sharing.

Therefore, for MZ twin pairs, the variance-covariance matrix is modeled as

$$\begin{bmatrix} \sigma_A^2 + \sigma_D^2 + \sigma_C^2 + \sigma_E^2 & \\ \sigma_A^2 + \sigma_D^2 + \sigma_C^2 & \sigma_A^2 + \sigma_D^2 + \sigma_C^2 + \sigma_E^2 \end{bmatrix}$$

whereas, for DZ twins, it is

$$\begin{bmatrix} \sigma_A^2 + \sigma_D^2 + \sigma_C^2 + \sigma_E^2 & \\ \dfrac{\sigma_A^2}{2} + \dfrac{\sigma_D^2}{4} + \sigma_C^2 & \sigma_A^2 + \sigma_D^2 + \sigma_C^2 + \sigma_E^2 \end{bmatrix}$$

These two matrices represent our model. Different values of σ_A^2, σ_D^2, σ_C^2, and σ_E^2 will result in different *expected* variance-covariance matrices. These matrices are "expected," in the sense that, *if* the values of the model parameters were true, then these are the averaged matrices we would expect to observe if we repeated the experiment a very large number of times.

As an example, consider a trait with a variance of 5. Imagine that variation in this trait was entirely due to an equal balance of additive genetic effects and nonshared environmental effects. In terms of the model, this assumption is equivalent to saying that σ_A^2 and σ_E^2 both equal 2.5, whereas σ_D^2 and σ_C^2 both equal 0. If this were true, then what variance-covariance matrices would we *expect* to observe for MZ and DZ twins? Simply substituting these values, we would expect to observe for MZ twins,

$$\begin{bmatrix} 2.5 + 0 + 0 + 2.5 & \\ 2.5 + 0 + 0 & 2.5 + 0 + 0 + 2.5 \end{bmatrix} = \begin{bmatrix} 5 & \\ 2.5 & 5 \end{bmatrix}$$

and for DZ twins,

$$\begin{bmatrix} 2.5 + 0 + 0 + 2.5 \\ \dfrac{2.5}{2} + \dfrac{0.0}{4} + 0 \quad 2.5 + 0 + 0 + 2.5 \end{bmatrix} = \begin{bmatrix} 5 \\ 1.25 \quad 5 \end{bmatrix}$$

To recap, we have seen how a specific set of parameter values will result in a certain expected set of variance-covariance matrices for twins. This result is, in itself, not very useful. We do not know the true values of these parameters—these are the very values we are trying to discover! Model fitting helps us to estimate the parameter values most likely to be true by evaluating the expected values produced by very many sets of parameter values. The set of parameter values that produces expected matrices that most closely match the observed matrices are selected as the *best-fit parameter estimates*. These represent the best estimates of the true parameter values. Because of the iterative nature of model fitting (evaluating very many different sets of parameter values), it is a computationally intensive technique that can only be performed by using computers.

2.2.5 An example of the model-fitting principle   Suppose that, for a certain trait, we observe the following variance-covariance matrices for MZ and DZ pairs, respectively (note that the observed variances are similar although not identical):

$$\begin{bmatrix} 2.81 \\ 2.13 \quad 3.02 \end{bmatrix}$$

$$\begin{bmatrix} 3.17 \\ 1.54 \quad 3.06 \end{bmatrix}$$

The model fitting would start by substituting *any* set of parameters to generate the expected matrices. Suppose we substituted $\sigma_A^2 = 0.7$, $\sigma_D^2 = 0.2$, $\sigma_C^2 = 1.2$, and $\sigma_E^2 = 0.8$. These values only represent a "first guess" that will be evaluated and improved on by the model-fitting process. These values imply that 24 percent $[0.7/(0.7 + 0.2 + 1.2 + 0.8)]$ of phenotypic variation is attributable to additive genetic effects. If these were the true values, the variance-covariance matrix we would expect to observe for MZ twins is

$$\begin{bmatrix} 0.7 + 0.2 + 1.2 + 0.8 \\ 0.7 + 0.2 + 1.2 \quad 0.7 + 0.2 + 1.2 + 0.8 \end{bmatrix} = \begin{bmatrix} 2.9 \\ 2.1 \quad 2.9 \end{bmatrix}$$

whereas, for DZ twins, it is

$$\begin{bmatrix} 0.7 + 0.2 + 1.2 + 0.8 \\ \dfrac{0.7}{2} + \dfrac{0.2}{4} + 1.2 \quad 0.7 + 0.2 + 1.2 + 0.8 \end{bmatrix} = \begin{bmatrix} 2.9 \\ 1.6 \quad 2.9 \end{bmatrix}$$

Comparing these expectations with the observed statistics, we can see that they are numerically similar but not exactly the same. We need an exact method for determining *how good* the fit between the expected and observed matrices is. Model fitting can therefore proceed, changing the parameter values to increase the *goodness of fit* between the model-dependent expected values and the sample-based observed values. When a set of values has been found that cannot be beaten for goodness of fit, these will be presented as the "output" from the model-fitting programs, the best-fit estimates. This process is called *optimization*. It would be very inefficient to evaluate *every* possible set of parameter values. For most models, evaluating every set would in fact be virtually impossible, given current computing technology. Rather, optimization will try to change the parameters in an intelligent way. One way of thinking about this process is as a form of a "hotter-colder" game: The aim is to increasingly refine your guess as to where the hidden object is, rather than exhaustively searching every inch of the room.

There are many indices of fit—one simple one is the chi-squared (χ^2, pronounced "ki," as in *kite*) goodness-of-fit statistic. This statistic essentially evaluates the magnitude of the discrepancies between expected and observed values by comparing how likely the observed data are under the model. The χ^2 goodness-of-fit statistic can be formally tested for significance in order to indicate whether or not the model provides a good approximation of the data. If the χ^2 goodness-of-fit statistic is low (i.e., nonsignificant), it indicates that the observed values *do not significantly deviate* from the expected values. However, a low χ^2 value does not necessarily mean that the parameter values being tested are the best-fit estimates. As we have mentioned, different values for the four parameters might provide a better fit (i.e., an even lower χ^2 goodness-of-fit statistic).

Just because we can write down a model that we believe to be an accurate description of the real-world processes affecting a trait, it does not necessarily mean that we can derive values for its parameters. In the preceding example, we would not be able to estimate the four parameters (additive and dominance genetic variances, shared and nonshared environmental variances) from our twin data. In simple terms, we are asking too many questions of too little information.

Consider what happens when we change the parameter values to see whether we can improve the fit of the model. Try substituting $\sigma_A^2 = 0.1$, $\sigma_D^2 = 0.6$, $\sigma_C^2 = 1.4$, and $\sigma_E^2 = 0.8$ instead, and you will notice that we obtain the same two expected variance-covariance matrices for both MZ and DZ twins as we did under the previous set of parameters. Both sets of parameters would therefore have an identical fit, so we would not be able to distinguish these two alternative explanations of the observations. This phenomenon can make model fitting very difficult or even impossible. This is an instance of a model not being *identified*.

2.2.6 The ACE model   Although we will not follow the proof here, researchers have demonstrated that we cannot ask about additive genetic effects, dominance genetic effects, *and* shared environmental effects simultaneously if the only information we have is from MZ and DZ twins reared together.

In virtually every circumstance, we will wish to retain the nonshared environmental variance component in the model. We wish to retain it partly because random measurement error is modeled as a nonshared environmental effect and we do not wish to have a model that assumes no measurement error (it is unlikely to fit very well). Most commonly, we would then model additive genetic variance and shared environmental variance. As mentioned earlier, such a model is called the ACE model.

If we had reason to suspect that dominance genetic variance might be affecting a trait, then we might fit an ADE model instead. If the MZ twin correlation is more than twice the DZ twin correlation, one explanation is that dominance genetic effects play a large role for that trait (an explanation that might suggest fitting an ADE model).

The ACE model (and the ADE model) is an identified model. That is, the best fit between the expected and observed matrices is produced by one and only one set of parameter values. As long as the twin covariances are both positive and the MZ covariance is not smaller than the DZ covariance (both of which are easily justified biologically as reasonable demands), the ACE model will always be able to select a unique set of parameters that best account for the observed statistics.

If we were to model standardized scores (so that differences in the observed variance elements could not reduce fit), then under the ACE model the best-fitting parameters will always have a χ^2 goodness of fit of precisely zero. Such a model is called a *saturated* model. Imagine that, for a standardized trait (i.e., one with a variance of 1), we found an MZ covariance of 0.6 (this can be considered as the MZ twin correlation, of course) and a DZ covariance of 0.4. There is, in fact, one and only one set of values for the three parameters of the ACE model that will produce expected values that exactly match these observed values. In this case, these are $\sigma_A^2 = 0.4$, $\sigma_C^2 = 0.2$, and $\sigma_E^2 = 0.4$. Substituting these into the model, we obtain for MZ twins,

$$\begin{bmatrix} 0.4 + 0.2 + 0.4 & \\ 0.4 + 0.2 & 0.4 + 0.2 + 0.4 \end{bmatrix} = \begin{bmatrix} 1.0 & \\ 0.6 & 1.0 \end{bmatrix}$$

and for DZ twins,

$$\begin{bmatrix} 0.4 + 0.2 + 0.4 & \\ \dfrac{0.4}{2} + 0.2 & 0.4 + 0.2 + 0.4 \end{bmatrix} = \begin{bmatrix} 1.0 & \\ 0.4 & 1.0 \end{bmatrix}$$

There are no other values that σ_A^2, σ_C^2, and σ_E^2 can take to produce the same expected variance-covariance matrices. This property does not mean that these values will necessarily reflect the true balance of genetic and environmental effects—they will only reflect the true values if the model (ACE or ADE or whatever) is a good one. All parameter estimates are model dependent: We can only conclude that, *if* the ACE model is a good model, then this result is the balance of genetic and environmental effects. We are able to test different models relative to one another, however, in order to get a sense of whether or not the model is a fair approximation of the underlying reality. We can only compare models if they are *nested*, however. A model is nested in another model if and only if that model results from constraining to zero one or more of the variance components in the larger model. For example, we may suspect that the shared environment plays no significant role for a given trait. We can test this supposition by fixing the shared environment variance component to zero and comparing the fit of the full model with the fit of this reduced model. Nesting is important because it forms the basis for testing and selecting between different models of our data.

A general principle of science is parsimony: to always prefer a simpler theory if it accounts equally well for the observations. This concept, often referred to as *Occam's razor*, is explicit in model fitting. Having derived estimates for genetic and environmental variance components under an ACE model, we might ask whether we could drop the shared environment term from the model. Might our simpler AE model provide a comparable fit to the data? Instead of estimating the shared environment variance component, we assume that it is zero (which is equivalent to ignoring it or removing it from the model). The AE model is therefore nested in the ACE model. We are able to calculate the goodness of fit of the ACE model, which estimates three parameters to explain the data, and the goodness of fit for the AE model, which only estimates two parameters to explain the same data. Any model with fewer parameters will not fit as well as a sensible model with more parameters. The question is whether or not the reduction in fit is *significantly* worse relative to the "advantage" of having fewer parameters in a more parsimonious model.

In our example, the ACE model will estimate $\sigma_A^2 = 0.4$, $\sigma_C^2 = 0.2$, and $\sigma_E^2 = 0.4$. As we saw earlier, substituting these values and *only* these values will produce expected variance-covariance matrices that match the observed perfectly (because we are modeling standardized scores, or correlations). In contrast, consider what happens under the AE model with the same data. Table A.2 shows that the AE model is unable to account for this particular set of observed values. Such a model is said to be *underidentified*. This condition is not necessarily problematic: In general, underidentified models are to be favored. Because a saturated model will *always* be able to fit the observed data perfectly, the goodness of fit does not really mean anything. However, if an underidentified model *does* fit the data, then we should take notice—it is not fitting out of mere statistical necessity. Perhaps it is a better, more parsimonious model of the data. Table A.2 represents three different sets of the values for the two parameters that attempt to explain the observed data. As the table shows, the AE model does not seem able to model our observed statistics quite as well as the ACE model.

If we run a model-fitting program such as *Mx*, we can formally determine which values for σ_A^2 and σ_E^2 give the best fit for the AE model and whether or not this fit is significantly worse than that of the saturated ACE model. Additionally, we can fit a CE

TABLE A.2

Fit of AE Model to Three Parameter Value Sets

Parameters			MZ	DZ
σ_A^2	σ_E^2	Variance	Covariance	Covariance
OBSERVED				
—	—	1.0	0.6	0.4
EXPECTED				
0.6	0.4	1.0	0.6	0.3
0.7	0.3	1.0	0.7	0.35
0.8	0.2	1.0	0.8	0.4

TABLE A.3

Best-Fit Univariate Parameter Estimates

Parameters		Variance	MZ Covariance	DZ Covariance	χ^2	df^a
AE Model						
σ_A^2	σ_E^2					
0.609	0.382	0.991	0.609	0.304	1.91	4
CE Model						
σ_C^2	σ_E^2					
0.5	0.5	1.000	0.500	0.500	6.75	4
E Model						
	σ_E^2					
	1.000	1.000	0.000	0.000	92.47	5

$^a df$, *degrees of freedom.*

model (which implies that any covariation between twins is not due to genetic factors) and an E model (which implies that there is no significant covariation between twins in any case). The results are presented in Table A.3, showing the optimized parameter values for the different models.

Because these models are not saturated, they cannot necessarily guarantee a perfect fit to the data. Adjusting one parameter to perfectly fit the MZ twin covariance pulls the DZ twin covariance or the variance estimate out of line, and vice versa. We see here that the AE model has estimated the variance and MZ covariance quite accurately in selecting the optimized parameters $\sigma_A^2 = 0.609$ and $\sigma_E^2 = 0.382$ but the expected DZ covariance departs substantially from the observed value of 0.4. But is this departure significant? The last two columns give the χ^2 and associated *degrees of freedom (df)* of the test. Because we have six observed statistics, from which we are estimating two parameters under the AE model, we say that we have $6 - 2 = 4$ degrees of freedom. The degrees of freedom therefore represent a measure of how simple or complex a model is—we need to know this when deciding which is the most parsimonious model. The E model, for example, estimates only one parameter and so has $6 - 1 = 5$ degrees of freedom.

The test of whether a nested, simpler model is more parsimonious is quite simple: We look at the difference in χ^2 goodness of fit between the two models. The difference in degrees of freedom between the two models is used to determine whether or not the difference in fit is significant. If the difference is significant, then we say that the nested submodel does *not* provide a good account of the data when compared with the goodness of fit of the fuller model. The χ^2 statistics calculated in our example in Table A.3 are dependent on sample size—these figures are based on 150 MZ twins and 150 DZ twins.

The ACE model estimates three parameters from the six observed statistics, so it has three degrees of freedom; the χ^2 is always 0.0 because the model is saturated. Therefore, the difference in fit between the ACE and AE models is $1.91 - 0 = 1.91$ with $4 - 3 = 1$

degree of freedom. Looking up this χ^2 value in significance tables tells us that it is not significant at the $p = 0.05$ level (in fact, $p = 0.17$). A p value lower than 0.05 indicates that the observed results would be expected to arise less than 5 percent of the time by chance alone, if there were in reality no effect. This is commonly accepted to be sufficient evidence to reject a null hypothesis, which states that no effect is present. Therefore, because the AE model does not show a significant reduction in fit relative to the ACE model, this result provides evidence that the shared environment is not important (i.e., that σ_C^2 is not substantially greater than 0.0) for this trait.

What about the CE and E models, though? The CE model fit is reduced by a χ^2 value of 6.75, also for a gain of one degree of freedom. This reduction in fit is significant at the $p = 0.05$ level ($p = 0.0093$). This significant reduction in goodness of fit suggests that additive genetic effects are important for this trait (i.e., that $\sigma_A^2 > 0.0$). Unsurprisingly, the E model shows an even greater reduction in fit ($\Delta\chi^2 = 92.47$ for two degrees of freedom: $p < 0.00001$), thus confirming the obvious fact that the members of both types of twins do in fact show a reasonable degree of resemblance to each other.

2.2.7 Path analysis   The kind of model fitting we have described so far is intimately related to a field of statistics called *path analysis*. Path analysis provides a visual and intuitive way to describe and explore any kind of model that describes some observed data. The *paths*, drawn as arrows, reflect the statistical effect of one variable on another, independent of all the other variables—what are called *partial regression coefficients*. The *variables* can be either measured traits (squares) or the *latent* (unmeasured; circles) variance components of our model. The twin ACE model can be represented as the path diagram in Figure A.9.

The curved, double-headed arrows between latent variables represent the covariance between them. The 1.0/0.5 on the covariance link between the two A latent variables indicates that for MZ twins, this covariance link is 1.0; for DZ twins, 0.5. The

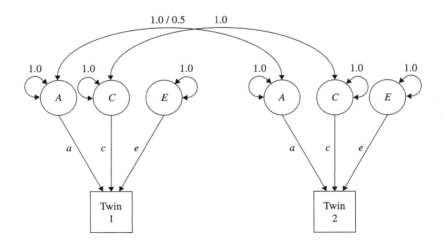

Figure A.9 ACE path diagram. This path diagram is equivalent to the matrix formulation of the ACE model. Path coefficients (*a, c,* and *e*) rather than variance components (which are assumed to be 1) are estimated.

covariance links between the C and E terms therefore represent the previously defined sharing of these variance components between twins (i.e., no link implies a 0 covariance). The double-headed arrow loops on each latent variable represent the variance of that variable. In our previous model fitting, we estimated the variances of these latent variables, calling them σ_A^2, σ_C^2, and σ_E^2. In our path diagram, we have fixed all the variances to 1.0. Instead, we estimate the path coefficients, which we have labeled as a, c, and e. The differences here are largely superficial: The diagram and the previous models are mathematically identical.

To understand a path diagram and how it relates to the kind of models we have discussed, we need to acquaint ourselves with a few basic rules of path analysis. The covariance between two variables is represented by tracing along all the paths that connect the two variables. There are certain rules about the directions in which paths can or cannot be traced, how loops in paths are dealt with, and so on, but the principle is simple. For each path, we multiply all the path coefficients together with the variances of any latent variables traced through. We sum these paths to calculate the expected covariance. The variance for the first twin is therefore a (up the first path) times 1.0 (the variance of latent variable A) times a (back down the path) plus the same for the paths to latent variables C and E. This equals $(a \times 1.0 \times a) + (c \times 1.0 \times c) + (e \times 1.0 \times e) = a^2 + c^2 + e^2$. So instead of estimating the variance components, we have written the model to estimate the path coefficients. This approach is used for practical reasons (e.g., it means that estimates of variance always remain positive, being the square of the path coefficient). The covariance between twins is derived in a similar way. When we trace the two paths between the twins, we get $(a \times 1.0 \times a) + (c \times 1.0 \times c)$ for MZ twins and $(a \times 0.5 \times a) + (c \times 1.0 \times c)$ for DZ twins. That is, $a^2 + c^2$ for MZ twins and $0.5a^2 + c^2$ for DZ twins, as before.

So we have seen how a properly constructed path diagram implies an expected variance-covariance (or correlation) matrix for the observed variables in the model. As noted, it is standard for the parameters in path diagrams to be path coefficients instead of variance components, although, for most basic purposes, this substitution makes very little difference. Any path diagram can be converted into a model that can be written down as algebraic terms in the elements of variance-covariance matrices, and vice versa.

2.2.8 Multivariate analysis

So far we have focused on the analysis of only one phenotype at a time. This method is often called a *univariate* approach—studying the genetic-environmental nature of the *variance of one trait*. If multiple measures have been assessed for each individual, however, a model-fitting approach easily extends to analyze the genetic-environmental basis of the *covariance between multiple traits*. Is, for example, the correlation between depression and anxiety due to genes that influence both traits, or is it largely due to environments that act as risk factors for both depression and anxiety? If we think of a correlation as essentially reflecting shared causes somewhere in the etiological pathways of the two traits, multivariate genetic analysis can tell us something about the nature of these shared causes. The development of multivariate quantitative genetics is one of the most important advances in behavioral genetics during the past two decades.

The essence of multivariate genetic analysis is the analysis of *cross-covariance* in relatives. That is, we can ask whether trait X is associated with another family member's trait Y. Path analysis provides an easy way to visualize multivariate analysis. The path diagram

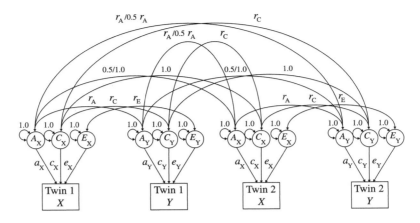

Figure A.10 Multivariate ACE path diagram. This path diagram represents a multivariate ACE model. The expected variance-covariance matrix (given in Table A.4) can be derived from this diagram by tracing the paths.

for a multivariate genetic analysis of two measures is shown in Figure A.10. The new parameters in this model are r_A, r_C, and r_E. These symbols represent the *genetic correlation*, the *shared environmental correlation*, and the *nonshared environmental correlation*, respectively. A genetic correlation of 1.0 would imply that all additive genetic influences on trait X also impact on trait Y. A shared environmental correlation of 0 would imply that the environmental influences that make twins more similar on measure X are independent of the environmental influences that make twins more similar on measure Y. The phenotypic correlation between X and Y can therefore be dissected into genetic and environmental constituents. A high genetic correlation implies that if a gene were found for one trait, there is a reasonable chance that this gene would also influence the second trait.

Multivariate analysis can model more than two variables—as many measures as we wish can be included. In matrix terms, instead of modeling a 2×2 matrix, we model a $2n \times 2n$ matrix, where n is the number of variables in the model. In a bivariate case, if we call the measures X and Y in Twins 1 and 2 (such that X_1 represents measure X for Twin 1), then the variance-covariance matrix would be

$$\begin{bmatrix} \text{Var}(X_1) & & & \\ \text{Cov}(X_1 X_2) & \text{Var}(X_2) & & \\ \text{Cov}(X_1 Y_1) & \text{Cov}(X_2 Y_1) & \text{Var}(Y_1) & \\ \text{Cov}(X_1 Y_2) & \text{Cov}(X_2 Y_2) & \text{Cov}(Y_1 Y_2) & \text{Var}(Y_2) \end{bmatrix}$$

giving us ten unique pieces of information. Along the diagonal, we have four variances—each measure in each twin. The terms $\text{Cov}(X_1 Y_1)$ and $\text{Cov}(X_2 Y_2)$ are the phenotypic covariances between X and Y for the first and second twin, respectively. The terms $\text{Cov}(X_1 X_2)$ and $\text{Cov}(Y_1 Y_2)$ are the univariate cross-twin covariances; the final two terms $\text{Cov}(X_1 Y_2)$ and $\text{Cov}(X_2 Y_1)$ are the cross-twin cross-trait covariances.

The corresponding multivariate ACE model for the expected variance-covariance matrix would be written in terms of univariate parameters as before (three parameters for measure X and three for measure Y) as well as three parameters for the genetic, shared environmental, and nonshared environmental correlations between the two measures (where G is the coefficient of relatedness; i.e., either 1.0 or 0.5 for MZ or DZ twins). Table A.4 presents the elements of this matrix in tabular form.

The shaded area in Table A.4 represents the cross-trait part of the model, which looks more complex than it really is. In path diagram terms, the phenotypic (within-individual) cross-trait covariance results from three paths. The first path includes the additive genetic path for measure X (a_X) multiplied by the genetic correlation between the two traits (r_A) and the additive genetic path for measure Y (a_Y). The shared environmental and nonshared environmental paths are constructed in a similar way. The cross-twin cross-trait correlations are identical, except that there are no nonshared environmental components (by definition) and there is a coefficient of relatedness, G, to determine the magnitude of shared additive genetic variance for MZ and DZ twins. Be careful in the interpretation of a *non-shared environmental correlation*; remember that this term means *nonshared* between family members, not *trait-specific*. Any environmental effect that family members do not have in common and that influences more than one trait will induce a nonshared environmental correlation between these traits.

Genetic, shared environmental, and nonshared environmental correlations are independent of univariate heritabilities. That is, two traits might both have low heritabilities but a high genetic correlation. This would mean that, although there are probably only a few genes of modest effect that influence both these traits, whichever gene influences one trait is very likely to influence the other trait also. In this way, the analysis of these three etiological correlations can begin to tell us not just *whether* two traits are correlated but also *why* they are correlated.

Imagine that we have measured three traits, X, Y, and Z, in a sample of MZ and DZ twins (400 MZ pairs, 400 DZ pairs). What might a multivariate genetic analysis be able

TABLE A.4

Variance-Covariance Matrix for a Multivariate Genetic Model

	Twin 1 Measure X	Twin 2 Measure X	Twin 1 Measure Y	Twin 2 Measure Y
Twin 1 Measure X	$a_X^2 + c_X^2 + e_X^2$			
Twin 2 Measure X	$Ga_X^2 + c_X^2$	$a_X^2 + c_X^2 + e_X^2$		
Twin 1 Measure Y	$r_A a_X a_Y +$ $r_C c_X c_Y +$ $r_E e_X e_Y$	$Gr_A a_X a_Y +$ $r_C c_X c_Y$	$a_Y^2 + c_Y^2 + e_Y^2$	
Twin 2 Measure Y	$Gr_A a_X a_Y$ $r_C c_X c_Y$	$r_A a_X a_Y +$ $r_C c_X c_Y +$ $r_E e_X e_Y$	$Ga_Y^2 + c_Y^2$	$a_Y^2 + c_Y^2 + e_Y^2$

to tell us about the relationships between these traits? Looking at the phenotypic correlations, we observe that each trait is moderately correlated with the other two:

$$\begin{bmatrix} 1.00 & & \\ .42 & 1.00 & \\ .30 & .45 & 1.00 \end{bmatrix}$$

Naturally, we would be interested in the twin correlations for these measures—both the univariate and cross-trait twin correlations. For MZ twins, we might observe

$$\begin{bmatrix} .78 & & \\ .44 & .91 & \\ .08 & .39 & .70 \end{bmatrix}$$

whereas for DZ twins, we might see

$$\begin{bmatrix} .40 & & \\ .23 & .61 & \\ .04 & .23 & .58 \end{bmatrix}$$

The twin correlations along the diagonal therefore represent univariate twin correlations. For example, we can see that the correlation between MZ twins for trait Y is .91. The off-diagonal elements represent the cross-twin cross-trait correlations. For example, the correlation between an individual's trait X with their co-twin's trait Y is .23 for DZ twins. Submitting our data to formal model-fitting analysis gives optimized estimates for the univariate parameters (heritability, proportion of variance attributable to shared environment, proportion of variance attributable to nonshared environment) shown in Table A.5.

That is, traits X and Y both appear to be strongly heritable. Trait Z appears less heritable, although one-fourth of the variation in the population of twins is still due to genetic factors. The more interesting results emerge when the multivariate structure of the data is examined. The best-fitting parameter estimates for the genetic correlation

TABLE A.5

Best-Fit Univariate Parameter Estimates

Trait	Optimized Estimate (%)[a]		
	h^2	c^2	e^2
X	74	4	22
Y	60	31	9
Z	23	47	30

[a]h^2, heritability or additive genetic variance; c^2, shared environmental variance; e^2, nonshared environmental variance.

matrix, the shared environment correlation matrix, and the nonshared environment correlation matrix, respectively, are presented in the following matrices:

$$
\begin{bmatrix}
1.00 & & \\
.44 & 1.00 & \\
.11 & .75 & 1.00
\end{bmatrix}
\quad
\begin{bmatrix}
1.00 & & \\
.98 & 1.00 & \\
.17 & .26 & 1.00
\end{bmatrix}
\quad
\begin{bmatrix}
1.00 & & \\
.10 & 1.00 & \\
.89 & .46 & 1.00
\end{bmatrix}
$$

Genetic correlation   Shared environmental   Nonshared environmental
matrix   correlation matrix   correlation matrix

These correlations tell an interesting story about the underlying nature of the association between the three traits. Although on the surface, traits X, Y, and Z appear to be all moderately intercorrelated, behavioral genetic analysis has revealed a nonuniform pattern of underlying genetic and environmental sources of association.

The genetic correlation between traits Y and Z is high ($r_A = .75$), so any genes impacting on Y are likely to also affect Z, and vice versa. The contribution of shared genetic factors to the phenotypic correlation between two traits is called the *bivariate heritability*. This statistic is calculated by tracing the genetic paths that contribute to the phenotypic correlation: in this case, a_Y and r_A (Y-Z correlation) and a_Z. In other words, the bivariate heritability is the product of the square root of both univariate heritabilities multiplied by the genetic correlation. In the case of traits Y and Z, this statistic equals $\sqrt{0.60} \times .75 \times \sqrt{0.23} = .28$. As shown in an earlier matrix, the phenotypic correlation between traits Y and Z is .45. Therefore, over half (62 percent $= .28/.45$) of the correlation between traits Y and Z can be explained by shared genes. Note that we take the square root of the univariate heritabilities because, in path analysis terms, we only trace up the path once—in calculating the univariate heritability, we would come back down that path, therefore squaring the estimate.

The same logic can be applied to the environmental influences. Focusing on traits Y and Z, tracing the paths for shared and nonshared environmental influences yields values of .10 ($\sqrt{0.31} \times .26 \times \sqrt{0.47}$) and .07 ($\sqrt{0.09} \times .46 \times \sqrt{0.30}$) for the bivariate estimates. Note that these add up to the phenotypic correlation, as expected (.28 + .10 + .07 = .45).

In contrast, the correlation between traits X and Z ($r = .30$) is not predominantly mediated by shared genetic influence: $\sqrt{0.74} \times .11 \times \sqrt{0.23} = .04$; only 13 percent of this phenotypic correlation is due to genes.

An interesting aspect of this kind of analysis is that it could potentially reveal a strong genetic overlap between two heritable traits even when the phenotypic correlation is near 0. This scenario could arise if there were, for example, a negative nonshared environmental correlation (i.e., certain environments [nonshared between family members] tend to make individuals dissimilar for two traits). Consider the following example: Two traits both have univariate heritabilities of 0.50 and no shared environmental influences, so the nonshared environment will account for the remaining 50 percent of the variance. If the traits had a genetic correlation of .75 but a nonshared environmental correlation of $-.75$, then the phenotypic correlation would be 0. The phenotypic correlation is the sum of the chains of paths ($\sqrt{0.5} \times .75 \times \sqrt{0.5}$) + ($\sqrt{0.5} \times -.75 \times \sqrt{0.5}$) = 0.0. This example shows that the phenotypic correlation by itself does not necessarily tell you very much about the shared etiologies of traits.

The preceding model is just one form of multivariate model. Different models that make different assumptions about the underlying nature of the traits can be fitted to test whether a more parsimonious explanation fits the data. For example, the *common-factor independent-pathway* model assumes that each measure has specific (subscript "S") genetic and environmental effects as well as general (subscript "C") genetic and environmental effects that create the correlations between all the measures. Figure A.11 shows a schematic path diagram for a three-trait version of this model. (*Note:* The diagram represents only one twin for convenience—the full model would have the three traits for both twins and the *A* and *C* latent variables would have the appropriate covariance links between twins.) In this path diagram, the general factors are at the bottom.

A similar but more restricted model, the *common-factor common-pathway* model, assumes that the common genetic *and* environmental effects load onto a latent variable, *L*, that in turn loads onto all the measures in the model. This model is said to be more restricted in that, because fewer parameters are estimated, the expected variance-covariance is not as free to model any pattern of phenotypic, cross-twin same-trait and cross-twin cross-trait, correlations. Figure A.12 represents this model (again, for only one twin).

The common-factor independent-pathway model is nested in the more general multivariate model presented earlier; the common-factor common-pathway model is nested in both. These models can therefore be tested against each other to see which provides the most parsimonious explanation of the observations. Note that these multivariate models can also vary in terms of whether they are ACE, ADE, CE, AE, or E models.

A more specific form of multivariate model that has received a lot of interest is the *longitudinal* model. This model is appropriate for designs that take repeated measures of a trait over a period of time (say, IQ at 5, 10, 15, and 20 years of age). Such models can

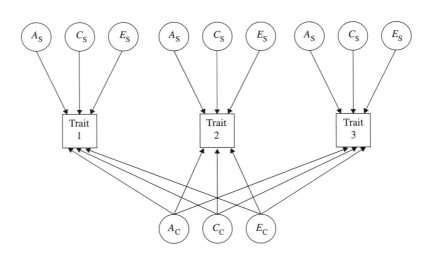

Figure A.11 Common-factor independent-pathway multivariate path diagram. This is a partial diagram, for one twin. *A*, additive genetic effects; *C*, shared environmental effects; *E*, nonshared environmental effects; S (subscript), specific effects; C (subscript), general effects.

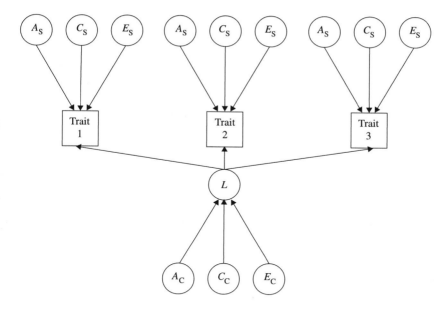

Figure A.12 Common-factor common-pathway multivariate path diagram. This is a partial diagram, for one twin. *A*, additive genetic effects; *C*, shared environmental effects; *E*, nonshared environmental effects; S (subscript), specific effects; C (subscript), general effects; *L*, latent variable.

be used to unravel the etiology of continuity and change in a trait over time and are especially powerful for studying the interaction of genetic makeup and environment.

2.2.9 Complex effects including gene-environment interaction

For the sake of simplicity (and parsimony), all the ACE-type models we have looked at so far have made various assumptions about the nature of the genetic and environmental influences that operate on the trait. Nature does not always conform to our expectations, however. In this section, we will briefly review some of the "complexities" that can be incorporated into models of genetic and environmental influence.

As mentioned earlier, an important feature of the model-fitting approach is that, as well as being flexible, it tends to make the assumptions of the model quite apparent. One such assumption is the *equal environments assumption* that MZ and DZ twins receive equally similar environments (see Chapter 5). The assumption is implicit in the model—we estimate the same parameter for shared environmental effects for MZ and DZ twins. This assumption might not always be true in practice. Can we account for potential inequalities of environment in our model? Unfortunately, not without collecting more information. The model-fitting approach is flexible, but it cannot do everything—this problem is an example of how experimental design and analysis should work hand-in-hand to tackle such questions. For example, research has compared MZ twins who have been mistakenly brought up as DZ twins, and vice versa, to study whether MZ twins are in fact treated more similarly, as indicated in Chapter 5.

Another assumption of the models used so far is random mating in the population. When nonrandom (or assortative) mating occurs (Chapter 8), then loci for a trait will be correlated between spouses. This unexpected correlation will lead to siblings and DZ twins sharing more than half their genetic variation, a situation that will bias the estimates derived from our models. In model fitting, the effects of assortative mating can be modeled (and therefore accounted for) if appropriate parental information is gathered.

Covariance between relatives on any trait can arise from a number of different sources that are not considered in our basic models. As mentioned earlier, shared causation is not the only process by which covariation can arise. The phenotype of one twin might *directly* influence the phenotype of the other, for example, because the co-twin is very much part of a twin's environment. Having an aggressive co-twin may influence levels of aggression as a result of the direct exposure to the co-twin's aggressive behavior. Such an effect is called *sibling interaction*. In the context of multivariate analysis, it is possible that trait X actually causes trait Y in the same individual, rather than a gene or environment impacting on both. These situations can be modeled by using fairly standard approaches. If such factors are important but are ignored in model fitting, they will bias estimates of genetic and environmental influence.

Another way in which the basic model might be extended is to account for possible *heterogeneity* in the sample. Genetic and environmental influences may be different for boys and girls on the same trait, or for young versus old people. Heritability is only a sample-based statistic: A heritability of 70 percent means that 70 percent of the variation *in the sample* can be accounted for by genetic effects. This outcome could be because the trait is completely heritable in 70 percent of the sample and not at all heritable in 30 percent. Such a sample would be called *heterogeneous*—there is something different and potentially interesting about the 30 percent that we may wish to study. The standard model-fitting approaches we have studied so far would leave the researcher oblivious to such effects.

To uncover heterogeneity, various approaches can be taken. Potential indices of heterogeneity (e.g., sex or age) can be incorporated into a model, for example. We could ask, Does heritability increase with age? Or we could test a model having separate parameter estimates for boys and girls for genetic effects against the nested model with only one parameter for both sexes. Same-sex and opposite-sex DZ twins can be modeled separately to test for quantitative and qualitative etiological differences between males and females. This design is called a *sex-limitation model*, and it can ask whether the magnitude of genetic and environmental effects are similar in males and females. Additionally, such designs are potentially able to test whether the *same* genes are important for both sexes, irrespective of magnitude of effect.

Other complications include *nonadditivity*, such as epistasis, gene-environment interaction, and gene-environment correlation. These three types of effects were defined under the preceding biometric model section. Epistasis is any gene-gene interaction; $G \times E$ interaction is the interaction between genetic effects and environments; G-E correlation occurs when certain genes are associated with certain environments. As an example of epistasis, imagine that an allele at locus A only predisposes toward depression if that individual also has a certain allele at locus B. As an example of gene-environment interaction, the allele at locus A may have an effect only for individuals living in deprived environments. These types of effects complicate model fitting because there are many forms in which they could occur. Normal twin designs do not offer much hope for identifying them. An MZ correlation that is much higher than twice the DZ correlation would be suggestive of epistasis, but the models cannot really go any further in quantifying such effects.

Although model fitting can often be extended to incorporate more complex effects, it is not generally possible to include *all* these "modifications" at the same time. Successful approaches will typically select specific types of models that should be fitted a priori, on the basis of existing etiological knowledge of the traits under study.

One exciting development in model fitting involves incorporating measured variables for individuals into the analysis. Measuring alleles at specific loci, or specific environmental variables, makes the detection of specific, complex, interactive effects feasible, as well as forming the basis for modern techniques for mapping genes, as we will review in the final section.

2.2.10 Environmental mediation   Behavioral genetic studies have convincingly demonstrated that genes play a significant role in many complex human traits and diseases. As a result, rather than just estimating heritability and other genetic quantities of interest, an increasingly important application of genetically informative designs, such as the twin study, is to shed light on the nature of *environmental effects.*

Although we might know that an environment and an outcome show a statistical correlation, we often do not understand the true nature of that association. For example, an association might be causal if the environment directly affects the outcome. Alternatively, the association might only arise as a reflection of some other underlying shared, possibly genetic, factor that influences both environment and the outcome. As illustrated in detail in Chapter 16, many "environmental" measures do indeed show genetic influence. By using a genetically informative design to control for genetic factors, researchers are able to make stronger inferences about environmental factors. A simple but powerful design is to focus on environmental measures that predict phenotypic differences between MZ twins.

2.2.11 Extremes analysis   When we partition the variance of a trait into portions attributable to genetic or environmental effects, we are analyzing the sources of *individual differences* across the entire range of the trait. When looking at a quantitative trait, we may be more interested in one end, or extreme, of that trait. Instead of asking what makes individuals different for a trait, we might want to ask what makes individuals score high on that trait.

Consider a trait such as reading ability. Low levels of reading ability have clinical significance; individuals scoring very low will tend to be diagnosed as having reading disability. We may want to ask what makes people reading disabled, rather than what influences individuals' reading ability. We could perform a qualitative analysis where the dependent variable is simply a *Yes* or a *No* to indicate whether or not individuals are reading disabled (i.e., low scoring). If we have used a quantitative trait measure (such as a score on a reading ability task) that we believe to be related to reading disability, we may wish to retain this extra information. Indeed, we can ask whether reading disability is etiologically related to the continuum of reading ability or whether it represents a distinct syndrome. In the latter case, the factors that tend to make individuals score lower on a reading ability task in the entire population will not be the same as the factors that make people reading disabled. A regression-based method for analyzing twin data, DF (DeFries-Fulker) extremes analysis, addresses such questions, by analyzing means as opposed to variances. The methodology for DF extremes analysis is described in Chapter 7.

3 Molecular Genetics

Mapping genes for quantitative traits (quantitative trait loci, or QTLs) and diseases is a fast-developing area in behavioral genetics. The goal is to identify either the chromosomal region in which a QTL resides (via *linkage analysis*) or to pinpoint the specific variants or genes involved (via *association analysis*). The starting point for both of these molecular genetic approaches is to collect DNA, either from families or samples of unrelated individuals, and directly measure the genotype (one or more variants) to study their relationship with the phenotype. The process of measuring genotypes is called *genotyping*, where we obtain the genotype for one or more *markers* (DNA variants) in each individual. Genotyping technology has evolved rapidly over the past few decades: Whereas early studies might have considered only a handful of markers, modern molecular genetic studies can now genotype half a million variants or more, in *genomewide association studies*, the current state-of-the-art.

Here we will briefly review the two complementary techniques of linkage and association analysis. Linkage tests whether or not the pattern of inheritance within families at a specific locus correlates with the pattern of trait similarity. Association, on the other hand, directly tests whether specific alleles at specific markers are correlated with increased or decreased scores on a trait or with prevalence of disease.

Although there are other molecular techniques that can be applied to complex behavioral traits, we restrict our focus in this section to approaches that correlate genotype marker data to phenotype. Other approaches not covered here include *expression analysis* using microarrays (to see whether patterns of gene expression, the amount of RNA produced in particular cell types, is related to phenotype), *DNA sequencing* (to study a region's entire DNA code for each individual, for example, to see whether rare mutations, that are not represented by common, polymorphic markers, are related to phenotype), and *epigenetics* (looking at features of the genome other than the standard inherited variation of DNA bases, such as methlylation patterns).

3.1 Linkage analysis

As described in Chapter 2, Mendel coined two famous "laws," based on his studies with garden peas. His first law, the "law of segregation," basically states that each person gets a paternal and a maternal copy of each gene, and which copy they pass on to each of their offspring is random. His second law, the "law of independent assortment," further states that which copy (i.e., the paternal or maternal) of a particular gene an individual passes on to his or her child does not depend on which copy of any other gene is passed on. In other words, Mendel believed that the transmission of any two genes is statistically independent, in the same way two coin tosses are, implying four equally likely possible combinations.

Mendel did not get it 100 percent right, however. There is an important exception, which is when the two genes, let's call them *A* and *B*, are close to each other on the same chromosome. In this case, we would say that *A* and *B* are *linked* or *in linkage*. Importantly, we can exploit the property of linkage (that nearby genes tend to be cotransmitted from a parent to its offspring) to localize genes that affect phenotypes, in *linkage analysis*, as described below.

3.1.1 Patterns of gene flow in families

If genes A and B were on different chromosomes, then Mendel's second law would hold. But consider what happens when they are not, as shown in Figure A.13. This figure shows a possible set of transmissions from a father and mother to their child for a stretch of this chromosome, which contains both genes A and B, very close to each other. For this whole region, the father transmits to his child the copy he received from his own father. In contrast, we see that during meiosis (the process of forming the sex cells) the mother's paternal and maternal chromosomes have experienced a *recombination event*, such that the mother transmits a mosaic of her own mother's and father's chromosomes.

Whether or not a recombination occurs at any one position is more or less a random process. Importantly, the farther away two points on a chromosome are, the more

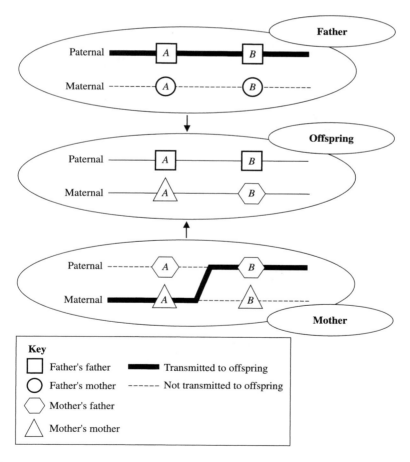

Figure A.13 Recombination of chromosomes during meiosis.

likely they are to be separated by a recombination event (technically, separated by *an odd number of recombination events*, as more than one can occur per chromosome). Two genes that are very close to each other on the same chromosome will tend not to be separated by recombination, however, and so they will tend to be cotransmitted from parent to offspring (i.e., either both are transmitted, or neither is). As mentioned above, this tendency is called linkage.

3.1.2 Genome scans using linkage

But what is the relevance of linkage to gene mapping? How does it help us find genes that influence particular phenotypes? First, linkage analysis has been centrally important in creating maps of the genome: By studying whether or not particular DNA variants are cotransmitted in families, researchers were able to infer the relative order and positions of these markers along each chromosome. Second, linkage analysis can help to detect genotype-phenotype correlations. Instead of considering markers at two genes, *A* and *B*, it is possible to consider linkage between a marker and a phenotype. If the marker and the phenotype are similarly cotransmitted in families, we can infer the presence of a phenotype-influencing gene that is linked to the marker.

A typical linkage analysis might involve genotyping a couple of hundred highly informative microsatellite markers (ones with many alleles) spaced across the genome, in a collection of families with multiple generations or multiple offspring. Often, the markers that are tested are not themselves assumed to be functional for the trait—they are merely selected because they are polymorphic in the population. The markers are used to statistically reconstruct the pattern of gene flow within these families for all positions along a chromosome. Such a study, often called a *genome scan*, provides an elegant way to search the entire genome for regions that might harbor phenotype-related genes. For disease traits, the simplest form of linkage analysis is to consider families with at least two affected siblings. If a region is linked with disease, we would expect the two siblings to have inherited the exact same stretch of chromosome from their parents more often than expected by chance, as a consequence of their sharing the same disease.

In practice, there are many complexities and many flavors of linkage analysis (e.g., for larger families, for continuous as well as disease traits, using different statistical models and assumptions, including variance components frameworks as described above that also incorporate marker data). Classical (parametric) linkage analysis relies on small numbers of large families (pedigrees) and explicitly models the distance between a test marker locus and a putative disease locus. The term *disease locus* (as opposed to QTL) reflects the fact that classical linkage is primarily concerned with mapping genes for dichotomous disease-like traits. Classical linkage requires that a model for the disease locus be specified a priori, in terms of allelic frequencies and mode of action (recessive or dominant). Figure A.14 shows an example of a pedigree in which a dominant gene is causing disease.

The approach of classical linkage is not so well suited to complex traits, however, for it is hard to specify any one model if we expect a large number of loci of small effect to impact on a trait. The alternative, nonparametric, or allele sharing, approach to linkage simply tests whether allele sharing at a locus correlates with trait similarity, as described above for affected sibling pairs. For quantitative traits, linkage analysis is often performed in nuclear families using a variance components framework similar to

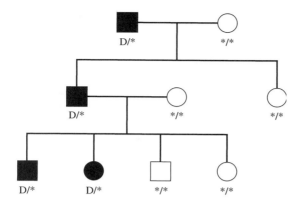

Figure A.14 A pedigree for a dominant disease (D) allele transmitted by the father. The asterisks refer to alleles other than D.

that described above for twin analysis. Using marker data, we can partition a sample of sibling pairs into those that share 0, 1, or 2 copies of the exact same parental DNA at any particular position along each chromosome. If the test locus is linked to the trait, then the sibling correlation should increase with the amount of sharing. Considering any one position, it is as if we are effectively splitting the siblings into unrelated pairs (those sharing 0 at that particular position), parent-offspring pairs (those sharing 1) and MZ twins (those sharing 2) and fitting the kind of quantitative genetic model described above, comparing these groups in the same way we compare MZ and DZ twins.

In general, linkage analysis has proven spectacularly successful in mapping many rare disease genes of large effect (for example, see Chapters 6 and 7). For many complex traits (which are often highly heritable but not influenced by only one or two major genes), linkage analysis has been less directly useful. Although linkage analysis can effectively search the entire genome with a relatively small number of markers, it lacks power to detect genes of small effect and has limited resolution. In many cases, collecting enough informative families might also be difficult.

3.2 Association analysis

Over the past decade, association analysis has become the approach of choice for many researchers attempting to map genes of small effect for complex traits. In many ways, association analysis asks a simpler question compared to linkage analysis. Whereas linkage analysis dissects patterns of genotypic and phenotypic sharing between related individuals, association analysis directly tests whether there is a genotype-phenotype correlation. Association is typically more powerful than linkage analysis to detect small effects, but it is necessary to genotype a much greater number of markers to cover the same genomic area. Traditionally, researchers would tend to restrict association analysis to a few "candidate" genes, or regions of the genome implicated by previous linkage studies. Modern advances in genotyping technology, which allow hundreds of thousands of markers to be genotyped per individual, have made very large scale studies feasible.

3.2.1 Population-based association analysis Imagine that a particular gene with two alleles, A_1 and A_2, is thought to be a QTL for a quantitative measure of cognitive ability. To test this hypothesis, a researcher might collect a sample of unrelated individuals, measured for this phenotype and genotyped for this particular locus (so we know whether an individual has A_1A_1, A_1A_2, or A_2A_2 genotype), and then ask whether the phenotype depends on genotype. The actual analysis might be a regression of phenotype (the dependent variable) on genotype (the independent variable, coded as the number of A_1 alleles an individual has, i.e., 0, 1, or 2). Similarly, if the phenotype was, instead, a disease, one might perform a *case-control study* in which a sample of cases (people with a particular disease, for example) are ascertained along with a control sample (people without the disease, but ideally who are otherwise well-matched to the case sample). If the frequency of a particular allele or genotype is significantly higher (or lower) in cases relative to controls, one would conclude that the gene shows an association with disease. For example, as discussed in Chapter 7, the frequency of the *ApoE4* allele of the gene that encodes apolipoprotein E is about 40 percent in individuals with Alzheimer's disease and about 15 percent in controls.

Consider the following example of a disease-based association analysis. The basic data for a single biallelic marker can be presented in a 3 × 2 contingency table of disease status by genotype. In this case, the cell counts refer to the number of individuals in each of the six categories.

	Case	Control
A_1A_1	64	41
A_1A_2	86	88
A_2A_2	26	42

One could perform a test of association based on a chi-squared test of independence for a contingency table. Often, however, such data are instead collapsed into allele counts, as opposed to genotype counts. In this case, each individual contributes twice (if the marker is autosomal): A_1A_1 individuals contribute two A_1 alleles, A_2A_2 individuals contribute two A_2 alleles, and A_1A_2 individuals contribute one of each. The 2 × 2 contingency table now represents the number of "case alleles" and "control alleles." A test based on this table implicitly assumes a simple dosage model for the effect of each allele, which will be more powerful, if true, than a genotypic analysis.

	Case	Control
A_1	$64 \times 2 + 86 = 214$	$41 \times 2 + 88 = 170$
A_2	$26 \times 2 + 86 = 138$	$42 \times 2 + 88 = 172$

Pearson's chi-squared statistic for this table is 8.63 (which has an associated *p*-value of 0.003, as this is a 1 degree of freedom test). Standard statistical software packages can be used to calculate this kind of association statistic. Often the effect will be described as an *odds ratio*, where a value of 1 indicates no effect, a value significantly greater than 1 represents a risk effect (of A_1 in this case), and a value significantly less than 1 represents a protective effect. If the four cells of a 2×2 table are labeled a, b, c and d :

	Case	Control
A_1	a	c
A_2	b	d

then the odds ratio is calculated ad/bc. In this example, the odds ratio is therefore $(214 \times 172)/(138 \times 170) = 1.57$, indicating that A_1 increases risk for disease. For many complex traits, researchers expect very small odds ratios, such as 1.2 or 1.1, for individuals markers; such small effects are statistically hard to detect. If the disease is rare, an odds ratio can be interpreted as a *relative risk*, meaning, in this example, that each extra copy of the A_1 allele an individual possesses increases his or her risk of disease by a factor of 1.57. So if A_2A_2 individuals have a baseline risk of disease of 1%, then A_1A_2 individuals would have an expected risk of 1.57% and A_1A_1 individuals would have a risk of $1.57\% \times 1.5\% = 2.46\%$.

3.2.2 Population stratification and family-based association

In the previous section, we noted that samples should be well-matched. In any association study, it is particularly critical that samples be well-matched in terms of ethnicity. Failure to adequately match can result in *population stratification* (a type of confounding) which causes spurious results in which between-group differences confound the search for biologically relevant within-group effects. For example, imagine a case/control study where the sample actually comes from two distinct ethnic groups. Further, imagine that one group is overrepresented in cases versus controls (this might be because the disease is more prevalent in one group, or it might just reflect differences in how cases and controls were ascertained). Any gene that is more common in one of the ethnic groups than in the other will now show an obligatory statistical association with disease because of this third, confounding variable, ethnicity. Almost always, these associations will be completely spurious (i.e., the gene has no causal association with disease).

That correlation does not imply causation is, of course, a maxim relevant to any epidemiological study. But often in genetics we are less concerned with proving causality, per se, than we are with having *useful* correlational evidence (i.e., that could be used in locating a nearby causal disease gene, as described below in the section on indirect association). The problem with population stratification is that it will tend to throw up a very large number of red herrings that have absolutely no useful interpretation.

Luckily there are a number of ways to avoid the possible confounding due to population stratification in association studies. The most obvious is to apply sound experimental and epidemiological principles of randomization and appropriate sampling protocols. Another alternative is to use families to test for association, as most family members are necessarily well-matched for ethnicity. For example, for siblings discordant for Alzheimer's disease, we would expect that the affected siblings would have a higher frequency of the *APoE4* allele of the gene encoding apolipoprotein E than would the unaffected members of the sibling pairs. Note that this is distinct from linkage analysis, which is based on sharing of chromosomal regions within families rather than testing the effects of specific alleles across families.

A common family-based association design is the *transmission/disequilibrium test* (TDT), which involves sampling affected individuals and their parents; in effect, the control individuals are created as "ghost-siblings" of cases, using the alleles that the *parents did not transmit* to their affected offspring. The test focuses only on parents who are heterozygous (e.g., have both an A_1 and an A_2 allele) and asks whether one allele was more often transmitted to affected offspring. If neither allele is associated with disease, we would expect 50:50 transmission of both alleles, as stated by Mendel's first law.

Although family-based association designs control against population stratification (and allow for some other specific hypotheses to be tested, for example, imprinting effects, in which the parental origin of an allele matters), they are in general less efficient, as more individuals must be sampled to achieve the same power as a population-based design. Recently, due to the increasing ability to genotype large numbers of markers, another approach to population stratification has emerged. By using markers randomly selected from across the genome, it is possible to empirically derive and control for ancestry in population-based studies using statistical methods.

3.2.3 *Indirect association and haplotype analysis*

In linkage analysis, the actual markers tested are not themselves assumed to be functional; they are merely proxies that provide information on the inheritance patterns of chromosomal regions. Similarly, in association analysis we do not necessarily assume the marker being tested is the functional, causal variant. This is because when we test any one marker, more often than not we are also implicitly testing the effects of surrounding markers, as alleles at nearby positions will be correlated at the population level. This phenomenon is closely related to linkage, described above, and is in fact called *linkage disequilibrium.*

A correlation between markers at the population level means that knowing a person's genotype at one marker tells you something about their genotype at a second marker. This correlation between markers, or linkage disequilibrium, actually reflects our shared ancestry. Over many generations, recombination has rearranged the genome, but like an imperfectly shuffled deck of cards, some traces of the previous order still exist. Because we inherit stretches of chromosomes that contain many alleles, certain strings of alleles will tend to be preserved by chance. Unless these strings are broken by recombination, the strings of alleles that sit on the same chromosomal stretch of DNA (called *haplotypes*) may become common at the population level. Considering three markers, A, B, and C (each with alleles coded 1 and 2), there may be only three common haplotypes in the population

$A_1 \, B_1 \, C_1$ 80%
$A_2 \, B_2 \, C_2$ 12%
$A_1 \, B_2 \, C_2$ 8%

In this example, possessing an A_2 allele makes you much more likely also to possess B_2 and C_2 alleles (100% of the time, in fact) than if you possess an A_1 allele (now only $8/(8 + 80) = 9\%$ of the time). We would, therefore, say that marker A is in linkage disequilibrium with B and C (and vice versa).

Linkage disequilibrium leads to indirect association; for example, if B were the true QTL, then performing an association analysis at A would still recover some of the true

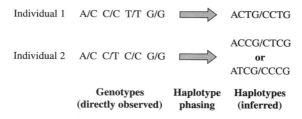

| | Genotypes (directly observed) | Haplotype phasing | Haplotypes (inferred) |

Figure A.15 Observed genotypes and inferred haplotypes.

signal, due to the correlation in alleles, although it would be somewhat attenuated. In contrast, genotyping C instead of B would recover all the information, as it is a perfect proxy for B.

It is possible to use haplotype information in association analysis, by testing haplotypes instead of genotypes. In the above example, we might ask whether the number of copies of the $A_2B_2C_2$ haplotype that an individual possesses predicts the phenotype. By combining multiple markers in this way (called *haplotype-based association analysis*), one can extract extra information without extra genotyping. For example, imagine a fourth, ungenotyped locus, D. In this case, the $A_1B_2C_2$ haplotype is a perfect proxy for D (as it is completely correlated with the D_2 allele) whereas none of the three original individual markers are.

$A_1\ B_1\ C_1\ D_1$ 80%
$A_2\ B_2\ C_2\ D_1$ 12%
$A_1\ B_2\ C_2\ D_2$ 8%

Any one individual will possess two of these haplotypes (one paternally, one maternally inherited), for example, $A_1B_1C_1$ and $A_2B_2C_2$ if we consider just the three genotyped markers. We do not usually observe haplotypes directly, however. Instead, we observe genotypes, which in this case would be A_1A_2 for the first marker, B_1B_2 for the second, and C_1C_2 for the last. As illustrated in Figure A.15, in themselves, genotypes do not contain information about haplotypes, so it might not always be possible to determine unambiguously which haplotypes an individual has. (A particular combination of genotypes might be compatible with more than one pair of haplotypes.) However, statistical techniques can be used to estimate the frequencies of the different possible haplotypes, which in turn can be used to guess which pair of haplotypes is most likely given the genotypes for an individual (this process is called *haplotype phasing*).

3.2.4 The HapMap and genomewide association studies In the example above, wouldn't it have been great if we knew in advance that markers B and C were perfect proxies for each other, or that marker D could be predicted by a haplotype of A, B and C? Knowing that, we would probably not want to waste money genotyping all the markers, when genotyping a subset would give exactly the same information. In fact, now we usually do know in advance, thanks to the HapMap Project (http://www.hapmap.org/). This was a large, international survey of patterns of linkage disequilibrium across the genome, performed in a number of different populations, focusing on single nucleotide polymorphisms (SNPs), the most common form of variation in the human genome. SNPs are biallelic markers, with the alleles being two of A, C, G and T (i.e., the four nucleotide basis of DNA).

For many common variants, the HapMap shows that there are lots of perfect proxies in the genome; that is, there is a lot of redundancy. This means that it is possible to measure almost all common variation in the human genome using a much smaller set of SNPs. This concept is called *tagging* and effectively determines how to optimally choose which markers to genotype.

Based on large-scale genomic efforts such as the HapMap and new genotyping technologies, association analysis has recently been taken to its logical conclusion: the *genomewide association study* (GWAS). As the name suggests, this involves genotyping hundreds of thousands of markers, usually in large case/control samples. The hope is that such studies combine the power of association analysis with the genomewide, unbiased coverage of the previous generation of linkage genome scans.

GLOSSARY

additive genetic variance Individual differences caused by the independent effects of alleles or loci that "add up." In contrast to *nonadditive genetic variance*, in which the effects of alleles or loci interact.

adoption studies A range of studies that use the separation of biological and social parentage brought about by adoption to assess the relative importance of genetic and environmental influences. Most commonly, the strategy involves a comparison of adoptees' resemblance to their biological parents, who did not rear them, and to their adoptive parents. May also involve the comparison of genetically related siblings and genetically unrelated (adoptive) siblings reared in the same family.

adoptive siblings Genetically unrelated children adopted by the same family and reared together.

affected sib-pair QTL linkage design A QTL linkage design that involves pairs of siblings who meet criteria for a disorder. Linkage with DNA markers is assessed by allele sharing within the pairs of siblings–whether they share 0, 1, or 2 alleles for a DNA marker. (See Box 6.1.)

allele An alternative form of a gene at a locus, for example, A_1 versus A_2.

allele sharing Presence of zero, one, or two of the parents' alleles in two siblings (a sibling pair, or sib pair).

allelic association An association between allelic frequencies and a phenotype. For example, the frequency of allele *4* of the gene that encodes apolipoprotein E is about 40 percent for individuals with Alzheimer's disease and 15 percent for control individuals who do not have the disorder.

allelic frequency Population frequency of an alternate form of a gene. For example, the frequency of the PKU allele is about 1 percent. (In contrast, see *genotypic frequency*.)

alternative splicing The process by which mRNA is reassembled to create different transcripts that are then translated into different proteins. More than half of human genes are alternatively spliced.

amino acid One of the 20 building blocks of proteins, specified by a triplet code of DNA.

amniocentesis A medical procedure used for prenatal diagnosis in which a small amount of amniotic fluid is extracted from the amnion surrounding a developing fetus. Because some of the fluid contains cells from the fetus, fetal chromosomes can be examined and fetal genes can be tested.

anticipation The severity of a disorder becomes greater or occurs at an earlier age in subsequent generations. In some disorders, this phenomenon is known to be due to the intergenerational expansion of DNA repeat sequences.

assortative mating Nonrandom mating that results in similarity between spouses. Assortative mating can be negative ("opposites attract") but is usually positive.

assortment Independent assortment is Mendel's second law of heredity. It states that the inheritance of one locus is not affected by the inheritance of another locus. Exceptions to the law occur when genes are inherited which are close together on the same chromosome. Such linkages make it possible to map genes to chromosomes.

autosome Any chromosome other than the X or Y sex chromosomes. Humans have 22 pairs of autosomal chromosomes and 1 pair of sex chromosomes.

balanced polymorphism Genetic variability that is maintained in a population, for example, by selecting against both dominant homozygotes and recessive homozygotes.

band (chromosomal) A chromosomal segment defined by staining characteristics.

base pair (bp) One step in the spiral staircase of the double helix of DNA, consisting of adenine bonded to thymine, or cytosine bonded to guanine.

behavioral genomics The study of how genes in the genome function at the behavioral level of analysis. In contrast to *functional genomics*, behavioral genomics is a top-down approach to understanding how genes work in terms of the behavior of the whole organism.

bioinformatics Techniques and resources to study the genome, transcriptome, and proteome, such as DNA sequences and functions, gene expression maps, and protein structures.

candidate gene A gene whose function suggests that it might be associated with a trait. For example, dopamine genes are considered as candidate genes for hyperactivity because the drug most commonly used to treat hyperactivity, methylphenidate, acts on the dopamine system.

carrier An individual who is heterozygous at a given locus, carrying both a normal allele and a mutant recessive allele, and who appears normal phenotypically.

centimorgan (cM) Measure of genetic distance on a chromosome. Two loci are 1 cM apart if there is a 1 percent chance of recombination due to crossover in a single generation. In humans, 1 cM corresponds to approximately 1 million base pairs.

centromere A chromosomal region without genes where the chromatids are held together during cell division.

chorion Sac within the placenta that surrounds the embryo. Two-thirds of the time, identical twins share the same chorion.

chromatid One member of one newly replicated chromosome in a chromosome pair, which may cross over with a chromatid from a homologous chromosome during meiosis.

chromosome A structure that is composed mainly of chromatin, which contains DNA, and resides in the nucleus of cells. Latin for "colored body" because chromosomes stain differently from the rest of the cell. (See also *autosome*.)

codon A sequence of three base pairs that codes for a particular amino acid or the end of a chain.

comorbidity The presence of more than one disorder or disease.

concordance Presence of a particular condition in two family members, such as twins.

copy number variant (CNV) A polymorphism that involves duplication of long stretches of DNA, often encompassing protein-coding genes as well as non-coding genes. Frequently used more broadly to refer to all structural variations in DNA, including insertions and deletions.

correlation An index of resemblance that ranges from .00, indicating no resemblance, to 1.00, indicating perfect resemblance.

crossover See *recombination.*

developmental genetic analysis Analysis of change and continuity of genetic and environmental parameters during development. Applied to longitudinal data, assesses genetic and environmental influences on age-to-age change and continuity.

DF extremes analysis An analysis of familial resemblance that takes advantage of quantitative scores of the relatives of probands rather than just assigning a dichotomous diagnosis to the relatives and assessing concordance. (In contrast, see *liability-threshold model.*)

diallel design Complete intercrossing of three or more inbred strains and comparing all possible F_1 crosses between them.

diathesis-stress A type of genotype-environment interaction in which individuals at genetic risk for a disorder (diathesis) are especially sensitive to the effects of risky (stress) environments.

dichotomous trait See *qualitative disorder.*

directional selection Natural selection operating against a particular allele, usually selection against a deleterious allele. (See also *stabilizing selection* and *balanced polymorphism.*)

dizygotic (DZ) Fraternal, or nonidentical, twins; literally, "two zygotes."

DNA (deoxyribonucleic acid) The double-stranded molecule that encodes genetic information. The two strands are held together by hydrogen bonds between two of the four bases, with adenine bonded to thymine, and cytosine bonded to guanine.

DNA marker A polymorphism in DNA itself, such as a restriction fragment length polymorphism (RFLP) or a simple sequence repeat (SSR) polymorphism.

DNA pooling Combining a few nanograms of DNA from each individual in a group in order to estimate allele frequencies for that group. Especially valuable in genomewide association studies with very large samples because instead of using a microarray for each individual in a group, one microarray can be used for the pooled DNA of all individuals in the group.

DNA sequence The order of base pairs on a single chain of the DNA double helix.

dominance An allele that produces a particular phenotype when present in the heterozygous state. The effect of one allele depends on that of another. (Compare with *epistasis,* which refers to nonadditive effects between genes at different loci.)

effect size The proportion of individual differences for the trait in the population accounted for by a particular factor. For example, heritability estimates the effect size of genetic differences among individuals.

electrophoresis A method used to separate DNA fragments by size. When an electrical charge is applied to DNA fragments in a gel, smaller fragments travel farther.

endophenotype An "inside" or intermediate phenotype that does not involve overt behavior.

epigenetics DNA modifications that affect gene expression without changing DNA sequence that can be "inherited" when cells divide. Can be involved in long-term developmental changes in gene expression.

epigenome Epigenetic events throughout the genome.

epistasis Nonadditive interaction between genes at different loci. The effect of one gene depends on that of another. (Compare with *dominance*, which refers to nonadditive effects between alleles at the same locus. See also *nonadditive genetic variance*.)

equal environments assumption In twin studies, the assumption that environments are similar for identical and fraternal twins.

evolutionary psychology and psychiatry Fields that focus on the adaptive value of behavior as a function of natural selection.

exon DNA sequence transcribed into messenger RNA and translated into protein. (Compare with *intron*.)

expanded triplet repeat A repeating sequence of three base pairs, such as the CGG repeat responsible for fragile X, that increases in number of repeats over several generations.

expression QTL (eQTL) Treating gene expression as a phenotype, QTLs can be identified that account for genetic influence on gene expression.

F_1, F_2 The offspring in the first and second generations following mating between two inbred strains.

familial Resemblance among family members.

family study Assessing the resemblance between genetically related parents and offspring, and between siblings living together. Resemblance can be due to heredity or to shared family environment.

first-degree relative See *genetic relatedness.*

fragile X Fragile sites are breaks in chromosomes that occur when chromosomes are stained or cultured. Fragile X is a fragile site on the X chromosome that is the second most important cause after Down syndrome of mental retardation in males, and is due to an expanded triplet repeat.

full siblings Individuals who have both biological (birth) parents in common.

functional genomics The study of how genes work in the sense of tracing pathways between genes, brain, and behavior. Usually implies a bottom-up approach that begins with molecules in a cell, in contrast to *behavioral genomics*.

gamete Mature reproductive cell (sperm or ovum) that contains a haploid (half) set of chromosomes.

gametic imprinting See *genomic imprinting.*

gene The basic unit of inheritance. A sequence of DNA bases that codes for a particular product. Includes DNA sequences that regulate transcription. (See also *allele; locus.*)

gene expression Transcription of DNA into mRNA.

gene expression profiling Using microarrays to assess the expression of all genes in the genome simultaneously.

gene frequency Can refer to either *allelic frequency* or *genotypic frequency*.

gene map Visual representation of the relative distances between genes or genetic markers on chromosomes.

gene targeting Mutations that are created in a specific gene and can then be transferred to an embryo.

genetical genomics Genes throughout the genome that affect gene expression. (See *transcriptome*.)

genetic anticipation See *anticipation*.

genetic counseling Conveys information about genetic risks and burdens, and helps individuals come to terms with the information and make their own decisions concerning actions.

genetic relatedness The extent or degree to which relatives have genes in common. *First-degree relatives* of the proband (parents and siblings) are 50 percent similar genetically. *Second-degree relatives* of the proband (grandparents, aunts, and uncles) are 25 percent similar genetically. *Third-degree relatives* of the proband (first cousins) are 12.5 percent similar genetically.

genome All the DNA sequences of an organism. The human genome contains about 3 billion DNA base pairs.

genomewide association An association study that assesses DNA variation throughout the genome.

genomic imprinting The process by which an allele at a given locus is expressed differently, depending upon whether it is inherited from the mother or the father.

genotype The genetic constitution of an individual, or the combination of alleles at a particular locus.

genotype-environment correlation Genetic influence on exposure to environment; experiences that are correlated with genetic propensities.

genotype-environment interaction Genetic sensitivity or susceptibility to environments. In quantitative genetics, genotype-environment interaction is usually limited to statistical interactions, such as genetic effects that differ in different environments.

genotypic frequency The frequency of alleles considered two at a time as they are inherited in individuals. The genotypic frequency of PKU individuals (homozygous for the recessive PKU allele) is 0.0001. The genotypic frequency of PKU carriers (who are heterozygous for the PKU allele) is 0.02. (See Box 2.2.)

half siblings Individuals who have just one biological (birth) parent in common.

haploid genotype (haplotype) The DNA sequence on one chromosome. In contrast to *genotype*, which refers to a pair of chromosomes, the DNA sequence on one chromosome is called a *haploid genotype*, which has been shortened to *haplotype*.

haplotype block A series of SNPs that are very highly correlated (i.e., seldom separated by recombination). The HapMap Project is systematizing haplotype blocks for several ethnic groups (http://www.hapmap.org/index.html.en).

Hardy-Weinberg equilibrium Allelic and genotypic frequencies remain the same, generation after generation, in the absence of forces such as natural selection that change these frequencies. If a two-allele locus is in Hardy-Weinberg equilibrium, the frequency of genotypes is $p^2 + 2pq + q^2$, where p and q are the frequencies of the two alleles.

heritability The proportion of phenotypic differences among individuals that can be attributed to genetic differences in a particular population. *Broad-sense heritability* involves all additive and nonadditive sources of genetic variance, whereas *narrow-sense heritability* is limited to additive genetic variance.

heterosis See *hybrid vigor.*

heterozygosity The presence of different alleles at a given locus on both members of a chromosome pair.

homeobox gene A class of genes that is highly conserved across animal species. These genes act as master switches that control the timing of development of different parts of the body, by coding for a protein that switches on cascades of other genes.

homozygosity The presence of the same allele at a given locus on both members of a chromosome pair.

hybrid vigor The increase in viability and fertility that can occur during outbreeding, for example, when inbred strains are crossed. The increase in heterozygosity masks the effects of deleterious recessive alleles.

imprinting See *genomic imprinting.*

inbred strain study Comparing inbred strains, created by mating brothers and sisters for at least 20 generations. Differences between strains can be attributed to their genetic differences when the strains are reared in the same laboratory environment. Differences within strains estimate environmental influences, because all individuals within an inbred strain are virtually identical genetically.

inbreeding Mating between genetically related individuals.

inbreeding depression A reduction in viability and fertility that can occur following inbreeding, which makes it more likely that offspring will have the same alleles at any locus and that deleterious recessive traits will be expressed.

inclusive fitness The reproductive fitness of an individual plus part of the fitness of kin that is genetically shared by the individual.

index case See *proband.*

indirect association An association between a trait and a DNA marker that is not itself the functional polymorphism that causes the association. In contrast to *direct association,* in which the DNA marker itself is the functional polymorphism.

innate Evolved capacities and constraints, not rigid hard-wiring that is impervious to experience.

instinct An innate behavioral tendency.

intergenic The 98% of the genome that is in between the coding regions of genes.

intron DNA sequence within a gene that is transcribed into messenger RNA but spliced out before the translation into protein. (Compare with *exon.*)

kilobase (kb) 1000 base pairs of DNA.

knock-out Inactivation of a gene by gene targeting.

latent class analysis A multivariate technique that clusters traits or symptoms into hypothesized underlying or latent classes.

liability-threshold model A model which assumes that dichotomous disorders are due to underlying genetic liabilities that are distributed normally. The disorder appears only when a threshold of liability is exceeded.

lifetime expectancy See *morbidity risk estimate.*

linkage Close proximity of loci on a chromosome. Linkage is an exception to Mendel's second law of independent assortment, because closely linked loci are not inherited independently within families.

linkage analysis A technique that detects linkage between DNA markers and traits, used to map genes to chromosomes. (See also *DNA marker; linkage; mapping.*)

linkage disequilibrium A violation of the Hardy-Weinberg equilibrium in which markers are uncorrelated. It is most frequently used to describe how close DNA markers are together on a chromosome; linkage disequilibrium of 1.0 means that the alleles of the DNA markers are perfectly correlated; 0.0 means that there is complete nonrandom association.

locus (plural, loci) The site of a specific gene on a chromosome. Latin for "place."

LOD score Log of the odds, a statistical term that indicates whether two loci are linked or unlinked. A LOD score of $+3$ or higher is commonly accepted as showing linkage, and a score of -2 excludes linkage.

major histocompatibility complex (MHC) A highly polymorphic region in a gene-dense chromosomal region (human chromosome 6) with genes particularly involved in immune functions.

mapping Linkage of DNA markers to a chromosome and to specific regions of chromosomes.

meiosis Cell division that occurs during gamete formation and results in halving the number of chromosomes, so that each gamete contains only one member of each chromosome pair.

messenger RNA (mRNA) Processed RNA that leaves the nucleus of the cell and serves as a template for protein synthesis in the cell body.

methylation An epigenetic process by which gene expression is inactivated by adding a methyl group.

microarray Commonly known as gene chips, microarrays are surfaces the size of a postage stamp with hundreds of thousands of DNA sequences that serve as probes to detect gene expression or single-nucleotide polymorphisms.

microRNA A class of non-coding RNA with just 21 base pairs.

microsatellite marker Two, three, or four DNA base pairs that are repeated up to a hundred times. Unlike SNPs which generally have just two alleles, microsatellite markers often have many alleles that are inherited in a Mendelian manner.

mitosis Cell division that occurs in somatic cells in which a cell duplicates itself and its DNA.

model fitting In quantitative genetics, a method to test the goodness of fit between a model of genetic and environmental relatedness and the observed data. Different models can be compared, and the best-fitting model is used to estimate genetic and environmental parameters.

molecular genetics The investigation of the effects of specific genes at the DNA level. In contrast to *quantitative genetics*, which investigates genetic and environmental components of variance.

monozygotic (MZ) Identical twins; literally, "one zygote."

morbidity risk estimate An incidence figure that is an estimate of the risk of being affected.

multiple-gene trait See *polygenic trait*.

multivariate genetic analysis Quantitative genetic analysis of the covariance between traits.

mutation A heritable change in DNA base-pair sequences.

natural selection The driving force in evolution in which individuals' alleles are spread, on the basis of the relative number of their surviving and reproducing offspring.

neurome Effects of the genome throughout the brain.

nonadditive genetic variance Individual differences due to the effects of alleles (dominance) or loci (epistasis) that interact with other alleles or loci. (In contrast, see *additive genetic variance*.)

non-coding RNA RNA that is transcribed but not translated into amino acid sequences. Can regulate gene expression by binding to mRNA.

nondisjunction Uneven division of members of a chromosome pair during meiosis.

nonshared environment Environmental influences that do not contribute to resemblance between family members.

nucleus The part of the cell that contains chromosomes.

odds ratio An effect size statistic for association calculated as the odds of an allele in cases divided by the odds of the allele in controls. An odds ratio of 1.0 means that there is no difference in allele frequency between cases and controls.

pedigree A family tree. Diagram depicting the genealogical history of a family, especially showing the inheritance of a particular condition in the family members.

pharmacogenetics and -genomics The genetics and genomics of responses to drugs.

phenotype An observed characteristic of an individual that results from the combined effects of genotype and environment.

pleiotropy Multiple effects of a gene.

polygenic trait A trait influenced by many genes.

polymerase chain reaction (PCR) A method to amplify a particular DNA sequence.

polymorphism A locus with two or more alleles. Greek for "multiple forms."

population genetics The study of allelic and genotypic frequencies in populations and the forces that change these frequencies, such as natural selection.

posttranslational modification Chemical change to polypeptides (amino acid sequences) after they have been translated from mRNA.

premutation Production of eggs or sperm with an unstable expanded number of repeats (up to 200 repeats for fragile X).

primer A short (usually 20-base) DNA sequence that marks the starting point for DNA replication. Primers on either side of a polymorphism mark the boundaries of a DNA sequence that is to be amplified by polymerase chain reaction (PCR).

proband The index case from whom other family members are identified.

proteome All the proteins translated from RNA (transcriptome).

psychopharmacogenetics and -genomics The genetics and genomics of behavioral responses to drugs.

QTL linkage analysis Linkage analysis that searches for linkages of small effect size, quantitative trait loci (QTLs). Most widely used is the affected sib-pair QTL linkage design.

qualitative disorder An either-or trait, usually a diagnosis.

quantitative dimension Psychological and physical traits that are continuously distributed within a population, for example, general cognitive ability, height, and blood pressure.

quantitative genetics A theory of multiple-gene influences that, together with environmental variation, result in quantitative (continuous) distributions of phenotypes. Quantitative genetic methods (such as the twin and adoption methods for human analysis, and inbred strain and selection methods for nonhuman analysis) estimate genetic and environmental contributions to phenotypic variance and covariance in a population.

quantitative trait loci (QTL) Genes of various effect sizes in multiple-gene systems that contribute to quantitative (continuous) variation in a phenotype.

recessive An allele that produces a particular phenotype only when present in the homozygous state.

recombinant inbred strains Inbred strains derived from brother-sister matings from an initial cross of two inbred progenitor strains. Called *recombinant* because, in the F$_2$ and subsequent generations, chromosomes from the progenitor strains recombine and exchange parts. Used to map genes.

recombination During meiosis, chromosomes exchange parts by a crossing over of chromatids.

recombinatorial hotspot Chromosomal location subject to much recombination. Often mark the boundaries of haplotype blocks.

repeat sequence Short sequences of DNA—two, three, or four nucleotide bases of DNA—that repeat a few times to a few dozen times. Used as DNA markers.

restriction enzyme Recognizes specific short DNA sequences and cuts DNA at that site.

restriction fragment length polymorphism (RFLP) DNA marker characterized by presence or absence of a particular sequence of DNA (restriction site) recognized by a specific restriction enzyme, which cuts the DNA at that site. Such sites are recognized by variation in the length of DNA fragments generated after DNA is digested with the particular restriction enzyme.

ribosome A small dense structure in the cell body (cytoplasm) that assembles amino acid sequences in the order dictated by mRNA.

RNA interference (RNAi) The use of double-stranded RNA to change the expression of the gene that shares its sequence. Also called *small interfering RNA* (siRNA) because it degrades complementary RNA transcripts.

second-degree relative See *genetic relatedness.*

segregation The process by which two alleles at a locus, one from each parent, separate during heredity. Mendel's law of segregation is his first law of heredity.

selection study Breeding for a phenotype over several generations by selecting parents with high scores on the phenotype, mating them, and assessing their offspring to determine the response to selection. Bidirectional selection studies also select in the other direction, that is, for low scores.

selective placement Adoption of children into families in which the adoptive parents are similar to the children's biological parents.

sex chromosome See *autosome.*

sex-linked trait See *X-linked trait.*

shared environment Environmental factors responsible for resemblance between family members.

simple sequence repeats (SSR) DNA markers that consist of two, three, or four DNA bases that repeat several times and are distributed throughout the genome for unknown reasons. The most common is the two-base (dinucleotide) repeat CA (cytosine followed by adenine).

single nucleotide polymorphism (SNP) The most common type of DNA polymorphism which involves a mutation in a single nucleotide. SNPs (pronounced "snips") can produce a change in an amino acid sequence (called *nonsynonymous,* i.e., not synonymous).

small interfering RNA (siRNA) See *RNA interference (RNAi).*

sociobiology An extension of evolutionary theory that focuses on inclusive fitness and kin selection.

somatic cells All cells in the body except gametes.

stabilizing selection Selection that maintains genetic variation within a population, for example, selection for intermediate phenotypic values.

synteny Loci on the same chromosome. *Synteny homology* refers to similar ordering of loci in chromosomal regions in different species.

targeted mutation A process by which a gene is changed in a specific way to alter its function, such as knock-outs. Called *transgenics* when the mutated gene is transferred from another species.

third-degree relative See *genetic relatedness.*

tiling microarray A microarray that covers, or "tiles," the entire genome for genomewide gene expression analysis.

transcription The synthesis of an RNA molecule from DNA in the cell nucleus.

transcriptome RNA transcribed from all the DNA in the genome.

transgenic Containing foreign DNA. For example, gene targeting can be used to replace a gene with a nonfunctional substitute in order to knock out the gene's functioning.

translation Assembly of amino acids into peptide chains on the basis of information encoded in messenger RNA. Occurs on ribosomes in the cell cytoplasm.

triplet code See *codon.*

triplet repeat See *expanded triplet repeat.*

trisomy Having three copies of a particular chromosome due to nondisjunction.

twin study Comparing the resemblance of identical and fraternal twins to estimate genetic and environmental components of variance.

variable expression A single genetic effect may result in variable manifestations in different individuals.

whole genome amplification Using a few restriction enzymes in polymerase chain reactions (PCRs) to chop up and amplify the entire genome. This makes microarrays possible.

X-linked trait A phenotype influenced by a gene on the X chromosome.

zygote The cell, or fertilized egg, resulting from the union of a sperm and an egg (ovum).

REFERENCES

Abecasis, G.R., Cardon, L.R., & Cookson, W.O. (2000). A general test of association for quantitative traits in nuclear families. *American Journal of Human Genetics, 66,* 279–292.

Abelson, J.F., Kwan, K.Y., O'Roak, B.J., Baek, D.Y., Stillman, A.A., Morgan, T.M., et al. (2005). Sequence variants in SLITRK1 are associated with Tourette's syndrome. *Science, 310,* 317–320.

Abelson, P., & Kennedy, D. (2004). The obesity epidemic. *Science, 304,* 1413.

Addington, A.M., Gornick, M., Duckworth, J., Sporn, A., Gogtay, N., Bobb, A., et al. (2005). GAD1 (2q31.1), which encodes glutamic acid decarboxylase (GAD67), is associated with childhood-onset schizophrenia and cortical gray matter volume loss. *Molecular Psychiatry, 10,* 581–588.

Aebersold, R., & Mann, M. (2003). Mass spectrometry-based proteomics. *Nature, 422,* 198–207.

Agrawal, A., Heath, A.C., Grant, J.D., Pergadia, M.L., Statham, D.J., Bucholz, K.K., et al. (2006). Assortative mating for cigarette smoking and for alcohol consumption in female Australian twins and their spouses. *Behavior Genetics, 36,* 553–566.

Agrawal, A., Jacobson, K.C., Prescott, C.A., & Kendler, K.S. (2002). A twin study of sex differences in social support. *Psychological Medicine, 32,* 1155–1164.

Agrawal, N., Sinha, S.N., & Jensen, A.R. (1984). Effects of inbreeding on Raven matrices. *Behavior Genetics, 14,* 579–585.

Ainsworth, M.D.S., Blehar, M.C., Waters, E., & Wall, S. (1978). *Patterns of attachment: A psychological study of the strange situation.* Hillsdale, NJ: Lawrence Erlbaum Associates.

Alarcón, M., DeFries, J.C., Light, J.G., & Pennington, B.F. (1997). A twin study of mathematics disability. *Journal of Learning Disabilities, 30,* 617–623.

Alarcón, M., Plomin, R., Fulker, D.W., Corley, R., & DeFries, J.C. (1998). Multivariate path analysis of specific cognitive abilities data at 12 years of age in the Colorado Adoption Project. *Behavior Genetics, 28,* 255–264.

Allen, M.G. (1976). Twin studies of affective illness. *Archives of General Psychiatry, 33,* 1476–1478.

Allison, D.B., Cui, X., Page, G.P., & Sabripour, M. (2006). Microarray data analysis: From disarray to consolidation and consensus. *Nature Reviews Genetics, 7,* 55–65.

Althoff, R.R., Faraone, S.V., Rettew, D.C., Morley, C.P., & Hudziak, J.J. (2005). Family, twin, adoption, and molecular genetic studies of juvenile bipolar disorder. *Bipolar Disorders, 7,* 598–609.

Altmuller, J., Palmer, L.J., Fischer, G., Scherb, H., & Wjst, M. (2001). Genomewide scans of complex human diseases: True linkage is hard to find. *American Journal of Human Genetics, 69,* 936–950.

Amir, R.E., Van den Veyer, I.B., Wan, M., Tran, C.Q., Francke, U., & Oghbi, H.Y. (1999). Rett syndrome is caused by mutations in X-linked MECP2, encoding methyl-CpG-binding protein 2. *Nature Genetics, 23,* 185–188.

Anderson, L.T., & Ernst, M. (1994). Self-injury in Lesch-Nyhan disease. *Journal of Autism and Developmental Disorders, 24,* 67–81.

Andreasen, N.C. (Ed.) (2005). *Research advances in genetics and genomics: Implications for psychiatry.* Washington, DC: American Psychiatric Publishing.

Andrews, G., Morris-Yates, A., Howie, P., & Martin, N. (1991). Genetic factors in stuttering confirmed. *Archives of General Psychiatry, 48,* 1034–1035.

Andrews, G., Stewart, G., Allen, R., & Henderson, A.S. (1990). The genetics of six neurotic disorders: A twin study. *Journal of Affective Disorders, 19,* 23–29.

Anholt, R.R., & Mackay, T.F. (2004). Quantitative genetic analyses of complex behaviours in *Drosophila. Nature Reviews Genetics, 5,* 838–849.

Antoch, M.P., Song, E.J., Chang, A.M., Vitaterna, M.H., Zhao, Y., Wilsbacher, L.D., et al. (1997). Functional identification of the mouse circadian clock gene by transgenic BAC rescue. *Cell, 89,* 655–667.

Antonarakis, S.E., & Epstein, C.J. (2006). The challenge of Down syndrome. *Trends in Molecular Medicine, 12,* 473–479.

Antonarakis, S.E., Lyle, R., Dermitzakis, E.T., Reymond, A., & Deutsch, S. (2004). Chromosome 21 and Down syndrome: From genomics to pathophysiology. *Nature Reviews Genetics, 5,* 725–738.

Arseneault, L., Moffitt, T.E., Caspi, A., Taylor, A., Rijsdijk, F.V., Jaffee, S.R., et al. (2003). Strong genetic effects on cross-situational antisocial behaviour among 5-year-old children according to mothers, teachers, examiner-observers, and twins' self-reports. *Journal of Child Psychology and Psychiatry and Allied Disciplines, 44,* 832–848.

Arvey, R.D., Bouchard, T.J., Jr., Segal, N.L., & Abraham, L.M. (1989). Job satisfaction: Environmental and genetic components. *Journal of Applied Psychology, 74,* 187–192.

Asbury, K., Dunn, J.F., Pike, A., & Plomin, R. (2003). Nonshared environmental influences on individual differences in early behavioral development: An MZ differences study. *Child Development, 74,* 933–943.

Asbury, K., Dunn, J.F., & Plomin, R. (2006a). The use of discordant MZ twins to generate hypotheses regarding non-shared environmental influence on anxiety in middle childhood. *Social Development, 15,* 564–570.

Asbury, K., Dunn, J.F., & Plomin, R. (2006b). Birthweight-discordance and differences in early parenting relate to monozygotic twin differences in behaviour problems and academic achievement at age 7. *Developmental Science, 9,* F22–F31.

Asbury, K., Wachs, T., & Plomin, R. (2005). Environmental moderators of genetic influence on verbal and nonverbal abilities in early childhood. *Intelligence, 33,* 643–661.

Atran, S. (2005). *In gods we trust: The evolutionary landscape of religion.* Oxford: Oxford University Press.

Bacchelli, E., & Maestrini, E. (2006). Autism spectrum disorders: Molecular genetic advances. *American Journal of Medical Genetics. Part C: Seminars in Medical Genetics, 142,* 13–23.

Badner, J.A., & Gershon, E.S. (2002). Meta-analysis of whole-genome linkage scans of bipolar disorder and schizophrenia. *Molecular Psychiatry, 7,* 405–411.

Bailey, A., Palferman, S., Heavey, L., & Le Couteur, A. (1998). Autism: The phenotype in relatives. *Journal of Autism and Developmental Disorders, 28,* 369–392.

Bailey, A., Phillips, W., & Rutter, M. (1996). Autism: Towards an integration of clinical, genetic, neuropsychological, and neurobiological perspectives. *Journal of Child Psychology and Psychiatry, 37,* 89–126.

Bailey, J.M., Dunne, M.P., & Martin, N.G. (2000). Genetic and environmental influences on sexual orientation and its correlates in an Australian twin sample. *Journal of Personality and Social Psychology, 78,* 524–536.

Bailey, J.M., & Pillard, R.C. (1991). A genetic study of male sexual orientation. *Archives of General Psychiatry, 48,* 1089–1096.

Bailey, J.M., Pillard, R.C., Dawood, K., Miller, M.B., Farrer, L.A., Trivedi, S., et al. (1999). A family history study of male sexual orientation using three independent samples. *Behavior Genetics, 29,* 79–86.

Bailey, J.M., Pillard, R.C., Neale, M.C., & Agyei, Y. (1993). Heritable factors influence sexual orientation in women. *Archives of General Psychiatry, 50,* 217–223.

Bailey, J.N., Breidenthal, S.E., Jorgensen, M.J., McCracken, J.T., & Fairbanks, L.A. (2007). The association of *DRD4* and novelty seeking is found in a nonhuman primate model. *Psychiatric Genetics, 17,* 23–27.

Bakermans-Kranenburg, M.J., van Uzendoorn, M.H., Bokhorst, C.L., & Schuengel, C. (2004). The importance of shared environment in infant-father attachment: A behavioral genetic study of the attachment q-sort. *Journal of Family Psychology, 18,* 545–549.

Bakwin, H. (1971). Enuresis in twins. *American Journal of Diseases in Children, 21,* 222–225.

Bakwin, H. (1973). Reading disability in twins. *Developmental Medicine and Child Neurology, 15,* 184–187.

Ball, D., & Collier, D. (2002). Substance misuse. In P. McGuffin, M.J. Owen, & I.I. Gottesman (Eds.), *Psychiatric genetics and genomics* (pp. 267–302). Oxford: Oxford University Press.

Baltes, P.B. (1993). The aging mind: Potential and limits. *Gerontologist, 33,* 580–594.

Barash, D.P., & Barash, N.R. (2005). *Madame Bovary's ovaries: A Darwinian look at literature.* New York: Delacorte Press.

Baron, M., Freimer, N.F., Risch, N., Lerer, B., & Alexander, J.R. (1993). Diminished support for linkage between manic depressive illness and X-chromosome markers in three Israeli pedigrees. *Nature Genetics, 3,* 49–55.

Baron, M., Gruen, R., Asnis, L., & Lord, S. (1985). Familial transmission of schizotypal and borderline personality disorder. *American Journal of Psychiatry, 142,* 927–934.

Bartels, M. (2007). An update on longitudinal twin and family studies. *Twin Research and Human Genetics, 10,* 3–12.

Bartels, M., Rietveld, M.J., van Baal, G.C., & Boomsma, D.I. (2002a). Genetic and environmental influences on the development of intelligence. *Behavior Genetics, 32,* 237–249.

Bartels, M., Rietveld, M.J., van Baal, G.C., & Boomsma, D.I. (2002b). Heritability of educational achievement in 12-year-olds and the overlap with cognitive ability. *Twin Research, 5,* 544–553.

Bashi, J. (1977). Effects of inbreeding on cognitive performance. *Nature, 266,* 440–442.

Bates, G.P. (2005). History of genetic disease: The molecular genetics of Huntington disease—A history. *Nature Reviews Genetics, 6,* 766–773.

Bates, T.C., Luciano, M., Castles, A., Coltheart, M., Wright, M.J., & Martin, N.G. (2007). Replication of reported linkages for dyslexia and spelling and suggestive evidence for novel regions on chromosomes 4 and 17. *European Journal of Human Genetics, 15,* 194–203.

Baum, A., & Posluszny, D.M. (1999). Health psychology: Mapping biobehavioral contributions to health and illness. *Annual Review of Psychology, 50,* 137–163.

Bearden, C.E., & Freimer, N.B. (2006). Endophenotypes for psychiatric disorders: Ready for primetime? *Trends in Genetics, 22,* 306–313.

Beaujean, A.A. (2005). Heritability of cognitive abilities as measured by mental chronometric tasks: A meta-analysis. *Intelligence, 33,* 187–201.

Bell, C.G., Walley, A.J., & Froguel, P. (2005). The genetics of human obesity. *Nature Reviews Genetics, 6,* 221–234.

Bellack, A.S. (2006). Scientific and consumer models of recovery in schizophrenia: Concordance, contrasts, and implications. *Schizophrenia Bulletin, 32,* 432–442.

Bender, B.G., Linden, M.G., & Robinson, A. (1993). Neuropsychological impairment in 42 adults with sex chromosome abnormalities. *American Journal of Medical Genetics. PartB: Neuropsychiatric Genetics, 48,* 169–173.

Benjamin, J., Ebstein, R., & Belmaker, R.H. (2002). *Molecular genetics and the human personality.* Washington, DC: American Psychiatric Publishing.

Benjamin, J., Li, L., Patterson, C., Greenburg, B.D., Murphy, D.L., & Hamer, D.H. (1996). Population and familial association between the D4 dopamine receptor gene and measures of novelty seeking. *Nature Genetics, 12,* 81–84.

Bennett, B., Carosone-Link, P., Zahniser, N.R., & Johnson, T.E. (2006). Confirmation and fine mapping of ethanol sensitivity quantitative trait loci, and candidate gene testing in the LXS recombinant inbred mice. *Journal of Pharmacological Experimental Theory, 319,* 299–307.

Bennett, B., Downing, C., Parker, C., & Johnson, T.E. (2006). Mouse genetic models in alcohol research. *Trends in Genetics, 22,* 367–374.

Benzer, S. (1973). Genetic dissection of behavior. *Scientific American, 229,* 24–37.

Berezikov, E., Thuemmler, F., van Laake, L.W., Kondova, I., Bontrop, R., Cuppen, E., et al. (2006). Diversity of microRNAs in human and chimpanzee brain. *Nature Genetics, 38,* 1375–1377.

Bergeman, C.S. (1997). *Aging: Genetic and environmental influences.* Newbury Park, CA: Sage.

Bergeman, C.S., Chipuer, H.M., Plomin, R., Pedersen, N.L., McClearn, G.E., Nesselroade, J.R., et al. (1993). Genetic and environmental effects on openness to experience, agreeableness, and conscientiousness: An adoption/twin study. *Journal of Personality, 61,* 159–179.

Bergeman, C.S., Plomin, R., McClearn, G.E., Pedersen, N.L., & Friberg, L. (1988). Genotype-environment interaction in personality development: Identical twins reared apart. *Psychology and Aging, 3,* 399–406.

Bergeman, C.S., Plomin, R., Pedersen, N.L., & McClearn, G.E. (1991). Genetic mediation of the relationship between social support and psychological well-being. *Psychology and Aging, 6,* 640–646.

Bergeman, C.S., Plomin, R., Pedersen, N.L., McClearn, G.E., & Nesselroade, J.R. (1990). Genetic and environmental influences on social support: The Swedish Adoption/Twin Study of Aging. *Journal of Gerontology, 45,* 101–106.

Bernards, R. (2006). Exploring the uses of RNAi-gene knockdown and the Nobel Prize. *The New England Journal of Medicine, 355,* 2391–2393.

Bertelsen, A. (1985). Controversies and consistencies in psychiatric genetics. *Acta Paediatrica Scandinavica, 71,* 61–75.

Bertram, L., McQueen, M.B., Mullin, K., Blacker, D., & Tanzi, R.E. (2007). Systematic meta-analyses of Alzheimer's disease genetic association studies: The AlzGene database. *Nature Genetics, 39,* 17–23.

Bertram, L., & Tanzi, R.E. (2004). Alzheimer's disease: One disorder, too many genes? *Human Molecular Genetics, 13,* R135–R141.

Bessman, S.P., Williamson, M.L., & Koch, R. (1978). Diet, genetics, and mental retardation interaction between phenylketonuric heterozygous mother and fetus to produce non-specific diminution of IQ: Evidence in support of the justification hypothesis. *Proceedings of the National Academy of Sciences (USA), 78,* 1562–1566.

Betsworth, D.G., Bouchard, T.J., Jr., Cooper, C.R., Grotevant, H.D., Hansen, J.I.C., Scarr, S., et al. (1994). Genetic and environmental influences on vocational interests assessed using adoptive and biological families and twins reared apart and together. *Journal of Vocational Behavior, 44,* 263–278.

Bickle, J. (2003). *Philosophy and neuroscience: A ruthlessly reductive account.* Boston: Kluwer Academic.

Biederman, J., Faraone, S.V., Keenan, K., Benjamin, J., Krifcher, B., Moore, C., et al. (1992). Further evidence for family-genetic risk factors in attention-deficit hyperactivity disorder. Patterns of comorbidity in probands and relatives psychiatrically and pediatrically referred samples. *Archives of General Psychiatry, 49,* 728–738.

Bienvenu, T., & Chelly, J. (2006). Molecular genetics of Rett syndrome: When DNA methylation goes unrecognized. *Nature Reviews Genetics, 7,* 415–426.

Biesecker, B.B., & Marteau, T. (1999). The future of genetic counseling: An international perspective. *Nature Genetics, 22,* 133–137.

Binder, E.B., & Holsboer, F. (2006). Pharmacogenomics and antidepressant drugs. *Annals of Medicine, 38,* 82–94.

Bishop, D.V. (2006). What causes specific language impairment in children? *Current Directions in Psychological Science, 15,* 217–221.

Blaser, R., & Gerlai, R. (2006). Behavioral phenotyping in zebrafish: Comparison of three behavioral quantification methods. *Behavior Research Methods, 38,* 456–469.

Blonigen, D.M., Hicks, B.M., Krueger, R.F., Patrick, C.J., & Iacono, W.G. (2006). Continuity and change in psychopathic traits as measured via normal-range personality: A longitudinal-biometric study. *Journal of Abnormal Psychology, 115,* 85–95.

Bloom, F.E., & Kupfer, D.J. (1995). *Psychopharmacology: A fourth generation of progress.* New York: Raven Press.

Bobb, A.J., Castellanos, F.X., Addington, A.M., & Rapoport, J.L. (2006). Molecular genetic studies of ADHD: 1991 to 2004. *American Journal of Medical Genetics. Part B: Neuropsychiatric Genetics, 141B,* 551–565.

Bock, G.R., & Goode, J.A. (1996). *Genetics of criminal and anti-social behaviour.* Chichester, UK: John Wiley & Sons.

Boehm, T., & Zufall, F. (2006). MHC peptides and the sensory evaluation of genotype. *Trends in Neuroscience, 29,* 100–107.

Bohman, M. (1996). Predisposition to criminality: Swedish adoption studies in retrospect. In G.R. Bock & J.A. Goode (Eds.), *Genetics of criminal and antisocial behaviour* (pp. 99–114). Chichester, UK: John Wiley & Sons.

Bohman, M., Cloninger, C.R., Sigvardsson, S., & von Knorring, A.L. (1982). Predisposition to petty criminals in Swedish adoptees. I. Genetic and environmental heterogeneity. *Archives of General Psychiatry, 39,* 1233–1241.

Bohman, M., Cloninger, C.R., von Knorring, A.L., & Sigvardsson, S. (1984). An adoption study of somatoform disorders. III. Cross-fostering analysis and genetic relationship to alcoholism and criminality. *Archives of General Psychiatry, 41,* 872–878.

Bokhorst, C.L., Bakermans-Kranenburg, M.J., Fearon, R.M., van Ijzendoorn, M.H., Fonagy, P., & Schuengel, C. (2003). The importance of shared environment in mother-infant attachment security: A behavioral genetic study. *Child Development, 74,* 1769–1782.

Bolinskey, P.K., Neale, M.C., Jacobson, K.C., Prescott, C.A., & Kendler, K.S. (2004). Sources of individual differences in stressful life event exposure in male and female twins. *Twin Research, 7,* 33–38.

Bolton, D., Eley, T.C., O'Connor, T.G., Perrin, S., Rabe-Hesketh, S., Rijsdijk, F.V., et al. (2006). Prevalence and genetic and environmental influences on anxiety disorders in 6-year-old twins. *Psychological Medicine, 36,* 335–344.

Bolton, D., & Hill, J. (2004). *Mind, meaning and mental disorder: The nature of causal explanation in psychology and psychiatry.* Oxford: Oxford University Press.

Böök, J.A. (1957). Genetical investigation in a north Swedish population: The offspring of first-cousin marriages. *Annals of Human Genetics, 21,* 191–221.

Boomsma, D., Busjahn, A., & Peltonen, L. (2002). Classical twin studies and beyond. *Nature Reviews Genetics, 3,* 872–882.

Borkenau, P., Riemann, R., Spinath, F.M., & Angleitner, A. (2006). Genetic and environmental influences on person × situation profiles. *Journal of Personality, 74,* 1451–1480.

Bouchard, C., Tremblay, A., Depres, J., Nadeau, A., Lupien, P.J., Theriault, G., et al. (1990). The response to long-term overfeeding in identical twins. *New England Journal of Medicine, 322,* 1477–1482.

Bouchard, T.J., Jr., & Loehlin, J.C. (2001). Genes, evolution, and personality. *Behavior Genetics, 31,* 243–273.

Bouchard, T.J., Jr., Lykken, D.T., McGue, M., Segal, N.L., & Tellegen, A. (1990). Sources of human psychological differences: The Minnesota Study of Twins Reared Apart. *Science, 250,* 223–228.

Bouchard, T.J., Jr., Lykken, D.T., McGue, M., Segal, N.L., & Tellegen, A. (1997). Genes, drives, environment, and experience: EPD theory revised. In C.P. Benbow & D. Lubinski (Eds.), *Intellectual talent: Psychometric and social issues* (pp. 5–43). Baltimore: John Hopkins University Press.

Bouchard, T.J., Jr., & McGue, M. (1981). Familial studies of intelligence: A review. *Science, 212,* 1055–1059.

Bouchard, T.J., Jr., & Propping, P. (Eds.). (1993). *Twins as a tool of behavioral genetics.* Chichester, UK: John Wiley & Sons.

Bovet, D. (1977). Strain differences in learning in the mouse. In A. Oliverio (Ed.), *Genetics, environment and intelligence* (pp. 79–92). Amsterdam: North-Holland.

Bovet, D., Bovet-Nitti, F., & Oliverio, A. (1969). Genetic aspects of learning and memory in mice. *Science, 163,* 139–149.

Braeckman, B.P., & Vanfleteren, J.R. (2007). Genetic control of longevity in *C. elegans. Experimental Gerontology, 42,* 90–98.

Brandenburg, N.A., Friedman, R.M., & Silver, S.E. (1990). The epidemiology of childhood psychiatric disorders: Prevalence findings from recent studies. *Journal of the American Academy of Child and Adolescent Psychiatry, 29,* 76–83.

Brandt, J., Welsh, K.A., Brietner, J.C., Folstein, M.F., Helms, M., & Christian, J.C. (1993). Hereditary influences on cognitive functioning in older men: A study of 4000 twin pairs. *Archives of Neurology, 50,* 599–603.

Bratko, D., (1997). Twin studies of verbal and spatial abilities. *Personality and Individual Differences, 23,* 365–369.

Bratko, D., & Butkovic, A. (2007). Stability of genetic and environmental effects from adolescence to young adulthood: Results of Croatian longitudinal twin study of personality. *Twin Research and Human Genetics, 10,* 151–157.

Braungart, J.M., Fulker, D.W., & Plomin, R. (1992). Genetic mediation of the home environment during infancy: A sibling adoption study of the HOME. *Developmental Psychology, 28,* 1048–1055.

Braungart, J.M., Plomin, R., DeFries, J.C., & Fulker, D.W. (1992). Genetic influence on tester-rated infant temperament as assessed by Bayley's Infant Behavior Record: Nonadoptive and adoptive siblings and twins. *Developmental Psychology, 28,* 40–47.

Bray, G.A. (1986). Effects of obesity on health and happiness. In K.E. Brownell & J.P. Foreyt (Eds.), *Handbook of eating disorders: Physiology, psychology and treatment of obesity, anorexia, and bulimia* (pp. 1–44). New York: Basic Books.

Breen, F.M., Plomin, R., & Wardle, J. (2006). Heritability of food preferences in young children. *Physiological Behavior, 88,* 443–447.

Breitner, J.C., Welsh, K.A., Gau, B.A., McDonald, W.M., Steffens, D.C., Saunders, A.M., et al. (1995). Alzheimer's disease in the National Academy of Sciences–National Research Council Registry of Aging Twin Veterans. III. Detection of cases, longitudinal results, and observations on twin concordance. *Archives of Neurology, 52,* 763–771.

Brennan, P.A., Mednick, S.A., & Jacobsen, B. (1996). Assessing the role of genetics in crime using adoption cohorts. In G.R. Bock & J.A. Goode (Eds.), *Genetics of criminal and anti-social behaviour* (pp. 115–128). Chichester, UK: John Wiley & Sons.

Brett, D., Pospisil, H., Valcárcel, J., Reich, J., & Bork, P. (2002). Alternative splicing and genome complexity. *Nature Genetics, 30,* 29–30.

Broadhurst, P.L. (1978). *Drugs and the inheritance of behaviour.* New York: Plenum.

Brody, N. (1992). *Intelligence* (2nd ed.). New York: Academic Press.

Broms, U., Silventoinen, K., Madden, P.A., Heath, A.C., & Kaprio, J. (2006). Genetic architecture of smoking behavior: A study of Finnish adult twins. *Twin Research and Human Genetics, 9,* 64–72.

Brouwer, S.I., van Beijsterveldt, T.C., Bartels, M., Hudziak, J.J., & Boomsma, D.I. (2006). Influences on achieving motor milestones: A twin-singleton study. *Twin Research and Human Genetics, 9,* 424–430.

Brown, S.D., Hancock, J.M., & Gates, H. (2006). Understanding mammalian genetic systems: The challenge of phenotyping in the mouse. *PLoS Genetics, 2,* E118.

Brunner, H.G. (1996). MAOA deficiency and abnormal behaviour. In G.R. Bock & J.A. Goode (Eds.), *Genetics of criminal and anti-social behaviour* (pp. 155–164). Chichester, UK: John Wiley & Sons.

Brunner, H.G., Nelen, M., Breakefield, X.O., Ropers, H.H., & van Oost, B.A. (1993). Abnormal behavior associated with a point mutation in the structural gene for monoamine oxidase A. *Science, 262,* 578–580.

Bruun, K., Markkananen, T., & Partanen, J. (1966). *Inheritance of drinking behaviour: A study of adult twins.* Helsinki, Finland: Finnish Foundation for Alcohol Research.

Buchwald, D., Herrell, R., Ashton, S., Belcourt, M., Schmaling, K., Sullivan, P., et al. (2001). A twin study of chronic fatigue. *Psychosomatic Medicine, 63,* 936–943.

Buck, K.J., Crabbe, J.C., & Belknap, J.K. (2000). Alcohol and other abused drugs. In D.W. Pfaff, W.H. Berrettini, T.H. Joh, & S.C. Maxson (Eds.), *Genetic influences on neural and behavioral functions* (pp. 159–183). Boca Raton, FL: CRC Press.

Buck, K.J., Rademacher, B.S., Metten, P., & Crabbe, J.C. (2002). Mapping murine loci for physical dependence on ethanol. *Psychopharmacology, 160,* 398–407.

Bulfield, G., Siller, W.G., Wight, P.A.L., & Moore, K.J. (1984). X chromosome-linked muscular dystrophy (mdx) in the mouse. *Proceedings of the National Academy of Sciences (USA), 81,* 1189–1192.

Bulik, C.M. (2005). Exploring the gene-environment nexus in eating disorders. *Journal of Psychiatry and Neuroscience, 30,* 335–339.

Bulik, C.M., Sullivan, P.F., Tozzi, F., Furberg, H., Lichtenstein, P., & Pedersen, N.L. (2006). Prevalence, heritability, and prospective risk factors for anorexia nervosa. *Archives of General Psychiatry, 63,* 305–312.

Bulik, C.M., Sullivan, P.F., Wade, T.D., & Kendler, K.S. (2000). Twin studies of eating disorders: A review. *International Journal of Eating Disorders, 27,* 1–20.

Buller, D.J. (2005a). *Adapting minds: Evolutionary psychology and the persistent quest for human nature.* London: Bradford Books.

Buller, D.J. (2005b). Evolutionary psychology: The emperor's new paradigm. *Trends in Cognitive Sciences, 9,* 277–283.

Bullock, B.M., Deater-Deckard, K., & Leve, L.D. (2006). Deviant peer affiliation and problem behavior: A test of genetic and environmental influences. *Journal of Abnormal Child Psychology, 34,* 29–41.

Burgess, R.L., & Drais, A.A. (1999). Beyond the "Cinderella effect"—Life history theory and child maltreatment. *Human Nature, 10,* 373–398.

Burke, K.C., Burke, J.D., Roe, D.S., & Regier, D.A. (1991). Comparing age at onset of major depression and other psychiatric disorders by birth cohorts in five U.S. community populations. *Archives of General Psychiatry, 48,* 789–795.

Burks, B. (1928). The relative influence of nature and nurture upon mental development: A comparative study on foster parent-foster child resemblance. *Yearbook of the National Society for the Study of Education. Part 1, 27,* 219–316.

Burt, S.A., Krueger, R.F., McGue, M., & Iacono, W. (2003). Parent-child conflict and the comorbidity among childhood externalizing disorders. *Archives of General Psychiatry, 60,* 505–513.

Burt, S.A., McGue, M., Carter, L.A., & Iacono, W.G. (2007). The different origins of stability and change in antisocial personality disorder symptoms. *Psychological Medicine, 37,* 27–38.

Burt, S.A., McGue, M., Iacono, W.G., & Krueger, R.F. (2006). Differential parent-child relationships and adolescent externalizing symptoms: Cross-lagged analyses within a monozygotic twin differences design. *Developmental Psychology, 42,* 1289–1298.

Burt, S.A., McGue, M., Krueger, R.F., & Iacono, W.G. (2005). How are parent-child conflict and childhood externalizing symptoms related over time? Results from a genetically informative cross-lagged study. *Development and Psychopathology, 17,* 145–165.

Busjahn, A. (2002). Twin registers across the globe: What's out there in 2002? *Twin Research, 5,* v–vi.

Buss, A.H., & Plomin, R. (1984). *Temperament: Early developing personality traits.* Hillsdale, NJ: Lawrence Erlbaum Associates.

Buss, D.M. (1994a). *The evolution of desire: Strategies of human mating.* New York: Basic Books.

Buss, D.M. (1994b). The strategies of human mating. *American Scientist, 82,* 238–249.

Buss, D.M. (1995). Evolutionary psychology—A new paradigm for psychological science. *Psychological Inquiry, 6*, 1–30.

Buss, D.M. (2003). *The evolution of desire: strategies of human mating*, revised edition. New York: Basic Books.

Buss, D.M. (Ed.). (2005a). *The handbook of evolutionary psychology*. New York: Wiley.

Buss, D.M. (2005b). *The murderer next door: Why the mind is designed to kill*. New York: Penguin Press.

Buss, D.M. (2007). *Evolutionary psychology: The new science of the mind* (3rd ed.). Boston: Allyn & Bacon.

Buss, D.M., & Greiling, H. (1999). Adaptive individual differences. *Journal of Personality, 67*, 209–243.

Butcher, L.M., Meaburn, E., Craig, I., Schalkwyk, L.C., & Plomin, R. (2006). A genomewide association scan for GE interplay in childhood. *American Journal of Medical Genetics. Part B: Neuropsychiatric Genetics, 141*, 722.

Butcher, L.M., Meaburn, E., Dale, P.S., Sham, P., Schalkwyk, L.C., Craig, I.W., et al. (2005). Association analysis of mild mental impairment using DNA pooling to screen 432 brain-expressed single-nucleotide polymorphisms. *Molecular Psychiatry, 10*, 384–392.

Butcher, L.M., Meaburn, E., Knight, J., Sham, P., Schalkwyk, L.C., Craig, I.W., et al. (2005). SNPs, microarrays, and pooled DNA: Identification of four loci associated with mild mental impairment in a sample of 6000 children. *Human Molecular Genetics, 14*, 1315–1325.

Butcher, L.M., Meaburn, E., Liu, L., Hill, L., Al-Chalabi, A., Plomin, R., et al. (2004). Genotyping pooled DNA on microarrays: A systematic genome screen of thousands of SNPs in large samples to detect QTLs for complex traits. *Behavior Genetics, 34*, 549–555.

Butler, R.J., Galsworthy, M.J., Rijsdijk, F., & Plomin, R. (2001). Genetic and gender influences on nocturnal bladder control: A study of 2900 3-year-old twin pairs. *Scandinavian Journal of Urology and Nephrology, 35*, 177–183.

Byrne, B., Samuelsson, S., Wadsworth, S., Hulslander, J., Corley, R., DeFries, J.C., et al. (2007). Longitudinal twin study of early literacy development: Preschool through grade 1. *Reading and Writing, 20*, 77–102.

Byrne, B., Wadsworth, S., Corley, R., Samuelsson, S., Quain, P., DeFries, J.C., et al. (2005). Longitudinal twin study of early literacy development: Preschool and kindergarten phases. *Scientific Studies of Reading, 9*, 219–235.

Cadoret, R.J., Cain, C.A., & Crowe, R.R. (1983). Evidence for gene-environment interaction in the development of adolescent antisocial behaviour. *Behavior Genetics, 13*, 301–310.

Cadoret, R.J., O'Gorman, T.W., Heywood, E., & Troughton, E. (1985). Genetic and environmental factors in major depression. *Journal of Affective Disorders, 9*, 155–164.

Cadoret, R.J., & Stewart, M.A. (1991). An adoption study of attention-deficit hyperactivity aggression and their relationship to adult antisocial personality. *Comprehensive Psychiatry, 32*, 73–82.

Cadoret, R.J., Yates, W.R., Troughton, E., & Woodworth, G. (1995). Genetic-environmental interaction in the genesis of aggressivity and conduct disorders. *Archives of General Psychiatry, 52*, 916–924.

Caldwell, B.M., & Bradley, R.H. (1978). *Home observation for measurement of the environment*. Little Rock: University of Arkansas.

Callinan, P.A., & Feinberg, A.P. (2006). The emerging science of epigenomics. *Human Molecular Genetics, 15,* R95–R101.

Camp, N.J., Lowry, M.R., Richards, R.L., Plenk, A.M., Carter, C., Hensel, C.H., et al. (2005). Genome-wide linkage analyses of extended Utah pedigrees identifies loci that influence recurrent, early-onset major depression and anxiety disorders. *American Journal of Medical Genetics. Part B: Neuropsychiatric Genetics, 135,* 85–93.

Cannon, T.D., Mednick, S.A., Parnas, J., Schulsinger, F., Praestholm, J., & Vestergaard, A. (1993). Developmental brain abnormalities in the offspring of schizophrenic mothers. I. Contributions of genetic and perinatal factors. *Archives of General Psychiatry, 50,* 551–564.

Canter, S. (1973). Personality traits in twins. In G. Claridge, S. Canter, & W.I. Hume (Eds.), *Personality differences and biological variations* (pp. 21–51). New York: Pergamon Press.

Cantor, C. (2005). *Evolution and posttraumatic stress.* London: Routledge.

Cantwell, D.P. (1975). Genetic studies of hyperactive children: Psychiatric illness in biological and adopting parents. In R.R. Fieve, D. Rosenthal, & H. Brill (Eds.), *Genetic research in psychiatry* (pp. 273–280). Baltimore: Johns Hopkins University Press.

Capecchi, M.R. (1994). Targeted gene replacement. *Scientific American, 270,* 52–59.

Capron, C., & Duyme, M. (1989). Assessment of the effects of socioeconomic status on IQ in a full cross-fostering study. *Nature, 340,* 552–554.

Capron, C., & Duyme, M. (1996). Effect of socioeconomic status of biological and adoptive parents on WISC-R subtest scores of their French adopted children. *Intelligence, 22,* 259–275.

Cardno, A.G., & Gottesman, I.I. (2000). Twin studies of schizophrenia: From bow-and-arrow concordances to Star Wars Mx and functional genomics. *American Journal of Medical Genetics, 97,* 12–17.

Cardno, A.G., Jones, L.A., Murphy, K.C., Sanders, R.D., Asherson, P.J., Owen, M.J., et al. (1998). Dimensions of psychosis in affected sibling pairs. *Schizophrenia Bulletin, 25,* 841–850.

Cardno, A.G., Rijsdijk, F.V., Sham, P.C., Murray, R.M., & McGuffin, P. (2002). A twin study of genetic relationships between psychotic symptoms. *American Journal of Psychiatry, 159,* 539–545.

Cardon, L.R. (1994a). Height, weight and obesity. In J.C. DeFries, R. Plomin, & D.W. Fulker (Eds.), *Nature and nurture during middle childhood* (pp. 165–172). Cambridge, MA: Blackwell.

Cardon, L.R. (1994b). Specific cognitive abilities. In J.C. DeFries, R. Plomin, & D.W. Fulker (Eds.), *Nature and nurture during middle childhood* (pp. 57–76). Oxford: Blackwell.

Cardon, L.R., & Abecasis, G.R. (2003). Using haplotype blocks to map human complex trait loci. *Trends in Genetics, 19,* 135–140.

Cardon, L.R., & Bell, J. (2001). Association study designs for complex diseases. *Nature Genetics, 2,* 91–99.

Cardon, L.R., & Fulker, D.W. (1993). Genetics of specific cognitive abilities. In R. Plomin & G.E. McClearn (Eds.), *Nature, nurture, and psychology* (pp. 99–120). Washington, DC: American Psychological Association.

Cardon, L.R., Smith, S.D., Fulker, D.W., Kimberling, W.J., Pennington, B.F., & DeFries, J.C. (1994). Quantitative trait locus for reading disability on chromosome 6. *Science, 266,* 276–279.

Carey, G. (1986). Sibling imitation and contrast effects. *Behavior Genetics, 16,* 319–341.

Carey, G. (1992). Twin imitation for anti-social behavior: Implications for genetic and family environmental research. *Journal of Abnormal Psychology, 101,* 18–25.

Carmelli, D., Swan, G.E., & Cardon, L.R. (1995). Genetic mediation in the relationship of education to cognitive function in older people. *Psychology and Aging, 10,* 48–53.

Carninci, P. (2006). Tagging mammalian transcription complexity. *Trends in Genetics, 22,* 501–510.

Caron, M.G. (1996). Images in neuroscience. Molecular biology, II. A dopamine transporter mouse knockout. *American Journal of Psychiatry, 153,* 1515.

Carroll, J.B. (1993). *Human cognitive abilities.* New York: Cambridge University Press.

Carroll, J.B. (1997). Psychometrics, intelligence, and public policy. *Intelligence, 24,* 25–52.

Caspi, A., McClay, J., Moffitt, T.E., Mill, J., Martin, J., Craig, I.W., et al. (2002). Role of genotype in the cycle of violence in maltreated children. *Science, 297,* 851–854.

Caspi, A., & Moffitt, T.E. (1995). The continuity of maladaptive behaviour: From description to understanding of antisocial behaviour. In D. Cicchetti & D.J. Cohen (Eds.), *Developmental psychopathology,* Volume 2: *Risk, disorder, and adaptation* (pp. 472–511). New York: Wiley.

Caspi, A., Moffitt, T.E., Cannon, M., McClay, J., Murray, R., Harrington, H., et al. (2005). Moderation of the effect of adolescent-onset cannabis use on adult psychosis by a functional polymorphism in the catechol-O-methyltransferase gene: Longitudinal evidence of a gene × environment interaction. *Biological Psychiatry, 57,* 1117–1127.

Caspi, A., Moffitt, T.E., Morgan, J., Rutter, M., Taylor, A., Arseneault, L., et al. (2004). Maternal expressed emotion predicts children's antisocial behavior problems: Using monozygotic-twin differences to identify environmental effects on behavioral development. *Developmental Psychology, 40,* 149–161.

Caspi, A., Sugden, K., Moffitt, T.E., Taylor, A., Craig, I.W., Harrington, H., et al. (2003). Influence of life stress on depression: Moderation by a polymorphism in the *5-HTT* gene. *Science, 301,* 386–389.

Cassidy, S.B., & Schwartz, S. (1998). Prader-Willi and Angelman syndromes: Disorders of genomic imprinting. *Medicine, 77,* 140–151.

Casto, S.D., DeFries, J.C., & Fulker, D.W. (1995). Multivariate genetic analysis of Wechsler Intelligence Scale for Children-Revised (WISC-R) factors. *Behavior Genetics, 25,* 25–32.

Cattell, R.B. (1982). *The inheritance of personality and ability.* New York: Academic Press.

Chapman, N.H., Igo, R.P., Thomson, J.B., Matsushita, M., Brkanac, Z., Holzman, T., et al. (2004). Linkage analyses of four regions previously implicated in dyslexia: Confirmation of a locus on chromosome 15*q. American Journal of Medical Genetics. Part B: Neuropsychiatric Genetics, 131,* 67–75.

Chawla, S. (1993). Demographic aging and development. *Generations, 17,* 20–23.

Chen, L., & Woo, S.L. (2005). Complete and persistent phenotypic correction of phenylketonuria in mice by site-specific genome integration of murine phenylalanine hydroxylase cDNA. *Proceedings of the National Academy of Sciences (USA), 102,* 15581–15586.

Cherniske, E.M., Carpenter, T.O., Klaiman, C., Young, E., Bregman, J., Insogna, K., et al. (2004). Multisystem study of 20 older adults with Williams syndrome. *American Journal of Medical Genetics: Part A, 131,* 255–264.

Cherny, S.S., Fulker, D.W., Emde, R.N., Robinson, J., Corley, R.P., Reznick, J.S., et al. (1994). Continuity and change in infant shyness from 14 to 20 months. *Behavior Genetics, 24,* 365–379.

Cherny, S.S., Fulker, D.W., & Hewitt, J.K. (1997). Cognitive development from infancy to middle childhood. In R.J. Sternberg & E.L. Grigorenko (Eds.), *Intelligence, heredity and environment* (pp. 463–482). Cambridge: Cambridge University Press.

Chesler, E.J., Lu, L., Shou, S., Qu, Y., Gu, J., Wang, J., et al. (2005). Complex trait analysis of gene expression uncovers polygenic and pleiotropic networks that modulate nervous system function. *Nature Genetics, 37,* 233–242.

Cheung, V.G., Conlin, L.K., Weber, T.M., Arcaro, M., Jen, K.Y., Morley, M., et al. (2003). Natural variation in human gene expression assessed in lymphoblastoid cells. *Nature Genetics, 33,* 422–425.

Chipuer, H.M., & Plomin, R. (1992). Using siblings to identify shared and non-shared HOME items. *British Journal of Developmental Psychology, 10,* 165–178.

Chipuer, H.M., Plomin, R., Pedersen, N.L., McClearn, G.E., & Nesselroade, J.R. (1993). Genetic influence on family environment: The role of personality. *Developmental Psychology, 29,* 110–118.

Chipuer, H.M., Rovine, M.J., & Plomin, R. (1990). LISREL modeling: Genetic and environmental influences on IQ revisited. *Intelligence, 14,* 11–29.

Christensen, K., Johnson, T.E., & Vaupel, J.W. (2006). The quest for genetic determinants of human longevity: Challenges and insights. *Nature Reviews Genetics, 7,* 436–448.

Christensen, K., Petersen, I., Skytthe, A., Herskind, A.M., McGue, M., & Bingley, P. (2006). Comparison of academic performance of twins and singletons in adolescence: Follow-up study. *British Medical Journal, 333,* 1095.

Christensen, K., Vaupel, J.W., Holm, N.V., & Yashlin, A.I. (1995). Mortality among twins after age 6: Foetal origins hypothesis versus twin method. *British Medical Journal, 310,* 432–436.

Christiansen, K.O. (1977). A preliminary study of criminality among twins. In S. Mednick & K.O. Christiansen (Eds.), *Biosocial bases of criminal behavior* (pp. 89–108). New York: Gardner Press, Inc.

Christiansen, L., Frederiksen, H., Schousboe, K., Skytthe, A., Wurmb-Schwark, N., Christensen, K., et al. (2003). Age- and sex-differences in the validity of questionnaire-based zygosity in twins. *Twin Research, 6,* 275–278.

Chua, S.C., Jr., Chung, W.K., Wu-Peng, X.S., Zhang, Y., Liu, S.M., Tartaglia, L., et al. (1996). Phenotypes of mouse diabetes and rat fatty due to mutations in the OB (leptin) receptor. *Science, 271,* 994–996.

Churchill, G.A., Airey, D.C., Allayee, H., Angel, J.M., Attie, A.D., Beatty, J., et al. (2004). The Collaborative Cross, a community resource for the genetic analysis of complex traits. *Nature Genetics, 36,* 1133–1137.

Cichowski, K., Shih, T.S., Schmitt, E., Santiago, S., Reilly, K., McLaughlin, M.E., et al. (1999). Mouse models of tumor development in neurofibromatosis type 1. *Science, 286,* 2172–2176.

Claridge, G., & Hewitt, J.K. (1987). A biometrical study of schizotypy in a normal population. *Personality and Individual Differences, 8,* 303–312.

Clementz, B.A., McDowell, J.E., & Zisook, S. (1994). Saccadic system functioning among schizophrenic patients and their first-degree biological relatives. *Journal of Abnormal Psychology, 103,* 277–287.

Cloninger, C.R. (1987). A systematic method for clinical description and classification of personality variants. A proposal. *Archives of General Psychiatry, 44,* 573–588.

Cloninger, C.R. (2002). The relevance of normal personality for psychiatrists. In J. Benjamin, R. Ebstein, & R.H. Belmaker (Eds.), *Molecular genetics and human personality.* New York: American Psychiatric Press.

Cloninger, C.R., Bohman, M., & Sigvardsson, S. (1981). Inheritance of alcohol abuse: Cross-fostering analysis of adopted men. *Archives of General Psychiatry, 38,* 861–868.

Cloninger, C.R., Sigvardsson, S., Bohman, M., & von Knorring, A.L. (1982). Predisposition to petty criminality in Swedish adoptees: II. Cross fostering analysis of gene-environment interaction. *Archives of General Psychiatry, 39,* 1242–1247.

Cloninger, C.R., Svrakic, D.M., & Przybeck, T.R. (1993). A psychobiological model of temperament and character. *Archives of General Psychiatry, 50,* 975–990.

Cobb, J.P., Mindrinos, M.N., Miller-Graziano, C., Calvano, S.E., Baker, H.V., Xiao, W., et al. (2005). Application of genome-wide expression analysis to human health and disease. *Proceedings of the National Academy of Sciences (USA), 102,* 4801–4806.

Cohen, B.H. (1964). Family patterns of mortality and life span. *Quarterly Review of Biology, 39,* 130–181.

Cohen, P., Cohen, J., Kasen, S., Velez, C.N., Hartmark, C., Johnson, J., et al. (1993). An epidemiological study of disorders in late childhood and adolescence. I. Age- and gender-specific prevalence. *Journal of Child Psychology and Psychiatry, 34,* 851–867.

Collier, D.A., Stober, G., Li, T., Heils, A., Catalano, M., Di Bella, D., et al. (1996). A novel functional polymorphism within the promoter of the serotonin transporter gene: Possible role in susceptibility to affective disorders. *Molecular Psychiatry, 1,* 453–460.

Collins, A.C. (1981). A review of research using short-sleep and long-sleep mice. In G.E. McClearn, R.A. Deitrich, & V.G. Erwin (Eds.), *Development of animal models as pharmocogenetic tools* (6th ed.). *USDHHS-NIAAA Research Monograph No.6* (pp. 161–170). Washington, DC: U.S. Government Printing Office.

Collins, F.S. (2006a). *The language of God: A scientist presents evidence for belief.* New York: Simon & Schuster.

Collins, F.S. (2006b). 2005 William Allan Award address. No longer just looking under the lamppost. *American Journal of Human Genetics, 79,* 421–426.

Collins, F.S., Green, E.D., Guttmacher, A.E., & Guyer, M.S. (2003). A vision for the future of genomics research. *Nature, 422,* 835–847.

Collins, F.S., & McKusick, V.A. (2001). Implications of the Human Genome Project for medical science. *Journal of the American Medical Association, 285,* 540–544.

Constantino, J.N., & Todd, R.D. (2003). Autistic traits in the general population: A twin study. *Archives of General Psychiatry, 60,* 524–530.

Cook, E.H., Jr., Courchesne, R.Y., Cox, N.J., Lord, C., Gonen, D., Guter, S.J., et al. (1998). Linkage-disequilibrium mapping of autistic disorder, with 15q11–13 markers. *American Journal of Human Genetics, 62,* 1077–1083.

Cooper, R.M., & Zubek, J.P. (1958). Effects of enriched and restricted early environments on the learning ability of bright and dull rats. *Canadian Journal of Psychology, 12,* 159–164.

Corder, E.H., Saunders, A.M., Risch, N.J., Strittmatter, W.J., Shmechel, D.E., Gaskell, P.C., Jr., et al. (1994). Protective effect of apolipoprotein E type 2 allele for late onset Alzheimer's disease. *Nature Genetics, 7,* 180–184.

Corder, E.H., Saunders, A.M., Strittmatter, W.J., Schmechel, D.E., Gaskell, P.C., Small, G.W., et al. (1993). Gene dose of apolipoprotein E type 4 allele and the risk of Alzheimer's disease in late onset families. *Science, 261,* 921–923.

Coren, S. (2005). *The intelligence of dogs: A guide to the thoughts, emotions, and inner lives of our canine companions.* New York: Simon & Schuster.

Correa, C.R., & Cheung, V.G. (2004). Genetic variation in radiation-induced expression phenotypes. *American Journal of Human Genetics, 75,* 885–890.

Costa, F.F. (2005). Non-coding RNAs: New players in eukaryotic biology. *Gene, 357,* 83–94.

Costa, P.T., & McCrae, R.R. (1994). Stability and change in personality from adolescence through adulthood. In C.F. Haverson, Jr., G.A. Kohnstamm, & R.P. Martin (Eds.), *The developing structure of temperament and personality from infancy to adulthood* (pp. 139–150). Hillsdale, NJ: Erlbaum.

Costa, R., & Kyriacou, C.P. (1998). Functional and evolutionary implications of variation in clock genes. *Current Opinion in Neurobiology, 8,* 659–664.

Cotton, N.S. (1979). The familial incidence of alcoholism. *Journal of Studies on Alcohol, 40,* 89–116.

Coude, F.X., Mignot, C., Lyonnet, S., & Munnich, A. (2006). Academic impairment is the most frequent complication of neurofibromatosis type-1 (NF1) in children. *Behavior Genetics, 36,* 660–664.

Crabbe, J.C., Belknap, J.K., & Buck, K.J. (1994). Genetic animal models of alcohol and drug abuse. *Science, 264,* 1715–1723.

Crabbe, J.C., & Harris, R.A. (1991). *The genetic basis of alcohol and drug actions.* New York: Plenum.

Crabbe, J.C., Kosobud, A., Young, E.R., Tam, B.R., & McSwigan, J.D. (1985). Bidirectional selection for susceptibility to ethanol withdrawal seizures in *Mus musculus. Behavior Genetics, 15,* 521–536.

Crabbe, J.C., Phillips, T.J., Buck, K.J., Cunningham, C.L., & Belknap, J.K. (1999). Identifying genes for alcohol and drug sensitivity: Recent progress and future directions. *Trends in Neurosciences, 22,* 173–179.

Crabbe, J.C., Phillips, T.J., Feller, D.J., Hen, R., Wenger, C.D., Lessov, C.N., et al. (1996). Elevated alcohol consumption in null mutant mice lacking 5-HT$_{1B}$ serotonin receptors. *Nature Genetics, 14,* 98–101.

Crabbe, J.C., Phillips, T.J., Harris, R.A., Arends, M.A., & Koob, G.F. (2006). Alcohol-related genes: Contributions from studies with genetically engineered mice. *Addiction Biology, 11,* 195–269.

Crabbe, J.C., Wahlsten, D., & Dudek, B.C. (1999). Genetics of mouse behavior: Interactions with laboratory environment. *Science, 284,* 1670–1672.

Craddock, N., & Forty, L. (2006). Genetics of affective (mood) disorders. *European Journal of Human Genetics, 14,* 660–668.

Craddock, N., O'Donovan, M.C., & Owen, M.J. (2005). The genetics of schizophrenia and bipolar disorder: Dissecting psychosis. *Journal of Medical Genetics, 42,* 193–204.

Craddock, N., & Owen, M.J. (2005). The beginning of the end for the Kraepelinian dichotomy. *British Journal of Psychiatry, 186,* 364–366.

Craddock, N., Owen, M.J., & O'Donovan, M.C. (2006). The catechol-O-methyl transferase (COMT) gene as a candidate for psychiatric phenotypes: Evidence and lessons. *Molecular Psychiatry, 11,* 446–458.

Crawford, C.B., & Salmon, C.A. (2004). *Evolutionary psychology, public policy and personal decisions.* Hillsdale, NJ: Lawrence Erlbaum Associates.

Crawford, D.C., Acuna, J.M., & Sherman, S.L. (2001). FMR1 and the fragile X syndrome: Human genome epidemiology review. *Genetics in Medicine, 3,* 359–371.

Crawley, J.N. (2003). Behavioral phenotyping of rodents. *Comparative Medicine, 53,* 140–146.

Crawley, J.N. (2007). *What's wrong with my mouse: Behavioral phenotyping of transgenic and knockout mice* (2nd ed.). Wilmington, DE: Wiley-Liss.

Cronk, N.J., Slutske, W.S., Madden, P.A., Bucholz, K.K., Reich, W., & Heath, A.C. (2002). Emotional and behavioral problems among female twins: An evaluation of the equal environments assumption. *Journal of the American Academy of Child and Adolescent Psychiatry, 41,* 829–837.

Crow, T.J. (1985). The two syndrome concept: Origins and current states. *Schizophrenia Bulletin, 11,* 471–486.

Crowe, R.R. (1972). The adopted offspring of women criminal offenders: A study of their arrest records. *Archives of General Psychiatry, 27,* 600–603.

Crowe, R.R. (1974). An adoption study of antisocial personality. *Archives of General Psychiatry, 31,* 785–791.

Cruts, M., van Duijn, C.M., Backhovens, H., van den Broeck, M., & Wehnert, A. (1998). Estimation of the genetic contribution of presenilin-1 and -2 mutations in a population-based study of presenile Alzheimer's disease. *Human Molecular Genetics, 71,* 43–51.

Cumings, J.L., & Benson, D.F. (1992). *Dementia: A clinical approach.* Boston: Butterworth.

Curry, J., Bebb, G., Moffat, J., Young, D., Khaidakov, M., Mortimer, A., et al. (1997). Similar mutant frequencies observed between pairs of monozygotic twins. *Human Mutation, 9,* 445–451.

Dale, P.S., Simonoff, E., Bishop, D.V.M., Eley, T.C., Oliver, B., Price, T.S., et al. (1998). Genetic influence on language delay in two-year-old children. *Nature Neuroscience, 1,* 324–328.

Daly, M., & Wilson, M. (1999). *The truth about Cinderella.* New Haven, CT: Yale University Press.

Daniels, J.K., Owen, M.J., McGuffin, P., Thompson, L., Detterman, D.K., Chorney, K., et al. (1994). IQ and variation in the number of fragile X CGG repeats: No association in a normal sample. *Intelligence, 19,* 45–50.

Darvasi, A. (1998). Experimental strategies for the genetic dissection of complex traits in animal models. *Nature Genetics, 18,* 19–24.

Darwin, C. (1859). *On the origin of species by means of natural selection, or the preservation of favoured races in the struggle for life.* London: John Murray.

Darwin, C. (1871). *The descent of man and selection in relation to sex.* London: John Murray.

Darwin, C. (1877). A biographical sketch of an infant. *Mind, 2,* 285–294.

Darwin, C. (1896). *Journal of researchers into the natural history and geology of the countries visited during the voyage of H.M.S. Beagle round the world under the command of Capt. Fitz Roy, T.N.* New York: Appleton.

Dasgupta, B., & Gutmann, D.H. (2003). Neurofibromatosis 1: Closing the GAP between mice and men. *Current Opinion in Genetics and Development, 13,* 20–27.

Davidson, J.R.T., Hughes, D., Blazer, D.G., & George, L. (1991). Posttraumatic stress disorder in the community: An epidemiological study. *Psychological Medicine, 21,* 713–721.

Davis, M.H., Luce, C., & Kraus, S.J. (1994). The heritability of characteristics associated with dispositional empathy. *Journal of Personality, 62,* 369–391.

Davis, R.L. (2004). Olfactory learning. *Neuron, 44,* 31–48.

Davis, R.L. (2005). Olfactory memory formation in *Drosophila:* From molecular to systems neuroscience. *Annual Review of Neuroscience, 28,* 275–302.

Dawkins, R. (1976). *The selfish gene.* New York: Oxford University Press.

Dawkins, R. (2006). *The God delusion.* Bantam Press.

Deary, I. J. (2000). *Looking down on human intelligence: From psychometrics to the brain.* Oxford: Oxford University Press.

Deary, I.J. (2001). Human intelligence differences: Towards a combined experimental-differential approach. *Trends in Cognitive Science, 5,* 164–170.

Deary, I.J., Pattie, A., Wilson, V., & Whalley, L.J. (2005). The cognitive cost of being a twin: Two whole-population surveys. *Twin Research and Human Genetics, 8,* 376–383.

Deary, I.J., Spinath, F.M., & Bates, T.C. (2006). Genetics of intelligence. *European Journal of Human Genetics, 14,* 690–700.

Deary, I.J., Whiteman, M.C., Pattie, A., Starr, J.M., Hayward, C., Wright, A.F., et al. (2002). Cognitive change and the APOE epsilon 4 allele. *Nature, 418,* 932.

Deary, I.J., Whiteman, M.C., Pattie, A., Starr, J.M., Hayward, C., Wright, A.F., et al. (2004). Apolipoprotein E gene variability and cognitive functions at age 79: A follow-up of the Scottish mental survey of 1932. *Psychological Aging, 19,* 367–371.

Deary, I.J., Whiteman, M.C., Starr, J.M., Whalley, L.J., & Fox, H.C. (2004). The impact of childhood intelligence on later life: Following up the Scottish mental surveys of 1932 and 1947. *Journal of Personality and Social Psychology, 86,* 130–147.

Deary, I.J., Wright, A.F., Harris, S.E., Whalley, L.J., & Starr, J.M. (2004). Searching for genetic influences on normal cognitive aging. *Trends in Cognitive Science, 8,* 178–184.

Deater-Deckard, K., & O'Connor, T.G. (2000). Parent-child mutuality in early childhood: Two behavioral genetic studies. *Developmental Psychology, 36,* 561–570.

Debener, S., Ullsperger, M., Siegel, M., & Engel, A.K. (2006). Single-trial EEG-fMRI reveals the dynamics of cognitive function. *Trends in Cognitive Sciences, 10,* 558–563.

de Castro, J.M. (1999). Behavioral genetics of food intake regulation in free-living humans. *Nutrition, 15,* 550–554.

Decker, S.N., & Vandenberg, S.G. (1985). Colorado twin study of reading disability. In D.B. Gray & J.F. Kavanagh (Eds.), *Biobehavioral measures of dyslexia* (pp. 123–135). Parkton, MD: York Press.

Deconinck, N., & Dan, B. (2007). Pathophysiology of Duchenne muscular dystrophy: Current hypotheses. *Pediatric neurology, 36,* 1–7.

DeFries, J.C., & Alarcón, M. (1996). Genetics of specific reading disability. *Mental Retardation and Developmental Disabilities Research Reviews, 2,* 39–47.

DeFries, J.C., & Fulker, D.W. (1985). Multiple regression analysis of twin data. *Behavior Genetics, 15,* 467–473.

DeFries, J.C., & Fulker, D.W. (1988). Multiple regression analysis of twin data: Etiology of deviant scores versus individual differences. *Acta Geneticae Medicae et Gemellologicae, 37,* 205–216.

DeFries, J.C., Fulker, D.W., & LaBuda, M.C. (1987). Evidence for a genetic aetiology in reading disability of twins. *Nature, 329,* 537–539.

DeFries, J.C., Gervais, M.C., & Thomas, E.A. (1978). Response to 30 generations of selection for open-field activity in laboratory mice. *Behavior Genetics, 8,* 3–13.

DeFries, J.C., & Gillis, J.J. (1993). Genetics and reading disability. In R. Plomin & G.E. McClearn (Eds.), *Nature, nurture, and psychology* (pp. 121–145). Washington, DC: American Psychological Association.

DeFries, J.C., Johnson, R.C., Kuse, A.R., McClearn, G.E., Polovina, J., Vandenberg, S.G., et al. (1979). Familial resemblance for specific cognitive abilities. *Behavior Genetics, 9,* 23–43.

DeFries, J.C., Knopik, V.S., & Wadsworth, S.J. (1999). Colorado Twin Study of reading disability. In D.D. Duane (Ed.), *Reading and attention disorders: Neurobiological correlates* (pp. 17–41). Baltimore: York Press.

DeFries, J.C., Vandenberg, S.G., & McClearn, G.E. (1976). Genetics of specific cognitive abilities. *Annual Review of Genetics, 10,* 179–207.

DeFries, J.C., Vogler, G.P., & LaBuda, M.C. (1986). Colorado Family Reading Study: An overview. In J.L. Fuller & E.C. Simmel (Eds.), *Perspectives in behavior genetics* (pp. 29–56). Hillsdale, NJ: Lawrence Erlbaum Associates.

de Geus, E.J., Wright, M.J., Martin, N.G., & Boomsma, D.I. (2001). Genetics of brain function and cognition. *Behavior Genetics, 31,* 489–495.

Delaval, K., & Feil, R. (2004). Epigenetic regulation of mammalian genomic imprinting. *Current Opinion in Genetic Development, 14,* 188–195.

DeLisi, L.E., Mirsky, A.F., Buchsbaum, M.S., van Kammen, D.P., Berman, K.F., Caton, C., et al. (1984). The Genain quadruplets 25 years later: A diagnostic and biochemical followup. *Psychiatric Research, 13,* 59–76.

Dennett, D.C. (2006). *Breaking the spell: Religion as a natural phenomenon.* St Albans, UK: Allen Lane.

Derks, E.M., Dolan, C.V., & Boomsma, D.I. (2006). A test of the equal environment assumption (EEA) in multivariate twin studies. *Twin Research and Human Genetics, 9,* 403–411.

Detterman, D.K. (1986). Human intelligence is a complex system of separate processes. In R.J. Sternberg & D.K. Detterman (Eds.), *What is intelligence? Contemporary viewpoints on its nature and measurement* (pp. 57–61). Norwood, NJ: Ablex.

Devlin, B., Daniels, M., & Roeder, K. (1997). The heritability of IQ. *Nature, 388,* 468–471.

de Vries, B.B.A., Pfundt, R., Leisink, M., Koolen, D.A., Vissers, L.E.L.M., Janssen, I.M., et al. (2005). Diagnostic genome profiling in mental retardation. *American Journal of Human Genetics, 77,* 606–616.

de Waal, F. (1996). *Good natured: The origins of right and wrong in humans and other animals.* Cambridge, MA: Harvard University Press.

de Waal, F. (2002). Evolutionary psychology: The wheat and the chaff. *Current Directions in Psychological Science, 11,* 187–191.

Dick, D.M., Agrawal, A., Schuckit, M.A., Bierut, L., Hinrichs, A., Fox, L., et al. (2006a). Marital status, alcohol dependence, and GABRA2: Evidence for gene-environment correlation and interaction. *Journal of Studies on Alcohol, 67,* 185–194.

Dick, D.M., Aliev, F., Bierut, L., Goate, A., Rice, J., Hinrichs, A., et al. (2006b). Linkage analyses of IQ in the collaborative study on the genetics of alcoholism (COGA) sample. *Behavior Genetics, 36,* 77–86.

Dick, D.M., & Bierut, L.J. (2006). The genetics of alcohol dependence. *Current Psychiatry Reports, 8,* 151–157.

Dick, D.M., Jones, K., Saccone, N., Hinrichs, A., Wang, J.C., Goate, A., et al. (2006c). Endophenotypes successfully lead to gene identification: Results from the collaborative study on the genetics of alcoholism. *Behavior Genetics, 36,* 112–126.

Dick, D.M., Rose, R.J., Viken, R.J., Kaprio, J., & Koskenvuo, M. (2001). Exploring gene-environment interactions: Socioregional moderation of alcohol use. *Journal of Abnormal Psychology, 110,* 625–632.

DiLalla, L.F., & Gottesman, I.I. (1989). Heterogeneity of causes for delinquency and criminality: Lifespan perspectives. *Development and Psychopathology, 1,* 339–349.

Dinwiddie, S.H., Hoop, J., & Gershon, E.S. (2004). Ethical issues in the use of genetic information. *International Review of Psychiatry, 16,* 320–328.

DiPetrillo, K., Wang, X., Stylianou, I.M., & Paigen, B. (2005). Bioinformatics toolbox for narrowing rodent quantitative trait loci. *Trends in Genetics, 21,* 683–692.

Dobzhansky, T. (1964). *Heredity and the nature of man.* New York: Harcourt, Brace & World.

Donarum, E.A., Halperin, R.F., Stephan, D.A., & Narayanan, V. (2006). Cognitive dysfunction in NF1 knock-out mice may result from altered vesicular trafficking of APP/DRD3 complex. *BMC Neuroscience, 7,* 22.

D'Onofrio, B.M., Turkheimer, E.N., Eaves, L.J., Corey, L.A., Berg, K., Solaas, M.H., et al. (2003). The role of the children of twins design in elucidating causal relations between parent characteristics and child outcomes. *Journal of Child Psychology and Psychiatry and Allied Disciplines, 44,* 1130–1144.

D'Onofrio, B.M., Turkheimer, E., Emery, R.E., Slutske, W.S., Heath, A.C., Madden, P.A., et al. (2006). A genetically informed study of the processes underlying the association between parental marital instability and offspring adjustment. *Developmental Psychology, 42,* 486–499.

Draghici, S., Khatri, P., Eklund, A.C., & Szallasi, Z. (2006). Reliability and reproducibility issues in DNA microarray measurements. *Trends in Genetics, 22,* 101–109.

Dubnau, J., & Tully, T. (1998). Gene discovery in *Drosophila:* New insights for learning and memory. *Annual Review of Neurosciences, 21,* 407–444.

Dunn, J.F., & Plomin, R. (1986). Determinants of maternal behavior toward three-year-old siblings. *British Journal of Developmental Psychology, 4,* 127–137.

Dunn, J.F., & Plomin, R. (1990). *Separate lives: Why siblings are so different.* New York: Basic Books.

Duyme, M., Dumaret, A.C., & Tomkiewicz, S. (1999). How can we boost IQs of "dull children"?: A late adoption study. *Proceedings of the National Academy of Sciences (USA), 96,* 8790–8794.

Dworkin, R.H. (1979). Genetic and environmental influences on person-situation interactions. *Journal of Research in Personality, 13,* 279–293.

Dworkin, R.H., & Lenzenweger, M.F. (1984). Symptoms and the genetics of schizophrenia: Implications for diagnosis. *American Journal of Psychiatry, 14,* 1541–1546.

Dykens, E.M., & Hodapp, R.M. (2001). Research in mental retardation: Toward an etiologic approach. *Journal of Child Psychology and Psychiatry, 42,* 49–71.

Dykens, E.M., Hodapp, R.M., & Leckman, J.F. (1994). *Behavior and development in fragile X syndrome.* London: Sage.

Eaves, L., Foley, D., & Silberg, J. (2003). Has the 'equal environments' assumption been tested in twin studies? *Twin Research, 6,* 486–489.

Eaves, L., Rutter, M., Silberg, J.L., Shillady, L., Maes, H., & Pickles, A. (2000). Genetic and environmental causes of covariation in interview assessments of disruptive behavior in child and adolescent twins. *Behavior Genetics, 30,* 321–334.

Eaves, L.J. (1976). A model for sibling effects in man. *Heredity, 36,* 205–214.

Eaves, L.J., D'Onofrio, B., & Russell, R. (1999). Transmission of religion and attitudes. *Twin Research, 2,* 59–61.

Eaves, L.J., & Eysenck, H.J. (1976). Genetical and environmental components of inconsistency and unrepeatability in twins' responses to a neuroticism questionaire. *Behavior Genetics, 6,* 145–160.

Eaves, L.J., Eysenck, H.J., & Martin, N.G. (1989). *Genes, culture, and personality: An empirical approach.* London: Academic Press.

Eaves, L.J., Heath, A.C., Martin, N.G., Maes, H., Neale, M., Kendler, K., et al. (1999). Comparing the biological and cultural inheritance of personality and social attitudes in the Virginia 30,000 study of twins and their relatives. *Twin Research, 2,* 62–80.

Eaves, L.J., Heath, A.C., Neale, M.C., Hewitt, J.K., & Martin, N.G. (1998). Sex differences and non-additivity in the effects of genes in personality. *Twin Research, 1,* 131–137.

Eaves, L.J., Kendler, K.S., & Schulz, S.C. (1986). The familial sporadic classification: Its power for the resolution of genetic and environmental etiological factors. *Journal of Psychiatric Research, 20,* 115–130.

Eaves, L.J., Silberg, J.L., Hewitt, J.K., Meyer, J., Rutter, M., Simonoff, E., et al. (1993). Genes, personality and psychopathology: A latent class analysis of liability to symptoms of attention-deficit hyperactivity disorder in twins. In R. Plomin & G.E. McClearn (Eds.), *Nature, nurture, and psychology* (pp. 285–303). Washington, DC: American Psychological Association.

Eaves, L.J., Silberg, J.L., Meyer, J.M., Maes, H.H., Simonoff, E., Pickles, A., et al. (1997). Genetics and developmental psychopathology: 2. The main effects of genes and environment on behavioral problems in the Virginia Twin Study of Adolescent Behavioral Development. *Journal of Child Psychology and Psychiatry, 38,* 965–980.

Ebstein, R.P. (2006). The molecular genetic architecture of human personality: Beyond self-report questionnaires. *Molecular Psychiatry, 11,* 427–445.

Ebstein, R.P., & Belmaker, R.H. (1997). Saga of an adventure gene: Novelty seeking, substance abuse and the dopamine D4 receptor (*D4DR*) exon III repeat polymorphism. *Molecular Psychiatry, 2,* 381–384.

Ebstein, R.P., & Kotler, M. (2002). Personality, substance abuse, and genes. In J. Benjamin, R. Ebstein, & R.H. Belmaker (Eds.), *Molecular genetics and human personality* (pp. 151–163). Washington, DC: American Psychiatric Press

Ebstein, R.P., Novick, O., Umansky, R., Priel, B., Osher, Y., Blaine, D., et al. (1996). Dopamine D4 receptor (*D4DR*) exon III polymorphism associated with the human personality trait of novelty seeking. *Nature Genetics, 12,* 78–80.

Eckman, P. (1973). *Darwin and facial expression: A century of research in review.* New York: Academic Press.

Edenberg, H.J., & Foroud, T. (2006). The genetics of alcoholism: Identifying specific genes through family studies. *Addiction Biology, 11,* 386–396.

Egeland, J.A., Gerhard, D.S., Pauls, D.L., Sussex, J.N., Kidd, K.K., Allen, C.R., et al. (1987). Bipolar affective disorders linked to DNA markers on chromosome 11. *Nature, 325*, 783–787.

Ehringer, M.A., Rhee, S.H., Young, S., Corley, R., & Hewitt, J.K. (2006). Genetic and environmental contributions to common psychopathologies of childhood and adolescence: A study of twins and their siblings. *Journal of Abnormal Child Psychology, 34*, 1–17.

Ehrman, L. (1972). A factor influencing the rare male mating advantage in *Drosophila. Behavior Genetics, 2*, 69–78.

Eisenstein, M. (2006). Protein arrays: Growing pains. *Nature, 444*, 959–962.

Eley, T.C., Bolton, D., O'Connor, T.G., Perrin, S., Smith, P., & Plomin, R. (2003). A twin study of anxiety-related behaviours in pre-school children. *Journal of Child Psychology and Psychiatry, 44*, 945–960.

Eley, T.C., Collier, D., & McGuffin, P. (2002). Anxiety and eating disorders. In P. McGuffin, M.J. Owen, & I.I. Gottesman (Eds.), *Psychiatric genetics and genomics* (pp. 303–340). Oxford: Oxford University Press.

Eley, T.C., Lichtenstein, P., & Stevenson, J. (1999). Sex differences in the aetiology of aggressive and non-aggressive antisocial behavior: Results from two twin studies. *Child Development, 70*, 155–168.

Elkins, I.J., McGue, M., & Iacono, W.G. (1997). Genetic and environmental influences on parent-son relationships: Evidence for increasing genetic influence during adolescence. *Developmental Psychology, 33*, 351–363.

Engle, S.J., Womer, D.E., Davies, P.M., Boivin, G., Sahota, A., Simmonds, H.A., et al. (1996). HPRT-APRT-deficient mice are not a model for Lesch-Nyhan syndrome. *Human Molecular Genetics, 5*, 1607–1610.

Erlenmeyer-Kimling, L. (1972). Gene-environment interactions and the variability of behavior. In L. Ehrman, G.S. Omenn, & E. Caspari (Eds.), *Genetics, environment, and behavior* (pp. 181–208). San Diego: Academic Press.

Erlenmeyer-Kimling, L., & Jarvik, L.F. (1963). Genetics and intelligence: A review. *Science, 142*, 1477–1479.

Erlenmeyer-Kimling, L., Squires-Wheeler, E., Adamo, U.H., Bassett, A.S., Cornblatt, B.A., Kestenbaum, C.J., et al. (1995). The New York high-risk project: Psychoses and cluster A personality disorders in offspring of schizophrenic parents at 23 years of follow-up. *Archives of General Psychiatry, 52*, 857–865.

Estourgie-van Burk, G.F., Bartels, M., van Beijsterveldt, T.C., Delemarre-van de Waal, H.A., & Boomsma, D.I. (2006). Body size in five-year-old twins: Heritability and comparison to singleton standards. *Twin Research and Human Genetics, 9*, 646–655.

Etcoff, N. (1999). *Survival of the prettiest: The science of beauty.* London: Little, Brown & Co.

Evans, W.E., & Relling, M.V. (2004). Moving towards individualized medicine with pharmacogenomics. *Nature, 429*, 464–468.

Eysenck, H.J. (1952). *The scientific study of personality.* London: Routledge & Kegan Paul.

Fagard, R., Bielen, E., & Amery, A. (1991). Heritability of aerobic power and anaerobic energy generation during exercise. *Journal of Applied Physiology, 70*, 357–362.

Fagard, R.H., Loos, R.J., Beunen, G., Derom, C., & Vlietinck, R. (2003). Influence of chorionicity on the heritability estimates of blood pressure: A study in twins. *Journal of Hypertension, 21*, 1313–1318.

Falconer, D.S. (1965). The inheritance of liability to certain diseases estimated from the incidence among relatives. *Annals of Human Genetics, 29,* 51–76.

Falconer, D.S., & Mackay, T.F.C. (1996). *Introduction to quantitative genetics* (4th ed.). Harlow, UK: Longman.

Fang, P., Lev-Lehman, E., Tsai, T.F., Matsuura, T., Benton, C.S., Sutcliffe, J.S., et al. (1999). The spectrum of mutations in UBE3A causing Angelman syndrome. *Human Molecular Genetics, 8,* 129–135.

Fantino, E., & Logan, C.A. (1979). *The experimental analysis of behavior.* San Francisco: Freeman.

Faraone, S.V. (2004). Genetics of adult attention-deficit hyperactivity disorder. *Psychiatric Clinics of North America, 27,* 303–321.

Faraone, S.V., Biederman, J., & Monuteaux, M.C. (2000). Attention-deficit disorder and conduct disorder in girls: Evidence for a familial subtype. *Biological Psychiatry, 48,* 21–29.

Faraone, S.V., Perlis, R.H., Doyle, A.E., Smoller, J.W., Goralnick, J.J., Holmgren, M.A., et al. (2005). Molecular genetics of attention-deficit hyperactivity disorder. *Biological Psychiatry, 57,* 1313–1323.

Faraone, S.V., Tsuang, M.T., & Tsuang, D.W. (2002). *Genetics of mental disorders: What practitioners and students need to know.* New York: Guilford Press.

Farmer, A., Elkin, A., & McGuffin, P. (2007). The genetics of bipolar affective disorder. *Current Opinion in Psychiatry, 20,* 8–12.

Farmer, A., Scourfield, J., Martin, N., Cardno, A., & McGuffin, P. (1999). Is disabling fatigue in childhood influenced by genes? *Psychological Medicine, 29,* 279–282.

Farmer, A.E., McGuffin, P., & Gottesman, I.I. (1987). Twin concordance for DSM-III schizophrenia: Scrutinzing the validity of the definition. *Archives of General Psychiatry, 44,* 634–641.

Fearon, R.M., van Ijzendoorn, M.H., Fonagy, P., Bakermans-Kranenburg, M.J., Schuengel, C., & Bokhorst, C.L. (2006). In search of shared and nonshared environmental factors in security of attachment: A behavior-genetic study of the association between sensitivity and attachment security. *Developmental Psychology, 42,* 1026–1040.

Federenko, I.S., Schlotz, W., Kirschbaum, C., Bartels, M., Hellhammer, D.H., & Wust, S. (2006). The heritability of perceived stress. *Psychological Medicine, 36,* 375–385.

Fehr, E., & Fischbacher, U. (2003). The nature of human altruism. *Nature, 425,* 785–791.

Feigon, S.A., Waldman, I.D., Levy, F., & Hay, D.A. (2001). Genetic and environmental influences on separation anxiety disorder symptoms and their moderation by age and sex. *Behavior Genetics, 31,* 403–411.

Felsenfeld, S. (1994). Developmental speech and language disorders. In J.C. DeFries, R. Plomin, & D.W. Fulker (Eds.), *Nature and nurture during middle childhood* (pp. 102–119). Oxford: Blackwell.

Felsenfeld, S., Kirk, K.M., Zhu, G., Statham, D.J., Neale, M.C., & Martin, N.G. (2000). A study of the genetic and environmental etiology of stuttering in a selected twin sample. *Behavior Genetics, 30,* 359–366.

Felsenfeld, S., & Plomin, R. (1997). Epidemiological and offspring analyses of developmental speech disorders using data from the Colorado Adoption Project. *Journal of Speech, Language, and Hearing Research, 40,* 778–791.

Ferner, R.E. (2007). Neurofibromatosis 1. *European Journal of Human Genetics, 15,* 131–138.

Finch, C.E., & Ruvkun, G. (2001). The genetics of aging. *Annual Review of Genomics and Human Genetics, 2,* 435–462.

Finkel, D., & McGue, M. (2007). Genetic and environmental influences on intraindividual variability in reaction time. *Experimental Aging Research, 33,* 13–35.

Finkel, D., Wille, D.E., & Matheny, A.P., Jr. (1998). Preliminary results from a twin study of infant-caregiver attachment. *Behavior Genetics, 28,* 1–8.

Finn, C.T., & Smoller, J.W. (2006). Genetic counseling in psychiatry. *Harvard Review of Psychiatry, 14,* 109–121.

Fischer, P.J., & Breakey, W.R. (1991). The epidemiology of alcohol, drug and mental disorders among homeless persons. *American Psychologist, 46,* 1115–1128.

Fisher, R.A. (1918). The correlation between relatives on the supposition of Mendelian inheritance. *Transactions of the Royal Society of Edinburgh, 52,* 399–433.

Fisher, R.A. (1922). On the mathematical foundations of theoretical statistics. *Philosophical Transactions of the Royal Society of London, 222,* 309–368.

Fisher, R.A. (1930). *The genetical theory of natural selection.* Oxford: Clarendon Press.

Fisher, R.C. (2004). *Why men won't ask for directions: The seductions of sociobiology.* Princeton, NJ: Princeton University Press.

Fisher, S.E., & DeFries, J.C. (2002). Developmental dyslexia: Genetic dissection of a complex cognitive trait. *Nature Reviews Neuroscience, 3,* 767–780.

Fisher, S.E., & Francks, C. (2006). Genes, cognition and dyslexia: Learning to read the genome. *Trends in Cognitive Science, 10,* 250–257.

Fisher, S.E., Francks, C., Marlow, A.J., MacPhie, I.L., Newbury, D.F., Cardon, L.R., et al. (2002). Independent genome-wide scans identify a chromosome 18 quantitative-trait locus influencing dyslexia. *Nature Genetics, 30,* 86–91.

Fisher, S.E., Marlow, A.J., Lamb, J., Maestrini, E., Williams, D.F., Richardson, A.J., et al. (1999). A quantitative-trait locus on chromosome 6p influences different aspects of developmental dyslexia. *American Journal of Human Genetics, 64,* 146–156.

Flaxman, S.M., & Sherman, P.W. (2000). Morning sickness: A mechanism for protecting mother and embryo. *The Quarterly Review of Biology, 75,* 113–148.

Fletcher, R. (1990). *The Cyril Burt scandal: Case for the defense.* New York: Macmillan.

Flint, J. (2004). The genetic basis of neuroticism. *Neurosciences and Biobehavioral Reviews, 28,* 307–316.

Flint, J., Corley, R., DeFries, J.C., Fulker, D.W., Gray, J.A., Miller, S., et al. (1995). A simple genetic basis for a complex psychological trait in laboratory mice. *Science, 269,* 1432–1435.

Flint, J., & Munafo, M.R. (2007). The endophenotype concept in psychiatric genetics. *Psychological Medicine, 37,* 163–180.

Flint, J., Valdar, W., Shifman, S., & Mott, R. (2005). Strategies for mapping and cloning quantitative trait genes in rodents. *Nature Reviews Genetics, 6,* 271–286.

Folstein, S., & Rutter, M. (1977). Infantile autism: A genetic study of 21 twin pairs. *Journal of Child Psychology and Psychiatry, 18,* 297–321.

Folstein, S.E., & Rosen-Sheidley, B. (2001). Genetics of autism: Complex aetiology for a heterogeneous disorder. *Nature Reviews Genetics, 2,* 943–955.

Foster, K., Foster, H., & Dickson, J.G. (2006). Gene therapy progress and prospects: Duchenne muscular dystrophy. *Gene Therapy, 13,* 1677–1685.

Fountoulakis, M., & Kossida, S. (2006). Proteomics-driven progress in neurodegeneration research. *Electrophoresis, 27,* 1556–1573.

Fountoulakis, M., Tsangaris, G.T., Maris, A., & Lubec, G. (2005). The rat brain hippocampus proteome. *Journal of Chromatography. B. Analytical Technologies in the Biomedical and Life Sciences, 819,* 115–129.

Fraga, M.F., Ballestar, E., Paz, M.F., Ropero, S., Setien, F., Ballestar, M.L., et al. (2005). Epigenetic differences arise during the lifetime of monozygotic twins. *Proceedings of the National Academy of Sciences (USA), 102,* 10604–10609.

Frayling, T.M., Timpson, N.J., Weedon, M.N., Zeggini, E., Freathy, R.M., Lindgren, C.M., et al. (2007). A common variant in the FTO gene is associated with body mass index and predisposes to childhood and adult obesity. *Science, 316,* 889–894.

Freeman, F.N., Holzinger, K.J., & Mitchell, B. (1928). The influence of environment on the intelligence, school achievement, and conduct of foster children. *Yearbook of the National Society for the Study of Education, 27,* 103–217.

Freitag, C.M. (2007). The genetics of autistic disorders and its clinical relevance: A review of the literature. *Molecular Psychiatry, 12,* 2–22.

Fulker, D.W. (1979). Nature and nurture: Heredity. In H.J. Eysenck (Ed.), *The structure and measurement of intelligence* (pp. 102–132). New York: Springer-Verlag.

Fulker, D.W., Cardon, L.R., DeFries, J.C., Kimberling, W.J., Pennington, B.F., & Smith, S.D. (1991). Multiple regression analysis of sib-pair data on reading to detect quantitative trait loci. *Reading and Writing: An Interdisciplinary Journal, 3,* 299–313.

Fulker, D.W., Cherny, S.S., & Cardon, L.R. (1993). Continuity and change in cognitive development. In R. Plomin & G.E. McClearn (Eds.), *Nature, nurture, and psychology* (pp. 77–97). Washington, DC: American Psychological Association.

Fulker, D.W., Cherny, S.S., Sham, P.C., & Hewitt, J.K. (1999). Combined linkage and association sib-pair analysis for quantitative traits. *American Journal of Human Genetics, 64,* 259–267.

Fulker, D.W., DeFries, J.C., & Plomin, R. (1988). Genetic influence on general mental ability increases between infancy and middle childhood. *Nature, 336,* 767–769.

Fulker, D.W., Eysenck, S.B.G., & Zuckerman, M. (1980). A genetic and environmental analysis of sensation seeking. *Journal of Research in Personality, 14,* 261–281.

Fuller, J.L., & Thompson, W.R. (1960). *Behavior genetics.* New York: Wiley.

Fuller, J.L., & Thompson, W.R. (1978). *Foundations of behavior genetics.* St Louis: Mosby.

Fullerton, J. (2006). New approaches to the genetic analysis of neuroticism and anxiety. *Behavior Genetics, 36,* 147–161.

Fullerton, J., Cubin, M., Tiwari, H., Wang, C., Bomhra, A., Davidson, S., et al. (2003). Linkage analysis of extremely discordant and concordant sibling pairs identifies quantitative-trait loci that influence variation in the human personality trait neuroticism. *American Journal of Human Genetics, 72,* 879–890.

Furnham, A., Moutafi, J., & Baguma, P. (2002). A cross-cultural study on the role of weight and waist-to-hip ratio on female attractiveness. *Personality and Individual Differences, 32,* 729–745.

Fyer, A.J., Hamilton, S.P., Durner, M., Haghighi, F., Heiman, G.A., Costa, R., et al. (2006). A third-pass genome scan in panic disorder: Evidence for multiple susceptibility loci. *Biological Psychiatry, 60,* 388–401.

Fyer, A.J., Mannuzza, S., Chapman, T.F., Martin, L.Y., & Klein, D.F. (1995). Specificity in familial aggregation of phobic disorders. *Archives of General Psychiatry, 52,* 564–573.

Galaburda, A.M., LoTurco, J., Ramus, F., Fitch, R.H., & Rosen, G.D. (2006). From genes to behavior in developmental dyslexia. *Nature Neuroscience, 9,* 1213–1217.

Galsworthy, M.J., Paya-Cano, J.L., Liu, L., Monleon, S., Gregoryan, G., Fernandes, C., et al. (2005). Assessing reliability, heritability and general cognitive ability in a battery of cognitive tasks for laboratory mice. *Behavior Genetics, 35,* 675–692.

Galton, F. (1865). Heredity talent and character. *Macmillan's Magazine, 12,* 157–166, 318–327.

Galton, F. (1869). *Hereditary genius: An enquiry into its laws and consequences.* Cleveland, OH: World, 1992.

Galton, F. (1876). The history of twins as a criterion of the relative powers of nature and nurture. *Royal Anthropological Institute of Great Britain and Ireland Journal, 6,* 391–406.

Galton, F. (1883). *Inquiries into human faculty and its development.* London: Macmillan.

Galton, F. (1889). *Natural inheritance.* London: Macmillan.

Gammie, S.C., Garland, T., Jr., & Stevenson, S.A. (2006). Artificial selection for increased maternal defense behavior in mice. *Behavior Genetics, 36,* 713–722.

Gangestad, S.W., Garver-Apgar, C.E., Simpson, J.A., & Cousins, A.J. (2007). Changes in women's mate preferences across the ovulatory cycle. *Journal of Personality and Social Psychology, 92,* 151–163.

Gangestad, S.W., Haselton, M.G., & Buss, D.M. (2006). Evolutionary foundations of cultural variation: Evoked culture and mate preferences. *Psychological Inquiry, 17,* 75–95.

Gangestad, S.W., & Scheyd, G.J. (2005). The evolution of human physical attractiveness. *Annual Review of Anthropology, 34,* 523–548.

Gangestad, S.W., & Simpson, J.A. (2007). *The evolution of mind: Fundamental questions and controversies.* New York: Guilford Publications.

Gangestad, S.W., Thornhill, R., & Garver-Apgar, C.E. (2005). Adaptations to ovulation—Implications for sexual and social behavior. *Current Directions in Psychological Science, 14,* 312–316.

Gao, W., Li, L., Cao, W., Zhan, S., Lv, J., Qin, Y., et al. (2006). Determination of zygosity by questionnaire and physical features comparison in Chinese adult twins. *Twin Research and Human Genetics, 9,* 266–271.

Garber, K., Smith, K.T., Reines, D., & Warren, S.T. (2006). Transcription, translation and fragile X syndrome. *Current Opinion in Genetics and Development, 16,* 270–275.

Gardenfors, P. (2006). *How homo became sapiens: On the evolution of thinking.* Oxford: Oxford University Press.

Gardner, H. (2006). *Multiple intelligences: New horizons in theory and practice.* New York: Basic Books.

Garrigan, D., & Hammer, M.F. (2006). Reconstructing human origins in the genomic era. *Nature Reviews Genetics, 7,* 669–680.

Garver-Apgar, C.E., Gangestad, S.W., Thornhill, R., Miller, R.D., & Olp, J.J. (2006). Major histocompatibility complex alleles, sexual responsivity, and unfaithfulness in romantic couples. *Psychological Science, 17,* 830–835.

Gatz, M., Pedersen, N.L., Berg, S., Johansson, B., Johansson, K., Mortimer, J.A., et al. (1997). Heritability for Alzheimer's disease: The study of dementia in Swedish twins. *Journals of Gerontology Series A: Biological Science and Medical Science, 52,* M117–M125.

Gatz, M., Pedersen, N.L., Plomin, R., Nesselroade, J.R., & McClearn, G.E. (1992). Importance of shared genes and shared environments for symptoms of depression in older adults. *Journal of Abnormal Psychology, 101,* 701–708.

Gatz, M., Reynolds, C.A., Fratiglioni, L., Johansson, B., Mortimer, J.A., Berg, S., et al. (2006). Role of genes and environments for explaining Alzheimer's disease. *Archives of General Psychiatry, 63,* 168–174.

Gayán, J., & Olson, R.K. (2003). Genetic and environmental influences on individual differences in printed word recognition. *Journal of Experimental Child Psychology, 84,* 97–123.

Gayán, J., Smith, S.D., Cherny, S.S., Cardon, L.R., Fulker, D.W., Brower, A.W., et al. (1999). Quantitative-trait locus for specific language and reading deficits on chromosome *6p. American Journal of Human Genetics, 64,* 157–164.

Gayán, J., Willcutt, E.G., Fisher, S.E., Francks, C., Cardon, L.R., Olson, R.K., et al. (2005). Bivariate linkage scan for reading disability and attention-deficit hyperactivity disorder localizes pleiotropic loci. *Journal of Child Psychology and Psychiatry, 46,* 1045–1056.

Ge, X., Conger, R.D., Cadoret, R.J., Neiderhiser, J.M., Yates, W., Troughton, E., et al. (1996). The developmental interface between nature and nurture: A mutual influence model of child antisocial behaviour and parenting. *Developmental Psychology, 32,* 574–589.

Geary, D.C. (2005). *The origin of mind: Evolution of brain, cognition, and general intelligence.* Washington, DC: American Psychological Association.

Geerts, M., Steyaert, J., & Fryns, J.P. (2003). The XYY syndrome: A follow-up study on 38 boys. *Genetic Counseling, 14,* 267–279.

Gelhorn, H., Stallings, M., Young, S., Corley, R., Rhee, S.H., Christian, H., et al. (2006). Common and specific genetic influences on aggressive and nonaggressive conduct disorder domains. *Journal of the American Academy of Child and Adolescent Psychiatry, 45,* 570–577.

Ghazalpour, A., Doss, S., Zhang, B., Wang, S., Plaisier, C., Castellanos, R., et al. (2006). Integrating genetic and network analysis to characterize genes related to mouse weight. *PLoS Genetics, 2,* E130.

Gibbs, R.A., Weinstock, G.M., Metzker, M.L., Muzny, D.M., Sodergren, E.J., Scherer, S., et al. (2004). Genome sequence of the Brown Norway rat yields insights into mammalian evolution. *Nature, 428,* 493–521.

Giedd, J.N., Clasen, L.S., Wallace, G.L., Lenroot, R.K., Lerch, J.P., Wells, E.M., et al. (2007). XXY (Klinefelter syndrome): A pediatric quantitative brain magnetic resonance imaging case-control study. *Pediatrics, 119,* E232–E240.

Gillespie, N.A., Whitfield, J.B., Williams, B., Heath, A.C., & Martin, N.G. (2005). The relationship between stressful life events, the serotonin transporter (5-HTTLPR) genotype and major depression. *Psychological Medicine, 35,* 101–111.

Gillespie, N.A., Zhu, G., Heath, A.C., Hickie, I.B., & Martin, N.G. (2000). The genetic aetiology of somatic distress. *Psychological Medicine, 30,* 1051–1061.

Gillis, J.J., Gilger, J.W., Pennington, B.F., & DeFries, J.C. (1992). Attention deficit disorder in reading-disabled twins: Evidence for a genetic etiology. *Journal of Abnormal Child Psychology, 20,* 303–315.

Giot, L., Bader, J.S., Brouwer, C., Chaudhuri, A., Kuang, B., Li, Y., et al. (2003). A protein interaction map of *Drosophila melanogaster. Science, 302,* 1727–1736.

Giros, B., Jaber, M., Jones, S.R., Wightman, R.M., & Caron, M.G. (1996). Hyperloco-motion and indifference to cocaine and amphetamine in mice lacking the dopamine transporter. *Nature, 379*, 606–612.

Gladkevich, A., Kauffman, H.F., & Korf, J. (2004). Lymphocytes as a neural probe: Potential for studying psychiatric disorders. *Progress in Neuro-Psychopharmacology and Biological Psychiatry, 28*, 559–576.

Glatt, S.J., Everall, I.P., Kremen, W.S., Corbeil, J., Sasik, R., Khanlou, N., et al. (2005). Comparative gene expression analysis of blood and brain provides con-current validation of SELENBP1 up-regulation in schizophrenia. *Proceedings of the National Academy of Sciences (USA), 102*, 15533–15538.

Godinho, S.I., & Nolan, P.M. (2006). The role of mutagenesis in defining genes in behaviour. *European Journal of Human Genetics, 14*, 651–659.

Goedert, M., & Spillantini, M.G. (2006). A century of Alzheimer's disease. *Science, 314*, 777–781.

Goetz, A.T., & Shackelford, T.K. (2006). Modern application of evolutionary theory to psychology: Key concepts and clarifications. *American Journal of Psychology, 119*, 567–584.

Goldberg, L.R. (1990). An alternative description of personality: The big five factor structure. *Journal of Personality and Social Psychology, 59*, 1216–1229.

Goldberg, T.E., & Weinberger, D.R. (2004). Genes and the parsing of cognitive processes. *Trends in Cognitive Sciences, 8*, 325–335.

Goldman, D., Oroszi, G., & Ducci, F. (2005). The genetics of addictions: Uncovering the genes. *Nature Reviews Genetics, 6*, 521–532.

Goldner, E.M., Hsu, L., Waraich, P., & Somers, J.M. (2002). Prevalence and incidence studies of schizophrenic disorders: A systematic review of the literature. *Canadian Journal of Psychiatry, 47*, 833–843.

Goldsmith, H.H. (1983). Genetic influences on personality from infancy to adulthood. *Child Development, 54*, 331–355.

Goldsmith, H.H. (1993). Nature-nurture and the development of personality: Intro-duction. In R. Plomin & G.E. McClearn (Eds.), *Nature, nurture, and psychology* (pp. 155–160). Washington, DC: American Psychological Association.

Goldsmith, H.H., Buss, K.A., & Lemery, K.S. (1997). Toddler and childhood tempera-ment: Expanded content, stronger genetic evidence, new evidence for the impor-tance of environment. *Developmental Psychology, 33*, 891–905.

Goldsmith, H.H., Buss, A.H., Plomin, R., Rothbart, M.K., Chess, S., Hinde, R.A., et al. (1987). Roundtable: What is temperament? Four approaches. *Child Develop-ment, 58*, 505–529.

Goldsmith, H.H., & Campos, J.J. (1986). Fundamental issues in the study of early development: The Denver twin temperament study. In M.E. Lamb, A.L. Brown, & B. Rogoff (Eds.), *Advances in developmental psychology* (pp. 231–283). Hillsdale, NJ: Lawrence Erlbaum Associates.

Goldstein, D.B., Tate, S.K., & Sisodiya, S.M. (2003). Pharmacogenetics goes genomic. *Nature Reviews Genetics, 4*, 937–947.

Goleman, D. (2005). *Emotional intelligence.* New York: Bantam Books.

Goncalves, M.A. (2005). A concise peer into the background, initial thoughts and practices of human gene therapy. *BioEssays, 27*, 506–517.

Goodman, R., & Stevenson, J. (1989). A twin study of hyperactivity II. The aetiologi-cal role of genes, family relationships and perinatal adversity. *Journal of Child Psy-chology and Psychiatry, 30*, 691–709.

Goodwin, F.K., & Jamison, K.R. (1990). *Manic-depressive illness*. New York: Oxford University Press.

Gornick, M.C., Addington, A.M., Sporn, A., Gogtay, N., Greenstein, D., Lenane, M., et al. (2005). Dysbindin (DTNBP1, 6p22.3) is associated with childhood-onset psychosis and endophenotypes measured by the Premorbid Adjustment Scale (PAS). *Journal of Autism and Developmental Disorders, 35*, 831–838.

Gottesman, I.I. (1991). *Schizophrenia genesis: The origins of madness*. New York: Freeman.

Gottesman, I.I., & Bertelsen, A. (1989). Confirming unexpressed genotypes for schizophrenia. *Archives of General Psychiatry, 46*, 867–872.

Gottesman, I.I., & Gould, T.D. (2003). The endophenotype concept in psychiatry: Etymology and strategic intentions. *American Journal of Psychiatry, 160*, 636–645.

Gottfredson, L.S. (1997). Why g matters: The complexity of everyday life. *Intelligence, 24*, 79–132.

Gottfredson, L.S. (1999). The nature and nurture of vocational interests. In M.L. Savickas & A.R. Spokane (Eds.), *Vocational interests: Meaning, measurement, and counseling use* (pp. 57–85). Palo Alto, CA: Davies-Black.

Gottschall, J., & Wilson, D.S. (2005). *The literary animal evolution and the nature of narrative*. Evanston, IL: Northwestern University Press.

Gould, S.J. (1996). *The mismeasure of man* (2nd ed.). New York: W.W. Norton.

Gould, S.J. (1999). *Rocks of ages: Science and religion in the fullness of life*. New York: Ballantine.

Gould, T.D., & Gottesman, I.I. (2006). Psychiatric endophenotypes and the development of valid animal models. *Genes, Brain, and Behavior, 5*, 113–119.

Grados, M.A., & Walkup, J.T. (2006). A new gene for Tourette's syndrome: A window into causal mechanisms? *Trends in Genetics, 22*, 291–293.

Grant, S.G., Marshall, M.C., Page, K.L., Cumiskey, M.A., & Armstrong, J.D. (2005). Synapse proteomics of multiprotein complexes: En route from genes to nervous system diseases. *Human Molecular Genetics, 14*, R225–R234.

Grant, S.G.N. (2003). An integrative neuroscience program linking genes to cognition and disease. In R. Plomin, J.C. DeFries, I.W. Craig, & P. McGuffin (Eds.), *Behavioral genetics in the postgenomic era* (pp. 123–138). Washington, DC: American Psychological Association.

Gray, J.R., & Thompson, P.M. (2004). Neurobiology of intelligence: Science and ethics. *Nature Reviews Neuroscience, 5*, 471–482.

Greenspan, R.J. (1995). Understanding the genetic construction of behavior. *Scientific American, 272*, 72–78.

Greer, M.K., Brown, F.R., Pai, G.S., Choudry, S.H., & Klein, A.J. (1997). Cognitive, adaptive, and behavioral characteristics of Williams syndrome. *American Journal of Medical Genetics. Part B: Neuropsychiatric Genetics, 74*, 521–525.

Gregory, S.G., Barlow, K.F., McLay, K.E., Kaul, R., Swarbreck, D., Dunham, A., et al. (2006). The DNA sequence and biological annotation of human chromosome 1. *Nature, 441*, 315–321.

Grigorenko, E.L., Naples, A., & Chang, J. (2007). Back to Africa: Tracing dyslexia genes in East Africa. *Reading and Writing, 20*, 27–49.

Grigorenko, E.L., Wood, F.B., Meyer, M.S., Hart, L.A., Speed, W.C., Shuster, A., et al. (1997). Susceptibility loci for distinct components of developmental dyslexia on chromosomes 6 and 15. *American Journal of Human Genetics, 60*, 27–39.

Grigorenko, E.L., Wood, F.B., Meyer, M.S., & Pauls, D.L. (2000). Chromosome *6p* influences on different dyslexia-related cognitive processes: Further confirmation. *American Journal of Human Genetics, 66,* 715–723.

Grilo, C.M., & Pogue-Geile, M.F. (1991). The nature of environmental influences on weight and obesity: A behavior genetic analysis. *Psychological Bulletin, 10,* 520–537.

Grof, P., Duffy, A., Cavazzoni, P., Grof, E., Garnham, J., MacDougall, M., et al. (2002). Is response to prophylactic lithium a familial trait? *Journal of Clinical Psychiatry, 63,* 942–947.

Gunderson, E.P., Tsai, A.L., Selby, J.V., Caan, B., Mayer-Davis, E.J., & Risch, N. (2006). Twins of mistaken zygosity (TOMZ): Evidence for genetic contributions to dietary patterns and physiologic traits. *Twin Research and Human Genetics, 9,* 540–549.

Guo, G. (2006). Genetic similarity shared by best friends among adolescents. *Twin Research and Human Genetics, 9,* 113–121.

Guo, S. (2004). Linking genes to brain, behavior and neurological diseases: What can we learn from zebrafish? *Genes, Brain, and Behavior, 3,* 63–74.

Gupta, A.R., & State, M.W. (2007). Recent advances in the genetics of autism. *Biological Psychiatry, 61,* 429–437.

Gusella, J.F., Tazi, R.E., Anderson, M.A., Hobbs, W., Gibbons, K., Raschtchian, R., et al. (1984). DNA markers for nervous system diseases. *Science, 225,* 1320–1326.

Gusella, J.F., Wexler, N.S., Conneally, P.M., Naylor, S.L., Anderson, M.A., Tanzi, R.E., et al. (1983). A polymorphic DNA marker genetically linked to Huntington's disease. *Nature, 306,* 234–238.

Guthrie, R. (1996). The introduction of newborn screening for phenylketonuria: A personal history. *European Journal of Pediatrics, 155,* 4–5.

Gutknecht, L., Spitz, E., & Carlier, M. (1999). Long-term effect of placental type on anthropometrical and psychological traits among monozygotic twins: A follow-up study. *Twin Research, 1999,* 212–217.

Guze, S.B. (1993). Genetics of Briquet's syndrome and somatization disorder: A review of family, adoption and twin studies. *Annals of Clinical Psychiatry, 5,* 225–230.

Guze, S.B., Cloninger, C.R., Martin, R.L., & Clayton, P.J. (1986). A follow-up and family study of Briquet's syndrome. *British Journal of Psychiatry, 149,* 17–23.

Haberstick, B.C., Timberlake, D., Ehringer, M.A., Lessem, J.M., Hopfer, C.J., Smolen, A., et al. (2007). Genes, time to first cigarette and nicotine dependence in a general population sample of young adults. *Addiction, 102,* 655–665.

Hagerman, R. (1995). Lessons from fragile X sydrome. In G.T. O'Brien & W. Yule (Eds.), *Behavioural phenotypes* (pp. 59–74). London: McKeith Press.

Halaas, J.L., Gajiwala, K.S., Maffei, M., Cohen, S.L., Chait, B.T., Rabinowitz, D., et al. (1995). Weight-reducing effects of the plasma protein encoded by the obese gene. *Science, 269,* 543–546.

Hallgren, B. (1957). Enuresis, a clinical and genetic study. *Acta Psychiatrica et Neurologica Scandinavica Supplementum, 114,* 1–159.

Hamer, D.H., & Copeland, P. (1998). *Living with our genes.* New York: Doubleday.

Hamer, D.H., Hu, S., Magnuson, V.L., Hu, N., & Pattatucci, A.M.L. (1993). A linkage between DNA markers on the X chromosome and male sexual orientation. *Science, 261,* 321–327.

Hamilton, A.S., Lessov-Schlaggar, C.N., Cockburn, M.G., Unger, J.B., Cozen, W., & Mack, T.M. (2006). Gender differences in determinants of smoking initiation and persistence in California twins. *Cancer Epidemiology Biomarkers and Prevention, 15,* 1189–1197.

Hamilton, W.D. (1964). Genetical evolution of social behaviour II. *Journal of Theoretical Biology, 7,* 17–52.

Hannon, G.J. (2002). RNA interference. *Nature, 418,* 244–251.

Hansell, N.K., Wright, M.J., Luciano, M., Geffen, G.M., Geffen, L.B., & Martin, N.G. (2005). Genetic covariation between event-related potential (ERP) and behavioral non-ERP measures of working-memory, processing speed, and IQ. *Behavior Genetics, 35,* 695–706.

Hanson, D.R., & Gottesman, I.I. (1976). The genetics, if any, of infantile autism and childhood schizophrenia. *Journal of Autism and Developmental Disorders, 6,* 209–234.

Happé, F., Ronald, A., & Plomin, R. (2006). Time to give up on a single explanation for autism. *Nature Neuroscience, 9,* 1218–1220.

Harden, K.P., Turkheimer, E., & Loehlin, J.C. (2007). Genotype by environment interaction in adolescents' cognitive aptitude. *Behavior Genetics, 37,* 273–283.

Hardy, J. (1997). Amyloid, the presenilins and Alzheimer's disease. *Trends in Neuroscience, 20,* 154–159.

Hardy, J., & Selkoe, D.J. (2002). The amyloid hypothesis of Alzheimer's disease: Progress and problems on the road to therapeutics. *Science, 297,* 353–356.

Hardy, J.A., & Hutton, M. (1995). Two new genes for Alzheimer's disease. *Trends in Neurosciences, 18,* 463.

Hare, R.D. (1993). *Without conscience: The disturbing world of psychopaths among us.* New York: Pocket Books.

Hariri, A.R., Drabant, E.M., Munoz, K.E., Kolachana, B.S., Mattay, V.S., Egan, M.F., et al. (2005). A susceptibility gene for affective disorders and the response of the human amygdala. *Archives of General Psychiatry, 62,* 146–152.

Hariri, A.R., & Holmes, A. (2006). Genetics of emotional regulation: The role of the serotonin transporter in neural function. *Trends in Cognitive Sciences, 10,* 182–191.

Hariri, A.R., Mattay, V.S., Tessitore, A., Kolachana, B., Fera, F., Goldman, D., et al. (2002). Serotonin transporter genetic variation and the response of the human amygdala. *Science, 297,* 400–403.

Harlaar, N. (2006). *Individual differences in early reading achievement: Developmental insights from a twin study.* Unpublished doctoral dissertation, University of London.

Harlaar, N., Butcher, L., Meaburn, E., Craig, I.W., & Plomin, R. (2005a). A behavioural genomic analysis of DNA markers associated with general cognitive ability in 7 year olds. *Journal of Child Psychology and Psychiatry, 46,* 1097–1107.

Harlaar, N., Hayiou-Thomas, M.E., & Plomin, R. (2005b). Reading and general cognitive ability: A multivariate analysis of 7-year-old twins. *Scientific Studies of Reading, 9,* 197–218.

Harlaar, N., Spinath, F.M., Dale, P.S., & Plomin, R. (2005c). Genetic influences on early word recognition abilities and disabilities: A study of 7-year-old twins. *Journal of Child Psychology and Psychiatry, 46,* 373–384.

Harris, C.R. (2003). A review of sex differences in sexual jealousy, including self-report data, psychophysiological responses, interpersonal violence, and morbid jealousy. *Personality and Social Psychology Review, 7,* 102–128.

Harris, J.A., Vernon, P.A., Johnson, A.M., & Jang, K.L. (2006). Phenotypic and genetic relationships between vocational interests and personality. *Personality and Individual Differences, 40,* 1531–1541.

Harris, J.R. (1998). *The nurture assumption: Why children turn out the way they do.* New York: The Free Press.

Harris, J.R., Pedersen, N.L., Stacey, C., McClearn, G.E., & Nesselroade, J.R. (1992). Age differences in the etiology of the relationship between life satisfaction and self-rated health. *Journal of Aging and Health, 4,* 349–368.

Harrison, P.J., & Law, A.J. (2006). Neuregulin 1 and schizophrenia: Genetics, gene expression, and neurobiology. *Biological Psychiatry, 60,* 132–140.

Harrison, P.J., & Weinberger, D.R. (2005). Schizophrenia genes, gene expression, and neuropathology: On the matter of their convergence. *Molecular Psychiatry, 10,* 40–68.

Harter, S. (1983). Developmental perspectives on the self-system. In E.M. Hetherington (Ed.), *Handbook of child psychology: Socialization, personality, and social development* (4th ed., pp. 275–385). New York: Wiley.

Hartl, D. (2004). *A primer of population genetics* (3rd ed.). Sunderland, MA: Sinauer Associates.

Hartl, D., & Clark, H.G. (2006). *Principles of population genetics* (4th ed.). Sunderland, MA: Sinauer Associates.

Hawkins, R.D., Kandel, E.R., & Bailey, C.H. (2006). Molecular mechanisms of memory storage in *Aplysia. The Biological Bulletin, 210,* 174–191.

Hearnshaw, L.S. (1979). *Cyril Burt, psychologist.* Ithaca, NY: Cornell University Press.

Heath, A.C., Bucholz, K.K., Madden, P.A.F., Dinwiddie, S.H., Slutski, W.S., Bierut, L.J., et al. (1997). Genetic and environmental contributions to alcohol dependence risk in a national twin sample: Consistency of findings in women and men. *Psychological Medicine, 27,* 1396.

Heath, A.C., Jardine, R., & Martin, N.G. (1989). Interactive effects of genotype and social environment on alcohol consumption. *Journal of Studies on Alcohol, 50,* 38–48.

Heath, A.C., & Madden, P.A.F. (1995). Genetic influences on smoking behavior. In J.R. Turner, L.R. Cardon, & J.K. Hewitt (Eds.), *Behavior genetic approaches in behavioral medicine* (pp. 45–66). New York: Plenum.

Heath, A.C., Madden, P.A.F., Bucholz, K.K., Nelson, E.C., Todorov, A., Price, R.K., et al. (2003). Genetic and environmental risks of dependence on alcohol, tobacco, and other drugs. In R. Plomin, J.C. DeFries, I.W. Craig, & P. McGuffin (Eds.), *Behavioral genetics in the postgenomic era* (pp. 309–334). Washington, DC: American Psychological Association.

Heath, A.C., & Martin, N. (1993). Genetic models for the natural history of smoking: Evidence for a genetic influence on smoking persistence. *Addictive Behaviours, 18,* 19–34.

Heath, A.C., Neale, M.C., Kessler, R.C., Eaves, L.J., & Kendler, K.S. (1992). Evidence for genetic influences on personality from self-reports and informant ratings. *Journal of Social and Personality Psychology, 63,* 85–96.

Hebb, D.O. (1949). *The organization of behavior.* New York: Wiley.

Hebebrand, J. (1992). A critical appraisal of X-linked bipolar illness: Evidence for the assumed mode of inheritance is lacking. *British Journal of Psychiatry, 160,* 7–11.

Heiman, N., Stallings, M.C., Young, S.E., & Hewitt, J.K. (2004). Investigating the genetic and environmental structure of Cloninger's personality dimensions in adolescence. *Twin Research, 7,* 462–470.

Heisenberg, M. (2003). Mushroom body memoir: From maps to models. *Nature Reviews Neuroscience, 4,* 266–275.

Heitmann, B.L., Kaprio, J., Harris, J.R., Rissanen, A., Korkeila, M., & Koskenvuo, M. (1997). Are genetic determinants of weight gain modified by leisure-time physical activity? A prospective study of Finnish twins. *American Journal of Clinical Nutrition, 66*, 672–678.

Hemmings, S.M., & Stein, D.J. (2006). The current status of association studies in obsessive-compulsive disorder. *The Psychiatric Clinics of North America, 29*, 411–444.

Henderson, N.D. (1967). Prior treatment effects on open field behaviour of mice—A genetic analysis. *Animal Behaviour, 15*, 365–376.

Henderson, N.D. (1972). Relative effects of early rearing environment on discrimination learning in house mice. *Journal of Comparative and Physiological Psychology, 72*, 505–511.

Henderson, N.D., Turri, M.G., DeFries, J.C., & Flint, J. (2004). QTL analysis of multiple behavioral measures of anxiety in mice. *Behavior Genetics, 34*, 267–293.

Herbert, A., Gerry, N.P., McQueen, M.B., Heid, I.M., Pfeufer, A., Illig, T., et al. (2006). A common genetic variant is associated with adult and childhood obesity. *Science, 312*, 279–283.

Herbert, A., Gerry, N.P., McQueen, M.B., Heid, I.M., Pfeufer, A., Illig, T., et al. (2007). Response to comments on "A common genetic variant is associated with adult and childhood obesity." *Science, 315*, 187.

Herrnstein, R.J., & Murray, C. (1994). *The bell curve: Intelligence and class structure in American life.* New York: Free Press.

Hershberger, S.L., Lichtenstein, P., & Knox, S.S. (1994). Genetic and environmental influences on perceptions of organizational climate. *Journal of Applied Psychology, 79*, 24–33.

Hershberger, S.L., Plomin, R., & Pedersen, N.L. (1995). Traits and metatraits: Their reliability, stability, and shared genetic influence. *Journal of Personality and Social Psychology, 69*, 673–684.

Herzog, E.D., Takahashi, J.S., & Block, G.D. (1998). Clock controls circadian period in isolated suprachiasmatic nucleus neurons. *Nature Neuroscience, 1*, 708–713.

Heston, L.L. (1966). Psychiatric disorders in foster home reared children of schizophrenic mothers. *British Journal of Psychiatry, 112*, 819–825.

Hetherington, E.M., & Clingempeel, W.G. (1992). Coping with marital transitions: A family systems perspective. *Monographs of the Society for Research in Child Development, 57*, 1–238.

Hettema, J.M., Annas, P., Neale, M.C., Kendler, K.S., & Fredrikson, M. (2003). A twin study of the genetics of fear conditioning. *Archives of General Psychiatry, 60*, 702–708.

Hettema, J.M., Neale, M.C., & Kendler, K.S. (2001). A review and meta-analysis of the genetic epidemiology of anxiety disorders. *American Journal of Psychiatry, 158*, 1568–1578.

Hettema, J.M., Neale, M.C., Myers, J.M., Prescott, C.A., & Kendler, K.S. (2006). A population-based twin study of the relationship between neuroticism and internalizing disorders. *American Journal of Psychiatry, 163*, 857–864.

Hettema, J.M., Prescott, C.A., & Kendler, K.S. (2001). A population-based twin study of generalized anxiety disorder in men and women. *Journal of Nervous and Mental Disease, 189*, 413–420.

Hettema, J.M., Prescott, C.A., Myers, J.M., Neale, M.C., & Kendler, K.S. (2005). The structure of genetic and environmental risk factors for anxiety disorders in men and women. *Archives of General Psychiatry, 62*, 182–189.

Higuchi, S., Matsushita, S., & Kashima, H. (2006). New findings on the genetic influences on alcohol use and dependence. *Current Opinion in Psychiatry, 19,* 253–265.

Hill, S.Y., Shen, S., Zezza, N., Hoffman, E.K., Perlin, M., & Allan, W. (2004). A genomewide search for alcoholism susceptibility genes. *American Journal of Medical Genetics. Part B: Neuropsychiatric Genetics, 128,* 102–113.

Hirschhorn, J.N., & Daly, M.J. (2005). Genomewide association studies for common diseases and complex traits. *Nature Reviews Genetics, 6,* 95–108.

Hirschhorn, J.N., Lohmueller, K., Byrne, E., & Hirschhorn, K. (2002). A comprehensive review of genetic association studies. *Genetics in Medicine, 4,* 45–61.

Hjelmborg, J., Iachine, I., Skytthe, A., Vaupel, J.W., McGue, M., Koskenvuo, M., et al. (2006). Genetic influence on human lifespan and longevity. *Human Genetics, 119,* 312–321.

Ho, H., Baker, L., & Decker, S.N. (1988). Covariation between intelligence and speed of cognitive processing: Genetic and environmental influences. *Behavior Genetics, 18,* 247–261.

Hobcraft, J. (2006). The ABC of demographic behaviour: How the interplays of alleles, brains, and contexts over the life course should shape research aimed at understanding population processes. *Population Studies, 60,* 153–187.

Hobert, O. (2003). Behavioral plasticity in *C. elegans:* Paradigms, circuits, genes. *Journal of Neurobiology, 54,* 203–223.

Hodgkinson, S., Mullan, M., & Murray, R.M. (1991). The genetics of vulnerability to alcoholism. In P. McGuffin & R. Murray (Eds.), *The new genetics of mental illness* (pp. 182–197). London: Mental Health Foundation.

Hofman, M.A., & Swaab, D.F. (2006). Living by the clock: The circadian pacemaker in older people. *Aging Research Reviews, 5,* 33–51.

Hollister, J.M., Mednick, S.A., Brennan, P., & Cannon, T.D. (1994). Impaired autonomic nervous system habituation in those at genetic risk for schizophrenia. *Archives of General Psychiatry, 51,* 552–558.

Holmans, P., Weissman, M.M., Zubenko, G.S., Scheftner, W.A., Crowe, R.R., Depaulo, J.R., Jr., et al. (2007). Genetics of recurrent early-onset major depression (GenRED): Final genome scan report. *American Journal of Psychiatry, 164,* 248–258.

Hotta, Y., & Benzer, S. (1970). Genetic dissection of the *Drosophila* nervous sytem by means of mosaics. *Proceedings of the National Academy of Sciences (USA), 67,* 1156–1163.

Howie, P. (1981). Concordance for stuttering in monozygotic and dizygotic twin pairs. *Journal of Speech and Hearing Research, 5,* 343–348.

Hrdy, S.B. (1999). *Mother nature: A history of mothers, infants and natural selection.* London: Pantheon/Chatto & Windus.

Hu, S., Pattatucci, A.M.L., Patterson, C., Li, L., Fulker, D.W., Cherny, S.S., et al. (1995). Linkage between sexual orientation and chromosome Xq28 in males but not in females. *Nature Genetics, 11,* 248–256.

Hublin, C., Kaprio, J., Partinen, M., & Koskenvuo, M. (1998). Nocturnal enuresis in a nationwide twin cohort. *Sleep, 21,* 579–585.

Hudziak, J.J., Derks, E.M., Althoff, R.R., Rettew, D.C., & Boomsma, D.I. (2005). The genetic and environmental contributions to attention-deficit hyperactivity disorder as measured by the Conners' Rating Scales-Revised. *American Journal of Psychiatry, 162,* 1614–1620.

Hudziak, J.J., Van Beijsterveldt, C.E., Althoff, R.R., Stanger, C., Rettew, D.C., Nelson, E.C., et al. (2004). Genetic and environmental contributions to the Child Behavior Checklist Obsessive-Compulsive Scale: A cross-cultural twin study. *Archives of General Psychiatry, 61*, 608–616.

Husén, T. (1959). *Psychological twin research.* Stockholm: Almqvist & Wiksell.

Huttenhofer, A., Schattner, P., & Polacek, N. (2005). Non-coding RNAs: Hope or hype? *Trends in Genetics, 21*, 289–297.

Hyman, S.L., Arthur, S.E., & North, K.N. (2006). Learning disabilities in children with neurofibromatosis type 1: Subtypes, cognitive profile, and attention-deficit hyperactivity disorder. *Developmental Medicine and Child Neurology, 48*, 973–977.

Hyman, S.L., Shores, A., & North, K.N. (2005). The nature and frequency of cognitive deficits in children with neurofibromatosis type 1. *Neurology, 65*, 1037–1044.

Iervolino, A.C., Pike, A., Manke, B., Reiss, D., Hetherington, E.M., & Plomin, R. (2002). Genetic and environmental influences in adolescent peer socialization: Evidence from two genetically sensitive designs. *Child Development, 73*, 162–175.

Inlow, J.K., & Restifo, L.L. (2004). Molecular and comparative genetics of mental retardation. *Genetics, 166*, 835–881.

International HapMap Consortium. (2005). A haplotype map of the human genome. *Nature, 437*, 1299–1320.

International Human Genome Sequencing Consortium. (2001). Initial sequencing and analysis of the human genome. *Nature, 409*, 860–921.

International Molecular Genetic Study of Autism Consortium (IMGSAC). (1998). A full genome screen for autism with evidence for linkage to a region on chromosome 7q. *Human Molecular Genetics, 7*, 571–578.

Ioannides, A.A. (2006). Magnetoencephalography as a research tool in neuroscience: State of the art. *Neuroscientist, 12*, 524–544.

Ioannidis, J.P., Ntzani, E.E., Trikalinos, T.A., & Contopoulos-Ioannidis, D.G. (2001). Replication validity of genetic association studies. *Nature Genetics, 29*, 306–309.

Ioannidis, J.P., Trikalinos, T.A., & Khoury, M.J. (2006). Implications of small effect sizes of individual genetic variants on the design and interpretation of genetic association studies of complex diseases. *American Journal of Epidemiology, 164*, 609–614.

Jacob, H.J., & Kwitek, A.E. (2002). Rat genetics: Attaching physiology and pharmacology to the genome. *Nature Reviews Genetics, 3*, 33–42.

Jacobs, N., Gestel, S.V., Derom, C., Thiery, E., Vernon, P.A., Derom, R., et al. (2001). Heritability estimates of intelligence in twins: Effect of chorion type. *Behavior Genetics, 31*, 209–217.

Jacobson, K.C. (2005). Genetic influence on the development of antisocial behavior. In K.S. Kendler & L.J. Eaves (Eds.), *Psychiatric genetics* (pp. 197–232). Arlington, VA: American Psychiatric Publishing.

Jacobson, K.C., Prescott, C.A., & Kendler, K.S. (2002). Sex differences in the genetic and environmental influences on the development of antisocial behavior. *Development and Psychopathology, 14*, 395–416.

Jacobson, K.C., & Rowe, D.C. (1999). Genetic and environmental influences on the relationships between family connectedness, school connectedness, and adolescent depressed mood: Sex differences. *Developmental Psychology, 35*, 926–939.

Jacobson, P., Torgerson, J.S., Sjostrom, L., & Bouchard, C. (2007). Spouse resemblance in body mass index: Effects on adult obesity prevalence in the offspring generation. *American Journal of Epidemiology, 165*, 101–108.

Jaenisch, R., & Bird, A. (2003). Epigenetic regulation of gene expression: How the genome integrates intrinsic and environmental signals. *Nature Genetics, 33*, 245–254.

Jaffee, S.R., Caspi, A., Moffitt, T.E., Dodge, K.A., Rutter, M., Taylor, A., et al. (2005). Nature × nurture: Genetic vulnerabilities interact with physical maltreatment to promote conduct problems. *Development and Psychopathology, 17*, 67–84.

Jaffee, S.R., Caspi, A., Moffitt, T.E., Polo-Tomas, M., Price, T.S., & Taylor, A. (2004). The limits of child effects: Evidence for genetically mediated child effects on corporal punishment but not on physical maltreatment. *Developmental Psychology, 40*, 1047–1058.

Jaffee, S.R., & Price, T. (2007). Gene-environment correlations: A review of the evidence and implications for prevention of mental illness. *Molecular Psychiatry, 12*, 432–442.

James, W. (1890). *Principles of psychology*. New York: Holt.

Jang, K.L. (1993). *A behavioral genetic analysis of personality, personality disorder, the environment, and the search for sources of nonshared environmental influences.* Unpublished doctoral dissertation, University of Western Ontario, London, Ontario.

Jang, K.L. (2005). *The behavioral genetics of psychopathology: A clinical guide.* Hillsdale, NJ: Lawrence Erlbaum Associates.

Jang, K.L., Lam, R.W., Livesley, W.J., & Vernon, P.A. (1997). The relationship between seasonal mood change and personality: More apparent than real? *Acta Psychiatrica Scandinavica, 95*, 539–543.

Jang, K.L., & Livesley, W.J. (1999). Why do measures of normal and disordered personality correlate? A study of genetic comorbidity. *Journal of Personality Disorders, 13*, 10–17.

Jang, K.L., Livesley, W.J., Ando, J., Yamagata, S., Suzuki, A., Angleitner, A., et al. (2006). Behavioral genetics of the higher-order factors of the Big Five. *Personality and Individual Differences, 41*, 261–272.

Jang, K.L., Livesley, W.J., & Vernon, P.A. (1996). Heritability of the Big Five dimensions and their facets: A twin study. *Journal of Personality, 64*, 577–591.

Jang, K.L., Livesley, W.J., Vernon, P.A., & Jackson, D.N. (1996). Heritability of personality disorder traits: A twin study. *Acta Psychiatrica Scandinavica, 94*, 438–444.

Jang, K.L., McCrae, R.R., Angleitner, A., Riemann, R., & Livesley, W.J. (1998). Heritability of facet-level traits in a cross-cultural twin sample: Support for a hierarchical model of personality. *Journal of Personality and Social Psychology, 74*, 1556–1565.

Jang, K.L., Woodward, T.S., Lang, D., Honer, W.G., & Livesley, W.J. (2005). The genetic and environmental basis of the relationship between schizotypy and personality: A twin study. *Journal of Nervous and Mental Disease, 193*, 153–159.

Jansen, R.C., & Nap, J.P. (2001). Genetical genomics: The added value from segregation. *Trends in Genetics, 17*, 388–391.

Jensen, A.R. (1978). Genetic and behavioural effects of nonrandom mating. In R.T. Osbourne, C.E. Noble, & N. Weyl (Eds.), *Human variation: The biopsychology of age, race, and sex* (pp. 51–105). New York: Academic Press.

Jensen, A.R. (1998). *The g factor: The science of mental ability*. Westport, CT: Praeger.

Jensen, A.R. (2006). *Clocking the mind: Mental chronometry and individual differences.* Oxford: Elsevier.

Jepsen, J.R.M., & Michel, M. (2006). ADHD and the symptom dimensions inattention, impulsivity, and hyperactivity—A review of aetiological twin studies from 1996 to 2004. *Nordic Psychology, 58*, 108–135.

Jinks, J.L., & Fulker, D.W. (1970). Comparison of the biometrical genetical, MAVA, and classical approaches to the analysis of human behavior. *Psychological Bulletin, 73*, 311–349.

Jinnah, H.A., Harris, J.C., Nyhan, W.L., & O'Neill, J.P. (2004). The spectrum of mutations causing HPRT deficiency: An update. *Nucleosides, Nucleotides and Nucleic Acid, 23,* 1153–1160.

Jinnah, H.A., Visser, J.E., Harris, J.C., Verdu, A., Larovere, L., Ceballos-Picot, I., et al. (2006). Delineation of the motor disorder of Lesch-Nyhan disease. *Brain, 129,* 1201–1217.

Johnson, C., Drgon, T., Liu, Q.R., Walther, D., Edenberg, H., Rice, J., et al. (2006). Pooled association genome scanning for alcohol dependence using 104,268 SNPs: Validation and use to identify alcoholism vulnerability loci in unrelated individuals from the collaborative study on the genetics of alcoholism. *American Journal of Medical Genetics. Part B: Neuropsychiatric Genetics, 141,* 844–853.

Johnson, J.M., Castle, J., Garrett-Engele, P., Kan, Z., Loerch, P.M., Armour, C.D., et al. (2003). Genome-wide survey of human alternative pre-mRNA splicing with exon junction microarrays. *Science, 302,* 2141–2144.

Johnson, J.M., Edwards, S., Shoemaker, D., & Schadt, E.E. (2005). Dark matter in the genome: Evidence of widespread transcription detected by microarray tiling experiments. *Trends in Genetics, 21,* 93–102.

Johnson, W., Krueger, R.F., Bouchard, T.J., Jr., & McGue, M. (2002). The personalities of twins: Just ordinary folks. *Twin Research, 5,* 125–131.

Johnson, W., McGue, M., Gaist, D., Vaupel, J.W., & Christensen, K. (2002). Frequency and heritability of depression symptomatology in the second half of life: Evidence from Danish twins over 45. *Psychological Medicine, 32,* 1175–1185.

Johnson, W., McGue, M., & Krueger, R.F. (2005). Personality stability in late adulthood: A behavioral genetic analysis. *Journal of Personality, 73,* 523–552.

Johnson, W., McGue, M., Krueger, R.F., & Bouchard, T.J., Jr. (2004). Marriage and personality: A genetic analysis. *Journal of Personality and Social Psychology, 86,* 285–294.

Jones, P.B., & Murray, R.M. (1991). Aberrant neurodevelopment as the expression of schizophrenia genotype. In P. McGuffin & R. Murray (Eds.), *The new genetics of mental illness* (pp. 112–129). Oxford: Butterworth-Heinemann.

Jones, S. (1999). *Almost like a whale: The origin of species, updated.* New York: Doubleday.

Jones, S.R., Gainetdinov, R.R., Jaber, M., Giros, B., Wightman, R.M., & Caron, M.G. (1998). Profound neuronal plasticity in response to inactivation of the dopamine transporter. *Proceedings of the National Academy of Sciences (USA), 95,* 4029–4034.

Jonsson, E.G., Kaiser, R., Brockmoller, J., Nimgaonkar, V.L., & Crocq, M.A. (2004). Meta-analysis of the dopamine D3 receptor gene (*DRD3*) Ser9Gly variant and schizophrenia. *Psychiatric Genetics, 14,* 9–12.

Jordan, K.W., Morgan, T.J., & Mackay, T.F. (2006). Quantitative trait loci for locomotor behavior in *Drosophila melanogaster. Genetics, 174,* 271–284.

Joynson, R.B. (1989). *The Burt affair.* London: Routledge.

Kafkafi, N., Benjamini, Y., Sakov, A., Elmer, G.I., & Golani, I. (2005). Genotype-environment interactions in mouse behavior: A way out of the problem. *Proceedings of the National Academy of Sciences (USA), 102,* 4619–4624.

Kallmann, F.J. (1952). Twin and sibship study of overt male homosexuality. *Journal of Human Genetics, 4,* 136–146.

Kallmann, F.J., & Kaplan, O.J. (1955). Genetic aspects of mental disorders in later life. In O.J. Kaplan (Ed.), *Mental disorders in later life* (pp. 26–46). Stanford, CA: Stanford University Press.

Kallmann, F.J., & Roth, B. (1956). Genetic aspects of preadolescent schizophrenia. *American Journal of Psychiatry, 112,* 599–606.

Kamakura, T., Ando, J., & Ono, Y. (2007). Genetic and environmental effects of stability and change in self-esteem during adolescence. *Personality and Individual Differences, 42*, 181–190.

Kamin, L.J. (1974). *The science and politics of IQ.* Potomac, MD: Lawrence Erlbaum Associates.

Kanner, L. (1943). Autistic disturbances of affective contact. *Nervous Child, 2*, 217–250.

Kaplan, D.E., Gayán, J., Ahn, J., Won, T.W., Pauls, D., Olson, R.K., et al. (2002). Evidence for linkage and association with reading disability on 6p21.3–22. *American Journal of Human Genetics, 70*, 1287–1298.

Karanjawala, Z.E., & Collins, F.S. (1998). Genetics in the context of medical practice. *Journal of the American Medical Association, 280*, 1533–1544.

Karmiloff-Smith, A., Grant, J., Berthoud, I., Davies, M., Howlin, P., & Udwin, O. (1997). Language and Williams syndrome: How intact is "intact"? *Child Development, 68*, 246–262.

Kato, K., & Pedersen, N.L. (2005). Personality and coping: A study of twins reared apart and twins reared together. *Behavior Genetics, 35*, 147–158.

Kato, T. (2007). Molecular genetics of bipolar disorder and depression. *Psychiatry and Clinical Neuroscience, 61*, 3–19.

Kaufmann, W.E. (1996). Mental retardation and learning disabilities: A neuropathological differentiation. In A.J. Capute & P.J. Accardo (Eds.), *Developmental disabilities in infancy and childhood* (pp. 49–70). Baltimore, MD: Paul H. Brookes Publishing.

Kaufmann, W.E., & Reiss, A.L. (1999). Molecular and cellular genetics of fragile X syndrome. *American Journal of Medical Genetics, 88*, 11–24.

Keebaugh, A.C., Sullivan, R.T., & Thomas, J.W. (2007). Gene duplication and inactivation in the HPRT gene family. *Genomics, 89*, 134–142.

Keller, L.M., Bouchard, T.J., Jr., Segal, N.L., & Dawes, R.V. (1992). Work values: Genetic and environmental influences. *Journal of Applied Psychology, 77*, 79–88.

Keller, M.C., & Miller, G. (2006). An evolutionary framework for mental disorders: Integrating adaptationist and evolutionary genetic models. *Behavioral and Brain Sciences, 29*, 429–441.

Keller, M.C., Coventry, W.L., Heath, A.C., & Martin, N.G. (2005). Widespread evidence for non-additive genetic variation in Cloninger's and Eysenck's personality dimensions using a twin plus sibling design. *Behavior Genetics, 35*, 707–721.

Kelsoe, J.R., Ginns, E.I., Egeland, J.A., Gerhard, D.S., Goldstein, A.M., Bale, S.J., et al. (1989). Re-evaluation of the linkage relationship between chromosome *11p* loci and the gene for bipolar affective disorder in the Old Order Amish. *Nature, 342*, 238–242.

Kendler, K.S. (1988). Familial aggregation of schizophrenia and schizophrenia spectrum disorder. *Archives of General Psychiatry, 45*, 377–383.

Kendler, K.S. (1996a). Major depression and generalised anxiety disorder. Same genes, (partly) different environments revisited. *British Journal of Psychiatry. Supplement 168*, 68–75.

Kendler, K.S. (1996b). Parenting: A genetic-epidemiologic perspective. *American Journal of Psychiatry, 153*, 11–20.

Kendler, K.S. (2001). Twin studies of psychiatric illness: An update. *Archives of General Psychiatry, 58*, 1005–1014.

Kendler, K.S. (2005). Toward a philosophical structure for psychiatry. *The American Journal of Psychiatry, 162*, 433–440.

Kendler, K.S., Aggen, S.H., Tambs, K., & Reichborn-Kjennerud, T. (2006). Illicit psychoactive substance use, abuse and dependence in a population-based sample of Norwegian twins. *Psychological Medicine, 36*, 955–962.

Kendler, K.S., & Baker, J.H. (2006). Genetic influences on measures of the environment: A systematic review. *Psychological Medicine, 19*, 1–12.

Kendler, K.S., Czajkowski, N., Tambs, K., Torgersen, S., Aggen, S.H., Neale, M.C., et al. (2006). Dimensional representations of DSM-IV cluster A personality disorders in a population-based sample of Norwegian twins: A multivariate study. *Psychological Medicine, 36*, 1583–1591.

Kendler, K.S., & Eaves, L.J. (1986). Models for the joint effects of genotype and environment on liability to psychiatric illness. *American Journal of Psychiatry, 143*, 279–289.

Kendler, K.S., & Eaves, L.J. (Eds.). (2005). *Psychiatric genetics.* Washington, DC: American Psychiatric Publishing.

Kendler, K.S., Gardner, C.O., & Prescott, C.A. (2001). Panic syndromes in a population-based sample of male and female twins. *Psychological Medicine, 31*, 989–1000.

Kendler, K.S., Gardner, C.O., & Prescott, C.A. (2003). Personality and the experience of environmental adversity. *Psychological Medicine, 33*, 1193–1202.

Kendler, K.S., Gatz, M., Gardner, C.O., & Pedersen, N.L. (2006a). A Swedish national twin study of lifetime major depression. *American Journal of Psychiatry, 163*, 109–114.

Kendler, K.S., Gatz, M., Gardner, C.O., & Pedersen, N.L. (2006b). Personality and major depression: A Swedish longitudinal, population-based twin study. *Archives of General Psychiatry, 63*, 1113–1120.

Kendler, K.S., & Greenspan, R.J. (2006). The nature of genetic influences on behavior: Lessons from "simpler" organisms. *The American Journal of Psychiatry, 163*, 1683–1694.

Kendler, K.S., Gruenberg, A.M., & Kinney, D.K. (1994). Independent diagnoses of adoptees and relatives, as defined by DSM-II, in the provincial and national samples of the Danish adoption study of schizophrenia. *Archives of General Psychiatry, 51*, 456–468.

Kendler, K.S., & Hewitt, J. (1992). The structure of self-report schizotypy in twins. *Journal of Personality Disorders, 6*, 1–17.

Kendler, K.S., Jacobson, K.C., Gardner, C.O., Gillespie, N., Aggen, S.A., & Prescott, C.A. (2007). Creating a social world: A developmental twin study of peer-group deviance. *Archives of General Psychiatry, 64*, 958–965.

Kendler, K.S., Karkowski, L.M., Neale, M.C., & Prescott, C.A. (2000). Illicit psychoactive substance use, heavy use, abuse, and dependence in a US population-based sample of male twins. *Archives of General Psychiatry, 57*, 261–269.

Kendler, K.S., & Karkowski-Shuman, L. (1997). Stressful life events and genetic liability to major depression: Genetic control of exposure to the environment. *Psychological Medicine, 27*, 539–547.

Kendler, K.S., Kessler, R.C., Walters, E.E., MacLean, C.J., Neale, M.C., Heath, A.C., et al. (1995). Stressful life events, genetic liability, and onset of an episode of major depression in women. *American Journal of Psychiatry, 152*, 833–842.

Kendler, K.S., Kuhn, J.W., Vittum, J., Prescott, C.A., & Riley, B. (2005). The interaction of stressful life events and a serotonin transporter polymorphism in the prediction of episodes of major depression: A replication. *Archives of General Psychiatry, 62,* 529–535.

Kendler, K.S., Myers, J., Prescott, C.A., & Neale, M.C. (2001). The genetic epidemiology of irrational fears and phobias in men. *Archives of General Psychiatry, 58,* 257–265.

Kendler, K.S., Neale, M.C., Kessler, R.C., Heath, A.C., & Eaves, L.J. (1992). Major depression and generalized anxiety disorder. Same genes, (partly) different environments? *Archives of General Psychiatry, 49,* 716–722.

Kendler, K.S., Neale, M.C., Kessler, R.C., Heath, A.C., & Eaves, L.J. (1993a). A test of the equal-environment assumption in twin studies of psychiatric illness. *Behavior Genetics, 23,* 21–27.

Kendler, K.S., Neale, M.C., Kessler, R.C., Heath, A.C., & Eaves, L.J. (1993b). A twin study of recent life events and difficulties. *Archives of General Psychiatry, 50,* 789–796.

Kendler, K.S., Neale, M.C., Kessler, R.C., Heath, A.C., & Eaves, L.J. (1994). Parental treatment and the equal environment assumption in twin studies of psychiatric illness. *Psychological Medicine, 24,* 579–590.

Kendler, K.S., & Prescott, C.A. (1998). Cannabis use, abuse, and dependence in a population-based sample of female twins. *American Journal of Psychiatry, 155,* 1016–1022.

Kendler, K.S., & Prescott, C.A. (2006). *Genes, environment, and psychopathology: Understanding the cause of psychiatric and substance use disorders.* New York: Guilford Press.

Kendler, K.S., Prescott, C.A., Myers, J., & Neale, M.C. (2003). The structure of genetic and environmental risk factors for common psychiatric and substance use disorders in men and women. *Archives of General Psychiatry, 60,* 929–937.

Kendler, K.S., Prescott, C.A., Neale, M.C., & Pedersen, N.L. (1997). Temperance Board registration for alcohol abuse in a national sample of Swedish male twins, born 1902 to 1949. *Archives of General Psychiatry, 54,* 178–184.

Kendler, K.S., Thornton, L.M., Gilman, S.E., & Kessler, R.C. (2000). Sexual orientation in a U.S. national sample of twin and nontwin sibling pairs. *American Journal of Psychiatry, 157,* 1843–1846.

Kenrick, D.T., & Funder, D.C. (1988). Profiting from controversy: Lessons from the person-situation debate. *American Psychologist, 43,* 23–34.

Kessler, R.C., McGonagle, K.A., Zhao, C.B., Nelson, C.B., Hughes, M., Eshleman, S., et al. (1994). Lifetime and 12-month prevalence of DSM-III-R psychiatric disorders in the United States: Results from the National Comorbidity Study. *Archives of General Psychiatry, 51,* 8–19.

Kessler, R.C., Berglund, P., Demler, O., Jin, R., Merikangas, K.R., & Walters, E.E. (2005). Lifetime prevalence and age-of-onset distributions of DSM-IV disorders in the National Comorbidity Survey Replication. *Archives of General Psychiatry, 62,* 593–602.

Kessler, R.C., Chiu, W.T., Demler, O., Merikangas, K.R., & Walters, E.E. (2005). Prevalence, severity, and comorbidity of 12-month DSM-IV disorders in the National Comorbidity Survey Replication. *Archives of General Psychiatry, 62,* 617–627.

Kessler, R.C., Kendler, K.S., Heath, A.C., Neale, M.C., & Eaves, L.J. (1992). Social support, depressed mood, and adjustment to stress: A genetic epidemiological investigation. *Journal of Personality and Social Psychology, 62,* 257–272.

Kety, S.S. (1987). The significance of genetic factors in the etiology of schizophrenia: Results from the national study of adoptees in Denmark. *Journal of Psychiatric Research, 21,* 423–430.

Kety, S.S., Wender, P.H., Jacobsen, B., Ingraham, L.J., Jansson, L., Faber, B., et al. (1994). Mental illness in the biological and adoptive relatives of schizophrenic adoptees: Replication of the Copenhagen study in the rest of Denmark. *Archives of General Psychiatry, 51*, 442–455.

Khan, A.A., Jacobson, K.C., Gardner, C.O., Prescott, C.A., & Kendler, K.S. (2005). Personality and comorbidity of common psychiatric disorders. *British Journal of Psychiatry, 186*, 190–196.

Khandjian, E.W., Bechara, E., Davidovic, L., & Bardoni, B. (2005). Fragile X mental retardation protein: Many partners and multiple targets for a promiscuous function. *Current Genomics, 6*, 515–522.

Kidd, K. (1983). Recent progress on the genetics of stuttering. In C. Ludlow & J. Cooper (Eds.), *Genetic aspects of speech and language disorders* (pp. 197–213). New York: Academic Press.

Kieseppa, T., Partonen, T., Haukka, J., Kaprio, J., & Lonnqvist, J. (2004). High concordance of bipolar I disorder in a nationwide sample of twins. *American Journal of Psychiatry, 161*, 1814–1821.

Kile, B.T., & Hilton, D.J. (2005). The art and design of genetic screens: Mouse. *Nature Reviews Genetics, 6*, 557–567.

Kim, D.H., & Rossi, J.J. (2007). Strategies for silencing human disease using RNA interference. *Nature Reviews Genetics, 8*, 173–184.

Kim-Cohen, J., Caspi, A., Taylor, A., Williams, B., Newcombe, R., Craig, I.W., et al. (2006). MAOA, maltreatment, and gene-environment interaction predicting children's mental health: New evidence and a meta-analysis. *Molecular Psychiatry, 11*, 903–913.

King, D.P., Zhao, Y., Sangoram, A.M., Wilsbacher, L.D., Tanaka, M., Antoch, M.P., et al. (1997). Positional cloning of the mouse circadian clock gene. *Cell, 89*, 641–653.

Kirov, G., Nikolov, I., Georgieva, L., Moskvina, V., Owen, M.J., & O'Donovan, M.C. (2006). Pooled DNA genotyping on Affymetrix SNP genotyping arrays. *BMC Genomics, 7*, 27.

Kishino, T., Lalande, M., & Wagstaff, J. (1997). UBE3A/E6-AP mutations cause Angelman syndrome. *Nature Genetics, 15*, 70–73.

Klein, R.G., & Mannuzza, S. (1991). Long-term outcome of hyperactive children—A review. *Journal of the American Academy of Child and Adolescent Psychiatry, 30*, 383–387.

Klose, J., Nock, C., Herrmann, M., Stuhler, K., Marcus, K., Bluggel, M., et al. (2002). Genetic analysis of the mouse brain proteome. *Nature Genetics, 30*, 385–393.

Knafo, A., & Plomin, R. (2006a). Prosocial behavior from early to middle childhood: Genetic and environmental influences on stability and change. *Developmental Psychology, 42*, 771–86.

Knafo, A., & Plomin, R. (2006b). Parental discipline and affection, and children's prosocial behavior: Genetic and environmental link. *Journal of Personality and Social Psychology, 90*, 147–164.

Knight, S.J., & Regan, R. (2006). Idiopathic learning disability and genome imbalance. *Cytogenetic and Genome Research, 115*, 215–224.

Knight, S.J., Regan, R., Nicod, A., Horsley, S.W., Kearney, L., Homfray, T., et al. (1999). Subtle chromosomal rearrangements in children with unexplained mental retardation. *Lancet, 354*, 1676–1681.

Knopik, V.S., Alarcón, M., & DeFries, J.C. (1997). Comorbidity of mathematics and reading deficits: Evidence for a genetic etiology. *Behavior Genetics, 27*, 447–453.

Knoppien, P. (1985). Rare male mating advantage—A review. *Biological Reviews of the Cambridge Philosophical Society, 60,* 81–117.

Ko, C.H., & Takahashi, J.S. (2006). Molecular components of the mammalian circadian clock. *Human Molecular Genetics, 15,* R271–R277.

Koenig, L.B., McGue, M., Krueger, R.F., & Bouchard, T.J., Jr. (2005). Genetic and environmental influences on religiousness: Findings for retrospective and current religiousness ratings. *Journal of Personality, 73,* 471–488.

Koeppen-Schomerus, G., Spinath, F.M., & Plomin, R. (2003). Twins and non-twin siblings: Different estimates of shared environmental influence in early childhood. *Twin Research, 6,* 97–105.

Koeppen-Schomerus, G., Wardle, J., & Plomin, R. (2001). A genetic analysis of weight and overweight in 4-year-old twin pairs. *International Journal of Obesity and Related Metabolic Disorders, 25,* 838–844.

Kohnstamm, G.A., Bates, J.E., & Rothbart, M.K. (1989). *Temperament in childhood.* New York: Wiley.

Kolevzon, A., Smith, C.J., Schmeidler, J., Buxbaum, J.D., & Silverman, J.M. (2004). Familial symptom domains in monozygotic siblings with autism. *American Journal of Medical Genetics. Part B: Neuropsychiatric Genetics, 129,* 76–81.

Konopka, R.J., & Benzer, S. (1971). Clock mutants of *Drosophila melanogaster. Proceedings of the National Academy of Sciences (USA), 68,* 2112–2116.

Konradi, C. (2005). Gene expression microarray studies in polygenic psychiatric disorders: Applications and data analysis. *Brain Research Reviews, 50,* 142–155.

Koopmans, J.R., Boomsma, D.I., Heath, A.C., & van Doornen, L.J.P. (1995). A multivariate genetic analysis of sensation seeking. *Behavior Genetics, 25,* 349–356.

Koopmans, J.R., Slutske, W.S., van Baal, G.C., & Boomsma, D.I. (1999). The influence of religion on alcohol use initiation: Evidence for genotype × environment interaction. *Behavior Genetics, 29,* 445–453.

Kosik, K.S. (2006). The neuronal microRNA system. *Nature Reviews Neuroscience, 7,* 911–920.

Kotler, M., Cohen, H., Segman, R., Gritsenko, I., Nemanov, L., Lerer, B., et al. (1997). Excess dopamine D4 receptor (*D4DR*) exon III seven repeat allele in opioid-dependent subjects. *Molecular Psychiatry, 2,* 251–254.

Koukoui, S.D., & Chaudhuri, A. (2007). Neuroanatomical, molecular genetic, and behavioral correlates of fragile X syndrome. *Brain Research Reviews, 53,* 27–38.

Kovas, Y., Harlaar, N., Petrill, S.A., & Plomin, R. (2005). 'Generalist genes' and mathematics in 7-year-old twins. *Intelligence, 5,* 473–489.

Kovas, Y., Haworth, C.M.A., Dale, P.S., & Plomin, R. (2007).The genetic and environmental origins of learning abilities and disabilities in the early school years. *Monographs of the Society for Research in Child Development, 72,* 1–144..

Kovas, Y., Haworth, C.M.A., Harlaar, N., Petrill, S.A., Dale, P.S., & Plomin, R. (2007). Overlap and specificity of genetic and environmental influences on mathematics and reading disability in 10-year-old twins. *Journal of Child Psychology and Psychiatry, 49,* 914–922.

Kovas, Y., Haworth, C.M.A., Petrill, S., & Plomin, R. (in press). Mathematical ability of 10-year-old boys and girls: Genetic and environmental etiology of normal and low performance. *Journal of Learning Disabilities.*

Kovas, Y., Petrill, S.A., & Plomin, R. (2007). The origins of diverse domains of mathematics: Generalist genes but specialist environments. *Journal of Educational Psychology, 99,* 128–139.

Kovas, Y., & Plomin, R. (2006). Generalist genes: Implications for cognitive sciences. *Trends in Cognitive Science, 10,* 198–203.

Kremen, W.S., Jacobson, K.C., Xian, H., Eisen, S.A., Waterman, B., Toomey, R., et al. (2005). Heritability of word recognition in middle-aged men varies as a function of parental education. *Behavior Genetics, 35,* 417–433.

Kringlen, E., & Cramer, G. (1989). Offspring of monozygotic twins discordant for schizophrenia. *Archives of General Psychiatry, 46,* 873–877.

Krueger, R.F. (1999). The structure of common mental disorders. *Archives of General Psychiatry, 56,* 921–926.

Krueger, R.F., Caspi, A., Moffitt, T.E., Silva, A., & McGee, R. (1996). Personality traits are differentially linked to mental disorders: A multitrait-multidiagnosis study of an adolescent birth cohort. *Journal of Abnormal Psychology, 105,* 299–312.

Krueger, R.F., Hicks, B.M., Patrick, C.J., Carlson, S.R., Iacono, W.G., & McGue, M. (2002). Etiologic connections among substance dependence, antisocial behavior, and personality: Modeling the externalizing spectrum. *Journal of Abnormal Psychology, 111,* 411–424.

Krueger, R.F., Johnson, W., & Caspi, A. (in press). Behavioral genetics and personality: A new look at the integration of nature and nurture. In L.A. Pervin & O.P. John (Eds.), *Handbook of personality: Theory and research* (3rd ed.). New York: Guilford.

Krueger, R.F., Markon, K.E., & Bouchard, T.J., Jr. (2003). The extended genotype: The heritability of personality accounts for the heritability of recalled family environments in twins reared apart. *Journal of Personality, 71,* 809–833.

Kuehn, M.R., Bradley, A., Robertson, E.J., & Evans, M.J. (1987). A potential animal model for Lesch-Nyhan syndrome through introduction of HPRT mutations into mice. *Nature, 326,* 295–298.

Kuntsi, J., Rijsdijk, F., Ronald, A., Asherson, P., & Plomin, R. (2005). Genetic influences on the stability of attention-deficit hyperactivity disorder symptoms from early to middle childhood. *Biological Psychiatry, 57,* 647–654.

Lack, D. (1953). *Darwin's finches.* Cambridge: Cambridge University Press.

Lai, C.S., Fisher, S.E., Hurst, J.A., Vargha-Khadem, F., & Monaco, A.P. (2001). A forkhead-domain gene is mutated in a severe speech and language disorder. *Nature, 413,* 519–523.

Lander, E.S. (1999). Array of hope. *Nature Genetics, 21,* 3–4.

Lanfranco, F., Kamischke, A., Zitzmann, M., & Nieschlag, E. (2004). Klinefelter's syndrome. *Lancet, 364,* 273–283.

Langlois, J.H., Ritter, J.M., Casey, R.J., & Sawin, D.B. (1995). Infant attractiveness predicts maternal behaviors and attitudes. *Developmental Psychology, 31,* 464–472.

Langlois, J.H., Ritter, J.M., Roggman, L.A., & Vaughn, L. (1991). Facial diversity and infant preferences for attractive faces. *Developmental Psychology, 27,* 79–84.

Larsson, H., Andershed, H., & Lichtenstein, P. (2006). A genetic factor explains most of the variation in the psychopathic personality. *Journal of Abnormal Psychology, 115,* 221–230.

Larsson, H., Lichtenstein, P., & Larsson, J.O. (2006). Genetic contributions to the development of ADHD subtypes from childhood to adolescence. *Journal of the American Academy of Child and Adolescent Psychiatry, 45,* 973–981.

Larsson, H., Tuvblad, C., Rijsdijk, F.V., Andershed, H., Grann, M., & Lichtenstein, P. (2007). A common genetic factor explains the association between psychopathic personality and antisocial behavior. *Psychological Medicine, 37,* 15–26.

Larsson, J.O., Larsson, H., & Lichtenstein, P. (2004). Genetic and environmental contributions to stability and change of ADHD symptoms between 8 and 13 years of age: A longitudinal twin study. *Journal of the American Academy of Child and Adolescent Psychiatry, 43,* 1267–1275.

Lau, B., Bretaud, S., Huang, Y., Lin, E., & Guo, S. (2006). Dissociation of food and opiate preference by a genetic mutation in zebrafish. *Genes, Brain, and Behavior, 5,* 497–505.

Leahy, A.M. (1935). Nature-nurture and intelligence. *Genetic Psychology Monographs, 17,* 236–308.

Le Couteur, A., Bailey, A., Goode, S., Pickles, A., Robertson, S., Gottesman, I.I., et al. (1996). A broader phenotype of autism: The clinical spectrum in twins. *Journal of Child Psychology and Psychiatry, 37,* 785–801.

LeDoux, J.E. (2000). Emotion circuits in the brain. *Annual Review of Neuroscience, 23,* 155–184.

Lee, P.J., Ridout, D., Walter, J.H., & Cockburn, F. (2005). Maternal phenylketonuria: Report from the United Kingdom Registry 1978–97. *Archives of Disease in Childhood, 90,* 143–146.

Legrand, L.N., McGue, M., & Iacono, W.G. (1999). A twin study of state and trait anxiety in childhood and adolescence. *Journal of Child Psychology and Psychiatry, 40,* 953–958.

Lein, E.S., Hawrylycz, M.J., Ao, N., Ayres, M., Bensinger, A., Bernard, A., et al. (2007). Genome-wide atlas of gene expression in the adult mouse brain. *Nature, 445,* 168–176.

Lerner, I.M. (1968). *Heredity, evolution and society.* San Francisco: Freeman.

Lesch, K.P., Bengel, D., Heils, A., Zhang Sabol, S., Greenburg, B.D., Petri, S., et al. (1996). Association of anxiety-related traits with a polymorphism in the serotonin transporter gene regulatory region. *Science, 274,* 1527–1531.

Letwin, N.E., Kafkafi, N., Benjamini, Y., Mayo, C., Frank, B.C., Luu, T., et al. (2006). Combined application of behavior genetics and microarray analysis to identify regional expression themes and gene-behavior associations. *The Journal of Neuroscience, 26,* 5277–5287.

Levinson, D.F., Evgrafov, O.V., Knowles, J.A., Potash, J.B., Weissman, M.M., Scheftner, W.A., et al. (2007). Genetics of recurrent early-onset major depression (GenRED): Significant linkage on chromosome 15q25–q26 after fine mapping with single nucleotide polymorphism markers. *American Journal of Psychiatry, 164,* 259–264.

Levy, D.L., Holzman, P.S., Matthysse, S., & Mendell, N.R. (1993). Eye tracking disfunction and schizophrenia: A critical perspective. *Schizophrenia Bulletin, 19,* 461–536.

Levy, R., Mirlesse, V., Jacquemard, F., & Daffos, F. (2002). Prenatal diagnosis of zygosity by fetal DNA analysis, a contribution to the management of multiple pregnancies. A series of 31 cases. *Fetal Diagnosis and Therapy, 17,* 339–342.

Lewin, B. (2004). *Genes VIII.* New York: Prentice Hall.

Lewis, C.M., Levinson, D.F., Wise, L.H., DeLisi, L.E., Straub, R.E., Hovatta, I., et al. (2003). Genome scan meta-analysis of schizophrenia and bipolar disorder, part II: Schizophrenia. *American Journal of Human Genetics, 73,* 34–48.

Li, D., Collier, D.A., & He, L. (2006). Meta-analysis shows strong positive association of the neuregulin 1 (*NRG1*) gene with schizophrenia. *Human Molecular Genetics, 15,* 1995–2002.

Li, D., Sham, P.C., Owen, M.J., & He, L. (2006). Meta-analysis shows significant association between dopamine system genes and attention-deficit hyperactivity disorder (ADHD). *Human Molecular Genetics, 15*, 2276–2284.

Li, J., & Burmeister, M. (2005). Genetical genomics: Combining genetics with gene expression analysis. *Human Molecular Genetics, 14*, R163–R169.

Li, L.L., Keverne, E.B., Aparicio, S.A., Ishino, F., Barton, S.C., & Surani, M.A. (1999). Regulation of maternal behavior and offspring growth by paternally expressed *Peg3*. *Science, 284*, 333.

Li, T., Xu, K., Deng, H., Cai, G., Liu, J., Liu, X., et al. (1997). Association analysis of the dopamine D4 gene exon III VNTR and heroin abuse in Chinese subjects. *Molecular Psychiatry, 2*, 413–416.

Liang, H., Masoro, E.J., Nelson, J.F., Strong, R., McMahan, C.A., & Richardson, A. (2003). Genetic mouse models of extended lifespan. *Experimental Gerontology, 38*, 1353–1364.

Lichtenstein, P., Harris, J.R., Pedersen, N.L., & McClearn, G.E. (1992). Socioeconomic status and physical health, how are they related? An empirical study based on twins reared apart and twins reared together. *Social Science and Medicine, 36*, 441–450.

Lichtenstein, P., Pedersen, N.L., & McClearn, G.E. (1992). The origins of individual differences in occupational status and educational level. *Acta Sociologica, 35*, 13–31.

Lidsky, A.S., Robson, K., Chandra, T., Barker, P., Ruddle, F., & Woo, S.L.C. (1984). The PKU locus in man is on chromosome 12. *American Journal of Human Genetics, 36*, 527–533.

Lilienfeld, S.O. (1992). The association between antisocial personality and somatization disorders: A review and integration of theoretical models. *Clinical Psychology Review, 12*, 641–662.

Lim, L.P., Lau, N.C., Garrett-Engele, P., Grimson, A., Schelter, J.M., Castle, J., et al. (2005). Microarray analysis shows that some microRNAs downregulate large numbers of target mRNAs. *Nature, 433*, 769–773.

Lindblad-Toh, K., Wade, C.M., Mikkelsen, T.S., Karlsson, E.K., Jaffe, D.B., Kamal, M., et al. (2005). Genome sequence, comparative analysis and haplotype structure of the domestic dog. *Nature, 438*, 803–819.

Linney, Y.M., Murray, R.M., Peters, E.R., MacDonald, A.M., Rijsdijk, F., & Sham, P.C. (2003). A quantitative genetic analysis of schizotypal personality traits. *Psychological Medicine, 33*, 803–816.

Lipovechaja, N.G., Kantonistowa, N.S., & Chamaganova, T.G. (1978). The role of heredity and environment in the determination of intellectual function. *Medicinskie, Probleing Formirovaniga Livenosti, 1*, 48–59.

Liu, Q.R., Drgon, T., Walther, D., Johnson, C., Poleskaya, O., Hess, J., et al. (2005). Pooled association genome scanning: Validation and use to identify addiction vulnerability loci in two samples. *Proceedings of the National Academy of Sciences (USA), 102*, 11864–11869.

Liu, X., & Davis, R.L. (2006). Insect olfactory memory in time and space. *Current Opinion in Neurobiology, 16*, 679–685.

Livesley, W.J., Jang, K.L., & Vernon, P.A. (1998). Phenotypic and genetic structure of traits delineating personality disorder. *Archives of General Psychiatry, 55*, 941–948.

Loehlin, J.C. (1989). Partitioning environmental and genetic contributions to behavioral development. *American Psychologist, 44*, 1285–1292.

Loehlin, J.C. (1992). *Genes and environment in personality development.* Newbury Park, CA: Sage Publications Inc.

Loehlin, J.C. (1997). Genes and environment. In D. Magnusson (Ed.), *The lifespan development of individuals: Behavioral, neurobiological, and psychosocial perspectives: A synthesis* (pp. 38–51). New York: Cambridge University Press.

Loehlin, J.C., Horn, J.M., & Willerman, L. (1989). Modeling IQ change: Evidence from the Texas Adoption Project. *Child Development, 60,* 993–1004.

Loehlin, J.C., Horn, J.M., & Willerman, L. (1990). Heredity, environment, and personality change: Evidence from the Texas Adoption Study. *Journal of Personality, 58,* 221–243.

Loehlin, J.C., Horn, J.M., & Willerman, L. (1997). Heredity, environment and IQ in the Texas Adoption Study. In E.M. Sternberg & E.L. Grigorenko (Eds.), *Intelligence, heredity and environment* (pp. 105–125). New York: Cambridge University Press.

Loehlin, J.C., & Martin, N.G. (2001). Age changes in personality traits and their heritabilities during the adult years: Evidence from Australian twin registry samples. *Personality and Individual Differences, 30,* 1147–1174.

Loehlin, J.C., Neiderhiser, J.M., & Reiss, D. (2003). The behavior genetics of personality and the NEAD study. *Journal of Research in Personality, 37,* 373–387.

Loehlin, J.C., & Nichols, J. (1976). *Heredity, environment and personality.* Austin: University of Texas Press.

Loehlin, J.C., Willerman, L., & Horn, J.M. (1982). Personality resemblances between unwed mothers and their adopted-away offspring. *Journal of Personality and Social Psychology, 42,* 1089–1099.

Lohmueller, K.E., Pearce, C.L., Pike, M., Lander, E.S., & Hirschhorn, J.N. (2003). Meta-analysis of genetic association studies supports a contribution of common variants to susceptibility to common disease. *Nature Genetics, 33,* 177–182.

Long, J., Knowler, W., Hanson, R., Robin, R., Urbanek, M., Moore, E., et al. (1998). Evidence for genetic linkage to alcohol dependence on chromosomes 4 and 11 from an autosome-wide scan in an American Indian population. *American Journal of Medical Genetics. Part B: Neuropsychiatric Genetics, 81,* 216–221.

Losoya, S.H., Callor, S., Rowe, D.C., & Goldsmith, H.H. (1997). Origins of familial similarity in parenting: A study of twins and adoptive siblings. *Developmental Psychology, 33,* 1012–1023.

Lovinger, D.M., & Crabbe, J.C. (2005). Laboratory models of alcoholism: Treatment target identification and insight into mechanisms. *Nature Neuroscience, 8,* 1471–1480.

Lucht, M., Barnow, S., Schroeder, W., Grabe, H.J., Finckh, U., John, U., et al. (2006). Negative perceived paternal parenting is associated with dopamine D2 receptor exon 8 and GABA(A) alpha 6 receptor variants: An explorative study. *American Journal of Medical Genetics. Part B: Neuropsychiatric Genetics, 141,* 167–172.

Luciano, M., Posthuma, D., Wright, M.J., de Geus, E.J., Smith, G.A., Geffen, G.M., et al. (2005). Perceptual speed does not cause intelligence, and intelligence does not cause perceptual speed. *Biological Psychology, 70,* 1–8.

Luciano, M., Wright, M.J., Duffy, D.L., Wainwright, M.A., Zhu, G., Evans, D.M., et al. (2006). Genome-wide scan of IQ finds significant linkage to a quantitative trait locus on 2q. *Behavior Genetics, 36,* 45–55.

Luciano, M., Wright, M.J., Geffen, G.M., Geffen, L.B., Smith, G.A., & Martin, N.G. (2004). A genetic investigation of the covariation among inspection time, choice reaction time, and IQ subtest scores. *Behavior Genetics, 34,* 41–50.

Luo, D., Petrill, S.A., & Thompson, L.A. (1994). An exploration of genetic *g:* Hierarchical factor analysis of cognitive data from the Western Reserve Twin Project. *Intelligence, 18,* 335–348.

Luo, D., Thompson, L.A., & Detterman, D.K. (2006). The criterion validity of tasks of basic cognitive processes. *Intelligence, 34,* 79–120.

Lykken, D.T. (1982). Research with twins: The concept of emergenesis. *Psychophysiology, 19,* 361–373.

Lykken, D.T. (2006). The mechanism of emergenesis. *Genes, Brain, and Behavior, 5,* 306–310.

Lykken, D.T., & Tellegen, A. (1993). Is human mating adventitious or the result of lawful choice? A twin study of mate selection. *Journal of Personality and Social Psychology, 65,* 56–68.

Lynch, M.A. (2004). Long-term potentiation and memory. *Physiologica, 84,* 87–136.

Lynch, S.K., Turkheimer, E., D'Onofrio, B.M., Mendle, J., Emery, R.E., Slutske, W.S., et al. (2006). A genetically informed study of the association between harsh punishment and offspring behavioral problems. *Journal of Family Psychology, 20,* 190–198.

Lyons, M.J. (1996). A twin study of self-reported criminal behaviour. In G.R. Bock & J.A. Goode (Eds.), *Genetics of criminal and antisocial behaviour* (pp. 1–75). Chichester, UK: John Wiley & Sons.

Lyons, M.J., Goldberg, J., Eisen, S.A., True, W., Tsuang, M.T., Meyer, J.M., et al. (1993). Do genes influence exposure to trauma: A twin study of combat. *American Journal of Medical Genetics. Part B: Neuropsychiatric Genetics, 48,* 22–27.

Lyons, M.J., True, W.R., Eisen, S.A., Goldberg, J., Meyer, J.M., Faraone, S.V., et al. (1995). Differential heritability of adult and juvenile antisocial traits. *Archives of General Psychiatry, 52,* 906–915.

Lytton, H. (1977). Do parents create or respond to differences in twins? *Developmental Psycholology, 13,* 456–459.

Lytton, H. (1980). *Parent-child interaction: The socialization process observed in twin and singleton families.* New York: Plenum.

Lytton, H. (1991). Different parental practices—Different sources of influence. *Behavioral and Brain Sciences, 14,* 399–400.

Ma, D.Q., Cuccaro, M.L., Jaworski, J.M., Haynes, C.S., Stephan, D.A., Parod, J., et al. (2007). Dissecting the locus heterogeneity of autism: Significant linkage to chromosome 12q14. *Molecular Psychiatry, 12,* 376–384.

Maat-Kievit, A., Vegter-van der Vlis, M., Zoeteweij, M., Losekoot, M., van Haeringen, A., & Roos, R. (2000). Paradox of a better test for Huntington's disease. *Journal of Neurology, Neurosurgery, and Psychiatry, 69,* 579–583.

MacBeath, G. (2002). Protein microarrays and proteomics. *Nature Genetics, 32. Supplement,* 526–532.

MacGillivray, I., Campbell, D.M., & Thompson, B. (1988). *Twinning and twins.* Chichester, UK: John Wiley & Sons.

Mack, K.J., & Mack, P.A. (1992). Introduction of transcription factors in somatosensory cortex after tactile stimulation. *Molecular Brain Research, 12,* 141–149.

Mackay, T.F., & Anholt, R.R. (2006). Of flies and man: *Drosophila* as a model for human complex traits. *Annual Review of Genomics and Human Genetics, 7,* 339–367.

Mackintosh, M.A., Gatz, M., Wetherell, J.L., & Pedersen, N.L. (2006). A twin study of lifetime generalized anxiety disorder (GAD) in older adults: Genetic and environmental influences shared by neuroticism and GAD. *Twin Research and Human Genetics, 9,* 30–37.

Mackintosh, N.J. (1995). *Cyril Burt: Fraud or framed?* Oxford: Oxford University Press.

Mackintosh, N.J. (1998). *IQ and human intelligence.* Oxford: Oxford University Press.

Macphail, E.M. (1993). *The neuroscience of animal intelligence: From the seahare to the seahorse.* New York: Columbia University Press.

Maes, H.H., Neale, M.C., & Eaves, L.J. (1997). Genetic and environmental factors in relative body weight and human adiposity. *Behavior Genetics, 27,* 325–351.

Maes, H.H., Neale, M.C., Kendler, K.S., Martin, N.G., Heath, A.C., & Eaves, L.J. (2006). Genetic and cultural transmission of smoking initiation: An extended twin kinship model. *Behavior Genetics, 36,* 795–808.

Maguire, E.A., Gadian, D.G., Johnsrude, I.S., Good, C.D., Ashburner, J., Frackowiak, R.S., et al. (2000). Navigation-related structural change in the hippocampi of taxi drivers. *Proceedings of the National Academy of Sciences (USA), 97,* 4398–4403.

Maguire, E.A., Woollett, K., & Spiers, H.J. (2006). London taxi drivers and bus drivers: A structural MRI and neuropsychological analysis. *Hippocampus, 16,* 1091–1101.

Mahowald, M.B., Verp, M.S., & Anderson, R.R. (1998). Genetic counseling: Clinical and ethical challenges. *Annual Review of Genetics, 32,* 547–559.

Malykh, S.B., Iskoldsky, N.V., & Gindina, E.V. (2005). Genetic analysis of IQ in young adulthood: A Russian twin study. *Personality and Individual Differences, 38,* 1475–1485.

Mandoki, M.W., Sumner, G.S., Hoffman, R.P., & Riconda, D.L. (1991). A review of Klinefelter's syndrome in children and adolescents. *Journal of the American Academy of Child and Adolescent Psychiatry, 30,* 167–172.

Manke, B., McGuire, S., Reiss, D., Hetherington, E.M., & Plomin, R. (1995). Genetic contributions to adolescents' extrafamilial social interactions: Teachers, best friends, and peers. *Social Development, 4,* 238–256.

Mao, R., & Pevsner, J. (2005). The use of genomic microarrays to study chromosomal abnormalities in mental retardation. *Mental Retardation and Developmental Disabilities Research Reviews, 11,* 279–285.

Margulies, C., Tully, T., & Dubnau, J. (2005). Deconstructing memory in *Drosophila. Current Biology, 15,* R700–R713.

Marino, C., Giorda, R., Vanzin, L., Nobile, M., Lorusso, M.L., Baschirotto, C., et al. (2004). A locus on 15q15-15qter influences dyslexia: Further support from a transmission/disequilibrium study in an Italian speaking population. *Journal of Medical Genetics, 41,* 42–46.

Marks, I.M., & Nesse, R.M. (1994). Fear and fitness: An evolutionary analysis of anxiety disorders. *Etiology and Sociobiology, 15,* 247–261.

Maron, E., Toru, I., Must, A., Tasa, G., Toover, E., Vasar, V., et al. (2007). Association study of tryptophan hydroxylase 2 gene polymorphisms in panic disorder. *Neuroscience Letters, 411,* 180–184.

Martin, N., Boomsma, D.I., & Machin, G. (1997). A twin-pronged attack on complex trait. *Nature Genetics, 17,* 387–392.

Matheny, A.P., Jr. (1980). Bayley's infant behavioral record: Behavioral components and twin analysis. *Child Development, 51,* 1157–1167.

Matheny, A.P., Jr. (1989). Children's behavioral inhibition over age and across situations: Genetic similarity for a trait during change. *Journal of Personality, 57,* 215–235.

Matheny, A.P., Jr. (1990). Developmental behavior genetics: Contributions from the Louisville Twin Study. In M.E. Hahn, J.K. Hewitt, N.D. Henderson, & R.H. Benno (Eds.), *Developmental behavior genetics: Neural, biometrical, and evolutionary approaches* (pp. 25–39). New York: Chapman & Hall.

Matheny, A.P., Jr., & Dolan, A.B. (1975). Persons, situations, and time: A genetic view of behavioral change in children. *Journal of Personality and Social Psychology, 14,* 224–234.

Mather, K., & Jinks, J.L. (1982). *Biometrical genetics: The study of continuous variation* (3rd ed.). New York: Chapman & Hall.

Mattay, V.S., & Goldberg, T.E. (2004). Imaging genetic influences in human brain function. *Current Opinion in Neurobiology, 14,* 239–247.

Matthews, K.A., Kaufman, T.C., & Gelbart, W.M. (2005). Research resources for *Drosophila:* The expanding universe. *Nature Reviews Genetics, 6,* 179–193.

Mattick, J.S. (2004). RNA regulation: A new genetics? *Nature Reviews Genetics, 5,* 316–323.

Mattick, J.S. (2005). The functional genomics of noncoding RNA. *Science, 309,* 1527–1528.

Mattick, J.S., & Makunin, I.V. (2006). Non-coding RNA. *Human Molecular Genetics, 15,* R17–R29.

Matzel, L.D., Townsend, D.A., Grossman, H., Han, Y.R., Hale, G., Zappulla, M., et al. (2006). Exploration in outbred mice covaries with general learning abilities irrespective of stress reactivity, emotionality, and physical attributes. *Neurobiology of Learning and Memory, 86,* 228–240.

Maxson, S.C., & Canastar, A. (2003). Conceptual and methodological issues in the genetics of mouse agonistic behavior. *Hormones and Behavior, 44,* 258–262.

Mayford, M., & Kandel, E.R. (1999). Genetic approaches to memory storage. *Trends in Genetics, 15,* 463–470.

McCartney, K., Harris, M.J., & Bernieri, F. (1990). Growing up and growing apart: A developmental meta-analysis of twin studies. *Psychological Bulletin, 107,* 226–237.

McClearn, G.E. (1963). The inheritance of behavior. In L.J. Postman (Ed.), *Psychology in the making* (pp. 144–252). New York: Knopf.

McClearn, G.E. (1976). Experimental behavioural genetics. In D. Barltrop (Ed.), *Aspects of genetics in paediatrics* (pp. 31–39). London: Fellowship of Postdoctorate Medicine.

McClearn, G.E., & DeFries, J.C. (1973). *Introduction to behavioral genetics.* San Francisco: Freeman.

McClearn, G.E., Johansson, B., Berg, S., Pedersen, N.L., Ahern, F., Petrill, S.A., et al. (1997). Substantial genetic influence on cognitive abilities in twins 80+ years old. *Science, 276,* 1560–1563.

McClearn, G.E., & Rodgers, D.A. (1959). Differences in alcohol preference among inbred strains of mice. *Quarterly Journal of Studies on Alcohol, 52,* 62–67.

McCourt, K., Bouchard, T.J., Jr., Lykken, D.T., Tellegen, A., & Keyes, M. (1999). Authoritarianism revisited: Genetic and environmental influences examined in twins reared apart and together. *Personality and Individual Differences, 27,* 985–1014.

McGowan, E., Eriksen, J., & Hutton, M. (2006). A decade of modeling Alzheimer's disease in transgenic mice. *Trends in Genetics, 22,* 281–289.

McGrath, L.M., Smith, S.D., & Pennington, B.F. (2006). Breakthroughs in the search for dyslexia candidate genes. *Trends in Molecular Medicine, 12,* 333–341.

McGrath, M., Kawachi, I., Ascherio, A., Colditz, G.A., Hunter, D.J., & De Vivo, I. (2004). Association between catechol-O-methyltransferase and phobic anxiety. *American Journal of Psychiatry, 161*, 1703–1705.

McGue, M. (1993). From proteins to cognitions: The behavioral genetics of alcoholism. In R. Plomin & G.E. McClearn (Eds.), *Nature, nurture, and psychology* (pp. 245–268). Washington, DC: American Psychological Association.

McGue, M. (2000). *Behavioral genetic models of alcoholism and drinking.* New York: Guilford Press.

McGue, M., Bacon, S., & Lykken, D.T. (1993). Personality stability and change in early adulthood: A behavioral genetic analysis. *Developmental Psychology, 29*, 96–109.

McGue, M., & Bouchard, T.J., Jr. (1989). Genetic and environmental determinants of information processing and special mental abilities: A twin analysis. In R.J. Sternberg (Ed.), *Advances in the psychology of human intelligence, 5* (pp. 7–45). Hillsdale, NJ: Lawrence Erlbaum Associates.

McGue, M., Bouchard, T.J., Jr., Iacono, W.G., & Lykken, D.T. (1993). Behavioral genetics of cognitive ability: A life-span perspective. In R. Plomin & G.E. McClearn (Eds.), *Nature, nurture, and psychology* (pp. 59–76). Washington, DC: American Psychological Association.

McGue, M., & Christensen, K. (2001). The heritability of cognitive functioning in very old adults: Evidence from Danish twins aged 75 years and older. *Psychology and Aging, 16*, 272–280.

McGue, M., Elkins, I., Walden, B., & Iacono, W.G. (2005). Perceptions of the parent-adolescent relationship: A longitudinal investigation. *Developmental Psychology, 41*, 971–984.

McGue, M., & Gottesman, I.I. (1989). Genetic linkage in schizophrenia: Perspectives from genetic epidemiology. *Schizophrenia Bulletin, 15*, 464.

McGue, M., Hirsch, B., & Lykken, D.T. (1993). Age and the self-perception of ability: A twin study analysis. *Psychology and Aging, 8*, 72–80.

McGue, M., & Lykken, D.T. (1992). Genetic influence on risk of divorce. *Psychological Science, 3*, 368–373.

McGue, M., Sharma, S., & Benson, P. (1996). Parent and sibling influences on adolescent alcohol use and misuse: Evidence from a U.S. adoption court. *Journal of Studies on Alcohol, 57*, 8–18.

McGuffin, P., Cohen, S., & Knight, J. (2007). Homing in on depression genes. *American Journal of Psychiatry, 164*, 195–197.

McGuffin, P., Farmer, A.E., & Gottesman, I.I. (1987). Is there really a split in schizophrenia? The genetic evidence. *British Journal of Psychiatry, 50*, 581–592.

McGuffin, P., & Gottesman, I.I. (1985). Genetic influences on normal and abnormal development. In M. Rutter & L. Hersov (Eds.), *Child and adolescent psychiatry: Modern approaches* (2nd ed.). (pp. 17–33). Oxford: Blackwell Scientific.

McGuffin, P., & Katz, R. (1986). Nature, nurture, and affective disorder. In J.W.F. Deakin (Ed.), *The biology of depression* (pp. 26–51). London: Gaskell Press.

McGuffin, P., Katz, R., & Rutherford, J. (1991). Nature, nurture and depression: A twin study. *Psychological Medicine, 21*, 329–335.

McGuffin, P., Katz, R., Watkins, S., & Rutherford, J. (1996). A hospital-based twin register of the heritability of DSM-IV unipolar depression. *Archives of General Psychiatry, 53*, 129–136.

McGuffin, P., Knight, J., Breen, G., Brewster, S., Boyd, P.R., Craddock, N., et al. (2005). Whole genome linkage scan of recurrent depressive disorder from the depression network study. *Human Molecular Genetics, 14*, 3337–3345.

McGuffin, P., Owen, M.J., & Gottesman, I.I. (2002). *Psychiatric genetics and genomics.* Oxford: Oxford University Press.

McGuffin, P., Owen, M.J., O'Donovan, M.C., Thapar, A., & Gottesman, I.I. (1994). *Seminars in psychiatric genetics.* London: Gaskell.

McGuffin, P., Rijsdijk, F., Andrew, M., Sham, P., Katz, R., & Cardno, A. (2003). The heritability of bipolar affective disorder and the genetic relationship to unipolar depression. *Archives of General Psychiatry, 60*, 497–502.

McGuffin, P., Sargeant, M., Hetti, G., Tidmarsh, S., Whatley, S., & Marchbanks, R.M. (1990). Exclusion of a schizophrenia susceptibility gene from the chromosome 5q11-q13 region. New data and a reanalysis of previous reports. *American Journal of Human Genetics, 47*, 534–535.

McGuffin, P., & Sturt, E. (1986). Genetic markers in schizophrenia. *Human Heredity, 16*, 461–465.

McGuire, M., & Troisi, A. (1998). *Darwinian psychiatry.* Oxford: Oxford University Press.

McGuire, S., Neiderhiser, J.M., Reiss, D., Hetherington, E.M., & Plomin, R. (1994). Genetic and environmental influences on perceptions of self-worth and competence in adolescence: A study of twins, full siblings, and step-siblings. *Child Development, 65*, 785–799.

McGuire, S.E., Deshazer, M., & Davis, R.L. (2005). Thirty years of olfactory learning and memory research in *Drosophila melanogaster. Progress in Neurobiology, 76*, 328–347.

McKie, R. (2007). *Face of Britain: How our genes reveal the history of Britain.* New York: Simon & Schuster.

McLoughlin, G., Ronald, A., Kuntsi, J., Asherson, P., & Plomin, R. (2007). Genetic support for the dual nature of attention-deficit hyperactivity disorder: Substantial genetic overlap between the inattentive and hyperactive-impulsive components. *Journal of Abnormal Child Psychology, 35*, 999–1008.

McMahon, R.C. (1980). Genetic etiology in the hyperactive child syndrome: A critical review. *American Journal of Orthopsychiatry, 50*, 145–150.

McQueen, M.B., Devlin, B., Faraone, S.V., Nimgaonkar, V.L., Sklar, P., Smoller, J.W., et al. (2005). Combined analysis from eleven linkage studies of bipolar disorder provides strong evidence of susceptibility loci on chromosomes 6q and 8q. *American Journal of Human Genetics, 77*, 582–595.

McRae, A.F., Matigian, N.A., Vadlamudi, L., Mulley, J.C., Mowry, B., Martin, N.G., et al. (2007). Replicated effects of sex and genotype on gene expression in human lymphoblastoid cell lines. *Human Molecular Genetics, 16*, 364–373.

Meaburn, E., Dale, P.S., Craig, I.W., & Plomin, R. (2002). Language-impaired children: No sign of the FOXP2 mutation. *NeuroReport, 13*, 1075–1077.

Meaburn, E.L., Butcher, L.M., Knight, J., Craig, I.C., Schalkwyk, L.C., & Plomin, R. (2005). QTLS for reading disability at 7 years: Genotyping pooled DNA on 100K SNP microarrays. *American Journal of Medical Genetics. Part B: Neuropsychiatric Genetics, 51*.

Medlund, P., Cederlof, R., Floderus-Myrhed, B., Friberg, L., & Sorensen, S. (1977). A new Swedish twin registry. *Acta Medica Scandinavica Supplementum, 60*, 1–11.

Mednick, S.A., Gabrielli, W.F., & Hutchings, B. (1984). Genetic factors in criminal behavior: Evidence from an adoption cohort. *Science, 224*, 891–893.

Mello, C.V., Vicario, D.S., & Clayton, D.F. (1992). Song presentation induces gene expression in the songbird forebrain. *Proceedings of the National Academy of Sciences (USA), 89,* 6818–6821.

Mendel, G.J. (1866). Versuche ueber Pflanzenhybriden. *Verhandlungen des Naturforschunden Vereines in Bruenn, 4,* 3–47.

Mendes Soares, L.M., & Valcárcel, J. (2006). The expanding transcriptome: The genome as the "Book of Sand." *The EMBO Journal, 25,* 923–931.

Mendle, J., Turkheimer, E., D'Onofrio, B.M., Lynch, S.K., Emery, R.E., Slutske, W.S., et al. (2006). Family structure and age at menarche: A children-of-twins approach. *Developmental Psychology, 42,* 533–542.

Mendlewicz, J., & Rainer, J.D. (1977). Adoption study supporting genetic transmission in manic-depressive illness. *Nature, 268,* 327–329.

Merikangas, K.R. (1990). The genetic epidemiology of alcoholism. *Psychological Medicine, 20,* 11–22.

Merikangas, K.R., Stolar, M., Stevens, D.E., Goulet, J., Preisig, M.A., Fenton, B., et al. (1998). Familial transmission of substance use disorders. *Archives of General Psychiatry, 55,* 973–979.

Merriman, C. (1924). The intellectual resemblance of twins. *Psychological Monographs, 33,* 1–58.

Merrow, M., Spoelstra, K., & Roenneberg, T. (2005). The circadian cycle: Daily rhythms from behaviour to genes. *EMBO Reports, 6,* 930–935.

Meyer, J.M. (1995). Genetic studies of obesity across the life span. In L.R. Cardon & J.K. Hewitt (Eds.), *Behavior genetic approaches to behavioral medicine* (pp. 145–166). New York: Plenum.

Middeldorp, C.M., Cath, D.C., Van Dyck, R., & Boomsma, D.I. (2005). The co-morbidity of anxiety and depression in the perspective of genetic epidemiology. A review of twin and family studies. *Psychological Medicine, 35,* 611–624.

Middeldorp, C.M., Cath, D.C., Vink, J.M., & Boomsma, D.I. (2005). Twin and genetic effects on life events. *Twin Research and Human Genetics, 8,* 224–231.

Miklos, G.L., & Maleszka, R. (2004). Microarray reality checks in the context of a complex disease. *Nature Biotechnology, 22,* 615–621.

Miles, D.R., Silberg, J.L., Pickens, R.W., & Eaves, L.J. (2005). Familial influences on alcohol use in adolescent female twins: Testing for genetic and environmental interactions. *Journal of Studies of Alcohol, 66,* 445–451.

Miller, G.F. (2000). *The mating mind.* New York: Doubleday.

Ming, J.E., Geiger, E., James, A.C., Ciprero, K.L., Nimmakayalu, M., Zhang, Y., et al. (2006). Rapid detection of submicroscopic chromosomal rearrangements in children with multiple congenital anomalies using high density oligonucleotide arrays. *Human Mutation, 27,* 467–473.

Moehring, A.J., & Mackay, T.F. (2004). The quantitative genetic basis of male mating behavior in *Drosophila melanogaster. Genetics, 167,* 1249–1263.

Moffitt, T.E. (1993). Adolescence-limited and life-course-persistent antisocial behavior: A developmental taxonomy. *Psychological Review, 100,* 674–701.

Moffitt, T.E. (2005). The new look of behavioral genetics in developmental psychopathology: Gene-environment interplay in antisocial behaviors. *Psychological Bulletin, 131,* 533–554.

Moffitt, T.E., Caspi, A., & Rutter, M. (2005). Strategy for investigating interactions between measured genes and measured environments. *Archives of General Psychiatry, 62,* 473–481.

Moldin, S. (1999). Attention-deficit hyperactivity disorder. *Biological Psychiatry, 45,* 599–602.

Monks, S.A., Leonardson, A., Zhu, H., Cundiff, P., Pietrusiak, P., Edwards, S., et al. (2004). Genetic inheritance of gene expression in human cell lines. *American Journal of Human Genetics, 75,* 1094–1105.

Montague, C.T., Farooqi, I.S., Whitehead, J.P., Soos, M.A., Rau, H., Wareham, N.J., et al. (1997). Congenital leptin deficiency is associated with severe early-onset obesity in humans. *Nature, 387,* 904–908.

Moog, U., Arens, Y.H., Lent-Albrechts, J.C., Huijts, P.E., Smeets, E.E., Schrander-Stumpel, C.T., et al. (2005). Subtelomeric chromosome aberrations: Still a lot to learn. *Clinical Genetics, 68,* 397–407.

Moore, R.Y. (1999). A clock for the ages. *Science, 284,* 2102–2103.

Moore, T., & Haig, D. (1991). Genomic imprinting in mammalian development: A parental tug-of-war. *Trends in Genetics, 7,* 45–49.

Morgan, T.H., Sturtevant, A.H., Muller, H.J., & Bridges, C.B. (1915). *The mechanism of Mendelian heredity.* New York: Holt.

Morison, I.M., Ramsay, J.P., & Spencer, H.G. (2005). A census of mammalian imprinting. *Trends in Genetics, 21,* 457–465.

Morley, M., Molony, C.M., Weber, T.M., Devlin, J.L., Ewens, K.G., Spielman, R.S., et al. (2004). Genetic analysis of genome-wide variation in human gene expression. *Nature, 430,* 743–747.

Morris, D.W., Ivanov, D., Robinson, L., Williams, N., Stevenson, J., Owen, M.J., et al. (2004). Association analysis of two candidate phospholipase genes that map to the chromosome 15q15.1–15.3 region associated with reading disability. *American Journal of Medical Genetics. Part B: Neuropsychiatric Genetics, 129,* 97–103.

Morris-Yates, A., Andrews, G., Howie, P., & Henderson, S. (1990). Twins: A test of the equal environments assumption. *Acta Psychiatrica Scandinavica, 81,* 322–326.

Mosher, L.R., Pollin, W., & Stabenau, J.R. (1971). Identical twins discordant for schizophrenia: Neurological findings. *Archives of General Psychiatry, 24,* 422–430.

Muhle, R., Trentacoste, S.V., & Rapin, I. (2004). The genetics of autism. *Pediatrics, 113,* e472–e486.

Munafo, M.R., Clark, T., & Flint, J. (2005). Does measurement instrument moderate the association between the serotonin transporter gene and anxiety-related personality traits? A meta-analysis. *Molecular Psychiatry, 10,* 415–419.

Munafo, M.R., Clark, T.G., Moore, L.R., Payne, E., Walton, R., & Flint, J. (2003). Genetic polymorphisms and personality in healthy adults: A systematic review and meta-analysis. *Molecular Psychiatry, 8,* 471–484.

Murray, R.M., Lewis, S.W., & Reveley, A.M. (1985). Towards an aetiological classification of schizophrenia. *Lancet, 1,* 1023–1026.

Mutsuddi, M., Morris, D.W., Waggoner, S.G., Daly, M.J., Scolnick, E.M., & Sklar, P. (2006). Analysis of high-resolution HapMap of DTNBP1 (Dysbindin) suggests no consistency between reported common variant associations and schizophrenia. *American Journal of Human Genetics, 79,* 903–909.

Nadder, T.S., Rutter, M., Silberg, J.L., Maes, H.H., & Eaves, L.J. (2002). Genetic effects on the variation and covariation of attention-deficit hyperactivity disorder (ADHD) and oppositional-defiant disorder/conduct disorder (Odd/CD) symptomatologies across informant and occasion of measurement. *Psychological Medicine, 32,* 39–53.

Nadler, J.J., Zou, F., Huang, H., Moy, S.S., Lauder, J., Crawley, J.N., et al. (2006). Large-scale gene expression differences across brain regions and inbred strains correlate with a behavioral phenotype. *Genetics, 174*, 1229–1236.

Nash, M.W., Huezo-Diaz, P., Williamson, R.J., Sterne, A., Purcell, S., Hoda, F., et al. (2004). Genome-wide linkage analysis of a composite index of neuroticism and mood-related scales in extreme selected sibships. *Human Molecular Genetics, 13*, 2173–2182.

National Foundation for Brain Research. (1992). *The care of disorders of the brain.* Washington, DC: National Foundation for Brain Research.

Neale, M.C. (2004). *Mx: Statistical modeling* [computer software].

Neale, M.C., & Maes, H.H.M. (2003). *Methodology for genetic studies of twins and families.* Dordrecht, The Netherlands: Kluwer Academic Publishers B.V.

Neale, M.C., & Stevenson, J. (1989). Rater bias in the EASI temperament scales: A twin study. *Journal of Personality and Social Psychology, 56*, 446–455.

Neher, A. (2006). Evolutionary psychology: Its programs, prospects, and pitfalls. *American Journal of Psychology, 119*, 517–566.

Neiderhiser, J.M., & McGuire, S. (1994). Competence during middle childhood. In J.C. DeFries, R. Plomin, & D.W. Fulker (Eds.), *Nature and nurture during middle childhood* (pp. 141–151). Cambridge, MA: Blackwell.

Neiderhiser, J.M., Reiss, D., Hetherington, E.M., & Plomin, R. (1999). Relationships between parenting and adolescent adjustment over time: Genetic and environmental contributions. *Developmental Psychology, 35*, 680–692.

Neiderhiser, J.M., Reiss, D., Pedersen, N.L., Lichtenstein, P., Spotts, E.L., Hansson, K., et al. (2004). Genetic and environmental influences on mothering of adolescents: A comparison of two samples. *Developmental Psychology, 40*, 335–351.

Neiss, M.B., Sedikides, C., & Stevenson, J. (2006). Genetic influences on level and stability of self-esteem. *Self and Identity, 5*, 247–266.

Neisser, U. (1997). Never a dull moment. *American Psychologist, 52*, 79–81.

Neisser, U., Boodoo, G., Bouchard, T.J., Jr., Boykin, A.W., Brody, N., Ceci, S.J., et al. (1996). Intelligence: Knowns and unknowns. *American Psychologist, 51*, 77–101.

Nelson, R.J., Demas, G.E., Huang, P.L., Fishman, M.C., Dawson, V.L., Dawson, T.M., et al. (1995). Behavioural abnormalities in male mice lacking neuronal nitric oxide synthase. *Nature, 378*, 383–386.

Nesse, R.M., & Williams, G.C. (1996). *Why we get sick.* New York: Vintage.

Nestadt, G., Samuels, J., Riddle, M.A., Liang, K.Y., Bienvenu, O.J., Hoehn-Saric, R., et al. (2001). The relationship between obsessive-compulsive disorder and anxiety and affective disorders: Results from the Johns Hopkins OCD Family Study. *Psychological Medicine, 31*, 481–487.

Nettle, D. (2006). The evolution of personality variation in humans and other animals. *American Psychologist, 61*, 622–631.

Neubauer, A.C., Sange, G., & Pfurtscheller, G. (1999). Psychometric intelligence and event-related desynchronisation during performance of a letter matching task. In G. Pfurtscheller & Lopes da Silva (Eds.), *Event-related desynchronisation (ERD)—And related oscillatory EEG-phenomena of the awake brain.* Amsterdam: Elsevier.

Neubauer, A.C., Spinath, F.M., Riemann, R., Borkenau, P., & Angleitner, A. (2000). Genetic (and environmental) influence on two measures of speed of information processing and their relation to psychometric intelligence: Evidence from the German Observational Study of adult twins. *Intelligence, 28*, 267–289.

Newbury, D.F., Bonora, E., Lamb, J.A., Fisher, S.E., Lai, C.S.L., Baird, G., et al. (2002). *FOXP2* is not a major susceptibility gene for autism or specific language impairment. *American Journal of Human Genetics, 70,* 1318–1327.

Newcomer, J.W., & Krystal, J.H. (2001). NMDA receptor regulation of memory and behavior in humans. *Hippocampus, 11,* 529–542.

Newson, A., & Williamson, R. (1999). Should we undertake genetic research on intelligence? *Bioethics, 13,* 327–342.

Nichols, P.L. (1984). Familial mental retardation. *Behavior Genetics, 14,* 161–170.

Nichols, R.C. (1978). Twin studies of ability, personality, and interests. *HOMO, 29,* 158–173.

Nicholson, A.C., Unger, E.R., Mangalathu, R., Ojaniemi, H., & Vernon, S.D. (2004). Exploration of neuroendocrine and immune gene expression in peripheral blood mononuclear cells. *Molecular Brain Research, 129,* 193–197.

Nicolson, R., Brookner, F.B., Lenane, M., Gochman, P., Ingraham, L.J., Egan, M.F., et al. (2003). Parental schizophrenia spectrum disorders in childhood-onset and adult-onset schizophrenia. *American Journal of Psychiatry, 160,* 490–495.

Nicolson, R., & Rapoport, J.L. (1999). Childhood-onset schizophrenia: Rare but worth studying. *Biological Psychiatry, 46,* 1418–1428.

Nievergelt, C.M., Kripke, D.F., Barrett, T.B., Burg, E., Remick, R.A., Sadovnick, A.D., et al. (2006). Suggestive evidence for association of the circadian genes PERIOD3 and ARNTL with bipolar disorder. *American Journal of Medical Genetics. Part B: Neuropsychiatric Genetics, 141,* 234–241.

Nigg, J.T., & Goldsmith, H.H. (1994). Genetics of personality disorders: Perspectives from personality and psychopathology research. *Psychological Bulletin, 115,* 346–380.

Norton, N., Williams, H.J., & Owen, M.J. (2006). An update on the genetics of schizophrenia. *Current Opinion in Psychiatry, 19,* 158–164.

Nuffield Council on Bioethics. (2002). *Genetics and human behaviour: The ethical context.* London: Nuffield Council on Bioethics.

Nyhan, W.L., & Wong, D.F. (1996). New approaches to understanding Lesch-Nyhan disease. *New England Journal of Medicine, 334,* 1602–1604.

O'Connor, S., Sorbel, J., Morxorati, S., Li, T.K., & Christian, J.C. (1999). A twin study of genetic influences on the acute adaptation of the EEG to alcohol. *Alcoholism: Clinical and Experimental Research, 23,* 494–501.

O'Connor, T.G., & Croft, C.M. (2001). A twin study of attachment in preschool children. *Child Development, 72,* 1501–1511.

O'Connor, T.G., Deater-Deckard, K., Fulker, D.W., Rutter, M., & Plomin, R. (1998). Genotype-environment correlations in late childhood and early adolescence: Antisocial behavioural problems in the Colorado Adoption Project. *Developmental Psychology, 34,* 970–981.

O'Connor, T.G., Hetherington, E.M., Reiss, D., & Plomin, R. (1995). A twin-sibling study of observed parent-adolescent interactions. *Child Development, 66,* 812–829.

Ogdie, M.N., Fisher, S.E., Yang, M., Ishii, J., Francks, C., Loo, S.K., et al. (2004). Attention-deficit hyperactivity disorder: Fine mapping supports linkage to $5p13$, $6q12$, $16p13$, and $17p11$. *American Journal of Human Genetics, 75,* 661–668.

Ohman, A., & Mineka, S. (2001). Fears, phobias, and preparedness: Toward an evolved module of fear and fear learning. *Psychological Review, 108,* 483–522.

Oliver, B., Harlaar, N., Hayiou-Thomas, M.E., Kovas, Y., Walker, S.O., Petrill, S.A., et al. (2004). A twin study of teacher-reported mathematics performance and low performance in 7-year-olds. *Journal of Educational Psychology, 96,* 504–517.

Olson, J.M., Vernon, P.A., Harris, J.A., & Jang, K.L. (2001). The heritability of attitudes: A study of twins. *Journal of Personality and Social Psychology, 80,* 845–860.

Olson, R.K. (2007). Introduction to the special issue on genes, environment, and reading. *Reading and Writing, 20,* 1–11.

Ooki, S. (2005). Genetic and environmental influences on stuttering and tics in Japanese twin children. *Twin Research and Human Genetics, 8,* 69–75.

Oppenheimer, S. (2006). *The origins of the British: A genetic detective story.* London: Constable & Robinson.

Ostrander, E.A., Giger, U., & Lindblad-Toh, K. (Eds.). (2006). *The dog and its genome.* New York: Cold Spring Harbor Laboratory Press.

Ostrander, E.A., & Wayne, R.K. (2005). The canine genome. *Genome Research, 15,* 1706–1716.

Owen, M.J., Craddock, N., & O'Donovan, M.C. (2005). Schizophrenia: Genes at last? *Trends in Genetics, 21,* 518–525.

Owen, M.J., Liddell, M.B., & McGuffin, P. (1994). Alzheimer's disease: An association with apolipoprotein e4 may help unlock the puzzle. *British Medical Journal, 308,* 672–673.

Pagan, J.L., Rose, R.J., Viken, R.J., Pulkkinen, L., Kaprio, J., & Dick, D.M. (2006). Genetic and environmental influences on stages of alcohol use across adolescence and into young adulthood. *Behavior Genetics, 36,* 483–497.

Pahl, A. (2005). Gene expression profiling using RNA extracted from whole blood: Technologies and clinical applications. *Expert Review of Molecular Diagnostics, 5,* 43–52.

Papassotiropoulos, A., Stephan, D.A., Huentelman, M.J., Hoerndli, F.J., Craig, D.W., Pearson, J.V., et al. (2006). Common Kibra alleles are associated with human memory performance. *Science, 314,* 475–478.

Paris, J. (1999). *Genetics and psychopathology: Predisposition-stress interactions.* Washington, DC: American Psychiatric Press.

Parker, H.G., Kim, L.V., Sutter, N.B., Carlson, S., Lorentzen, T.D., Malek, T.B., et al. (2004). Genetic structure of the purebred domestic dog. *Science, 304,* 1160–1164.

Parnas, J., Cannon, T.D., Jacobsen, B., Schulsinger, H., Schulsinger, F., & Mednick, S.A. (1993). Lifetime DSM-II-R diagnostic outcomes in the offspring of schizophrenic mothers: Results from the Copenhagen high-risk study. *Archives of General Psychiatry, 50,* 707–714.

Patterson, D., & Costa, A.C. (2005). Down syndrome and genetics—A case of linked histories. *Nature Reviews Genetics, 6,* 137–147.

Pauler, F.M., & Barlow, D.P. (2006). Imprinting mechanisms—It only takes two. *Genes and Development, 20,* 1203–1206.

Pauls, D.L. (1990). Genetic influences on child psychiatric conditions. In M. Lewis (Ed.), *Child and Adolescent Psychiatry: A Comprehensive Textbook* (pp. 351–353). Baltimore, MD: Williams & Wilkins.

Pauls, D.L. (2003). An update on the genetics of Gilles de la Tourette's syndrome. *Journal of Psychosomatic Research, 55,* 7–12.

Pauls, D.L., Leckman, J.F., & Cohen, D.J. (1993). Familial relationship between Gilles de la Tourette's syndrome, attention-deficit disorder, learning difficulties, speech disorders, and stuttering. *Journal of the American Academy of Child and Adolescent Psychiatry, 32,* 1044–1050.

Pauls, D.L., Towbin, K.E., Leckman, J.F., Zahner, G.E.P., & Cohen, D.J. (1986). Gilles de la Tourette's syndrome and obsessive compulsive disorder. *Archives of General Psychiatry, 43,* 1182.

Pavuluri, M.N., Birmaher, B., & Naylor, M.W. (2005). Pediatric bipolar disorder: A review of the past 10 years. *Journal of the American Academy of Child and Adolescent Psychiatry, 44,* 846–871.

Payton, A. (2006). Investigating cognitive genetics and its implications for the treatment of cognitive deficit. *Genes, Brain and Behavior, 5,* 44–53.

Pearson, J.V., Huentelman, M.J., Halperin, R.F., Tembe, W.D., Melquist, S., Homer, N., et al. (2007). Identification of the genetic basis for complex disorders by use of pooling-based genomewide single-nucleotide polymorphism association studies. *American Journal of Human Genetics, 80,* 126–139.

Pedersen, N.L. (1996). Gerontological behavioral genetics. In J.E. Birren & K.W. Schaie (Eds.), *Handbook of the psychology of aging* (4th ed.). (pp. 59–77). San Diego: Academic Press.

Pedersen, N.L., Gatz, M., Plomin, R., Nesselroade, J.R., & McClearn, G.E. (1989). Individual differences in locus of control during the second half of the life span for identical and fraternal twins reared apart and reared together. *Psychosomatic Medicine, 51,* 428–440.

Pedersen, N.L., Lichtenstein, P., Plomin, R., DeFaire, U., McClearn, G.E., & Matthews, K.A. (1989). Genetic and environmental influences for type A-like measures and related traits: A study of twins reared apart and twins reared together. *Psychosomatic Medicine, 51,* 440.

Pedersen, N.L., McClearn, G.E., Plomin, R., & Nesselroade, J.R. (1992a). Effects of early rearing environment on twin similarity in the last half of the life span. *British Journal of Developmental Psychology, 10,* 255–267.

Pedersen, N.L., McClearn, G.E., Plomin, R., & Nesselroade, J.R. (1992b). A quantitative genetic analysis of cognitive abilities during the second half of the life span. *Psychological Science, 3,* 346–353.

Pedersen, N.L., Plomin, R., & McClearn, G.E. (1994). Is there *G* beyond *g*? (Is there genetic influence on specific cognitive abilities independent of genetic influence on general cognitive ability?). *Intelligence, 18,* 133–143.

Peirce, J.L., Li, H., Wang, J., Manly, K.F., Hitzemann, R.J., Belknap, J.K., et al. (2006). How replicable are mRNA expression QTL? *Mammalian Genome, 17,* 643–656.

Pergadia, M.L., Heath, A.C., Martin, N.G., & Madden, P.A. (2006). Genetic analyses of DSM-IV nicotine withdrawal in adult twins. *Psychological Medicine, 36,* 963–972.

Pergadia, M.L., Madden, P.A., Lessov, C.N., Todorov, A.A., Bucholz, K.K., Martin, N.G., et al. (2006). Genetic and environmental influences on extreme personality dispositions in adolescent female twins. *Journal of Child Psychology and Psychiatry, 47,* 902–909.

Pervin, L.A., & John, O.P. (in press). *Handbook of personality: Theory and research* (3rd ed.). New York: Guilford Press.

Peters, L.L., Robledo, R.F., Bult, C.J., Churchill, G.A., Paigen, B.J., & Svenson, K.L. (2007). The mouse as a model for human biology: A resource guide for complex trait analysis. *Nature Reviews Genetics, 8,* 58–69.

Peto, R., Lopez, A.D., Boreham, J., Thun, M., & Heath, C. (1992). Mortality from tobacco in developed countries: Indirect estimation from national vital statistics. *Lancet, 339,* 1268–1278.

Petretto, E., Mangion, J., Dickens, N.J., Cook, S.A., Kumaran, M.K., Lu, H., et al. (2006). Heritability and tissue specificity of expression quantitative trait loci. *PLoS Genetics, 2,* E172.

Petrill, S.A. (1997). Molarity versus modularity of cognitive functioning? A behavioral genetic perspective. *Current Directions in Psychological Science, 6,* 96–99.

Petrill, S.A. (2002). The case for general intelligence: A behavioral genetic perspective. In R.J. Sternberg & E.L. Grigorenko (Eds.), *The general factor of intelligence: How general is it?* (pp. 281–298). Hillsdale, NJ: Lawrence Erlbaum Associates.

Petrill, S.A., Ball, D.M., Eley, T.C., Hill, L., & Plomin, R. (1998). Failure to replicate a QTL association between a DNA marker identified by EST00083 and IQ. *Intelligence, 25,* 179–184.

Petrill, S.A., Deater-Deckard, K., Thompson, L.A., Schatschneider, C., DeThorne, L.S., & Vandenbergh, D.J. (2007). Longitudinal genetic analysis of early reading: The Western Reserve Reading Project. *Reading and Writing, 20,* 127–146.

Petrill, S.A., Lipton, P.A., Hewitt, J.K., Plomin, R., Cherny, S.S., Corley, R., et al. (2004). Genetic and environmental contributions to general cognitive ability through the first 16 years of life. *Developmental Psychology, 40,* 805–812.

Petrill, S.A., Plomin, R., DeFries, J.C., & Hewitt, J.K. (2003). *Nature, nurture, and the transition to early adolescence.* Oxford: Oxford University Press.

Petrill, S.A., Saudino, K.J., Cherny, S.S., Emde, R.N., Hewitt, J.K., Fulker, D.W., et al. (1997). Exploring the genetic etiology of low general cognitive ability from 14 to 36 months. *Developmental Psychology, 33,* 544–548.

Petrill, S.A., Thompson, L.A., & Detterman, D.K. (1995). The genetic and environmental variance underlying elementary cognitive tasks. *Behavior Genetics, 25,* 199–209.

Petronis, A. (2006). Epigenetics and twins: Three variations on the theme. *Trends in Genetics, 22,* 347–350.

Phelps, E.A., & LeDoux, J.E. (2005). Contributions of the amygdala to emotion processing: From animal models to human behavior. *Neuron, 48,* 175–187.

Phelps, J.A., Davis, O.J., & Schwartz, K.M. (1997). Nature, nurture and twin research strategies. *Current Directions in Psychological Science, 6,* 117–121.

Phillips, D.I.W. (1993). Twin studies in medical research: Can they tell us whether diseases are genetically determined? *Lancet, 341,* 1008–1009.

Phillips, K., & Matheny, A.P., Jr. (1995). Quantitative genetic analysis of injury liability in infants and toddlers. *American Journal of Medical Genetics. Part B: Neuropsychiatric Genetics, 60,* 64–71.

Phillips, K., & Matheny, A.P., Jr. (1997). Evidence for genetic influence on both cross-situation and situation-specific components of behavior. *Journal of Personality and Social Psychology, 73,* 129–138.

Phillips, T.J., Belknap, J.K., Buck, K.J., & Cunningham, C.L. (1998a). Genes on mouse chromosomes 2 and 9 determine variation in ethanol consumption. *Mammalian Genome, 9,* 936–941.

Phillips, T.J., Brown, K.J., Burkhart-Kasch, S., Wenger, C.D., Kelly, M.A., Rubinstein, M., et al. (1998b). Alcohol preference and sensitivity are markedly reduced in mice lacking dopamine D2 receptors. *Nature Neuroscience, 1,* 610–615.

Phillips, T.J., & Crabbe, J.C. (1991). Behavioral studies of genetic differences in alcohol action. In J.C. Crabbe & R.A. Harris (Eds.), *The genetic basis of alcohol and drug actions* (pp. 25–104). New York: Plenum.

Pietilainen, K.H., Kaprio, J., Rissanen, A., Winter, T., Rimpela, A., Viken, R.J., et al. (1999). Distribution and heritability of BMI in Finnish adolescents aged 16y and 17y: A study of 4884 twins and 2509 singletons. *International Journal of Obesity and Related Metabolic Disorders: Journal of the International Association for the Study of Obesity, 23*, 107–115.

Pike, A., McGuire, S., Hetherington, E.M., Reiss, D., & Plomin, R. (1996). Family environment and adolescent depressive symptoms and antisocial behavior: A multivariate genetic analysis. *Developmental Psychology, 32*, 590–603.

Pike, A., Reiss, D., Hetherington, E.M., & Plomin, R. (1996). Using MZ differences in the search for nonshared environmental effects. *Journal of Child Psychology and Psychiatry, 37*, 695–704.

Pillard, R.C., & Bailey, J.M. (1998). Human sexual orientation has a heritable component. *Human Biology, 70*, 347–365.

Pinker, S. (1994). *The language instinct: The new language of science in mind.* London: Penguin.

Pinker, S. (2002). *The blank slate: The modern denial of human nature.* New York: Penguin.

Platek, S., & Shackelford, T. (Eds.). (2006). *Female infidelity and paternal uncertainty: Evolutionary perspectives on male anti-cuckoldry tactics.* Cambridge: Cambridge University Press.

Plomin, R. (1986). *Development, genetics, and psychology.* Hillsdale, NJ: Erlbaum.

Plomin, R. (1987). Developmental behavioral genetics and infancy. In J. Osofsky (Ed.), *Handbook of infant development* (2nd ed.). (pp. 363–417). New York: Interscience.

Plomin, R. (1988). The nature and nurture of cognitive abilities. In R.J. Sternberg (Ed.), *Advances in the psychology of human intelligence, 4* (pp. 1–33). Hillsdale, NJ: Lawrence Erlbaum Associates.

Plomin, R. (1991). Genetic risk and psychosocial disorders: Links between the normal and abnormal. In M. Rutter & P. Casaer (Eds.), *Biological risk factors for psychosocial disorders* (pp. 101–138). Cambridge: Cambridge University Press.

Plomin, R. (1993). Nature and nurture: Perspective and prospective. In R. Plomin & G.E. McClearn (Eds.), *Nature, nurture, and psychology* (pp. 457–483). Washington, DC: American Psychological Association.

Plomin, R. (1994). *Genetics and experience: The interplay between nature and nurture.* Thousand Oaks, California: Sage Publications Inc.

Plomin, R. (1999a). Genetic research on general cognitive ability as a model for mild mental retardation. *International Review of Psychiatry, 11*, 34–36.

Plomin, R. (1999b). Genetics and general cognitive ability. *Nature, 402*, C25–C29.

Plomin, R. (2001). The genetics of g in human and mouse. *Nature Reviews Neuroscience, 2*, 136–141.

Plomin, R., Asbury, K., & Dunn, J. (2001). Why are children in the same family so different? Nonshared environment a decade later. *Canadian Journal of Psychiatry, 46*, 225–233.

Plomin, R., & Bergeman, C.S. (1991). The nature of nurture: Genetic influences on "environmental" measures. *Behavioral and Brain Sciences, 14*, 373–385.

Plomin, R., & Caspi, A. (1999). Behavioral genetics and personality. In L.A. Pervin & O.P. John (Eds.), *Handbook of personality: Theory and research* (2nd ed.). (pp. 251–276). New York: Guilford Press.

Plomin, R., Chipuer, H.M., & Loehlin, J.C. (1990). Behavioral genetics and personality. In L.A. Pervin (Ed.), *Handbook of personality: Theory and research* (pp. 225–243). New York: Guilford.

Plomin, R., Coon, H., Carey, G., DeFries, J.C., & Fulker, D.W. (1991). Parent-offspring and sibling adoption analyses of parental ratings of temperament in infancy and childhood. *Journal of Personality, 59,* 705–732.

Plomin, R., Corley, R., Caspi, A., Fulker, D.W., & DeFries, J.C. (1998). Adoption results for self-reported personality: Evidence for nonadditive genetic effects? *Journal of Personality and Social Psychology, 75,* 211–218.

Plomin, R., & Crabbe, J.C. (2000). DNA. *Psychological Bulletin, 126,* 806–828.

Plomin, R., & Daniels, D. (1987). Why are children in the same family so different from each other? *Behavioral and Brain Sciences, 10,* 1–16.

Plomin, R., & Davis, O.S.P. (2006). Gene-environment interactions and correlations in the development of cognitive abilities and disabilities. In J. MacCabe, O. O'Daly, R.M. Murray, P. McGuffin, & P. Wright (Eds.), *Beyond nature and nurture: Genes, environment and their interplay in psychiatry* (pp. 35–45). Andover, UK: Thomson Publishing Services.

Plomin, R., & DeFries, J.C. (1985). A parent-offspring adoption study of cognitive abilities in early childhood. *Intelligence, 9,* 341–356.

Plomin, R., & DeFries, J.C. (1998). The genetics of cognitive abilities and disabilities. *Scientific American, 278,* 62–69.

Plomin, R., DeFries, J.C., & Fulker, D.W. (1988). *Nature and nurture during infancy and early childhood.* Cambridge: Cambridge University Press.

Plomin, R., DeFries, J.C., & Loehlin, J.C. (1977a). Assortative mating by unwed biological parents of adopted children. *Science, 196,* 499–450.

Plomin, R., DeFries, J.C., & Loehlin, J.C. (1977b). Genotype-environment interaction and correlation in the analysis of human behaviour. *Psychological Bulletin, 84,* 309–322.

Plomin, R., DeFries, J.C., & McClearn, G.E. (1980). *Behavioral genetics: A primer.* New York: W.H. Freeman.

Plomin, R., DeFries, J.C., McClearn, G.E., & Rutter, M. (1997a). *Behavioral genetics* (3rd ed.). New York: W.H. Freeman.

Plomin, R., Emde, R.N., Braungart, J.M., Campos, J., Corley, R., Fulker, D.W., et al. (1993). Genetic change and continuity from fourteen to twenty months: The MacArthur Longitudinal Twin Study. *Child Development, 64,* 1354–1376.

Plomin, R., & Foch, T.T. (1980). A twin study of objectively assessed personality in childhood. *Journal of Personality and Social Psychology, 39,* 680–688.

Plomin, R., Foch, T.T., & Rowe, D.C. (1981). Bobo clown aggression in childhood: Environment not genes. *Journal of Research in Personality, 14,* 331–342.

Plomin, R., Fulker, D.W., Corley, R., & DeFries, J.C. (1997b). Nature, nurture and cognitive development from 1 to 16 years: A parent-offspring adoption study. *Psychological Science, 8,* 442–447.

Plomin, R., Hill, L., Craig, I., McGuffin, P., Purcell, S., Sham, P., et al. (2001). A genome-wide scan of 1842 DNA markers for allelic associations with general cognitive ability: A five-stage design using DNA pooling and extreme selected groups. *Behavior Genetics, 31,* 497–509.

Plomin, R., Kennedy, J.K.J., & Craig, I.W. (2006). The quest for quantitative trait loci associated with intelligence. *Intelligence, 34,* 513–526.

Plomin, R., & Kovas, Y. (2005). Generalist genes and learning disabilities. *Psychological Bulletin, 131,* 592–617.

Plomin, R., Loehlin, J.C., & DeFries, J.C. (1985). Genetic and environmental components of "environmental" influences. *Developmental Psychology, 21,* 391–402.

Plomin, R., & McClearn, G.E. (1990). Human behavioral genetics of aging. In J.E. Birren & K.W. Schaie (Eds.), *Handbook of the psychology of aging* (pp. 66–77). New York: Academic Press.

Plomin, R., & McClearn, G.E. (1993a). *Nature, nurture, and psychology.* Washington, DC: American Psychological Association.

Plomin, R., & McClearn, G.E. (1993b). Quantitative trait loci (QTL) analyses and alcohol-related behaviors. *Behavior Genetics, 23,* 197–211.

Plomin, R., McClearn, G.E., Smith, D.L., Skuder, P., Vignetti, S., Chorney, M.J., et al. (1995). Allelic associations between 100 DNA markers and high versus low IQ. *Intelligence, 21,* 31–48.

Plomin, R., Pedersen, N.L., Lichtenstein, P., & McClearn, G.E. (1994a). Variability and stability in cognitive abilities are largely genetic later in life. *Behavior Genetics, 24,* 207–215.

Plomin, R., Reiss, D., Hetherington, E.M., & Howe, G.W. (1994b). Nature and nurture: Genetic contributions to measures of the family environment. *Developmental Psychology, 30,* 32–43.

Plomin, R., & Schalkwyk, L.C. (2007). Microarrays. *Developmental Science, 10,* 19–23.

Plomin, R., & Spinath, F.M. (2002). Genetics and general cognitive ability (*g*). *Trends in Cognitive Science, 6,* 169–176.

Plomin, R., & Walker, S.O. (2003). Genetics and educational psychology. *British Journal of Educational Psychology, 73,* 3–14.

Pogue-Geile, M.F., & Rose, R.J. (1985). Developmental genetic studies of adult personality. *Developmental Psychology, 21,* 547–557.

Poinar, G. (1999). Ancient DNA. *American Scientist, 87,* 446–457.

Poirier, L., & Seroude, L. (2005). Genetic approaches to study aging in *Drosophila melanogaster. Age, 27,* 165–182.

Pollak, D.D., John, J., Hoeger, H., & Lubec, G. (2006). An integrated map of the murine hippocampal proteome based upon five mouse strains. *Electrophoresis, 27,* 2787–2798.

Pollak, D.D., John, J., Schneider, A., Hoeger, H., & Lubec, G. (2006). Strain-dependent expression of signaling proteins in the mouse hippocampus. *Neuroscience, 138,* 149–158.

Pollen, D.A. (1993). *Hannah's heirs: The quest for the genetic origins of Alzheimer's disease.* Oxford: Oxford University Press.

Posthuma, D., de Geus, E.J., Baare, W.F., Pol, H.E.H., Kahn, R.S., & Boomsma, D.I. (2002). The association between brain volume and intelligence is of genetic origin. *Nature Neuroscience, 5,* 83–84.

Posthuma, D., Luciano, M., de Geus, E.J., Wright, M.J., Slagboom, P.E., Montgomery, G.W., et al. (2005). A genomewide scan for intelligence identifies quantitative trait loci on 2*q* and 6*p. American Journal of Human Genetics, 77,* 318–326.

Prescott, C.A., Madden, P.A., & Stallings, M.C. (2006). Challenges in genetic studies of the etiology of substance use and substance use disorders: Introduction to the special issue. *Behavior Genetics, 36,* 473–482.

Prescott, C.A., Sullivan, P.F., Kuo, P.H., Webb, B.T., Vittum, J., Patterson, D.G., et al. (2006). Genomewide linkage study in the Irish affected sib pair study of alcohol dependence: Evidence for a susceptibility region for symptoms of alcohol dependence on chromosome 4. *Molecular Psychiatry, 11*, 603–611.

Price, D.L., Sisodia, S.S., & Borchelt, D.R. (1998). Alzheimer's disease—When and why? *Nature Genetics, 19*, 314–316.

Price, R.A., Kidd, K.K., Cohen, D.J., Pauls, D.L., & Leckman, J.F. (1985). A twin study of Tourette's syndrome. *Archives of General Psychiatry, 42*, 815–820.

Price, T.S., Simonoff, E., Kuntsi, J., Curran, S., Asherson, P., Waldman, I., et al. (2005). Continuity and change in preschool hyperactive behaviors: Longitudinal genetic analysis with contrast effects. *Behavior Genetics, 35*, 121–132.

Profet, M. (1992). Pregnancy sickness as adaptation: A deterrent to maternal ingestion on teratogens. In J. Barkow, L. Cosmides, & J. Tooby (Eds.), *The adapted mind* (pp. 327–366). New York: Oxford University Press.

Propping, P. (1987). Single gene effects in psychiatric disorders. In F. Vogel & K. Sperling (Eds.), *Human genetics: Proceedings of the 7th International Congress, Berlin* (pp. 452–457). New York: Springer.

Purcell, S. (2002). Variance components models for gene-environment interaction in twin analysis. *Twin Research, 5*, 554–571.

Raiha, I., Kaprio, J., Koskenvuo, M., Rajala, T., & Sourander, L. (1996). Alzheimer's disease in Finnish twins. *Lancet, 347*, 573–578.

Raine, A. (1993). *The psychopathology of crime: Criminal behavior as a clinical disorder.* San Diego: Academic Press.

Rakyan, V.K., & Beck, S. (2006). Epigenetic variation and inheritance in mammals. *Current Opinion in Genetic Development, 16*, 573–577.

Ralph, M.R., & Menaker, M. (1988). A mutation of the ciradian system in golden hamsters. *Science, 241*, 1225–1227.

Ramus, F. (2006). Genes, brain, and cognition: A roadmap for the cognitive scientist. *Cognition, 101*, 247–269.

Rankin, C.H. (2002). From gene to identified neuron to behaviour in *Caenorhabditis elegans. Nature Reviews Genetics, 3*, 622–630.

Rankinen, T., Zuberi, A., Chagnon, Y.C., Weisnagel, S.J., Argyropoulos, G., Walts, B., et al. (2006). The human obesity gene map: The 2005 update. *Obesity, 14*, 529–644.

Rasmussen, E.R., Neuman, R.J., Heath, A.C., Levy, F., Hay, D.A., & Todd, R.D. (2004). Familial clustering of latent class and DSM-IV defined attention-deficit hyperactivity disorder (ADHD) subtypes. *Journal of Child Psychology and Psychiatry, 45*, 589–598.

Rasmussen, S.A., & Tsuang, M.T. (1984). The epidemiology of obsessive compulsive disorder. *Journal of Clinical Psychiatry, 45*, 450–457.

Ratcliffe, S.G. (1994). The psychological and psychiatric consequences of sex chromosome abnormalities in children, based on population studies. In F. Poustka (Ed.), *Basic approaches to genetic and molecular-biological developmental psychiatry* (pp. 92–122). Berlin: Quintessenz Library of Psychiatry.

Raymond, F.L., & Tarpey, P. (2006). The genetics of mental retardation. *Human Molecular Genetics, 15*, R110–R116.

Read, S., Vogler, G.P., Pedersen, N.L., & Johansson, B. (2006). Stability and change in genetic and environmental components of personality in old age. *Personality and Individual Differences, 40*, 1637–1647.

Redon, R., Ishikawa, S., Fitch, K.R., Feuk, L., Perry, G.H., Andrews, T.D., et al. (2006). Global variation in copy number in the human genome. *Nature, 444*, 444–454.

Reed, E.W., & Reed, S.C. (1965). *Mental retardation: A family study*. Philadelphia: Saunders.

Reich, T., & Cloninger, R. (1990). Time-dependent model of the familial transmission of alcoholism. In *Banbury report 33: Genetics and biology of alcoholism* (pp. 55–73). Cold Spring Harbor, NY: Cold Spring Harbor Laboratory Press.

Reich, T., Edenberg, H.J., Goate, A., Williams, J., Rice, J., Van Eerdewegh, P., et al. (1998). Genome-wide search for genes affecting the risk for alcohol dependence. *American Journal of Medical Genetics, 81*, 207–215.

Reich, T., Hinrichs, A., Culverhouse, R., & Bierut, L. (1999). Genetics studies of alcoholism and substance dependence. *American Journal of Human Genetics, 65*, 599–605.

Reik, W., & Walter, J. (2001). Genomic imprinting: Parental influence on the genome. *Nature Reviews Genetics, 2*, 21–32.

Reiss, D., Neiderhiser, J.M., Hetherington, E.M., & Plomin, R. (2000). *The relationship code: Deciphering genetic and social patterns in adolescent development*. Cambridge, MA: Harvard University Press.

Rettew, D.C., Vink, J.M., Willimsen, G., Doyle, A., Hudziak, J.J., & Boomsa, D.I. (2006). The genetic architecture of neuroticism in 3301 Dutch adolescent twins as a function of age and sex: A study from the Dutch Twins Register. *Twin Research and Human Genetics, 9*, 24–29.

Rexbye, H., Petersen, I., Iachina, M., Mortensen, J., McGue, M., Vaupel, J.W., et al. (2005). Hair loss among elderly men: Etiology and impact on perceived age. *Journal of Gerontology Series A—Biological Sciences and Medical Sciences, 60*, 1077–1082.

Rhee, S.H., Hewitt, J.K., Young, S.E., Corley, R.P., Crowley, T.J., & Stallings, M.C. (2003). Genetic and environmental influences on substance initiation, use, and problem use in adolescents. *Archives of General Psychiatry, 60*, 1256–1264.

Rhee, S.H., & Waldman, I.D. (2002). Genetic and environmental influences on antisocial behavior: A meta-analysis of twin and adoption studies. *Psychological Bulletin, 128*, 490–529.

Rhodes, G. (2006). The evolutionary psychology of facial beauty. *Annual Review of Psychology, 57*, 199–226.

Rhodes, J.S., & Crabbe, J.C. (2003). Progress towards finding genes for alcoholism in mice. *Clinical Neuroscience Research, 3*, 315–323.

Rice, G., Anderson, C., Risch, N., & Ebers, G. (1999). Male homosexuality: Absence of linkage to microsatellite markers at Xq28. *Science, 284*, 665–667.

Richards, E.J. (2006). Inherited epigenetic variation—Revisiting soft inheritance. *Nature Reviews Genetics, 7*, 395–401.

Riemann, R., Angleitner, A., & Strelau, J. (1997). Genetic and environmental influences on personality: A study of twins reared together using the self and peer report NEO-FFI scales. *Journal of Personality, 65*, 449–476.

Riese, M.L. (1990). Neonatal temperament in monozygotic and dizygotic twin pairs. *Child Development, 61*, 1230–1237.

Riese, M.L. (1999). Effects of chorion type on neonatal temperament differences in monozygotic pairs. *Behavior Genetics, 29*, 87–94.

Rietveld, M.J., Dolan, C.V., van Baal, G.C., & Boomsma, D.I. (2003). A twin study of differentiation of cognitive abilities in childhood. *Behavior Genetics, 33*, 367–381.

Rietveld, M.J., Hudziak, J.J., Bartels, M., Van Beijsterveldt, C.E., & Boomsma, D.I. (2004). Heritability of attention problems in children: Longitudinal results from a study of twins, age 3 to 12. *Journal of Child Psychology and Psychiatry, 45*, 577–588.

Rietveld, M.J., Posthuma, D., Dolan, C.V., & Boomsma, D.I. (2003). ADHD: Sibling interaction or dominance. An evaluation of statistical power. *Behavior Genetics, 33*, 247–255.

Rijsdijk, F.V., & Boomsma, D.I. (1997). Genetic mediation of the correlation between peripheral nerve conduction velocity and IQ. *Behavior Genetics, 27*, 87–98.

Rijsdijk, F.V., Boomsma, D.I., & Vernon, P.A. (1995). Genetic analysis of peripheral nerve conduction velocity in twins. *Behavior Genetics, 25*, 341–348.

Rijsdijk, F.V., Vernon, P.A., & Boomsma, D.I. (2002). Application of hierarchical genetic models to Raven and WAIS subtests: A Dutch twin study. *Behavior Genetics, 32*, 199–210.

Riley, B., & Kendler, K.S. (2006). Molecular genetic studies of schizophrenia. *European Journal of Human Genetics, 14*, 669–680.

Risch, N.J. (2000). Searching for genetic determinants in the new millennium. *Nature, 405*, 847–856.

Roberts, C.A., & Johansson, C.B. (1974). The inheritance of cognitive interest styles among twins. *Journal of Vocational Behavior, 4*, 237–243.

Robins, L.N. (1978). Sturdy childhood predictors of adult antisocial behaviour: Replications from longitudinal analyses. *Psychological Medicine, 8*, 611–622.

Robins, L.N., & Price, R.K. (1991). Adult disorders predicted by childhood conduct problems: Results from the NIMH epidemiologic catchment area project. *Psychiatry, 54*, 116–132.

Robins, L.N., & Regier, D.A. (1991). *Psychiatric disorders in America*. New York: Free Press.

Robinson, D.G., Woerner, M.G., McMeniman, M., Mendelowitz, A., & Bilder, R.M. (2004). Symptomatic and functional recovery from a first episode of schizophrenia or schizoaffective disorder. *American Journal of Psychiatry, 161*, 473–479.

Robinson, J.L., Kagan, J., Reznick, J.S., & Corley, R. (1992). The heritability of inhibited and uninhibited behavior: A twin study. *Developmental Psychology, 28*, 1030–1037.

Rocha, B.L., Scearce-Levie, K., Lucas, J.J., Hiroi, N., Castanon, N., Crabbe, J.C., et al. (1998). Increased vulnerability to cocaine in mice lacking the serotonin-1B receptor. *Nature, 393*, 175–178.

Rockman, M.V., & Kruglyak, L. (2006). Genetics of global gene expression. *Nature Reviews Genetics, 7*, 862–872.

Rogaeva, E., Meng, Y., Lee, J.H., Gu, Y., Kawarai, T., Zou, F., et al. (2007). The neuronal sortilin-related receptor SORL1 is genetically associated with Alzheimer's disease. *Nature Genetics, 39*, 168–177.

Roisman, G.I., & Fraley, R.C. (2006). The limits of genetic influence: A behavior-genetic analysis of infant-caregiver relationship quality and temperament. *Child Development, 77*, 1656–1667.

Roizen, N.J., & Patterson, D. (2003). Down syndrome. *Lancet, 361*, 1281–1289.

Romeis, J.C., Grant, J.D., Knopik, V.S., Pedersen, N.L., & Heath, A.C. (2004). The genetics of middle-age spread in middle-class males. *Twin Research, 7*, 596–602.

Ronald, A., Happé, F., & Plomin, R. (2005). The genetic relationship between individual differences in social and nonsocial behaviours characteristic of autism. *Developmental Science, 8*, 444–458.

Ronald, A., Happé, F., Price, T.S., Baron-Cohen, S., & Plomin, R. (2006). Phenotypic and genetic overlap between autistic traits at the extremes of the general population. *Journal of the American Academy of Child and Adolescent Psychiatry, 45*, 1206–1214.

Ronalds, G.A., De Stavola, B.L., & Leon, D.A. (2005). The cognitive cost of being a twin: Evidence from comparisons within families in the Aberdeen children of the 1950s cohort study. *British Medical Journal, 331,* 1306.

Rooms, L., Reyniers, E., & Kooy, R.F. (2005). Subtelomeric rearrangements in the mentally retarded: A comparison of detection methods. *Human Mutation, 25,* 513–524.

Ropers, H.H. (2006). X-linked mental retardation: Many genes for a complex disorder. *Current Opinion in Genetics and Development, 16,* 260–269.

Ropers, H.H., & Hamel, B.C. (2005). X-linked mental retardation. *Nature Reviews Genetics, 6,* 46–57.

Rosanoff, A.J., Handy, L.M., & Plesset, I.R. (1937). The etiology of mental deficiency with special reference to its occurrence in twins. *Psychological Monographs, 216,* 1–137.

Rosato, E., Tauber, E., & Kyriacou, C.P. (2006). Molecular genetics of the fruit-fly circadian clock. *European Journal of Human Genetics, 14,* 729–738.

Rosenthal, D., Wender, P.H., Kety, S.S., & Schulsinger, F. (1971). The adopted-away offspring of schizophrenics. *American Journal of Psychiatry, 128,* 307–311.

Rosenthal, D., Wender, P.H., Kety, S.S., Schulsinger, F., Welner, J., & Ostergaard, L. (1968). Schizophrenics' offspring reared in adoptive homes. *Journal of Psychiatric Research, 6,* 377–391.

Rosenthal, N.E., Sack, D.A., Gillin, J.C., Lewy, A.J., Goodwin, F.K., Davenport, Y., et al. (1984). Seasonal affective disorder. A description of the syndrome and preliminary findings with light therapy. *Archives of General Psychiatry, 41,* 72–80.

Roses, A.D. (2000). Pharmacogenetics and the practice of medicine. *Nature, 405,* 857–865.

Ross, C.A. (2004). Huntington's disease: New paths to pathogenesis. *Cell, 118,* 4–7.

Ross, M.T., Grafham, D.V., Coffey, A.J., Scherer, S., McLay, K., Muzny, D., et al. (2005). The DNA sequence of the human X chromosome. *Nature, 434,* 325–337.

Rothe, C., Koszycki, D., Bradwejn, J., King, N., Deluca, V., Tharmalingam, S., et al. (2006). Association of the Val158Met catechol-O-methyltransferase genetic polymorphism with panic disorder. *Neuropsychopharmacology, 31,* 2237–2242.

Rothstein, M.A. (2005). Science and society: Applications of behavioural genetics: Outpacing the science? *Nature Reviews Genetics, 6,* 793–798.

Roush, W. (1995). Conflict marks crime conference. *Science, 269,* 1808–1809.

Rowe, D.C. (1981). Environmental and genetic influences on dimensions of perceived parenting: A twin study. *Developmental Psychology, 17,* 203–208.

Rowe, D.C. (1983a). A biometrical analysis of perceptions of family environment: A study of twin and singleton sibling relationships. *Child Development, 54,* 416–423.

Rowe, D.C. (1983b). Biometrical genetic models of self-reported delinquent behavior: A twin study. *Behavior Genetics, 13,* 473–489.

Rowe, D.C. (1987). Resolving the person-situation debate: Invitation to an interdisciplinary dialogue. *American Psychologist, 42,* 218–227.

Rowe, D.C. (1994). *The limits of family influence: Genes, experience, and behaviour.* New York: Guilford Press.

Rowe, D.C., Jacobson, K.C., & van den Oord, E.J. (1999). Genetic and environmental influences on vocabulary IQ: Parental education level as moderator. *Child Development, 70,* 1151–1162.

Rowe, D.C., & Linver, M.R. (1995). Smoking and addictive behaviors: Epidemiological, individual, and family factors. In J.R. Turner, L.R. Cardon, & J.K. Hewitt (Eds.), *Behavior genetic approaches in behavioral medicine* (pp. 67–84). New York: Plenum.

Rowe, D.C., Vesterdal, W.J., & Rodgers, J.L. (1999). Herrnstein's syllogism: Genetic and shared environmental influences on IQ, education, and income. *Intelligence, 26*, 405–423.

Roy, M.A., Neale, M.C., & Kendler, K.S. (1995). The genetic epidemiology of self-esteem. *British Journal of Psychiatry, 166*, 813–820.

Royce, T.E., Rozowsky, J.S., Bertone, P., Samanta, M., Stolc, V., Weissman, S., et al. (2005). Issues in the analysis of oligonucleotide tiling microarrays for transcript mapping. *Trends in Genetics, 21*, 466–475.

Rubanyi, G.M. (2001). The future of human gene therapy. *Molecular Aspects of Medicine, 22*, 113–142.

Rubin, J.B., & Gutmann, D.H. (2005). Neurofibromatosis type 1—A model for nervous system tumour formation? *Nature Reviews Cancer, 5*, 557–564.

Rubinstein, M., Phillips, T.J., Bunzow, J.R., Falzone, T.L., Dziewczapolski, G., Zhang, G., et al. (1997). Mice lacking dopamine D4 receptors are supersensitive to ethanol, cocaine, and methamphetamine. *Cell, 90*, 991–1001.

Rush, A.J., & Weissenburger, J.E. (1994). Melancholic symptom features and DSM-IV. *American Journal of Psychiatry, 151*, 489–498.

Rushton, J.P. (2002). New evidence on Sir Cyril Burt: His 1964 speech to the association of educational psychologists. *Intelligence, 30*, 555–567.

Rushton, J.P. (2004). Genetic and environmental contributions to pro-social attitudes: A twin study of social responsibility. *Proceedings of the Royal Society of London Series B, 271*, 2583–2585.

Rushton, J.P., & Bons, T.A. (2005). Mate choice and friendship in twins: Evidence for genetic similarity. *Psychological Science, 16*, 555–559.

Rushton, J.P., & Jensen, A.R. (2005). Thirty years of research on race differences in cognitive ability. *Psychology, Public Policy and Law, 11*, 235–294.

Rutherford, J., McGuffin, P., Katz, R.J., & Murray, R.M. (1993). Genetic influences on eating attitudes in a normal female twin population. *Psychological Medicine, 23*, 425–436.

Rutter, M. (1996a). Concluding remarks. In G.R. Bock & J.A. Goode (Eds.), *Genetics of criminal and antisocial behaviour* (pp. 265–271). Chichester, UK: John Wiley & Sons.

Rutter, M. (1996b). Introduction: Concepts of antisocial behavior, of cause, and of genetic influences. In G.R. Bock & J.A. Goode (Eds.), *Genetics of criminal and antisocial behaviour* (pp. 1–15). Chichester, UK: John Wiley & Sons.

Rutter, M. (2005a). Aetiology of autism: Findings and questions. *Journal of Intellectual Disability Research, 49*, 231–238.

Rutter, M. (2005b). Environmentally mediated risks for psychopathology: Research strategies and findings. *Journal of the American Academy of Child and Adolescent Psychiatry, 44*, 3–18.

Rutter, M. (2006). *Genes and behavior: Nature-nurture interplay explained.* Oxford: Blackwell Publishing.

Rutter, M. (2007). Gene-environment interdependence. *Developmental Science, 10*, 12–18.

Rutter, M., Maughan, B., Meyer, J., Pickles, A., Silberg, J., Simonoff, E., et al. (1997). Heterogeneity of antisocial behavior: Causes, continuities, and consequences. In R. Dienstbier & D.W. Osgood (Eds.), *Nebraska Symposium on Motivation: Vol. 44. Motivation and delinquency* (pp. 45–118). Lincoln: University of Nebraska Press.

Rutter, M., Moffitt, T.E., & Caspi, A. (2006). Gene-environment interplay and psychopathology: Multiple varieties but real effects. *Journal of Child Psychology and Psychiatry, 47,* 226–261.

Rutter, M., & Plomin, R. (1997). Opportunities for psychiatry from genetic findings. *British Journal of Psychiatry, 171,* 209–219.

Rutter, M., & Redshaw, J. (1991). Annotation: Growing up as a twin: Twin-singleton differences in psychological development. *Journal of Child Psychology and Psychiatry, 32,* 885–895.

Rutter, M., Silberg, J., O'Connor, T.G., & Simonoff, E. (1999). Genetics and child psychiatry. II. Empirical research findings. *Journal of Child Psychology and Psychiatry, 40,* 19–55.

Saad, G. (2007). *The evolutionary bases of consumption.* Hillsdale, NJ: Lawrence Erlbaum Associates.

Saetre, P., Strandberg, E., Sundgren, P.E., Pettersson, U., Jazin, E., & Bergstrom, T.F. (2006). The genetic contribution to canine personality. *Genes, Brain and Behavior, 5,* 240–248.

Sakai, T., & Ishida, N. (2001). Circadian rhythms of female mating activity governed by clock genes in *Drosophila. Proceedings of the National Academy of Sciences (USA), 98,* 9221–9225.

Sakai, T., Tamura, T., Kitamoto, T., & Kidokoro, Y. (2004). A clock gene, period, plays a key role in long-term memory formation in *Drosophila. Proceedings of the National Academy of Sciences (USA), 101,* 16058–16063.

Salahpour, A., Medvedev, I.O., Beaulieu, J.M., Gainetdinov, R.R., & Caron, M.G. (2007). Local knockdown of genes in the brain using small interfering RNA: A phenotypic comparison with knockout animals. *Biological Psychiatry, 61,* 65–69.

Sambandan, D., Yamamoto, A., Fanara, J.J., Mackay, T.F., & Anholt, R.R. (2006). Dynamic genetic interactions determine odor-guided behavior in *Drosophila melanogaster. Genetics, 174,* 1349–1363.

Samuelsson, S., Olson, R., Wadsworth, S., Corley, R., DeFries, J.C., Willcutt, E., et al. (2007). Genetic and environmental influences on prereading skills and early reading and spelling development in the United States, Australia, and Scandinavia. *Reading and Writing, 20,* 51–75.

Saudino, K.J., & Eaton, W.O. (1991). Infant temperament and genetics: An objective twin study of motor activity level. *Child Development, 62,* 1167–1174.

Saudino, K.J., McGuire, S., Reiss, D., Hetherington, E.M., & Plomin, R. (1995). Parent rating of EAS temperaments in twins, full siblings, half siblings, and step siblings. *Journal of Personality and Social Psychology, 68,* 723–733.

Saudino, K.J., Pedersen, N.L., Lichtenstein, P., McClearn, G.E., & Plomin, R. (1997). Can personality explain genetic influences on life events? *Journal of Personality and Social Psychology, 72,* 196–206.

Saudino, K.J., & Plomin, R. (1997). Cognitive and temperamental mediators of genetic contributions to the home environment during infancy. *Merrill-Palmer Quarterly, 43,* 1–23.

Saudino, K.J., Plomin, R., & DeFries, J.C. (1996). Tester-rated temperament at 14, 20, and 24 months: Environmental change and genetic continuity. *British Journal of Developmental Psychology, 14,* 129–144.

Saudino, K.J., Plomin, R., Pedersen, N.L., & McClearn, G.E. (1994). The etiology of high and low cognitive ability during the second half of the life span. *Intelligence, 19,* 353–371.

Saudino, K.J., Ronald, A., & Plomin, R. (2005). Rater effects in the etiology of behavior problems in 7-year-old twins: Parent ratings and ratings by same and different teachers. *Journal of Abnormal Child Psychology, 33*, 113–130.

Saudino, K.J., Wertz, A.E., Gagne, J.R., & Chawla, S. (2004). Night and day: Are siblings as different in temperament as parents say they are? *Journal of Personality and Social Psychology, 87*, 698–706.

Saudou, F., Amara, D.A., Dierich, A., LeMeur, M., Ramboz, S., Segu, L., et al. (1994). Enhanced aggressive behavior in mice lacking 5-HT$_{1B}$ receptor. *Science, 265*, 1875–1878.

Savitz, J., Solms, M., & Ramesar, R. (2006). Apolipoprotein E variants and cognition in healthy individuals: A critical opinion. *Brain Research Reviews, 51*, 125–135.

Scarr, S. (1992). Developmental theories for the 1990s: Development and individual differences. *Child Development, 63*, 1–19.

Scarr, S., & Carter-Saltzman, L. (1979). Twin method: Defense of a critical assumption. *Behavior Genetics, 9*, 527–542.

Scarr, S., & McCartney, K. (1983). How people make their own environments: A theory of genotype → environmental effects. *Child Development, 54*, 424–435.

Scarr, S., & Weinberg, R.A. (1978a). Attitudes, interests, and IQ. *Human Nature, 1*, 29–36.

Scarr, S., & Weinberg, R.A. (1978b). The influence of "family background" on intellectual attainment. *American Sociological Review, 43*, 674–692.

Scarr, S., & Weinberg, R.A. (1981). The transmission of authoritarianism in families: Genetic resemblance in social-political attitudes? In S. Scarr (Ed.), *Race, social class, and individual differences in IQ* (pp. 399–427). Hillsdale, NJ: Lawrence Erlbaum Associates.

Schadt, E.E. (2006). Novel integrative genomics strategies to identify genes for complex traits. *Animal Genetics, 37*, 18–23.

Schafer, W.R. (2005). Deciphering the neural and molecular mechanisms of *C. elegans* behavior. *Current Biology, 15*, R723–R729.

Scherrer, J.F., True, W.R., Xian, H., Lyons, M.J., Eisen, S.A., Goldberg, J., et al. (2000). Evidence for genetic influences common and specific to symptoms of generalized anxiety and panic. *Journal of Affective Disorders, 57*, 25–35.

Schmidt, F.L., & Hunter, J. (2004). General mental ability in the world of work: Occupational attainment and job performance. *Journal of Personality and Social Psychology, 86*, 162–173.

Schmitt, J.E., Prescott, C.A., Gardner, C.O., Neale, M.C., & Kendler, K.S. (2005). The differential heritability of regular tobacco use based on method of administration. *Twin Research and Human Genetics, 8*, 60–62.

Schmitz, S. (1994). Personality and temperament. In J.C. DeFries & R. Plomin (Eds.), *Nature and nurture during middle childhood* (pp. 120–140). Cambridge, MA: Blackwell.

Schmitz, S., Saudino, K.J., Plomin, R., Fulker, D.W., & DeFries, J.C. (1996). Genetic and environmental influences on temperament in middle childhood: Analyses of teacher and tester ratings. *Child Development, 67*, 409–422.

Schoenfeldt, L.F. (1968). The hereditary components of the project TALENT two-day test battery. *Measurement and Evaluation in Guidance, 1*, 130–140.

Schousboe, K., Willemsen, G., Kyvik, K.O., Mortensen, J., Boomsma, D.I., Cornes, B.K., et al. (2003). Sex differences in heritability of BMI: A comparative study of results from twin studies in eight countries. *Twin Research, 6*, 409–421.

Schulsinger, F. (1972). Psychopathy: Heredity and environment. *International Journal of Mental Health, 1*, 190–206.

Schulte-Körne, G., Grimm, T., Nothen, M.M., Muller-Myhsok, B., Cichon, S., Vogt, I.R., et al. (1998). Evidence for linkage of spelling disabilty to chromosome 15. *American Journal of Human Genetics, 63*, 279–282.

Schulze, T.G., Hedeker, D., Zandi, P., Rietschel, M., & McMahon, F.J. (2006). What is familial about familial bipolar disorder? Resemblance among relatives across a broad spectrum of phenotypic characteristics. *Archives of General Psychiatry, 63*, 1368–1376.

Schumacher, J., Hoffmann, P., Schmael, C., Schulte-Korne, G., & Nothen, M.M. (2007). Genetics of dyslexia: The evolving landscape. *Journal of Medical Genetics, 44*, 289–297.

Schutzwohl, A. (2006). Judging female figures: A new methodological approach to male attractiveness judgments of female waist-to-hip ratio. *Biological Psychology, 71*, 223–229.

Schweizer, J., Zynger, D., & Francke, U. (1999). In vivo nuclease hypersensitivity studies reveal mutliple sites of parental origin-dependent differential chromatin conformation in the 150 kb SNRPN transcription unit. *Human Molecular Genetics, 8*, 555–566.

Scott, J.P., & Fuller, J.L. (1965). *Genetics and the social behavior of the dog.* Chicago: University of Chicago Press.

Scriver, C.R., & Waters, P.J. (1999). Monogenetic traits are not simple: Lessons from phenylketonuria. *Trends in Genetics, 15*, 267–272.

Seale, T.W. (1991). Genetic differences in response to cocaine and stimulant drugs. In J.C. Crabbe & R.A. Harris (Eds.), *The genetic basis of alcohol and drug actions* (pp. 279–321). New York: Plenum.

Segal, N.L. (1999). *Entwined lives: Twins and what they tell us about human behavior.* New York: Dutton.

Segal, N.L., & MacDonald, K.B. (1998). Behavioral genetics and evolutionary psychology: Unified perspective on personality research. *Human Biology, 70*, 159–184.

Segurado, R., Detera-Wadleigh, S.D., Levinson, D.F., Lewis, C.M., Gill, M., Nurnberger, J.I., Jr., et al. (2003). Genome scan meta-analysis of schizophrenia and bipolar disorder, part III: Bipolar disorder. *American Journal of Human Genetics, 73*, 49–62.

Seregaza, Z., Roubertoux, P.L., Jamon, M., & Soumireu-Mourat, B. (2006). Mouse models of cognitive disorders in trisomy 21: A review. *Behavior Genetics, 36*, 387–404.

Serretti, A., Calati, R., Mandelli, L., & De Ronchi, D. (2006). Serotonin transporter gene variants and behavior: A comprehensive review. *Current Drug Targets, 7*, 1659–1669.

Service, R.F. (2006). Gene sequencing. The race for the $1000 genome. *Science, 311*, 1544–1546.

Sesardic, N. (2005). *Making sense of heritability.* Cambridge: Cambridge University Press.

Sham, P., Bader, J.S., Craig, I., O'Donovan, M., & Owen, M. (2002). DNA pooling: A tool for large-scale association studies. *Nature Reviews Genetics, 3*, 862–871.

Sham, P.C., Cherny, S.S., Purcell, S., & Hewitt, J. (2000). Power of linkage versus association analysis of quantitative traits, by use of variance-components models, for sibship data. *American Journal of Human Genetics, 66*, 1616–1630.

Sharma, A., Sharma, V.K., Horn-Saban, S., Lancet, D., Ramachandran, S., & Brahmachari, S.K. (2005). Assessing natural variations in gene expression in humans by comparing with monozygotic twins using microarrays. *Physiological Genomics, 21,* 117–123.

Sharp, F.R., Xu, H., Lit, L., Walker, W., Apperson, M., Gilbert, D.L., et al. (2006). The future of genomic profiling of neurological diseases using blood. *Archives of Neurology, 63,* 1529–1536.

Sher, L., Goldman, D., Ozaki, N., & Rosenthal, N.E. (1999). The role of genetic factors in the etiology of seasonal affective disorder and seasonality. *Journal of Affective Disorders, 53,* 203–210.

Sherman, S.L., DeFries, J.C., Gottesman, I.I., Loehlin, J.C., Meyer, J.M., Pelias, M.Z., et al. (1997). Recent developments in human behavioral genetics: Past accomplishments and future directions. *American Journal of Human Genetics, 60,* 1265–1275.

Sherrington, R., Brynjolfsson, J., Petursson, H., Potter, M., Dudleston, K., Barraclough, B., et al. (1988). Localisation of susceptibility locus for schizophrenia on chromosome 5. *Nature, 336,* 164–167.

Sherrington, R., Rogaev, E.I., Liang, Y., Rogaeva, E.A., Levesque, G., Ikeda, M., et al. (1995). Cloning of a gene bearing missense mutation in early-onset familial Alzheimer's disease. *Nature, 375,* 754–760.

Shih, R.A., Belmonte, P.L., & Zandi, P.P. (2004). A review of the evidence from family, twin and adoption studies for a genetic contribution to adult psychiatric disorders. *International Review of Psychiatry, 16,* 260–283.

Shmookler Reis, R.J., Kang, P., & Ayyadevara, S. (2006). Quantitative trait loci define genes and pathways underlying genetic variation in longevity. *Experimental Gerontology, 41,* 1046–1054.

Siever, L.J., Silverman, K.M., Horvath, T.B., Klar, H., Coccaro, E., Keefe, R.S.E., et al. (1990). Increased morbid risk for schizophrenia-related disorders in relatives of schizotypal personality disordered patients. *Archives of General Psychiatry, 47,* 634–640.

Sigvardsson, S., Bohman, M., & Cloninger, C.R. (1996). Replication of Stockholm adoption study of alcoholism. *Archives of General Psychiatry, 53,* 681–687.

Silberg, J., Pickles, A., Rutter, M., Hewitt, J., Simonoff, E., Maes, H., et al. (1999). The influence of genetic factors and life stress on depression among adolescent girls. *Archives of General Psychiatry, 56,* 225–232.

Silberg, J.L., Rutter, M., & Eaves, L. (2001). Genetic and environmental influences on the temporal association between earlier anxiety and later depression in girls. *Biological Psychiatry, 49,* 1040–1049.

Silberg, J.L., Rutter, M.L., Meyer, J., Maes, H., Hewitt, J., Simonoff, E., et al. (1996). Genetic and environmental influences on the covariation between hyperactivity and conduct disturbance in juvenile twins. *Journal of Child Psychology and Psychiatry, 37,* 803–816.

Silva, A.J., Kogan, J.H., Frankland, P.W., & Kida, S. (1998). CREB and memory. *Annual Review of Neuroscience, 21,* 127–148.

Silva, A.J., Paylor, R., Wehner, J.M., & Tonegawa, S. (1992). Impaired spatial learning in α-calcium-calmodulin kinase mutant mice. *Science, 257,* 206–211.

Silva, A.J., Smith, A.M., & Giese, K.P. (1997). Gene targeting and the biology of learning and memory. *Annual Review of Genetics, 31,* 527–546.

Simm, P.J., & Zacharin, M.R. (2006). The psychosocial impact of Klinefelter syndrome— A 10 year review. *Journal of Pediatric Endocrinology and Metabolism, 19,* 499–505.

Singer, J.J., Falchi, M., MacGregor, A.J., Cherkas, L.F., & Spector, T.D. (2006). Genome-wide scan for prospective memory suggests linkage to chromosome 12q22. *Behavior Genetics, 36*, 18–28.

Singer, J.J., MacGregor, J.J., Cherkas, L.F., & Spector, T.D. (2006). Genetic influences on cognitive function using the Cambridge neuropsychological test automated battery. *Intelligence, 34*, 421–428.

Singh, D. (1993). Adaptive significance of waist-to-hip ratio and female physical attractiveness. *Journal of Personality and Social Psychology, 65*, 293–307.

Singh, D. (2004). Mating strategies of young women: Role of physical attractiveness. *Journal of Sex Research, 41*, 43–54.

Sison, M., Cawker, J., Buske, C., & Gerlai, R. (2006). Fishing for genes influencing vertebrate behavior: Zebrafish making headway. *Lab Animal, 35*, 33–39.

Sitskoorn, M.M., Aleman, A., Ebisch, S.J., Appels, M.C., & Kahn, R.S. (2004). Cognitive deficits in relatives of patients with schizophrenia: A meta-analysis. *Schizophrenia Research, 71*, 285–295.

Skodak, M., & Skeels, H.M. (1949). A final follow-up on one hundred adopted children. *Journal of Genetic Psychology, 75*, 84–125.

Skoog, I., Nilsson, L., Palmertz, B., Andreasson, L.A., & Svanborg, A. (1993). A population-based study of dementia in 85-year-olds. *New England Journal of Medicine, 328*, 153–158.

Skoulakis, E.M., & Grammenoudi, S. (2006). Dunces and da Vincis: The genetics of learning and memory in *Drosophila*. *Cellular and Molecular Life Sciences, 63*, 975–988.

Slater, E., & Cowie, V. (1971). *The genetics of mental disorders*. London: Oxford University Press.

SLI Consortium (SLIC). (2004). Highly significant linkage to the SLI1 locus in an expanded sample of individuals affected by specific language impairment. *American Journal of Human Genetics, 74*, 1225–1238.

Slof-Op 't Landt, M.C., van Furth, E.F., Meulenbelt, I., Slagboom, P.E., Bartels, M., Boomsma, D.I., et al. (2005). Eating disorders: From twin studies to candidate genes and beyond. *Twin Research and Human Genetics, 8*, 467–482.

Small, B.J., Rosnick, C.B., Fratiglioni, L., & Backman, L. (2004). Apolipoprotein E and cognitive performance: A meta-analysis. *Psychology and Aging, 19*, 592–600.

Smalley, S.L., Asarnow, R.F., & Spence, M.A. (1988). Autism and genetics: A decade of research. *Archives of General Psychiatry, 45*, 953–961.

Smeeth, L., Cook, C., Fombonne, E., Heavey, L., Rodrigues, L.C., Smith, P.G., et al. (2004). MMR vaccination and pervasive developmental disorders: A case-control study. *Lancet, 364*, 963–969.

Smith, C. (1974). Concordance in twins: Methods and interpretation. *American Journal of Human Genetics, 26*, 454–466.

Smith, D.L. (2004). *Why we lie: The evolutionary roots of deception and the unconscious mind*. New York: St Martin's Press.

Smith, E.M., North, C.S., McColl, R.E., & Shea, J.M. (1990). Acute postdisaster psychiatric disorders: Identification of persons at risk. *American Journal of Psychiatry, 147*, 202–206.

Smith, I., Beasley, M.G., Wolff, O.H., & Ades, A.E. (1991). Effect on intelligence of relaxing the low phenylalanine diet in phenylketonuria. *Archives of Disease in Childhood, 66*, 311–316.

Smith, S.D., Kimberling, W.J., & Pennington, B.F. (1991). Screening for multiple genes influencing dyslexia. *Reading and Writing, 3*, 285–298.

Smith, S.D., Kimberling, W.J., Pennington, B.F., & Lubs, H.A. (1983). Specific reading disability: Identification of an inherited form through linkage analysis. *Science, 219*, 1345–1347.

Smits, B.M., & Cuppen, E. (2006). Rat genetics: The next episode. *Trends in Genetics, 22*, 232–240.

Smoller, J.W., & Finn, C.T. (2003). Family, twin, and adoption studies of bipolar disorder. *American Journal of Medical Genetics. Part C: Seminars in Medical Genetics, 123*, 48–58.

Snyderman, M., & Rothman, S. (1988). *The IQ controversy, the media and publication.* New Brunswick, NJ: Transaction.

Sokol, D.K., Moore, C.A., Rose, R.J., Williams, C.J., Reed, T., & Christian, J.C. (1995). Intrapair differences in personality and cognitive ability among young monozygotic twins distinguished by chorion type. *Behavior Genetics, 25*, 457–466.

Sokolowski, M.B. (2001). *Drosophila:* Genetics meets behaviour. *Nature Reviews Genetics, 2*, 879–890.

Speakman, J., Hambly, C., Mitchell, S., & Krol, E. (2007). Animal models of obesity. *Obesity Reviews, 8* (Suppl.), 55–61.

Spearman, C. (1904). "General intelligence," objectively determined and measured. *American Journal of Psychology, 15*, 201–292.

Spinath, F.M., Harlaar, N., Ronald, A., & Plomin, R. (2004). Substantial genetic influence on mild mental impairment in early childhood. *American Journal of Mental Retardation, 109*, 34–43.

Spitz, H.H. (1988). Wechsler subtest patterns of mentally retarded groups: Relationship to *g* and to estimates of heritability. *Intelligence, 12*, 279–297.

Spotts, E.L., Pederson, N.L., Neiderhiser, J.M., Reiss, D., Lichtenstein, P., Hansson, K., et al. (2005). Genetic effects on women's positive mental health: Do marital relationships and social support matter? *Journal of Family Psychology, 19*, 339–349.

Spotts, E.L., Prescott, C.A., & Kendler, K. (2006). Examining the origins of gender differences in marital quality: A behavior genetic analysis. *Journal of Family Psychology, 20*, 605–613.

Sprott, R.L., & Staats, J. (1975). Behavioral studies using genetically defined mice—A bibliography. *Behavior Genetics, 5*, 27–82.

Spuhler, J.N. (1968). Assortative mating with respect to physical characteristics. *Eugenics Quarterly, 15*, 128–140.

St. George-Hyslop, P., Haines, J., Rogaev, E., Mortilla, M., Vaula, G., Pericak-Vance, M., et al. (1992). Genetic evidence for a novel familial Alzheimer's disease locus on chromosome 14. *Nature Genetics, 2*, 330–334.

Stallings, M.C., Corley, R.P., Dennehey, B., Hewitt, J.K., Krauter, K.S., Lessem, J.M., et al. (2005). A genome-wide search for quantitative trait loci that influence antisocial drug dependence in adolescence. *Archives of General Psychiatry, 62*, 1042–1051.

Stallings, M.C., Corley, R.P., Hewitt, J.K., Krauter, K.S., Lessem, J.M., Mikulich, S.K., et al. (2003). A genome-wide search for quantitative trait loci influencing substance dependence vulnerability in adolescence. *Drug and Alcohol Dependence, 70*, 295–307.

Stallings, M.C., Hewitt, J.K., Cloninger, C.R., Heath, A.C., & Eaves, L.J. (1996). Genetic and environmental structure of the Tridimensional Personality Questionnaire: Three or four temperament dimensions? *Journal of Personality and Social Psychology, 70*, 127–140.

Stead, J.D., Clinton, S., Neal, C., Schneider, J., Jama, A., Miller, S., et al. (2006). Selective breeding for divergence in novelty-seeking traits: Heritability and enrichment in spontaneous anxiety-related behaviors. *Behavior Genetics, 36,* 697–712.

Stefansson, H., Sigurdsson, E., Steinthorsdottir, V., Bjornsdottir, S., Sigmundsson, T., Ghosh, S., et al. (2002). Neuregulin 1 and susceptibility to schizophrenia. *American Journal of Human Genetics, 71,* 877–892.

Stein, M.B., Chartier, M.J., Hazen, A.L., Kozak, M.V., Tancer, M.E., Lander, S., et al. (1998). A direct-interview family study of generalized social phobia. *American Journal of Psychiatry, 155,* 90–97.

Stent, G.S. (1963). *Molecular biology of bacterial viruses.* New York: W.H. Freeman.

Sternberg, R.J., & Grigorenko, E.L. (1997). *Intelligence: Heredity and environment.* Cambridge: Cambridge University Press.

Stevens, A., & Price, J. (2004). *Evolutionary psychiatry: A new beginning* (2nd ed.). London: Routledge.

Stevenson, J., Graham, P., Fredman, G., & McLoughlin, V. (1987). A twin study of genetic influences on reading and spelling ability and disability. *Journal of Child Psychology and Psychiatry, 28,* 229–247.

Stewart, S.E., Platko, J., Fagerness, J., Birns, J., Jenike, E., Smoller, J.W., et al. (2007). A genetic family-based association study of OLIG2 in obsessive-compulsive disorder. *Archives of General Psychiatry, 64,* 209–214.

Stoolmiller, M. (1999). Implications of the restricted range of family environments for estimates of heritability and nonshared environment in behavior-genetic adoption studies. *Psychological Bulletin, 125,* 392–409.

Stratakis, C.A., & Rennert, O.M. (2005). Turner syndrome—An update. *Endocrinologist, 15,* 27–36.

Straub, R.E., Jiang, Y., MacLean, C.J., Ma, Y., Webb, B.T., Myakishev, M.V., et al. (2002). Genetic variation in the *6p22.3* gene DTNBP1, the human ortholog of the mouse dysbindin gene, is associated with schizophrenia. *American Journal of Human Genetics, 71,* 337–348.

Stromswold, K. (2001). The heritability of language: A review and meta-analysis of twin, adoption and linkage studies. *Language, 77,* 647–723.

Stromswold, K. (2006). Why aren't identical twins linguistically identical? Genetic, prenatal and postnatal factors. *Cognition, 101,* 333–384.

Stunkard, A.J., Foch, T.T., & Hrubec, Z. (1986). A twin study of human obesity. *Journal of the American Medical Association, 256,* 51–54.

Sullivan, P.F., Evengard, B., Jacks, A., & Pedersen, N.L. (2005). Twin analyses of chronic fatigue in a Swedish national sample. *Psychological Medicine, 35,* 1327–1336.

Sullivan, P.F., Fan, C., & Perou, C.M. (2006). Evaluating the comparability of gene expression in blood and brain. *American Journal of Medical Genetics. Part B: Neuropsychiatric Genetics, 141,* 261–268.

Sullivan, P.F., Kendler, K.S., & Neale, M.C. (2003). Schizophrenia as a complex trait: Evidence from a meta-analysis of twin studies. *Archives of General Psychiatry, 60,* 1187–1192.

Sullivan, P.F., Kovalenko, P., York, T.P., Prescott, C.A., & Kendler, K.S. (2003). Fatigue in a community sample of twins. *Psychological Medicine, 33,* 263–281.

Sullivan, P.F., Neale, M.C., & Kendler, K.S. (2000). Genetic epidemiology of major depression: Review and meta-analysis. *American Journal of Psychiatry, 157,* 1552–1562.

Suresh, R., Ambrose, N., Roe, C., Pluzhnikov, A., Wittke-Thompson, J.K., Ng, M.C., et al. (2006). New complexities in the genetics of stuttering: Significant sex-specific linkage signals. *American Journal of Human Genetics, 78,* 554–563.

Sykes, B. (2007). *Saxons, Vikings and Celts: The genetic roots of Britain and Ireland.* New York: W.W. Norton.

Symons, D. (1979). *The evolution of human sexuality.* New York: Oxford University Press.

Syvanen, A.C. (2005). Toward genome-wide SNP genotyping. *Nature Genetics, 37,* S5–S10.

Szatmari, P., Paterson, A.D., Zwaigenbaum, L., Roberts, W., Brian, J., Liu, X.Q., et al. (2007). Mapping autism risk loci using genetic linkage and chromosomal rearrangements. *Nature Genetics, 39,* 319–328.

Tabor, H.K., Risch, N.J., & Myers, R.M. (2002). Opinion: Candidate-gene approaches for studying complex genetic traits: Practical considerations. *Nature Reviews Genetics, 3,* 391–397.

Taipale, M., Kaminen, N., Nopola-Hemmi, J., Haltia, T., Myllyluoma, B., Lyytinen, H., et al. (2003). A candidate gene for developmental dyslexia encodes a nuclear tetratricopeptide repeat domain protein dynamically regulated in brain. *Proceedings of the National Academy of Sciences (USA), 100,* 11553–11558.

Talbot, C.J., Nicod, A., Cherny, S.S., Fulker, D.W., Collins, A.C., & Flint, J. (1999). High-resolution mapping of quantitative trait loci in outbred mice. *Nature Genetics, 21,* 305–308.

Talkowski, M.E., Seltman, H., Bassett, A.S., Brzustowicz, L.M., Chen, X., Chowdari, K.V., et al. (2006). Evaluation of a susceptibility gene for schizophrenia: Genotype based meta-analysis of RGS4 polymorphisms from thirteen independent samples. *Biological Psychiatry, 60,* 152–162.

Tambs, K., Sundet, J.M., & Magnus, P. (1986). Genetic and environmental contribution to the covariation between the Wechsler Adult Intelligence Scale (WAIS) subtests: A study of twins. *Behavior Genetics, 16,* 475–491.

Tambs, K., Sundet, J.M., Magnus, P., & Berg, K. (1989). Genetic and environmental contributions to the covariance between occupational status, educational attainment, and IQ: A study of twins. *Behavior Genetics, 19,* 209–222.

Tang, Y.P., Shimizu, E., Dube, G.R., Rampon, C., Kerchner, G.A., Zhuo, M., et al. (1999). Genetic enhancement of learning and memory in mice. *Nature, 401,* 63–69.

Tanzi, R.E., & Bertram, L. (2005). Twenty years of the Alzheimer's disease amyloid hypothesis: A genetic perspective. *Cell, 120,* 545–555.

Taubman, P. (1976). The determinants of earnings: Genetics, family and other environments: A study of white male twins. *American Economic Review, 66,* 858–870.

Taylor, E. (1995). Dysfunctions of attention. In D. Cicchetti & D.J. Cohen (Eds.), *Developmental psychopathology: Vol. 2. Risk, disorder, and adaptation* (pp. 243–273). New York: Wiley.

Tellegen, A., Lykken, D.T., Bouchard, T.J., Jr., Wilcox, K., Segal, N., & Rich, A. (1988). Personality similarity in twins reared together and apart. *Journal of Personality and Social Psychology, 54,* 1031–1039.

Tennyson, C., Klamut, H.J., & Worton, R.G. (1995). The human dystrophin gene requires 16 hours to be transcribed and is contrascriptionally spliced. *Nature Genetics, 9,* 184–190.

Tesser, A. (1993). On the importance of heritability in psychological research: The case of attitudes. *Psychological Review, 100,* 129–142.

Tesser, A., Whitaker, D., Martin, L., & Ward, D. (1998). Attitude heritability, attitude change and physiological responsivity. *Personality and Individual Differences, 24,* 89–96.

Thakker, D.R., Hoyer, D., & Cryan, J.F. (2006). Interfering with the brain: Use of RNA interference for understanding the pathophysiology of psychiatric and neurological disorders. *Pharmacology and Therapeutics, 109,* 413–438.

Thapar, A., Harold, G., & McGuffin, P. (1998). Life events and depressive symptoms in childhood—Shared genes or shared adversity? A research note. *Journal of Child Psychology and Psychiatry, 39,* 1153–1158.

Thapar, A., Langley, K., O'Donovan, M., & Owen, M. (2006). Refining the attention-deficit hyperactivity disorder phenotype for molecular genetic studies. *Molecular Psychiatry, 11,* 714–720.

Thapar, A., & McGuffin, P. (1996). Genetic influences on life events in childhood. *Psychological Medicine, 26,* 813–820.

Thapar, A., O'Donovan, M., & Owen, M.J. (2005). The genetics of attention-deficit hyperactivity disorder. *Human Molecular Genetics, 14,* R275–R282.

Thapar, A., & Rice, F. (2006). Twin studies in pediatric depression. *Child and Adolescent Psychiatric Clinics of North America, 15,* 869–881.

Thapar, A., & Scourfield, J. (2002). Childhood disorders. In P. McGuffin, M.J. Owen, & I.I. Gottesman (Eds.), *Psychiatric genetics and genomics* (pp. 147–180). Oxford: Oxford University Press.

Theis, S.V.S. (1924). *How foster children turn out* (Publication No. 165). New York: State Charities Aid Association.

Tholin, S., Rasmussen, F., Tynelius, P., & Karlsson, J. (2005). Genetic and environmental influences on eating behavior: The Swedish Young Male Twins Study. *American Journal of Clinical Nutrition, 81,* 564–569.

Thomas, R.K. (1996). Investigating cognitive abilities in animals: Unrealized potential. *Cognitive Brain Research, 3,* 157–166.

Thompson, P.M., Cannon, T.D., Narr, K.L., van Erp, T., Poutanen, V.P., Huttunen, M., et al. (2001). Genetic influences on brain structure. *Nature Neuroscience, 4,* 1253–1258.

Tienari, P., Wynne, L.C., Sorri, A., Lahti, I., Laksy, K., Moring, J., et al. (2004). Genotype-environment interaction in schizophrenia-spectrum disorder. Long-term follow-up study of Finnish adoptees. *British Journal of Psychiatry, 184,* 216–222.

Toga, A.W., & Thompson, P.M. (2005). Genetics of brain structure and intelligence. *Annual Review of Neuroscience, 28,* 1–23.

Toma, D.P., White, K.P., Hirsch, J., & Greenspan, R.J. (2002). Identification of genes involved in *Drosophila melanogaster* geotaxis, a complex behavioral trait. *Nature Genetics, 31,* 349–353.

Tooley, G.A., Karakis, M., Stokes, M., & Ozanne-Smith, J. (2006). Generalising the Cinderella Effect to unintentional childhood fatalities. *Evolution and Human Behavior, 27,* 224–230.

Torgersen, S. (1983). Genetic factors in anxiety disorders. *Archives of General Psychiatry, 40,* 1085–1089.

Torgersen, S., Edvardsen, J., Oien, P.A., Onstad, S., Skre, I., Lygren, S., et al. (2002). Schizotypal personality disorder inside and outside the schizophrenic spectrum. *Schizophrenia Research, 54,* 33–38.

Torgersen, S., Lygren, S., Oien, P.A., Skre, I., Onstad, S., Edvardsen, J., et al. (2000). A twin study of personality disorders. *Comprehensive Psychiatry, 41,* 416–425.

Torgersen, S. (1980). The oral, obsessive, and hysterical personality syndromes: A study of hereditary and environmental factors by means of the twin method. *Archives of General Psychiatry, 37,* 1272–1277.

Torrey, E.F. (1990). Offspring of twins with schizophrenia. *Archives of General Psychiatry, 47,* 976–977.

Torrey, E.F., Bowler, A.E., Taylor, E.H., & Gottesman, I.I. (1994). *Schizophrenia and manic-depressive disorder.* New York: Basic Books.

The Tourette Syndrome Association International Consortium for Genetics (2007). Genome scan for Tourette's disorder in affected-sibling-pair and multigenerational families. *American Journal of Human Genetics, 80,* 265–272.

Trikalinos, T.A., Karvouni, A., Zintzaras, E., Ylisaukko-Oja, T., Peltonen, L., Jarvela, I., et al. (2006). A heterogeneity-based genome search meta-analysis for autism-spectrum disorders. *Molecular Psychiatry, 11,* 29–36.

Trivers, R.L. (1972). Parental investment and sexual selection. In B. Campbell (Ed.), *Sexual selection and the descent of man: 1871–1971* (pp. 136–179). Chicago: Aldine.

Trivers, R.L. (1985). *Social evolution.* Menlo Park, CA: Benjamin/Cummings.

Trovo-Marqui, A.B., & Tajara, E.H. (2006). Neurofibromin: A general outlook. *Clinical Genetics, 70,* 1–13.

True, W.R., Rice, J., Eisen, S.A., Heath, A.C., Goldberg, J., Lyons, M.J., et al. (1993). A twin study of genetic and environmental contributions to liability for post-traumatic stress symptoms. *Archives of General Psychiatry, 50,* 257–264.

Trut, L.N. (1999). Early canid domestication: The fox farm experiment. *American Scientist, 87,* 160–169.

Tsuang, M.T., & Faraone, S.D. (1990). *The genetics of mood disorders.* Baltimore: John Hopkins University Press.

Tsuang, M.T., Bar, J.L., Harley, R.M., & Lyons, M.J. (2001). The Harvard Twin Study of Substance Abuse: What we have learned. *Harvard Review of Psychiatry, 9,* 267–279.

Tsuang, M.T., Lyons, M.J., Eisen, S.A., True, W.T., Goldberg, J., & Henderson, W. (1992). A twin study of drug exposure and initiation of use. *Behavior Genetics, 22,* 756 (abstract).

Tsutsumi, T., Holmes, S.E., McInnis, M.G., Sawa, A., Callahan, C., DePaulo, J.R., et al. (2004). Novel CAG/CTG repeat expansion mutations do not contribute to the genetic risk for most cases of bipolar disorder or schizophrenia. *American Journal of Medical Genetics. Part B: Neuropsychiatric Genetics, 124,* 15–19.

Turic, D., Robinson, L., Duke, M., Morris, D.W., Webb, V., Hamshere, M., et al. (2003). Linkage disequilibrium mapping provides further evidence for a gene for reading disability on chromosome 6p21. *Molecular Psychiatry, 8,* 176–185.

Turkheimer, E., Haley, A., Waldron, M., D'Onofrio, B., & Gottesman, I.I. (2003). Socioeconomic status modifies heritability of IQ in young children. *Psychological Science, 14,* 623–628.

Turkheimer, E., & Waldron, M. (2000). Nonshared environment: A theoretical, methodological, and quantitative review. *Psychological Bulletin, 126,* 78–108.

Turner, J.R., Cardon, L.R., & Hewitt, J.K. (1995). *Behavior genetic approaches in behavioral medicine.* New York: Plenum.

Turri, M.G., Henderson, N.D., DeFries, J.C., & Flint, J. (2001). Quantitative trait locus mapping in laboratory mice derived from a replicated selection experiment for open-field activity. *Genetics, 158,* 1217–1226.

Tuvblad, C., Grann, M., & Lichtenstein, P. (2006). Heritability for adolescent antisocial behavior differs with socioeconomic status: Gene-environment interaction. *Journal of Child Psychology and Psychiatry, 47,* 734–743.

U.S. Bureau of the Census. (1995). *Sixty-five plus in America.* Washington, DC: U.S. Government Printing Office.

Ustun, T.B., Rehm, J., Chatterji, S., Saxena, S., Trotter, R., Room, R., et al. (1999). Multiple-informant ranking of the disabling effects of different health conditions in 14 countries. WHO/NIH Joint Project CAR Study Group. *Lancet, 354,* 111–115.

Valdar, W., Solberg, L.C., Gauguier, D., Burnett, S., Klenerman, P., Cookson, W.O., et al. (2006a). Genome-wide genetic association of complex traits in heterogeneous stock mice. *Nature Genetics, 38,* 879–887.

Valdar, W., Solberg, L.C., Gauguier, D., Cookson, W.O., Rawlins, J.N., Mott, R., et al. (2006b). Genetic and environmental effects on complex traits in mice. *Genetics, 174,* 959–984.

van Baal, G., de Geus, E.J., & Boomsma, D.I. (1998). Longitudinal study of genetic influences on ERP-P3 during childhood. *Developmental Neuropsychology, 14,* 19–45.

van Beijsterveldt, C.E., Molenaar, P.C., de Geus, E.J., & Boomsma, D.I. (1998). Genetic and environmental influences on EEG coherence. *Behavior Genetics, 28,* 443–453.

Vandenberg, S.G. (1971). What do we know today about the inheritance of intelligence and how do we know it? In R. Cancro (Ed.), *Genetic and environmental influences* (pp. 182–218). New York: Grune & Stratton.

Vandenberg, S.G. (1972). Assortative mating, or who marries whom? *Behavior Genetics, 2,* 127–157.

van den Oord, J.C.G., & Rowe, D.C. (1997). Continuity and change in children's social maladjustment: A developmental behavior genetic study. *Developmental Psychology, 33,* 319–332.

van der Sluis, S., Dolan, D.V., Neale, M.C., Boomsma, D.I., & Posthuma, D. (2006). Detecting genotype-environment interaction in monozygotic twin data: Comparing the Jinks and Fulker test and a new test based on marginal maximum likelihood estimation. *Twin Research and Human Genetics, 9,* 377–392.

van Grootheest, D.S., Cath, D.C., Beekman, A.T., & Boomsma, D.I. (2005). Twin studies on obsessive-compulsive disorder: A review. *Twin Research and Human Genetics, 8,* 450–458.

van Ijzendoorn, M.H., Moran, G., Belsky, J., Pederson, D., Bakermans-Kranenburg, M.J., & Fisher, K. (2000). The similarity of siblings' attachments to their mother. *Child Development, 71,* 1086–1098.

Venter, J.C., Adams, M.D., Myers, E.W., Li, P.W., Mural, R.J., Sutton, G.G., et al. (2001). The sequence of the human genome. *Science, 291,* 1304–1351.

Verkerk, A.J., Cath, D.C., van der Linde, H.C., Both, J., Heutink, P., Breedveld, G., et al. (2006). Genetic and clinical analysis of a large Dutch Gilles de la Tourette family. *Molecular Psychiatry, 11,* 954–964.

Verkerk, A.J., Pieretti, M., Sutcliffe, J.S., Fu, Y.-H., Kuhl, D.P.A., Pizzuti, A., et al. (1991). Identification of a gene (FMR-1) containing a CGG repeat coincident with a breakpoint cluster region exhibiting length variation in fragile X syndrome. *Cell, 65,* 905–914.

Vernon, P.A. (1989). The heritability of measures of speed of information-processing. *Personality and Individual Differences, 10,* 575–576.

Vernon, P.A., Jang, K.L., Harris, J.A., & McCarthy, J.M. (1997). Environmental predictors of personality differences: A twin and sibling study. *Journal of Personality and Social Psychology, 72,* 177–183.

Viding, E. (2004). Annotation: Understanding the development of psychopathy. *Journal of Child Psychology and Psychiatry, 45,* 1329–1337.

Viding, E., Blair, R.J.R., Moffitt, T.E., & Plomin, R. (2005). Evidence for substantial genetic risk for psychopathy in 7 year olds. *Journal of Child Psychology and Psychiatry, 46,* 592–597.

Villafuerte, S., & Burmeister, M. (2003). Untangling genetic networks of panic, phobia, fear and anxiety. *Genome Biology, 4,* 224.

Visser, B.A., Ashton, M.C., & Vernon, P.A. (2006). Beyond *g:* Putting multiple intelligences theory to the test. *Intelligence, 34,* 487–502.

Vitaterna, M.H., King, D.P., Chang, A.M., Kornhauser, J.M., Lowrey, P.L., McDonald, J.D., et al. (1994). Mutagenesis and mapping of a mouse gene, Clock, essential for circadian behavior. *Science, 264,* 719–725.

Vitaterna, M.H., Pinto, L.H., & Takahashi, J.S. (2006). Large-scale mutagenesis and phenotypic screens for the nervous system and behavior in mice. *Trends in Neurosciences, 29,* 233–240.

Vogel, K.S., Klesse, L.J., Velasco-Miguel, S., Meyers, K., Rushing, E.J., & Parada, L.F. (1999). Mouse tumor model for neurofibromatosis type 1. *Science, 286,* 2176–2179.

von Gontard, A., Schaumburg, H., Hollmann, E., Eiberg, H., & Rittig, S. (2001). The genetics of enuresis: A review. *Journal of Urology, 166,* 2438–2443.

von Knorring, A.L., Cloninger, C.R., Bohman, M., & Sigvardsson, S. (1983). An adoption study of depressive disorders and substance abuse. *Archives of General Psychiatry, 40,* 943–950.

Waddell, S., & Quinn, W.G. (2001). What can we teach *Drosophila?* What can they teach us? *Trends in Genetics, 17,* 719–726.

Wade, C.H., & Wilfond, B.S. (2006). Ethical and clinical practice considerations for genetic counselors related to direct-to-consumer marketing of genetic tests. *American Journal of Medical Genetics. Part C: Seminars in Medical Genetics, 142,* 284–292.

Wahlsten, D. (1999). Single-gene influences on brain and behavior. *Annual Review of Psychology, 50,* 599–624.

Wahlsten, D., Bachmanov, A., Finn, D.A., & Crabbe, J.C. (2006). Stability of inbred mouse strain differences in behavior and brain size between laboratories and across decades. *Proceedings of the National Academy of Sciences (USA), 103,* 16364–16369.

Wahlsten, D., Metten, P., Phillips, T.J., Boehm, S.L., Burkhart-Kasch, S., Dorow, J., et al. (2003). Different data from different labs: Lessons from studies of gene-environment interaction. *Journal of Neurobiology, 54,* 283–311.

Wainwright, M.A., Wright, M.J., Luciano, M., Geffen, G.M., & Martin, N.G. (2005). Multivariate genetic analysis of academic skills of the Queensland core skills test and IQ highlight the importance of genetic *g. Twin Research and Human Genetics, 8,* 602–608.

Wainwright, M.A., Wright, M.J., Luciano, M., Montgomery, G.W., Geffen, G.M., & Martin, N.G. (2006). A linkage study of academic skills defined by the Queensland core skills test. *Behavior Genetics, 36,* 56–64.

Waldman, I.D., & Gizer, I.R. (2006). The genetics of attention-deficit hyperactivity disorder. *Clinical Psychology Review, 26,* 396–432.

Walker, S.O., & Plomin, R. (2005). The nature-nurture question: Teachers' perceptions of how genes and the environment influence educationally relevant behavior. *Educational Psychology, 25,* 509–516.

Walker, S.O., & Plomin, R. (2006). Nature, nurture, and perceptions of the classroom environment as they relate to teacher assessed academic achievement: A twin study of 9-year-olds. *Educational Psychology, 26,* 541–561.

Wallace, G.L., Eric, S.J., Lenroot, R., Viding, E., Ordaz, S., Rosenthal, M.A., et al. (2006). A pediatric twin study of brain morphometry. *Journal of Child Psychology and Psychiatry, 47,* 987–993.

Waller, N.G., & Shaver, P.R. (1994). The importance of nongenetic influence on romantic love styles: A twin-family study. *Psychological Science, 5,* 268–274.

Walsh, C.M., Zainal, N.Z., Middleton, S.J., & Paykel, E.S. (2001). A family history study of chronic fatigue syndrome. *Psychiatric Genetics, 11,* 123–128.

Wang, Y., Barbacioura, C., Hyland, F., Xiao, W., Hunkapiller, K.L., Blake, J., et al. (2006). Large scale real-time PCR validation on gene expression measurements from two commercial long-oligonucleotide microarrays. *BMC Genomics, 21,* 59.

Ward, B.A., & Gutmann, D.H. (2005). Neurofibromatosis 1: From lab bench to clinic. *Pediatric Neurology, 32,* 221–228.

Ward, M.J., Vaughn, B.E., & Robb, M.D. (1988). Social-emotional adaptation and infant-mother attachment in siblings: Role of the mother in cross-sibling consistency. *Child Development, 59,* 643–651.

Waters, P.J. (2003). How PAH gene mutations cause hyper-phenylalaninemia and why mechanism matters: Insights from in vitro expression. *Human Mutation, 21,* 357–369.

Watson, J.B. (1930). *Behaviorism.* New York: Norton.

Watson, J.D., Baker, T.A., Bell, S.P., Gann, A., Levine, M., & Losick, R. (2004). *Molecular biology of the gene* (5th ed.). New York: Addison Wesley.

Weaving, L.S., Ellaway, C.J., Gecz, J., & Christodoulou, J. (2005). Rett syndrome: Clinical review and genetic update. *Journal of Medical Genetics, 42,* 1–7.

Wedekind, C., Seebeck, T., Bettens, F., & Paepke, A.J. (1995). MHC-dependent mate preferences in humans. *Proceedings. Clinical Biological Research, 260,* 245–249.

Wedell, N., Kvarnemo, C., Lessels, C.K.M., & Tregenza, T. (2006). Sexual conflict and life histories. *Animal Behavior, 71,* 999–1011.

Weeden, J., & Sabini, J. (2005). Physical attractiveness and health in Western societies: A review. *Psychological Bulletin, 131,* 635–653.

Weiner, J. (1994). *The beak of the finch.* New York: Vintage Books.

Weiner, J. (1999). *Time, love and memory: A great biologist and his quest for the origins of behavior.* New York: Alfred A. Knopf.

Weir, B.S., Anderson, A.D., & Hepler, A.B. (2006). Genetic relatedness analysis: Modern data and new challenges. *Nature Reviews Genetics, 7,* 771–780.

Weiss, D.S., Marmar, C.R., Schlenger, W.E., Fairbank, J.A., Jordan, B.K., Hough, R.L., et al. (1992). The prevalence of lifetime and partial posttraumatic stress disorder in Vietnam theater veterans. *Journal of Traumatic Stress, 5,* 365–376.

Weiss, P. (1982). *Psychogenetik: Humangenetik in psychologie and psychiatrie.* Jena, Germany: Risher.

Wells, R.D., & Warren, S.T. (1998). *Genetic instabilities and hereditary neurological diseases.* San Diego, CA: Academic Press.

Wender, P.H., Kety, S.S., Rosenthal, D., Schulsinger, F., Ortmann, J., & Lunde, I. (1986). Psychiatric disorders in the biological and adoptive families of adopted individuals with affective disorders. *Archives of General Psychiatry, 43*, 923–929.

Wender, P.H., Rosenthal, D., Kety, S.S., Schulsinger, F., & Welner, J. (1974). Cross-fostering: A research strategy for clarifying the role of genetic and experimental factors in the etiology of schizophrenia. *Archives of General Psychiatry, 30*, 121–128.

Wheeler, D.A., Kyriacou, C.P., Greenacre, M.L., Yu, Q., Rutila, J.E., Rosbash, M., et al. (1991). Molecular transfer of a species-specific courtship behavior from *Drosophila simulans* to *Drosophila melanogaster. Science, 251*, 1082–1085.

Whittington, J.E., Butler, J.V., & Holland, A.J. (2007). Changing rates of genetic sub-types of Prader-Willi syndrome in the UK. *European Journal of Human Genetics, 15*, 127–130.

Widiger, T.A., & Trull, T.J. (2007). Plate tectonics in the classification of personality disorder: Shifting to a dimensional model. *American Psychologist, 62*, 71–83.

Wilkins, J.F., & Haig, D. (2003). What good is genomic imprinting: The function of parent-specific gene expression. *Nature Reviews Genetics, 4*, 359–368.

Willcutt, E.G., Pennington, B.F., & DeFries, J.C. (2000). Twin study of the etiology of comorbidity between reading disability and attention-deficit hyperactivity disor-der. *American Journal of Medical Genetics, 12*, 293–301.

Williams, C.A., Beaudet, A.L., Clayton-Smith, J., Knoll, J.H., Kyllerman, M., Laan, L.A., et al. (2006). Angelman syndrome 2005: Updated consensus for diagnostic criteria. *American Journal of Medical Genetics: Part A, 140*, 413–418.

Williams, J., & O'Donovan, M.C. (2006). The genetics of developmental dyslexia. *European Journal of Human Genetics, 14*, 681–689.

Williams, R.W. (2006). Expression genetics and the phenotype revolution. *Mammalian Genome, 17*, 496–502.

Willis-Owen, S.A., & Flint, J. (2006). The genetic basis of emotional behaviour in mice. *European Journal of Human Genetics, 14*, 721–728.

Willis-Owen, S.A., & Flint, J. (2007). Identifying the genetic determinants of emo-tionality in humans; insights from rodents. *Neuroscience and Biobehavioral Reviews, 31*, 115–124.

Willis-Owen, S.A., Turri, M.G., Munafo, M.R., Surtees, P.G., Wainwright, N.W., Brixey, R.D., et al. (2005). The serotonin transporter length polymorphism, neu-roticism, and depression: A comprehensive assessment of association. *Biological Psychiatry, 58*, 451–456.

Wilson, E.O. (1975). *Sociobiology: The new synthesis.* Cambridge, MA: Belknap Press.

Wilson, R.K. (1999). How the worm was won: The *C.elegans* genome sequencing proj-ect. *Trends in Genetics, 15*, 51–58.

Wilson, R.S. (1983). The Louisville Twin Study: Developmental synchronies in behav-ior. *Child Development, 54*, 298–316.

Wilson, R.S., & Matheny, A.P., Jr. (1976). Retardation and twin concordance in infant mental development: A reassessment. *Behavior Genetics, 6*, 353–356.

Wilson, R.S., & Matheny, A.P., Jr. (1986). Behavior genetics research in infant tem-perament: The Louisiville Twin Study. In R. Plomin & J.F. Dunn (Eds.), *The study of temperament: Changes, continuities, and challenges* (pp. 81–97). Hillsdale, NJ: Lawrence Erlbaum Associates.

Winterer, G., & Goldman, D. (2003). Genetics of human prefrontal function. *Brain Research Reviews, 43*, 134–163.

Winterer, G., Hariri, A.R., Goldman, D., & Weinberger, D.R. (2005). Neuroimaging and human genetics. *International Review of Neurobiology, 67*, 325–383.

Wolf, N. (1992). *The beauty myth: How images of beauty are used against women.* New York: Anchor.

Wolpert, L. (2007). *Six impossible things before breakfast.* London: Faber & Faber.

Wong, K.K., deLeeuw, R.J., Dosanjh, N.S., Kimm, L.R., Cheng, Z., Horsman, D.E., et al. (2007). A comprehensive analysis of common copy-number variations in the human genome. *American Journal of Human Genetics, 80*, 91–104.

Wood, W., & Eagly, A.H. (2002). A cross-cultural analysis of the behavior of women and men: Implications for the origins of sex differences. *Psychological Bulletin, 128*, 699–727.

Wooldridge, A. (1994). *Measuring the mind: Education and psychology in England, c. 1860–c. 1990.* Cambridge: Cambridge University Press.

Wright, S. (1921). Systems of mating. *Genetics, 6*, 111–178.

Yairi, E., Ambrose, N., & Cox, N. (1996). Genetics of stuttering: A critical review. *Journal of Speech, Language, and Hearing Research, 39*, 771–784.

Yalcin, B., Willis-Owen, S.A., Fullerton, J., Meesaq, A., Deacon, R.M., Rawlins, J.N., et al. (2004). Genetic dissection of a behavioral quantitative trait locus shows that Rgs2 modulates anxiety in mice. *Nature Genetics, 36*, 1197–1202.

Yamasaki, C., Koyanagi, K.O., Fujii, Y., Itoh, T., Barrero, R., Tamura, T., et al. (2005). Investigation of protein functions through data-mining on integrated human transcriptome database, H-Invitational database (H-InvDB). *Gene, 364*, 99–107.

Yin, J.C.P., Vecchio, M.D., Zhou, H., & Tully, T. (1995). CRAB as a memory modulator: Induced expression of a dCREB2 activator isoform enhances long-term memory in *Drosophila. Cell, 8*, 107–115.

Young, J.P.R., Fenton, G.W., & Lader, M.H. (1971). The inheritance of neurotic traits: A twin study of the Middlesex Hospital Questionnaire. *British Journal of Psychiatry, 119*, 393–398.

Young, S.E., Rhee, S.H., Stallings, M.C., Corley, R.P., & Hewitt, J.K. (2006). Genetic and environmental vulnerabilities underlying adolescent substance use and problem use: General or specific? *Behavior Genetics, 36*, 603–615.

Yuferov, V., Nielsen, D., Butelman, E., & Kreek, M.J. (2005). Microarray studies of psychostimulant-induced changes in gene expression. *Addiction Biology, 10*, 101–118.

Zagreda, L., Goodman, J., Druin, D.P., McDonald, D., & Diamond, A. (1999). Cognitive deficits in a genetic mouse model of the most common biochemical cause of human mental retardation. *The Journal of Neuroscience, 19*, 6175–6182.

Zahn-Waxler, C., Robinson, J., & Emde, R.N. (1992). The development of empathy in twins. *Developmental Psychology, 28*, 1038–1047.

Zalsman, G., Huang, Y.Y., Oquendo, M.A., Burke, A.K., Hu, X.Z., Brent, D.A., et al. (2006). Association of a triallelic serotonin transporter gene promoter region (5-HTTLPR) polymorphism with stressful life events and severity of depression. *American Journal of Psychiatry, 163*, 1588–1593.

Zhang, Y., Proenca, R., Maffei, M., Barone, M., Leopold, L., & Friedman, J.M. (1994). Positional cloning of the mouse obese gene and its human homologue. *Nature, 372,* 425–432.

Zhang, Y.Q., & Broadie, K. (2005). Fathoming fragile X in fruit flies. *Trends in Genetics, 21,* 37–45.

Zhu, Y., Ghosh, P., Charnay, P., Burns, D.K., & Parada, L.F. (2002). Neurofibromas in NF1: Schwann cell origin and role of tumor environment. *Science, 296,* 920–922.

Zondervan, K.T., & Cardon, L.R. (2004). The complex interplay among factors that influence allelic association. *Nature Reviews Genetics, 5,* 89–100.

Zschocke, J. (2003). Phenylketonuria mutations in Europe. *Human Mutation, 21,* 345–356.

Zubenko, G.S., Hughes, H.B., III, Zubenko, W.N., & Maher, B.S. (2007). Genome survey for loci that influence successful aging: Results at 10-cM resolution. *American Journal of Geriatric Psychiatry, 15,* 184–193.

Zucker, R.A. (2006). The developmental behavior genetics of drug involvement: Overview and comments. *Behavior Genetics, 36,* 616–625.

Zuckerman, M. (1994). *Behavioral expression and biosocial bases of sensation seeking.* New York: Cambridge University Press.

Associations

The Behavior Genetics Association, with links to its journal, *Behavior Genetics*
   http://www.bga.org/

The International Society for Twin Studies is an international, multidisciplinary scientific organization whose purpose is to further research and public education in all fields related to twins and twin studies. Its Web site is linked to the society's journal, *Twin Research and Human Genetics.*
   http://www.ists.qimr.edu.au/

The International Society of Psychiatric Genetics is a worldwide organization that aims to promote and facilitate research in the genetics of psychiatric disorders, substance use disorders, and allied traits. With links to associated journals, *Psychiatric Genetics* and *Neuropsychiatric Genetics.*
   http://www.ispg.net/

The American Society of Human Genetics, with links to its journal, *American Journal of Human Genetics*
   http://www.ashg.org/

The European Society of Human Genetics, with links to its journal, *European Journal of Human Genetics*
   http://www.eshg.org/

The American Psychological Association "genetics in psychology" page
   http://www.apa.org/science/genetics/

The Human Genome Organization (HUGO), the international organization of scientists involved in human genetics
   http://www.hugo-international.org/

The International Behavioral and Neural Genetics Society (IBANGS) works to promote the field of neurobehavioral genetics. With links to its journal *Genes, Brain and Behavior.*
   http://www.ibngs.org/

Databases and Genome Browsers

EMBL-EBI, the European Molecular Biology Laboratory's (EMBL) European Bioinformatics Institute, is the European node for globally coordinated efforts to collect and disseminate biological data.

http://www.ebi.ac.uk/

NCBI, the National Center for Biotechnology Information, is the U.S. node.

http://www.ncbi.nlm.nih.gov/

Ensembl, the EBI and Wellcome Trust Sanger Institute's genome browser

http://www.ensembl.org/

University of California Santa Cruz (UCSC) genome browser

http://genome.ucsc.edu/

NCBI, Online Mendelian Inheritance in Man (OMIM). This database is a catalog of human genes and genetic disorders. The database contains textual information, pictures, and reference material.

http://www.ncbi.nlm.nih.gov/omim/

Resources

Behavioral Genetic Interactive Modules based on the Appendix to this text

http://statgen.iop.kcl.ac.uk/bgim/

Mx is freely available software widely used in quantitative genetic analysis.

http://www.vcu.edu/mx/

Mx script library

http://www.psy.vu.nl/mxbib/

The Jackson Laboratory, Mouse Genome Informatics, is an excellent resource for mouse genetics.

http://www.informatics.jax.org/

NIH Science provides the latest information on animal models used in genetic research.

http://www.nih.gov/science/models/

NCBI online bookshelf: search a growing collection of biomedical textbooks.

http://eutils.ncbi.nlm.nih.gov/entrez/query.fcgi?db=Books

The Gene Almanac includes an animated primer on the basics of DNA, genes, and heredity, as well as links to the Cold Spring Harbor Laboratory Eugenics Archive and other related information.

http://www.dnalc.org/home.html

Neuroscience Gateway, the latest research, news, and events in neuroscience and genomics research, developed collaboratively by the Allen Institute for Brain Science and Nature Publishing Group

http://www.brainatlas.org/aba/

Background on the Human Genome Project from the U.S. National Human Genome Research Institute

http://www.genome.gov/Education/

Statement from the American Society of Human Genetics on the bright future of behavioral genetics

http://www.faseb.org/genetics/ashg/policy/pol-28.htm

R is a free software environment for statistical computing and graphics. *Bioconductor* is an open source and open development software project for the analysis and comprehension of genomic data using the *R* environment.

http://www.r-project.org/

http://www.bioconductor.org/

Microarray Technology

Affymetrix and Illumina are two of the leading suppliers of microarray technology.

http://www.affymetrix.com/
http://www.illumina.com/

Public Understanding of Genetics

Genetics Resources on the Web (GROW) is dedicated to optimization of the use of the Web to provide health professionals and the public with high-quality information related to human genetics, with a particular focus on genetic medicine and health.

http://www.geneticsresources.org/

The Genetic Science Learning Center is an outreach education program at the University of Utah. Its aim is to help people understand how genetics affects their lives and society. This is an introductory guide to molecular genetics.

http://learn.genetics.utah.edu/units/basics/

The Genetics Home Reference provides consumer-friendly information about the effects of genetic variations on human health.

http://ghr.nlm.nih.gov/

NAME INDEX

SUBJECT INDEX